The Cell Cycle

Regulators, Targets, and Clinical Applications

GWUMC Department of Biochemistry
Annual Spring Symposia
Series Editors:
Allan L. Goldstein, Ajit Kumar, and J. Martyn Bailey
The George Washington University Medical Center

Recent volumes in this series:

ADVANCES IN MOLECULAR BIOLOGY AND TARGETED TREATMENT FOR AIDS
Edited by Ajit Kumar

BIOLOGY OF CELLULAR TRANSDUCING SIGNALS
Edited by Jack Y. Vanderhoek

BIOMEDICAL ADVANCES IN AGING
Edited by Allan L. Goldstein

CARDIOVASCULAR DISEASE
Molecular and Cellular Mechanisms, Prevention, and Treatment
Edited by Linda L. Gallo

CELL CALCIUM METABOLISM
Physiology, Biochemistry, Pharmacology, and Clinical Implications
Edited by Gary Fiskum

THE CELL CYCLE
Regulators, Targets, and Clinical Applications
Edited by Valerie W. Hu

GROWTH FACTORS, PEPTIDES, AND RECEPTORS
Edited by Terry W. Moody

NEURAL AND ENDOCRINE PEPTIDES AND RECEPTORS
Edited by Terry W. Moody

PROSTAGLANDINS, LEUKOTRIENES, AND LIPOXINS
Biochemistry, Mechanism of Action, and Clinical Applications
Edited by J. Martyn Bailey

PROSTAGLANDINS, LEUKOTRIENES, LIPOXINS, AND PAF
Mechanism of Action, Molecular Biology, and Clinical Applications
Edited by J. Martyn Bailey

THYMIC HORMONES AND LYMPHOKINES
Basic Chemistry and Clinical Applications
Edited by Allan L. Goldstein

The Cell Cycle

Regulators, Targets, and Clinical Applications

Edited by

Valerie W. Hu

The George Washington University Medical Center
Washington, D.C.

Springer Science+Business Media, LLC

Library of Congress Cataloging-in-Publication Data

The Cell cycle : regulators, targets, and clinical applications / edited by Valerie W. Hu.
p. cm. -- (GWUMC Department of Biochemistry annual spring symposia)
"Proceedings of the thirteenth Washington International Spring Symposium at the George Washington University, held May 10-14, 1993, in Washington, D.C."--T.p. verso.
Includes bibliographical references and index.
ISBN 978-0-306-44672-6 ISBN 978-1-4615-2421-2 (eBook)
DOI 10.1007/978-1-4615-2421-2
1. Cell cycle--Congresses. 2. Cells--Growth--Regulation--Congresses. 3. Carcinogenesis--Congresses. 4. Cellular control mechanisms--Congresses. I. Hu, Valerie W. II. International Washington Spring Symposium (13th : 1993 : Washington, D.C.) III. Series.
QH605.C45 1994
599'.087623--dc20 94-957
CIP

Proceedings of the Thirteenth Washington International Spring Symposium at The George Washington University, held May 10–14, 1993, in Washington, D.C.

ISBN 978-0-306-44672-6

PREFACE

Interest in the cell cycle has grown explosively in recent years as a result of the identification of key cell cycle regulators and their substrates. Aside from enhancing our understanding of normal cellular growth controls, this new knowledge has also been valuable in elucidating mechanisms of growth deregulation which occur in diseased states, such as cancer and, in some instances, viral or parasitic infections.

The Thirteenth Washington International Spring Symposium was organized with the intention of bringing together scientists working on different aspects of the cell cycle. Scientific topics presented ranged from molecular regulators and effectors to mitosis-specific changes in cell architecture to the role of the cell cycle in development and disease. The goal of this gathering was to help formulate a more comprehensive and integrated picture of events driving and being driven by the cell cycle, as well as to evaluate the possibilities for clinical application of this knowledge. This symposium, held in Washington, D.C. from May 10-14, 1993, was attended by more than 400 scientists from 20 countries, including many of the scientific leaders in this field. This volume contains most of the papers presented at the seven plenary sessions in addition to selected contributions from a total of nine special oral and poster sessions.

The book is divided into seven sections which are reflective of the seven plenary sessions. Part I focusses on the regulators of the cell cycle and opens with the keynote address presented by Dr. James Maller, co-recipient along with Dr. Tim Hunt of the 1993 Abraham White Distinguished Scientist Award. Drs. Maller and Hunt were honored at the symposium for their significant contributions to the advancement of cell cycle research. In his keynote address, Dr. Maller discussed highlights of studies on oocyte maturation, the resolution of questions raised by these studies, and new questions which have arisen from recent findings, including auto-amplification of MPF. Other papers in this section expand upon the role of protein phosphatases and the D-type G1 cyclins in cell cycle progression.

Part II is concerned with control of cell proliferation and the regulatory mechanisms for entry into and exit from the cell cycle. The chapter begins with a discussion of ERKs, protein kinases which are involved in numerous signal transduction pathways, including transition of cells from G_0 to G_1. The association of statin with the non-proliferating states of cells, specifically G_0 and cellular senescence, is also described along with a proposed model for the role of statin in the hypophosphorylation of Rb in both these states. Cell senescence is further characterized as a programmed event in which multiple senescence genes are activated and the expression of other genes, such as cdk2 and cyclins A and B, altered. The ability of DNA tumor viruses to trick a quiescent cell into entering S-phase in order to replicate viral DNA is also described and the roles of Rb and E2F are discussed in this context. In the final two papers of this section, a link between a GTPase cycle and the cell cycle is proposed and the novel idea of a sphingolipid cycle is advanced as a component of the signal transduction pathway regulating cell growth.

In part III, the cell cycle-dependent regulation of gene expression and repair is discussed. Several papers in this section describe the role of both cis- and trans-acting elements in driving transcription at the G_1/S border while others address mechanisms for sensing DNA damage and incomplete DNA replication. A novel method for detecting preferential repair of genes is also described.

Papers in Part IV deal with the regulation of cell entry into M-phase and the associated changes in cell architecture which must occur in preparation for cell division. The roles of the nim 1/cdr 1 kinase and p65 phosphatase as mitotic inducers are described and mitosis-specific changes in protein structure and localization as well as nuclear envelope assembly are discussed.

Perhaps the least developed area in cell cycle research concerns the involvement of cell cycle regulation in development. Part V contains papers that describe the evolution and control of the cell cycle at two levels of development: that of cellular aging and senescence and that of organismic aging and development. Within these contexts, the role of *mos*, G1 cyclins, and heat shock proteins are discussed and the significance of cell proliferation indices with respect to biological aging is addressed. The chapter closes with a discussion of apoptosis and its inverse relationship to cell proliferation.

Parts VI and VII focus on cell cycle regulation and deregulation in disease and explore clinical strategies involving cell cycle arrest or intervention. The aberrant expression of cyclin (E, D_1, A, B) genes in cancer cells is a recurrent theme in these papers, as is the down-regulation of tumor suppressor genes. In one study uniting the two themes, the increased expression of $p34^{cdc2}$ and cyclin A in leukemic cells is suggested to be related to deficient transcriptional regulation by Rb and p53, respectively. The respective roles of genetic damage, faulty DNA repair, and apoptosis are also discussed in relation to carcinogenesis. The last paper in Part VI describes a mechanism for HIV-induced T cell death involving apoptosis following G_2/M arrest, possibly as a result of hyperphosphorylation of $p34^{cdc2}$.

The involvement of the cell cycle in drug-induced apoptosis is further considered in Part VII and a case is made for altered pH regulation as an important component in this process. In terms of a therapeutic strategy, the importance of determining the molecular basis of specific tumors is stressed in order to utilize approaches based on challenging the defective cell cycle checkpoint in that given tumor. An alternative strategy would be to employ multiple checkpoint override to elicit inappropriate cell cycle progression, and subsequently apoptosis, in tumor cells. Yet another therapeutic approach might be to utilize protease inhibitors to block cyclin degradation leading, in some cases, to catastrophic mitosis. Along another vein, delayed cyclin B expression has been associated with the mechanism for the radioresistance of some tumor cells. Finally, the use of cell cycle analysis by single cell DNA cytophotometry as a prognostic tool for transitional cell carcinomas of the renal pelvis and ureter is described.

The articles in this volume highlight the multidisciplinary nature of current research on the cell cycle and brings together topics that sometimes are excluded from a meeting on cell cycle regulation, such as those related to cellular senescence, development, cell architecture, and clinical applications. This volume should thus be of interest to biochemists, cell, molecular, and developmental biologists, oncologists, pharmacologists, as well as other scientists interested in the regulation of cell proliferation.

Valerie W. Hu

Washington, D.C.

CONTENTS

PART I - REGULATORS OF THE CELL CYCLE

PART II - CONTROL OF CELL PROLIFERATION

PART III - CELL CYCLE CONTROL OF GENE EXPRESSION AND REPAIR

PART IV - MITOSIS: INDUCTION AND MECHANICS

PART V - CELL CYCLE REGULATION IN DEVELOPMENT

PART VI - ROLE OF THE CELL CYCLE IN DISEASE

PART VII - CLINICAL APPLICATIONS

*Presenting Authors at the 1993 Washington International Spring Symposium

PART I

REGULATORS OF THE CELL CYCLE

1

PROTEIN PHOSPHORYLATION AND THE REGULATION OF KEY EVENTS IN OOCYTE AND EGG CELL CYCLES

James L. Maller

Howard Hughes Medical Institute
and Department of Pharmacology
University of Colorado School of Medicine
4200 East Ninth Avenue
Denver, Colorado, 80262, USA

The aim of this chapter is to provide a description of the *Xenopus* oocyte and egg system in terms of the major experimental questions that existed in the mid-1970s, then to cover in brief form the resolution of those experimental questions as they now appear, and in so doing, to comment on how phosphorylation as a regulatory mechanism turns out to be the fundamental reaction that governs progression through the cell cycle in higher eucaryotes. In 1971, when the author was a second year graduate student, a paper appeared by Yoshi Masui at the University of Toronto in which he reported the results of experiments transferring cytoplasm from metaphase-arrested *Rana* unfertilized eggs into either resting G_2 oocytes or into S-phase blastomeres of a newly fertilized egg.[1] Masui found that the injection of metaphase cytoplasm into resting oocytes caused them to enter meiosis ($G_2 \Rightarrow M$ transition of oocyte maturation) in the absence of the normal mitogen (progesterone), and that it would do so even in the absence of protein synthesis. He gave the name to this activity of maturation-promoting factor, or MPF. In the same paper, Dr. Masui transferred metaphase cytoplasm into one blastomere of a two-cell embryo. He observed that the blastomere arrested in metaphase of the next mitotic division. He gave the name to this activity of cytostatic factor, or CSF, and postulated that it was responsible for the meiotic metaphase arrest that is characteristic of vertebrate unfertilized eggs.

Shortly after the existence of MPF was described by Dr. Masui, the author repeated the experiments with *Xenopus* eggs,[2] and other experiments were subsequently carried out by many investigators demonstrating that all types of cells in metaphase, from yeast to man, contained MPF activity.[3-6] Moreover, the biochemical properties of crude MPF extracts from various sources were remarkably similar. This led to the idea that MPF was a universal regulator of mitosis and meiosis and could be predicted to consist of evolutionarily-conserved components. While these early biological experiments clearly established the ubiquitous nature of MPF in M-phase, there were really no clues to the nature of MPF itself nor how it caused the events of nuclear envelope breakdown. The first clues in this regard came during the latter part of the author's thesis work that discovered an immediate effect of microinjection of crude preparations of MPF into recipient oocytes was a large, 3-5 fold, increase in total protein phosphorylation, which

The Cell Cycle: Regulators, Targets, and Clinical Applications
Edited by V.W. Hu, Plenum Press, New York, 1994

later became known as the "burst" of phosphorylation accompanying MPF action.[7] This burst of phosphorylation was non-cyclic AMP (cAMP)-dependent as judged by its insensitivity to co-injection of the heat stable inhibitor of cAMP protein kinase. Subsequent work showed that the burst of phosphorylation also accompanied MPF appearance in a variety of other systems, indicating it was a general feature of its action.[8]

The discovery that phosphorylation was an immediate biochemical consequence of MPF led to the hypothesis by the author that MPF might be a protein kinase or a regulator of a protein kinase,[7,9] and initiated the idea, which ultimately proved to be true, that protein phosphorylation underlay the biochemical basis of MPF action in the cell cycle. It was because of this discovery of protein phosphorylation as a biochemical effect of MPF that the author initiated post-doctoral studies with Dr. Krebs at the University of California at Davis in 1975. One of the first ideas that developed at Davis was the concept that phosphorylation of substrates following microinjection of MPF might be related to the amplification of MPF that occurred in recipient oocytes. It had been shown by the author's work and by others that after recipient oocytes received MPF injections and underwent meiosis, these matured oocytes now could serve as a cytoplasm donor for a second recipient oocyte, which also underwent meiosis.[1,2,10-13] This amplification could be carried through several serial transfers of oocytes, including some in the presence of cycloheximide, indicating there was a maternal store of MPF in each oocyte that could be activated by a post-translational mechanism.

One of the major problems under study in Dr. Krebs' laboratory at the time concerned mechanisms that regulate the activity of phosphorylase b kinase. It had been shown in the lab recently by Dr. Jerry Wang that autophosphorylation and autoactivation of phosphorylase b kinase could, under certain experimental conditions, cause a greater enzyme activation than that exhibited by phosphorylation by the cAMP-dependent protein kinase.[14] This led to the hypothesis by Maller and Krebs that perhaps the autoamplification of MPF activity in injected oocytes was a consequence of autoactivation of a protein kinase. As will be shown later in this chapter, this hypothesis ultimately turned out to be correct. However, at the time experiments were carried out in which calcium-activated phosphorylase b kinase was microinjected into oocytes and, in some cases, these microinjections led to the induction of maturation ($G_2 \Rightarrow M$ transitions). These inductions of maturation occurred only when phosphorylase b kinase had been pre-treated with calcium although calcium alone had no effect. The percentage of oocytes that matured in these experiments in response to phosphorylase b kinase injection was always rather small, and the response was highly variable in batches of oocytes from different animals. In later years it became clear that this response was in fact due to the calmodulin molecule bound to phosphorylase b kinase, since it is now well established that under certain experimental conditions calmodulin alone is capable of inducing oocyte maturation.[9,15,16] Nevertheless, the fact that even a modest response was obtained by microinjection of a purified protein kinase into the oocyte reinforced the hypothesis MPF might be a protein kinase or a regulator of a protein kinase. In addition, the phosphorylase b kinase experiments led to discovery of conditions for extraction and stabilization of MPF, which had been notoriously difficult to extract in active form. In particular, a modified phosphorylase kinase buffer, consisting of 80 mM β-glycerol phosphate (βGP), 20 mM EGTA, 15 mM $MgCl_2$. 1 mM DTT, was found to be the best yet for extraction of MPF,[2,7,17] and continues to be used today. It should not be used in MPF kinase assays, however since it is inhibitory.[18] At the same time the marked effect of phosphorylase b kinase buffer on MPF extraction reinforced the idea that the enzyme could be related to MPF. βGP had historically been used in phosphorylase kinase work because it appeared to stabilize the enzyme, and there was difficulty in WWII of obtaining any other buffer besides Pi; the latter could not be used because it interfered with the assay of phosphorylase a by generation of inorganic phosphate.[19]

The phosphorylase kinase connection also impacted the field by suggesting cAMP had a role in the cell cycle. Because Dr. Krebs' lab had become prominent in the field of protein phosphorylation by establishing the target of cyclic AMP was the cAMP-dependent protein kinase for promotion of phosphorylase b kinase activation,[20] experiments were carried out to see whether microinjection of the catalytic subunit of cAMP-dependent protein kinase would mimic the limited ability of phosphorylase b kinase to induce oocyte maturation. It was also of interest to determine if any component of the phosphorylation burst normally induced by microinjection of crude extracts of MPF could be mimicked by cAMP. Since these experiments were designed to see if cAMP-dependent protein kinase was a positive regulator of maturation, we were greatly surprised to discover that microinjection of the catalytic subunit of cAMP-dependent protein kinase actually led to inhibition of oocyte maturation.[21]

Shortly after these experiments were completed, a paper appeared by Speaker and Butcher in which *Rana* oocytes treated with progesterone were observed to undergo a transient decrease in the level of cAMP.[22] The results that we had obtained suggested this decrease in cAMP was in fact required, since raising the level of cAMP-dependent protein kinase catalytic subunit was able to completely block the hormone response.[21] To examine this situation further, experiments were initiated in which both the regulatory subunit and the heat stable inhibitor of cAMP-dependent protein kinase were microinjected into oocytes. Both these components were found to induce oocyte maturation in the complete absence of hormone.[21] This was an exciting result at the time because it indicated that the decrease in cAMP was in fact sufficient to induce the oocyte maturation response. Subsequent experiments correlated a decrease in cAMP levels in *Xenopus* oocytes with the period of effective inhibition of maturation by the C subunit.[9,21,23]

One general consequence of this work at the time was that it strongly reinforced the idea that changes in the level of cAMP mediated effects largely through changes in the activity of cAMP-dependent protein kinase. For the first time all the effects of a hormone that had been shown to change cAMP levels could be reproduced merely by changing the intracellular concentration of protein kinase subunits. This concept has stood the test of time with few exceptions noted to date.

A second consequence was that it defined a second period of oocyte maturation in which protein phosphorylation was important; in particular, it implicated protein phosphorylation as underlying the basis for the G_2-arrested state of the oocyte. In this case phosphorylation prevents cell cycle transit, in contrast to the role of protein phosphorylation in stimulating cell cycle traverse that was evident in the MPF microinjection experiments. These two mechanisms of protein phosphorylation control could be clearly dissociated in time as well, because experiments showed that the period of sensitivity to cAMP-dependent protein kinase was only during about the first third of the oocyte maturation time and not related temporally to the burst of protein phosphorylation that accompanies the action of MPF, which begins at about 0.6 of the way to nuclear envelope breakdown in oocytes (0.6 $GVBD_{50}$).[7,9,21,23]

These two similar findings defined the major questions about the control of oocyte cell cycles that would occupy many investigators for the next 15 years. The questions raised by these results focussed on elucidating the structure of MPF and determining how it stimulates protein phosphorylation in recipient oocytes, and secondly, determining what mechanism accounts for the arrest of the cell cycle in G_2 by elevation of cAMP levels. In the light of eighteen years later, it is possible to provide at least partial answers to these questions although much work remains to be done.

PURIFICATION AND CHARACTERIZATION OF MPF

The problem of the structure of MPF proved to be an extraordinarily difficult question to resolve. Numerous laboratories made efforts to purify MPF without notable success, and resolution of the MPF question became the great unsolved problem of oocyte maturation and the biochemistry of the $G_2 \Rightarrow M$ transition generally. One of the reasons for the great difficulty is the fact that the microinjection assay for MPF was a rather insensitive one. This is the case because only a single nucleus is present in the oocyte and so it either breaks down or it does not. This all or none assay means no intermediate effects of activity can be judged, and makes it difficult to detect changes in MPF activity that fall below the threshold amount required to give the response. Another reason for difficulty with the microinjection assay was that it proved to have a great volume dependence, that is, an important variable in the assay was not only how much MPF was injected, but how concentrated that activity was when it was microinjected.[17] The basis for this was not clear but seemed to suggest that there was a mechanism that titrated low amounts of MPF activity ("anti-MPF") and prevented the autocatalytic amplification of MPF activity after MPF injection.

The fundamental breakthrough that led to progress in cell cycle research and elucidation of MPF was the development of a cell-free extract from *Xenopus* eggs that would faithfully reproduce the features of cell cycle arrest either in M-phase or in S-phase (interphase) and importantly, would do so in nuclei that were simply incubated in the extract.[24,25] This system was developed by Fred Lohka, a postdoc in the lab, and was an extension of work he had begun in Yoshi Masui's laboratory in Toronto as a graduate student, where he had made extracts from *Rana* eggs that exhibited some of the properties he was able to develop in *Xenopus* egg extracts.[24] Surprisingly, the conditions required to get functional extracts in *Xenopus* were significantly different than with *Rana*, and nearly a year was required to get the system going.[25] Other laboratories also made efforts to develop functional *Xenopus* extracts, with varying degrees of success.[26,27] Using this assay Fred was able to show that when crude preparations of MPF were added to interphase egg extracts, the number of nuclei that broke down was directly proportional to the amount of MPF added. Importantly the concentration in which MPF was added to the extract was less important than the final concentration in the assay. Using then a simple morphological assay based on the fluorescent breakdown of nuclei that had been synthesized *in vitro*, Fred spent a year in the cold room and was able to purify MPF to near homogeneity, ultimately showing that it consisted of two bands on a silver-stained SDS-polyacrylamide gel.[28] Fractions active in the cell-free assay were also active in the classical assay of inducing GVBD in injected oocytes in the presence of cycloheximide. The two bands in the purified preparation were of equal staining intensity and exhibited M_rs of 34 kilodaltons and 45 kilodaltons. In considering the seminal importance of this landmark purification, it is important to point out that Dr. Lohka had essentially no prior experience in protein purification, and no assumptions were made about the nature of MPF during this purification. It is clear that numerous previous attempts by others to purify MPF had failed in part because of assumptions that had been made about the nature of MPF. Despite great joy when these two bands consistently appeared on a gel at the end of the purification, it was not immediately evident how these two bands were able to cause the events of MPF.

To evaluate the hypothesis that had been developed nearly 15 years earlier that MPF was a protein kinase, we incubated the purified fractions with [γ-^{32}P]ATP and discovered that the 45 kDa subunit became phosphorylated in fractions which also contained the 32 kDa subunit.[28] Similar results were obtained when exogenous substrates, such as histone H1, phosphatase inhibitor-1, and casein were used in the assays. This finding was the first evidence that MPF possessed histone H1 kinase activity, leading to direct verification of the initial hypothesis[9] that MPF was a protein kinase. A striking finding at the time was that there were some biologically active fractions of MPF that contained only the 45 kDa substrate and not any apparent kinase activity.[28] Resolution of this paradox came later when the identity of the 45 kDa component was determined. At this point the purification of MPF (August 1987) identified its subunit composition and, in so doing, resolved one of the great unsolved problems of oocyte maturation.

The next step was to try to purify sufficiently large amounts of MPF that partial protein sequence data could be obtained on the subunits. This presented a formidable challenge because relatively small amounts of MPF could be obtained through our purification procedure even starting with kilograms of unfertilized eggs. However, we were committed to this approach as the most straightforward one required to obtain absolute identification of the nature of the subunits and to clone the genes for these components.

At about this time (January 1988) a collaboration was initiated with Dr. Paul Nurse at the University of Oxford that was designed to evaluate whether either of the subunits of MPF were related to the products of genes that had been identified as being important in cell cycle control in fission yeast. The fission yeast, Schizosaccharomyces pombe, is a genetically tractable organism in which the major cell cycle control point is in the G_2 phase, in contrast to the largely G_1 phase control that has been studied extensively in the budding yeast, Saccharomyces cerevisiae. One of the genes that Dr. Nurse and others had identified in fission yeast as being crucial for transit from $G_2 \Rightarrow M$ restriction point in the cell cycle is the product of the cdc2 gene.[29-31] This gene had been identified genetically as causing a cell division cycle (cdc) phenotype in fission yeast, and the function of cdc2 had been shown to be conserved in the related gene product in budding yeast, CDC28, as well as in human cells.[32,33] The striking aspect of the cdc2 gene product that led to the collaboration between the Maller and Nurse laboratories was that the molecular weight of the cdc2 gene product was 34,000 daltons, and it was predicted to encode a serine/threonine protein kinase activity. It had not escaped our attention that one subunit of MPF appeared to have a molecular weight on SDS gels of 34,000 daltons and was correlated with the incorporation of phosphate into serine and threonine residues in the 45 kDa subunit as well as in the substrate proteins that had been identified. The major reagent that had been used in various laboratories to study the cdc2 gene product for many years was an antibody against a 16 amino acid sequence that appeared to be uniquely present in cdc2 homologs. This 16 amino acid sequence contained the acronym, PSTAIR, amongst its amino acids. Antibodies against this peptide were able to immunoblot and immunoprecipitate cdc2 from a variety of cells.[29,32,34] In what is now a landmark experiment in this field, we were able to show that the Nurse PSTAIR antibody was able to both immunoblot and immunoprecipitate the 34 kDa component of purified MPF, and, moreover, these immunocomplexes contained a protein kinase activity against histone H1.[35] This was a striking finding at the time because it indicated that people working in two completely unrelated fields had suddenly found themselves to be working on precisely the same protein, that is, the cdc2 protein. This connection was confirmed by ongoing work in other laboratories that showed depletion of cdc2 from M-phase by egg extracts depleted MPF activity, although a specific effect on cdc2 could not be demonstrated.[36] In addition, MPF purification from starfish eggs gave similar results.[37] This convergence of genetics and biochemistry led to an extraordinary burst of vitality in the study of cell cycle control and illustrated that the important breakthroughs in biology incorporate data from diverse fields. This was made even more clear later when the 45 kDa subunit of MPF was identified as a B-type cyclin by our laboratory in collaboration with Tim Hunt.[38] In a now familiar motif the cyclin molecules were homologs of a yeast cell cycle control gene known as cdc13. Cyclins had been discovered in invertebrate eggs of marine organisms as proteins that came and went in a cyclical fashion[39] but had not really been connected to the process of the cell cycle as important regulatory components of early development. The ability of cyclin to form kinase-active complexes with monomeric cdc2 explains the ability of cyclin mRNA injection to cause GVBD[40] as well as the activity of purified MPF fractions that contained only the 45 kDa component,[28] cyclin B. The purification of MPF thus identified the precise role and the context in which the cdc2 protein kinase and the cyclins intertwine with the cell cycle.

In retrospect it is now clear that the purification of MPF was the seminal event that led to the convergence of three different fields: protein synthesis in marine invertebrates, actual catalysis of mitosis, and cdc mutants in fission yeast. It is clear that no one of these approaches was capable of producing the breakthrough that came when all three approaches were combined. In particular, it is clear that without the purification of MPF the exact role of cdc2 in the control of mitosis might still be obscure, and the biochemical study of the protein kinase activity of cdc2 would not have occurred. The work focussed attention on the G_2 restriction point in the cell cycle, which could be studied genetically in yeast and biochemically in *Xenopus*. Prior to this time, "cell cycle control" had referred largely to studies of quiescent cells stimulated to re-enter G_1 and S-phase by serum and growth factors. Because of the great advance in understanding that occurred from this convergence of genetics and biochemistry, a large number of laboratories entered the cell cycle field, and as a consequence the pace of research in the field has quickened considerably. The complexity of the regulation of cdc2 has increased dramatically in the last few years.

The resolution of MPF structure raised two general questions that still occupy many laboratories. One question concerns the mechanism by which MPF actually induces the events of mitosis; since MPF is a protein kinase, this clearly relates to the question of what substrates MPF phosphorylates in order to carry out the $G_2 \rightarrow M$ transition. There are a number of substrates that have been identified. These include structural proteins such as histone H1, lamins, and caldesmon as well as enzymatic substrates that lead to pleiotrophic changes in cellular function, such as pp60^{src} and the Abelson gene product.[41-45]

The mix of substrates identified so far suggests that MPF has a role not only as a "workhorse" for induction of mitotic events directly,[41] but could also act as an upstream activator of a variety of kinases to initiate a more pleiotrophic range of responses that could be achieved by the activation of a single kinase alone. Indeed, a major protein kinase cascade involving receptor tyrosines, MAP kinase and S6 kinase was identified through experiments studying phosphorylation induced by MPF in oocytes.[46] Identification of substrates for cdc2 was enhanced by identification of the sites in histone H1 that were phosphorylated by cdc2. These sites turn out to encode a consensus sequence K S T P X K which form the basis for identifying many other substrates of cdc2.[45,47] At the present time all but one or two of the sites identified as being phosphorylated by cdc2 fit this consensus.[41-45,47] This represents a prediction index of something greater than 95% which is an extraordinarily high number compared to most other protein kinases.[48] The high degree of substrate specificity of cdc2 allowed the identification of substrates in a relatively short period of time and provided insight into the mechanism by which it carried out mitotic events. The sites in histone H1 that were phosphorylated were previously attributed to an enzyme known as growth-associated histone H1 kinase, or by some individuals, mitosis-specific histone H1 kinase.[45] That kinase is now known to be either cdc2 itself or else, most likely, a related enzyme known as cdk2.

Although there is interest in identification of further substrates of MPF, the major focus of current research in the cell cycle field has been on the mechanisms that regulate the activity of MPF, and in particular, mechanisms that the couple the onset of mitosis with the completion of DNA synthesis. This coupling represents the focus of attention because the ability of the cell in G_2 phase to monitor its DNA for the fidelity of replication and for possible mutagenic damage by ionizing radiation or mutagens is central to ensuring that the genome is accurately partitioned and segregated in mitosis. Defects in tumor cell checkpoint control may also underlie the efficacy of radiation therapy and chemotherapy in the treatment of human cancer.[49]

Several genes have already been identified in fission yeast as affecting this G_2 restriction point decision to initiate the onset of M-phase. Two important ones are the cdc25 gene and the wee1 gene, which cause dose-dependent activation or inhibition of mitotic onset.[50,51] The importance of these genes biochemically for MPF regulation did not really become apparent until a discovery by ourselves and others that the phosphorylation state of cdc2 oscillated in the cell cycle. In particular the phosphorylation state of cdc2 was high during interphase when H1 kinase activity was low, and low in mitosis when the H1 kinase activity of cdc2 was high.[52-55] Phosphoamino acid analysis of cdc2 showed that the phosphorylation occurring during the low activity state in interphase was on threonine and tyrosine residues. This was an exciting link between a tyrosine kinase signal and a serine/threonine kinase signal. At the time these results were reported, it had been thought that phosphotyrosine did not exist in either budding yeast or fission yeast, but close inspection of the cdc2 gene product in fission yeast revealed that in fact it was phosphorylated on tyrosine and the location of the tyrosine residue was at position 15, which turns out to be directly in the ATP-binding site.[55]

Initially it was thought that this phosphate would prevent ATP-binding and that would account for inhibition of mitosis in interphase, but recent evidence has indicated that ATP-binding is largely unaffected by tyrosine phosphorylation; most likely the transfer of the γ phosphate is inhibited.[56] The importance of this tyrosine phosphorylation in keeping cdc2 inactive during interphase was emphasized by the finding that when DNA synthesis was inhibited, the tyrosine phosphorylation of cdc2 was enhanced.[57] Moreover, when the cdc25 gene was overexpressed in fission yeast, cdc2 tyrosine phosphorylation declined and premature mitosis was initiated, which leads to mitotic catastrophe and death in fission yeast.[58]

About this time Andrew Murray at the University of California developed a modification of the *Xenopus* extract system of Lohka and Maller.[59,60] This modification involved largely the preparation of extracts from eggs that had been passed through an oil layer to remove all extracellular fluid. This resulted in a higher concentration of cytoplasmic protein and such extracts were able to oscillate spontaneously between S-phase and mitosis *in vitro*. It is important to note such oscillatory activity would not have been useful for MPF purification in the original assays. These more concentrated extracts were used by Murray and Kirschner to show that a single form of sea urchin cyclin B was capable of driving several cycles of mitosis and DNA synthesis in these extracts, indicating that cyclin synthesis alone was sufficient to drive the early embryonic cell cycle.[59] Dunphy and co-workers[57] and Solomon et al[61] were able to show that during these spontaneous oscillations there were changes in the phosphorylation on tyrosine 15 of the cdc2 molecule, and in particular if DNA synthesis was inhibited by aphidicolin the tyrosine phosphorylation accumulated.[57] Addition of recombinant cdc25 to the extract caused dephosphorylation and activation of cdc2.

About the same time work carried out by Helen Piwnica-Worms and her collaborators, and Featherstone and Russell, demonstrated that the wee1 protein kinase could directly phosphorylate tyrosine 15 in cdc2 and inhibit its protein kinase activity.[62,63] Together these findings indicated that the two genetically defined components in fission yeast mitotic controls, wee1 and cdc25, actually function at a biochemical level by directly controlling the phosphorylation state of cdc2. It is now clear that cdc25 is a tyrosine phosphatase and wee1 a tyrosine kinase that together control the balance between phosphorylation and dephosphorylation of cdc2 providing a switch that regulates the activation of MPF and the onset of mitosis. Because of their pivotal role in this activation, much attention is currently being directed to elucidation of regulatory inputs into both wee1 and cdc25. The latter has received considerably more attention, and the cdc25 gene has now been cloned from a number of organisms, including *Xenopus*. Studies on the *Xenopus* enzyme indicate that it shares a catalytic site motif conserved in all other known tyrosine phosphatases.[64,65]

Perhaps the most exciting result accompanying the cloning and expression of cdc25 was the discovery that it is activated by phosphorylation on serine and threonine residues.[64,65] This phosphorylation causes a dramatic electrophoretic shift of about 18 kDa on SDS gels and can be reversed completely by protein phosphatase 1 or protein phosphatase 2A. In cell cycles in which onset of mitosis is blocked by the incomplete replication of DNA, the phosphorylation of cdc25 does not occur, suggesting that its phosphorylation is regulated by the mechanism that prevents the onset of mitosis when DNA synthesis is incomplete. This situation has led to an intensive search for kinases and phosphatases that act on cdc25 to control its activity.

A variety of kinases involved in cell cycle control have been investigated for activity towards cdc25. These include cAMP-dependent protein kinase, MAP kinase, S6 kinase II, and MPF itself. Of all of these kinases, the only one that is able to phosphorylate, electrophoretically shift and activate cdc25 is, in fact, the MPF kinase itself, ie, cdc2/cyclin B.[66] This seems at first glance to be an apparent paradox. How can the product of the reaction be the initiator of the reaction? The answer to this is not clear. One possibility is that there is a kinase which initiates the activation of cdc25 by phosphorylation and that once sufficient MPF has been activated to begin the amplification of the system, then cdc2-dependent activation of cdc25 can cause a feed-forward amplification of MPF to maintain the metaphase state. Evidence against existence of another kinase has been the finding that the preparation of extracts in which cyclins A and B are absent and in which H1 kinase activity of cdc2 is very low also have very low cdc25 kinase activity,[66] suggesting that no other kinase, other than a cyclin-dependent kinase, is involved in this process. Moreover, other workers reported that in HeLa cells immunodepletion of cyclin A and cyclin B-associated H1 kinase activities depleted all detectable cdc25 activity.[67]

Of course the phosphorylation state of any protein is a balance between the rate of phosphorylation and the rate of dephosphorylation. Since no other protein kinases except cdc2 were found, at least *in vitro*, to act as cdc25 kinases, we investigated protein phosphatase regulation in this system. Interest in the phosphatase possibility came initially from a discovery by Duncan Walker in the lab that a block to the onset of mitosis due to inhibition of DNA replication by aphidicolin[68] could be overcome by adding the protein phosphatase inhibitor okadaic acid.[69] Okadaic acid is known to inhibit both protein phosphatases 1 and 2A, and therefore experiments were carried out to test the effects of protein phosphatase inhibitor 2, a heat-stable inhibitor protein that is specific for protein phosphatase 1 isozymes. Data demonstrated that inhibitor 2 could also overcome an aphidicolin/DNA synthesis block to the onset of mitosis.[69] These results suggested that protein phosphatase 1 was capable of regulating the feedback control that prevents mitosis until DNA synthesis is complete.

The next question was whether there actually are changes in protein phosphatase 1 activity during a normal cell cycle. Accordingly, egg extracts were prepared that oscillate between M-phase and S-phase upon addition of calcium, and the level of protein phosphatase 1 activity was evaluated. It was discovered that protein phosphatase 1 activity exhibited two peaks in the cell cycle, one in S-phase declining just as MPF became activated, and a second in mid-M-phase when mitosis was well under way.[69] These two biochemically defined peaks of protein phosphatase 1 activity fit well with genetic work in fission yeast, which indicate phenotypes affected by the protein phosphatase 1 gene in both S-phase and in mitosis.[70-74] When the block to mitosis was established by aphidicolin, there was a persistent activation of protein phosphatase 1, providing further evidence it is indeed a component of the pathway linking mitosis and DNA synthesis.[69]

In order to understand the molecular basis for the oscillations in PP1 during the cell cycle, Duncan Walker in the lab cloned the *Xenopus* homolog of PP1 γ-1 from an oocyte cDNA library. He was then able to express this protein in bacteria in a soluble enzymatically active form with a histidine tag, allowing for purification of the protein in one pass over a nickel-agarose column.[74] It was of some interest that the specific activity of this enzyme against phosphorylase a and para-nitrophenylphosphate was identical to that of mammalian skeletal muscle protein phosphatase 1, which is a mixture of pools of catalytic subunits, but had approximately 50 times higher histone H1 phosphatase activity than the mammalian enzyme. Upon addition of the recombinant enzyme to *Xenopus* egg extracts, Dr. Walker observed that the activity of the exogenously added protein phosphatase 1 oscillated in a manner superimposable on that of the endogenous enzyme, even though total protein phosphatase 1 activity in the extract had been elevated.[74]

The C-terminus of PP1 γ-1 is the only portion of the molecule that is significantly different from other PP1 isozymes.[75] However, a distinguishing consensus feature of this sequence in many isozymes is the presence of a cdc2 consensus site, T P P R, in the C-terminal domain. Dr. Walker was able to show that this domain was a substrate for MPF *in vitro*, whereas a C-terminally truncated mutant that lacks the phosphorylation site was not a substrate *in vitro*.[74] Phosphorylation of this molecule had no apparent effect on total enzymatic activity *in vitro*, but the mutant construct, which could not be phosphorylated when added to extracts, caused arrest of the extract in interphase with a persistently high level of protein phosphatase activity. Other data show that PP1γ-1 is phosphorylated in M-phase but not in interphase.[74] Although no effect of phosphorylation *in vitro* on catalytic activity can be detected, these results suggest that the unique C-terminus of the enzyme may be important in directing changes in activity of PP1 γ-1. Based on motifs described in other systems, it seems likely that phosphorylation of the T P residue of the T P P R consensus is important in targeting the phosphatase to specific cellular locations or possibly changing its binding certain substrates over others.

These results define essentially a signal transduction pathway from unreplicated DNA to MPF, and we believe that by working backwards, we will eventually be able to reconstruct the entire sequence of events that constitute the mechanism that links the completion of DNA synthesis to the onset of mitosis. This will be then a biochemical and molecular description of a crucial cell cycle restriction point in G_2 that is of great importance in the efficacy of chemotherapeutic agents in the treatment of human cancer.[49] Table I summarizes the key events in the development of the oocyte and egg cell cycle model that demonstrate the importance of protein phosphorylation as a regulatory mechanism.

SUMMARY

It is clear that protein phosphorylation is the dominant regulatory motif governing both the decision to transit cell cycle restriction points and to execute the events associated with M-phase itself. The feedback control mechanism that operates to link the onset of mitosis to completion of DNA synthesis appears to involve a complex network of protein kinases and phosphatases that operates in a signal transduction pathway from replicating DNA to the activation of cyclin B/cdc2 (MPF). A somewhat different pathway is evident in the meiotic cell cycles where proto-oncogene kinases and cAMP-dependent protein kinase have opposing effects via phosphorylation networks to control activation of MPF via signals originating at the cell surface. Study of the mechanism of action of MPF led to the identification of S6 kinase and MAP kinase as key elements in signal transduction by tyrosine kinase receptors and other mitogenic agents. The past and current work on the cell cycle and on mitogenic signalling reinforces the conclusion

TABLE I

LANDMARKS IN OOCYTE MATURATION

Year(s)	Landmark	Lab(s)
1971	Discovery of MPF and CSF Activity	Masui
1971-1975	Autoamplification of MPF Activity in Recipient Oocytes	Several
1975	Synthesis of Initiator Proteins Required	Masui
1975	Identification of Phosphorylation Burst by MPF-Protein Kinase Hypothesis	Maller
1977	cAMP Decrease Necessary and Sufficient for Release of G_2 Arrest	Maller
1983	Cyclins Identified in Sea Urchin Eggs	Hunt
1983-1985	Development of Cell-Free System for Mitotic Events & Cell Cycle Transitions	Lohka
1986	Clam Cyclins Induce Frog Oocyte Maturation	Ruderman
1986	1st S6 Kinase Purified (Rsk)	Maller
1988	MAP Kinase/S6 Kinase Cascade Identified	Sturgill Maller
1988	MPF Purified and Identified as Heterodimer of Cyclin B and cdc2	Maller
1988	c-mos Identified as Candidate Initiator	Vande Woude
1989	Regulation of cdc2 by Phosphorylation	Several
1989	c-mos Identified as Component of CSF	Vande Woude
1992	Regulation of cdc25 by Phosphorylation	Maller Dunphy
1992	Cdk2 Identified as Component of CSF	Maller

The table lists the key discoveries in oocyte maturation that in the author's opinion identified protein kinases as crucial regulators on the process. The work in the labs cited for each discovery is discussed in the text.

originating from the pioneering concept of Krebs and Fischer that protein phosphorylation is the major regulatory mechanism governing important cellular decisions.

REFERENCES

1. Y. Masui and C.L. Markert, Cytoplasmic control of nuclear behavior during meiotic maturation of frog oocytes. *J Exp Zool* 177:129, (1971).
2. J.L. Maller, Studies on the mechanism of cytoplasmic control of meiotic maturation of *Xenopus* oocytes. Ph.D. thesis, University of California, Berkeley, (1974).
3. T. Kishimoto, R. Kuriyama, H. Kondo, and H. Kanatani, Generality of the action of various maturation-promoting factors. *Exp Cell Res* 137:121, (1982).
4. T. Kishimoto, K. Yamazaki, Y. Kato, S.S. Koide, and H. Kanatani, Induction of starfish oocyte maturation by maturation-promoting factor of mouse and surf clam oocytes. *J Exp Zool* 231:293, (1984).
5. H. Weintraub, M. Buscaglia, M. Ferrez, S. Weiller, A. Boulet, R. Fabré, and E.E. Baulieu, Mise en évidence d'une activité "MPF" chez Saccharomyces cerevisae. *C R Seances Acad Sci Paris Ser III* 295:787, (1982).
6. P.S. Sunkara, D.A. Wright, and P.N. Rao, Mitotic factors from mammalian cells induce germinal vesicle breakdown and chromosome condensation in amphibian oocytes. *Proc Natl Acad Sci USA* 76:2799, (1979).
7. J.L. Maller, M. Wu, and J.C. Gerhart, Changes in protein phosphorylation accompanying maturation of *Xenopus laevis* oocytes. *Dev Biol* 58:295, (1977).
8. M. Dorée, G. Peaucellier, and A. Picard, Activity of the maturation-promoting factor and the extent of protein phosphorylation oscillate simultaneously during meiotic maturation of starfish oocytes. *Dev Biol* 99:489, (1983).
9. J.L. Maller and E.G. Krebs, Regulation of oocyte maturation. *In Current Topics in Cell Regulation*, B. Horecker and E. Stadtman (eds), Academic Press, New York, pp 272, (1980).
10. K.C. Drury and S. Schorderet-Slatkine, Effects of cycloheximide on the "autocatalytic" nature of the maturation-promoting factor (MPF) in oocytes of *Xenopus laevis*. *Cell* 4:268, (1975).
11. W.J. Wasserman and Y. Masui, Effects of cycloheximide on a cytoplasmic factor initiating meiotic maturation in *Xenopus laevis* oocytes. *Exp Cell Res* 91:381, (1975b).
12. L.D. Smith and R.E. Ecker, The interaction of steroids with *Rana pipiens* oocytes in the induction of maturation. *Dev Biol* 25:233, (1971).
13. J.K. Reynhout and L.D. Smith, Studies on the appearance and nature of a maturation-inducing factor in the cytoplasm of amphibian oocytes exposed to progesterone. *Dev Biol* 38:394, (1974).
14. J.H. Wang, J.T. Stull, T.S. Huang, and E.G. Krebs, A study on the autoactivation of rabbit muscle phosphorylase kinase. *J Biol Chem* 251:4521, (1976).
15. O. Mulner, J. Tso, D. Huchon, and R. Ozon, Calmodulin modulates the cyclic AMP levels in *Xenopus* oocytes. *Cell Differ* 12:211, (1983).
16. W.J. Wasserman and L.D. Smith, Calmodulin triggers the resumption of meiosis in amphibian oocytes. *J Cell Biol* 89:389, (1981).
17. M. Wu and J.C. Gerhart, Partial purification and characterization of the maturation-promoting factor from eggs of *Xenopus laevis*. *Dev Biol* 79:465, (1980).
18. E. Erikson and J.L. Maller, Biochemical characterization of the $p34^{cdc2}$ protein kinase component of purified maturation-promoting factor from *Xenopus* eggs. *J Biol Chem* 264:19577, (1989).
19. E.G. Krebs, personal communication.
20. D.A. Walsh, J.P. Perkins, and E.G. Krebs, An adenosine 3′:5′-phosphate-dependent protein kinase from rabbit skeletal muscle. *J Biol Chem* 243.3763, (1968).
21. J.L. Maller and E.G. Krebs, Progesterone-stimulated meiotic cell division in *Xenopus* oocytes: induction by regulatory subunit and inhibition by catalytic subunit of adenosine 3′:5′ monophosphate dependent protein kinase. *J Biol Chem* 252:1712, (1977).
22. M.G. Speaker and F.R. Butcher, Cyclic nucleotide fluctuations during steroid-induced meiotic maturation of frog oocytes. *Nature* 267:848, (1977).
23. J.L. Maller, F.R. Butcher, and Krebs EG, Early effect of progesterone on levels of cyclic adenosine 3′:5′ monophosphate in *Xenopus* oocytes. *J Biol Chem* 254:579, (1979).
24. M.J. Lohka and Y. Masui, Formation *in vitro* of sperm pronuclei and mitotic chromosomes induced by amphibian ooplasmic components. *Science* 220:719, (1983).
25. M.J. Lohka and J.L. Maller, Induction of nuclear envelope breakdown, chromosome condensation, and spindle formation in cell-free extracts. *J Cell Biol* 101:518, (1985).

26. R. Miake-Lye and M.W. Kirschner, Induction of early mitotic events in a cell-free system. *Cell* 41:165, (1985).
27. C.J. Hutchison, R. Cox, R.S. Drepaul, M. Gomperts,and C.C. Ford, Periodic DNA synthesis in cell-free extracts of *Xenopus* eggs. *EMBO J* 6:2003, (1987).
28. M.J. Lohka, M. Hayes, and J.L. Maller, Purification of maturation-promoting-factor, an intracellular regulator of early mitotic events. *Proc Natl Acad Sci USA* 85:3009, (1988).
29. P. Nurse, Cell cycle control genes in yeast. *Trends Genet* 1:51, (1985).
30. V. Simanis and P. Nurse, The cell cycle control gene $cdc2^{+}$ of fission yeast encodes a protein kinase potentially regulated by phosphorylation. *Cell* 45:261, (1986).
31. P. Nurse and P. Thuriaux, Regulatory genes controlling mitosis in the fission yeast *Schizosaccharomyces pombe. Genetics* 96:627, (1980).
32. M. Lee and P. Nurse, Cell cycle control genes in fission yeast and mammalian cells. *Trends Genet* 4:287, (1988).
33. D. Beach, B. Durkacz, and P. Nurse, Functionally homologous cell cycle control genes in fission yeast and budding yeast. *Nature* 300:706, (1982).
34. G. Draetta, L. Brizuela, J. Potashkin, and D. Beach, Identification of p34 and p13, human homologs of the cell cycle regulators of fission yeast encoded by $cdc2^{+}$ and $suc1^{+}$. *Cell* 50:319, (1987).
35. J. Gautier, C. Norbury, M. Lohka, P. Nurse, and J.L. Maller, Purified maturation-promoting factor contains the product of a *Xenopus* homolog of the fission yeast cell cycle control gene cdc 2^{+}. *Cell* 54:433, (1988).
36. W.G. Dunphy, L. Brizuela, D. Beach, and J. Newport, The *Xenopus* cdc2 protein is a component of MPF, a cytoplasmic regulator of mitosis. *Cell* 54:423, (1988).
37. J.C. Labbé, J.-P. Capony, D. Caput, J.C. Cavadore, J. Derancourt, M. Kaghad, J.-M. Lelias, A. Picard, and M. Dorée, MPF from starfish oocytes at first meiotic metaphase is a heterodimer containing one molecule of cdc2 and one molecule of cyclin B. *EMBO J* 8:3053, (1989).
38. J. Gautier, J. Minshull, M. Lohka, M. Glotzer, T. Hunt, J.L. Maller, Cyclin is a component of MPF from *Xenopus*. *Cell* 60:487, (1990).
39. T. Evans, E.T. Rosenthal, J. Youngblom, D. Distel, and T. Hunt, Cyclin: a protein specified by maternal mRNA in sea urchin eggs that is destroyed at each cleavage division. *Cell* 54, 433, (1983).
40. K.I. Swenson, K.M. Farrell, and J.V. Ruderman, The clam embryo protein cyclin A induces entry into M-phase and the resumption of meiosis in *Xenopus* oocytes. *Cell* 47:861, (1986).
41. E.A. Nigg, The substrates of the cdc2 kinase. *Sem Cell Biol* 2:261, (1990).
42. S. Shenoy, J.-P. Choi, S. Bagrodia, T.D. Copeland, J.L. Maller, and D. Shalloway, Purified maturation-promoting factor phosphorylates $pp60^{c\text{-}src}$ at the sites phosphorylated during fibroblast mitosis. *Cell* 57:763, (1989).
43. E.T. Kipreos, J.Y.J. Wang, Differential phosphorylation of c-abl in cell cycle determined by cdc2 kinase and phosphatase activity. *Science* 248:217, (1990).
44. L. Satterwhite, J. Minshull, T. Hunt, L. Cisek, J. Corden, and T. Pollard, Phosphorylation of myosin regulatory light chain by cyclin-p34 kinase provides a mechanism for the timing of cytokinesis. *J Cell Biol* 111:135a, (1990).
45. T.A. Langan, J. Gautier, M. Lohka, R. Hollingsworth, P.N. Moreno, P.N. Nurse, J.L. Maller and R. Sclafani, Mammalian growth-associated H1 histone kinase: a homolog of $cdc2^{+}$/CDC28 protein kinase controlling mitotic entry in yeast and frog cells. *Mol Cell Biol* 9:3860, (1989).
46. T.W. Sturgill, L.B. Ray, E. Erikson and J.L. Maller, Insulin-stimulated MAP-2 kinase phosphorylates and activates ribosomal protein S6 kinase II. *Nature* 334:715, (1988).
47. B.T.Y. Lin, S. Gruewald, A.O. Morla, W.H. Lee, and J.Y.J. Wang, Retinoblastoma cancer suppressor gene product is a substrate of the cell cycle regulator cdc2 kinase. *EMBO J* 10:857, (1991).
48. B.E. Kemp and R.B. Pearson, Protein kinase recognition sequence motifs. *Trends Biochem Sci* 1:342, (1990).
49. A.W. Murray, Creative blocks: cell cycle checkpoints and feedback controls. *Nature* 39:599, (1992).
50. P. Russell and P. Nurse, $Cdc25^{+}$ functions as an inducer in the mitotic control of fission yeast. *Cell* 45:145, (1986).
51. P. Russell and P. Nurse, Negative regulation of mitosis by $wee1^{+}$, a gene encoding a protein kinase homolog. *Cell* 49:559, (1987).
52. J. Gautier, T. Matsukawa, P. Nurse, and J.L. Maller, Dephosphorylation and activation of *Xenopus* $p34^{cdc2}$ protein kinase during the cell cycle. *Nature* 339:626, (1989).
53. C. Norbury, J. Blow, and P. Nurse, Phosphorylation of the $p34^{cdc2}$ protein kinase in vertebrates. *EMBO J* 10:3321, (1991).
54. W. Krek and E.A. Nigg, Differential phosphorylation of vertebrate $p34^{cdc2}$ kinase at the G_1/S and G_2/M transitions of the cell cycle: identification of major sites. *EMBO J* 10:305, (1991).

55. K. Gould and P. Nurse, Tyrosine phosphorylation of the fission yeast cdc2+ protein kinase regulates entry into mitosis. *Nature* 342:39, (1989).
56. S. Atherton-Fessler, L.L. Parker, R.L. Geahlen, and H. Piwnica-Worms, Mechanisms of $p34^{cdc2}$ regulation. *Mol Cell Biol* 13:1675, (1993).
57. A. Kumagai and W.G. Dunphy, The cdc25 protein controls tyrosine dephosphorylation of the cdc2 protein in a cell-free system. *Cell* 64:903, (1991).
58. Y. Enoch and P. Nurse, Mutation of fission yeast cell cycle control genes abolishes dependence of mitosis on DNA replication. *Cell* 60:665, (1990).
59. A.W. Murray and Kirshner, Cyclin synthesis drives the early embryonic cell cycle. *Nature* 339:275, (1989).
60. A.W. Murray, M.J. Solomon, and M.W. Kirschner, The role of cyclin synthesis and degradation in the control of MPF activity. *Nature* 339:280, (1989).
61. M.J. Solomon, M. Glotzer, T. Lee, M. Philippe, and M.W. Kirschner, Cyclin activation of $p34^{cdc2}$. *Cell* 63:1013, (1991).
62. C. Featherstone and P. Russell, Fission yeast $p107^{wee1}$ mitotic inhibitor is a tyrosine/serine kinase. *Nature* 349:808, (1991).
63. L. Parker, S. Atherton-Fessler, M.S. Lee, S. Ogg S, J.L. Faulk, K.I. Swenson, and H. Piwnica-Worms, Cyclin promotes the tyrosine phosphorylation of $p34^{cdc2}$ in a wee1-dependent manner. *EMBO J* 10:1255, (1991).
64. T. Izumi, D.H. Walker, and J.L. Maller, Periodic changes in phosphorylation of the *Xenopus* cdc25 phosphatase regulates its activity. *Mol Biol Cell* 3:927, (1992).
65. A. Kumagai and G. Dunphy, Regulation of the cdc25 protein during the cell cycle in *Xenopus* extracts. *Cell* 70:139, (1992).
66. T. Izumi, J.L. Maller, Manuscript in preparation.
67. I. Hoffmann, P.R. Clarke, M.J. Marcoti, E. Karsenti, and G. Draetta, Phosphorylation and activation of human cdc25-C by cdc2-cyclin B and its involvement in the self-amplification of MPF at mitosis. *EMBO J* 12:53, (1993).
68. D.H. Walker and J.L. Maller, Role for cyclin A in the dependence of mitosis on completion of DNA replication. *Nature* 354:314, (1991).
69. D.H. Walker, A.A. DePaoli-Roach, and J.L. Maller, Multiple roles for protein phosphatase 1 in regulating the *Xenopus* early embryonic cell cycle. *Mol Biol Cell* 3:687, (1992).
70. J.M. Axton, V. Dombradi, P.T.W. Cohen, and D.M. Glover, One of the protein phosphatase 1 isoenzymes in Drosophila is essential for mitosis. *Cell* 63:33, (1990).
71. J.H. Doonan and N.R. Morris, The *BimG* gene of *Aspergillus nidulans*, required for completion of anaphase, encodes a homolog of mammalian protein phosphatase 1. *Cell* 54:17, (1989).
72. N. Kinoshita, H. Ohkura, and M. Yanagida, Distinct, essential roles of type 1 and 2A protein phosphatases in the control of the fission yeast cell division cycle. *Cell* 63:405, (1990).
73. R. Booher and D. Beach, Involvement of a type 1 protein phosphatase encoded by bws1+ in fission yeast mitotic control. *Cell* 57:1009, (1989).
74. D.H. Walker, R.E. Rempel, and J.L. Maller, The c-terminal sequence of *Xenopus* phosphatase 1-γ1 determines it cell cycle-dependent regulation and phosphorylation, submitted.
75. K. Sasaki, H. Shima, Y. Kitagawa, S. Irino, T. Sugimura, and M. Nagao, Identification of members of the protein phosphatase 1α gene in rat hepatocellular carcinomas. *Jpn J Cancer Res* 81:1272, (1990).

2

CONTROL OF G1 PROGRESSION BY MAMMALIAN D-TYPE CYCLINS

Charles J. Sherr,[1,2] Hitoshi Matsushime,[2] Jun-ya Kato,[2]
Dawn E. Quelle,[2] and Martine F. Roussel[2]

[1] Howard Hughes Medical Institute
[2] Department of Tumor Cell Biology
St. Jude Children's Research Hospital
Memphis, TN 38105

INTRODUCTION

Progression through the first gap phase (G1) of the mammalian cell cycle is regulated by growth factors, but once cells commit to replicate their cellular DNA, they can undergo mitosis even if deprived of growth factors during ensuing cell cycle intervals.[1] Recently isolated mammalian cyclins, including cyclins D1, D2, D3, and E,[2-6] appear likely to play central roles in integrating growth factor-induced signals with the cell cycle clock, thereby driving cells into S phase. To date, the hypothesis that D-type cyclins and cyclin E function to govern G1 transitions has rested largely on circumstantial evidence, but recent data now indicate that their activities are rate limiting for G1 progression and required for the entry of cells into the DNA synthetic (S) phase of the cell cycle.

G1 CONTROL IN BUDDING YEAST

In *Saccharomyces cerevisiae*, the commitment to replicate chromosomal DNA is made at START, an execution point late in G1 where critical nutrients and α-mating factors no longer exert their influence on cell cycle progression.[7] The transition through START is governed by the accumulation of G1 cyclins (encoded by three *CLN* genes), which act to regulate the activity of the 34 kilodalton cyclin-dependent kinase, p34^{CDC28}, to drive entry into S phase (reviewed in 8). The activity of any one of the three *CLN* genes is sufficient for cell growth, and mutations in all three are required to induce G1 arrest at the same position in the cell cycle as mutants defective in *CDC28*.[9] Because both dominant and loss of function mutations in *CLN* genes can affect cell size, *CLN* function appears to be responsive to extracellular nutrients that determine when cell size is sufficient for START to occur.[10-13] Conversely, the

The Cell Cycle: Regulators, Targets, and Clinical Applications
Edited by V.W. Hu, Plenum Press, New York, 1994

α-mating factor arrests G1 progression before START by triggering signals that interfere with G1 progression.[11,12,14-16] Because START occurs before the actual onset of DNA synthesis, the activities of the *CDC28* holoenzymes are rate limiting for a late G1 commitment step, but not for the G1/S transition *per se*.

G1 CONTROL IN MAMMALIAN CELLS

In animal cells, a late G1 restriction (or R) point defines a time at which cells no longer require growth factors for their subsequent progression through the cell cycle.[1] The R point may be analogous to START in yeast in the sense that it appears to involve the commitment to enter S phase, which follows several hours later. A reasonable hypothesis is that labile mammalian G1 cyclins together with their cyclin-dependent kinases (cdks) might act as growth factor sensors in controlling this step, much in the same way as *CLN*s and *CDC28* exert this function in budding yeast. Although cyclins identical to products of *CLN* genes have yet to be isolated from mammalian cells, the D-type cyclins and cyclin E are now known to be expressed during mid to late G1, corresponding to the time when passage through the R point occurs.

Cyclin E is expressed periodically, peaking near the G1/S transition,[5] and it regulates the activity of a $p34^{CDC28}$-like kinase, cdk2,[17,18] whose activity is necessary for S phase entry.[19] The recent demonstration that cyclin E expression is rate limiting for the G1/S transition in mammalian cells[20] has provided the first direct evidence that it must regulate critical events before or during this interval.

The D-type cyclins generally appear earlier during G1 than cyclin E, and they are differentially regulated in various cell lineages with no one of them being universally expressed in all proliferating cells.[3] In general, the induction and continued synthesis of D-type cyclins during G1 depends upon stimulation of cells by growth factors, and because the cyclins are highly unstable, they are degraded rapidly following growth factor withdrawal.[3,21] If this occurs during mid G1, the disappearance of D-type cyclins correlates with the cell's failure to enter S phase. In contrast, if growth factors are withdrawn after cells have entered S phase, the ensuing depletion of D-type cyclins has no obvious effect on progression through the remaining phases of the cell cycle.[3,22]

Microinjection of antisense plasmids or antibodies to cyclin D1 into serum-stimulated human or rodent fibroblasts during mid-G1 can prevent their entry into S phase, but injections performed near the G1/S transition are without effect[23] (Quelle, D.E., Ashmun, R.A., Shurtleff, S.A., Kato, J-Y., Bar-Sagi, D., Roussel, M.F., and Sherr, C.J., manuscript submitted). The ability of such reagents to block host DNA synthesis when injected several hours after the G_0/G1 transition, but not after G1/S, argues that cyclin D1 function is required during mid to late G1 to advance the cell cycle. Overexpression of cyclins D1 or D2 in rodent fibroblasts accelerates their G1 transit by several hours and leads to an equivalent shortening of their generation time (Quelle et al., manuscript submitted). Although such cells remain contact inhibited and anchorage dependent, they have a reduced serum requirement for cell growth, and those expressing D1 become smaller in size than their normal counterparts. Therefore, in fibroblasts, cyclin D1 and most likely D2 are rate limiting for G1 progression.

The cell cycle phenotypes observed when cyclin D1 and D2 expression are enforced are quite similar to that induced by ectopic overexpression of cyclin E, except that in the latter case, there is a compensatory prolongation of S phase, so that the overall doubling time of the cells is unchanged.[20] Moreover, cyclins D1 and D2 may differ from one another in their ability to affect cell size. Assuming that these apparent differences prove to be significant, cyclins D1, D2, and E may regulate distinct events in G1. Alternatively, ectopic overexpression of cyclins might result in a loss of their specificity, so that in circumstances where their levels are significantly elevated, one cyclin might replace the normal function of another. The latter possibility is unlikely to explain the effects of cyclins D1 and E, based on accumulating evidence that these interact with different cdk partners (see below). Still, it will be important to test the effects of cyclin D and E overexpression under conditions where they are regulated by the same vectors in a single cell type, as well as to determine whether their abilities to contract the G1 interval are additive or complementary.

CYCLIN D-DEPENDENT KINASES

The D-type cyclins interact with an atypical catalytic partner, cdk4,[21,24,25] but some associate with as yet other cdks, including cdk2 and cdk5.[24,26] When inserted into a baculovirus vector and expressed in insect Sf9 cells, cdk4 is completely devoid of protein kinase activity. However, when the cells are programmed to coexpress cdk4 and any one of the three D-type cyclins, cdk4 is activated as a kinase.[25] The interactions between D-type cyclins and cdk4 appear to be specific, in the sense that this catalytic subunit cannot be activated by cyclins A, B, or E. However, cyclins D2 and D3, but not D1, are also able to activate cdk2 in this system,[26] and because cdk2 can productively partner with cyclins A, E, D2, or D3, it is a more promiscuous catalytic subunit than cdk4. The differences between cyclins D2 and D3 versus D1 in activating cdk2 are intriguing, because they suggest that the functions of D-type cyclins are not strictly redundant.

The combinatorial interactions of the cyclins and cdks, and the rapid turnover of the cyclins (but not the cdks) in such complexes[21] suggest that cdks might exchange cyclin partners as cells progress through the cell cycle. If the holoenzyme complexes target different substrates that are important during various cell cycle transitions, inhibition of cyclin D or E degradation might arrest cell cycle progression before G1/S, as occurs at metaphase when mitotic cyclin B is not degraded.[27,28] A prediction, then, is that microinjection of stabilized cyclin mutants into appropriate target cells might prevent G1 exit.

INTERACTIONS OF CYCLINS WITH THE RETINOBLASTOMA PROTEIN

The D-type cyclins (but not cyclins A, B, C, or E) contain a Leu-X-Cys-X-Glu (LXCXE) motif at their aminotermini which they share with DNA tumor virus oncoproteins (T antigen, E1A, and E7) that bind to the retinoblastoma gene product (pRb) and to pRb-related proteins (e.g., p107, p130). These oncoproteins bind specifically to the underphosphorylated forms of pRb, which are present only during the G1 interval, and thereby appear to inactivate pRb's growth suppressive function (reviewed in 29). The D-type cyclins can also bind directly to pRb, either *in vitro*[26,30] or in intact insect Sf9 cells programmed to express both classes of genes.[25] Cyclins D2 and D3 appear to form more stable complexes with pRb or p107 than does cyclin D1,[25,26] and disruption of their LXCXE segments greatly diminishes the stability of these interactions. Conversely, pRb mutants that are defective in preventing G1 exit do not bind to the D-type cyclins.[26,30]

When introduced into the *RB*-negative, human osteosarcoma cell line, Saos-2, pRb remains underphosphorylated and induces G1 arrest.[31,32] However, cotransfection of *RB* together with expression vectors encoding cyclins A, E, or D2 induces pRb hyperphosphorylation and concommitantly overrides its ability to arrest G1 progression.[26,33] Cyclin D2 mutants lacking the LXCXE motif are much less active in this assay.[26] In insect Sf9 cells coinfected with vectors encoding pRb, cdk4, and cyclins D2 or D3, pRb undergoes hyperphosphorylation at physiologically relevant sites, and complexes between pRb and the cyclins are destabilized.[25] The simplest interpretation, then, is that cyclins D2, E, and A can each activate cdk(s) which phosphorylate pRb, resulting in pRb inactivation, disruption of pRb complexes, and G1 exit. Note that cdk2 can be activated by each of these cyclins in insect Sf9 cells, but not by cyclin D1 (see above). In spite of the intriguing nature of these observations, an important caveat is that pRb has not been demonstrated to form stable complexes with any cyclin in normal, untransfected mammalian cells, and the identity of the physiological pRb kinase(s) remains unknown. It is therefore possible that the targets of cyclin D- and/or E-associated cdk(s) do not include pRb itself but, rather, other pRb-related proteins.

Cyclin D1 does not induce pRb hyperphosphorylation in Saos-2 cells, and its effects on pRb's growth suppressive function are far less dramatic.[26,30] Here, however, cyclin D1 mutants disrupted in their LXCXE motif were *more* active than the wild-type D1 gene in overriding pRb function, leading to speculation that cyclin D1, unlike D2 and D3, is itself regulated by pRb and acts "downstream" of pRb in controlling cell proliferation.[30] Whatever the interpretation, the data underscore the concept that cyclins D2 and D3 differ functionally from D1. Moreover, the opposing effects of the cyclins and a known tumor suppressor gene imply that, at least under certain circumstances, D-type cyclins might themselves contribute to tumor formation (see below).

D-TYPE CYCLINS CAN AFFECT DIFFERENTIATION

A possible consequence of ectopic D-type cyclin overexpression in certain cells might be a loss of their ability to differentiate. For example, growth arrest in G1 might be required to execute certain differentiation-specific programs, and cyclin D overexpression might override such decisions by driving cells into S phase. To test this possibility, D-type cyclins were overexpressed in immature 32Dcl3 myeloid cells, which self renew in interleukin-3 (IL-3) but differentiate to mature neutrophils in medium containing granulocyte colony-stimulating factor (G-CSF).[34] These cells are strictly growth factor dependent and die rapidly in the absence of hemopoietins. When grown in IL-3, parental 32Dcl3 cells normally synthesize cyclins D2 and D3, and not D1, but when shifted to medium containing G-CSF, cell proliferation stops, and cyclin D synthesis terminates. It is important to note that G-CSF does not provide a mitogenic signal, but instead supports the survival of growth arrested cells that commit to differentiation in the granulocyte lineage. Overexpression of cyclins D2 and D3 in 32Dcl3 cells had no detectable effects on their proliferation in IL-3 but inhibited differentiation in response to G-CSF and led to a loss in cell viability. In contrast, cyclin D1 was without effect (Kato, J-Y. and Sherr, C.J., manuscript submitted). Quite possibly, 32Dcl3 cells are refractory to effects of ectopically expressed cyclin D1, because they normally depend only upon cyclins D2 and D3 for proliferation.

Cyclin D2 mutants disrupted in their LXCXE motif were unable to inhibit differentiation in this system, implicating pRb or pRb-like proteins in the process. Conversely, introduction of catalytically inactive cdk4 into cells overexpressing cyclin D2 counteracted the effect of the cyclin and restored their ability to differentiate in G-CSF. Because their failure to respond to G-CSF could be reversed, the observed phenotypic effects must have been due to the cyclin and not to the fortuitous selection of clonal variants that had lost the ability to differentiate.

As yet, methods for measuring cyclin D-dependent kinase activity *in vitro* or in Sf9 cells have not proven sufficiently sensitive or specific to quantitate such activities in lysates of mammalian cells. However, we presume that in 32Dcl3 cells, the persistent activation of cdks by ectopically expressed cyclins D2 and D3 results in phosphorylation events that drive cells into S phase. Because the cyclins neither render the cells G-CSF responsive nor independent of hemopoietins for survival, such cells cannot proliferate and eventually die.

CYCLIN D1 IN TUMORIGENESIS

Cyclin D1 was simultaneously discovered by three independent routes: namely, as a gene induced in macrophages by colony-stimulating factor 1,[3] as a gene able to complement *CLN*-deficiency in yeast,[4] and as a genetic target of chromosomal inversion in parathyroid adenomas.[2] The latter findings imply that cyclin D1 might act as a proto-oncogene, with its deregulated expression contributing to tumor formation. Consistent with this finding, the cyclin D1 locus on human chromosome 11q13 is also the target of a translocation [t(11;14)] commonly observed in centrocytic lymphomas, where the immunoglobulin heavy chain enhancer is fused to undisrupted cyclin D1 coding sequences.[35,36] The cyclin D1 locus is also commonly amplified in breast and esophageal carcinomas and in squamous cell tumors of the head and neck,[37,38] suggesting that it may be frequently involved in tumorigenesis. Because overexpression of cyclin D1 does not in itself lead to oncogenic transformation, at least in fibroblasts (Quelle et al., manuscript submitted), it seems likely that secondary collaborating events are required for tumorigenesis. It would be interesting to know, for example, whether cases of centrocytic lymphoma and B cell chronic lymphocytic leukemia that exhibit a loss of RB function[39] are also those that have translocations affecting cyclin D1. The possibility that deregulated cyclin D1 expression might perturb the differentiation of epithelial cells also merits consideration and might be most readily tested in appropriate cell lines or in transgenic mice. Whatever the outcome, the involvement of the cyclin D1 locus in a variety of common human malignancies reinforces the view that the gene acts to positively regulate cell proliferation.

Acknowledgments

We thank the other members of our laboratory, who contributed to various aspects of this work, and Drs. Mark E. Ewen and David M. Livingston of the Dana Farber Cancer Institute (Boston, MA), who were responsible for many of the critical experiments involving pRb. This work was supported in part by NIH grant CA-47064 (to C.J.S.), by Cancer Center Core grant CA-21765, and by the American Lebanese Syrian Associated Charities of St. Jude Children's Research Hospital.

REFERENCES

1. A.B. Pardee, G1 events and regulation of cell proliferation, *Science* 246:603 (1989).
2. T. Motokura, T. Bloom, H.G. Kim, H. Juppner, J.V. Ruderman, H.M. Kronenberg, and A. Arnold, A novel cyclin encoded by a *bcl*1-linked candidate oncogene, *Nature (London)* 350:512 (1991).

3. H. Matsushime, M.F. Roussel, R.A. Ashmun, and C.J. Sherr, Colony-stimulating factor 1 regulates novel cyclins during the G1 phase of the cell cycle, *Cell* 65:701 (1991).
4. Y. Xiong, T. Connolly, B. Futcher, and D. Beach, Human D-type cyclin, *Cell* 65:691 (1991).
5. D.J. Lew, V. Dulic, and S.I. Reed, Isolation of three novel human cyclins by rescue of G1 cyclin (Cln) function in yeast, *Cell* 66:1197 (1991).
6. A. Koff, F. Cross, A. Fisher, J. Schumacher, K. Leguellec, M. Philippe, and J.M. Roberts, Human cyclin E, a new cyclin that interacts with two members of the CDC2 gene family, *Cell* 66:1217 (1991).
7. L.H. Hartwell, *Saccharomyces cerevisiae* cell cycle, *Bact Rev* 38:164 (1974).
8. S.I. Reed, G1-specific cyclins: In search of an S-phase promoting factor, *Trends Genet* 7:95 (1991).
9. H.E. Richardson, C. Wittenberg, F. Cross, and S.I. Reed, An essential G1 function for cyclin-like proteins in yeast, *Cell* 59:1127 (1989).
10. P.E. Sudbery, A.R. Goodey, and B.L.A. Carter, Genes which control cell division in the yeast *Saccharomyces cerevisiae*, *Nature* 288:401 (1980).
11. R. Nash, G. Tokiwa, S. Anand, K. Erickson, and A.B. Futcher, The $WH1^+$ gene of *Saccharomyces cerevesiae* tethers cell division to cell size and is a cyclin homolog, *EMBO J* 7:4335 (1988).
12. F. Cross, *DAF*1, a mutant gene affecting size control, pheromone arrest, and cell cycle kinetics of *Saccharomyces cerevesiae*, *Mol Cell Biol* 8:4675 (1988).
13. J.A. Hadwiger, C. Wittenberg, H.E. Richardson, M. de Barros Lopes, and S.I. Reed, A novel family of cyclin homologs that control G1 in yeast, *Proc Natl Acad Sci USA* 86:6255 (1989).
14. F. Chang, and I. Herskowitz, Identification of a gene necessary for cell cycle arrest by a negative growth factor of yeast: *FAR*1 is an inhibitor of a G1 cyclin, *CLN*2, *Cell* 63:999 (1990).
15. E.A. Elion, P.L. Grisafi, and G.R. Fink, FUS3 encodes a $cdc2^+$/CDC28-related kinase required for the transition from mitosis into conjugation, *Cell* 60:649 (1990).
16. C. Wittenberg, K. Sugimoto, and S.I. Reed, G1-specific cyclins of *S. cerevisiae*: Cell cycle periodicity, regulation by mating pheromone, and association with the $p34^{CDC28}$ protein kinase, *Cell* 62:225 (1990).
17. A. Koff, A. Giordano, D. Desai, K. Yamashita, J.W. Harper, S. Elledge, T. Nishimoto, D.O. Morgan, B.R. Franza, and J.M. Roberts, Formation and activation of a cyclin E-cdk2 complex during the G1 phase of the human cell cycle, *Science* 257:1689 (1992).
18. V. Dulic, E. Lees, and S.I. Reed, Association of human cyclin E with a periodic G1-S phase protein kinase, *Science* 257:1958 (1992).
19. L-H. Tsai, E. Lees, B. Faha, E. Harlow, and K. Riabowol, The cdk2 kinase is required for the G1-to-S transition in mammalian cells, *Oncogene*, in press.
20. M. Ohtsubo, and J.M. Roberts, Cyclin-dependent regulation of G1 in mammalian fibroblasts, *Science* 259:1908 (1993).
21. H. Matsushime, M.E. Ewen, D.K. Strom, J-Y. Kato, S.K. Hanks, M.F. Roussel, and C.J. Sherr, Identification and properties of an atypical catalytic subunit ($p34^{PSKJ3}$/CDK4) for mammalian D-type G1 cyclins, *Cell* 71:323 (1992).
22. H. Matsushime, M.F. Roussel, and C.J. Sherr, Novel mammalian cyclin (CYL) genes expressed during G_1, *in:* The Cell Cycle, Cold Spring Harbor, NY, Cold Spring Harbor Symp Quant Biol, p 69 (1991).

23. V. Baldin, J. Likas, M.J. Marcote, M. Pagano, J. Bartek, and G. Draetta, Cyclin D1 is a nuclear protein required for cell cycle progression in G1, *Genes & Devel*, in press.
24. Y. Xiong, H. Zhang, and D. Beach, D-type cyclins associate with multiple protein kinases and the DNA replication and repair factor PCNA, *Cell* 71:505 (1992).
25. J-Y. Kato, H. Matsushime, S.W. Hiebert, M.E. Ewen, and C.J. Sherr, Direct binding of cyclin D to the retinoblastoma gene product (pRb) and pRb phosphorylation by the cyclin D-dependent kinase, CDK4, *Genes & Devel* 7:331 (1993).
26. M.E. Ewen, H.K. Sluss, C.J. Sherr, H. Matsushime, J-Y. Kato, and D.M. Livingston, Functional interactions of the retinoblastoma protein with mammalian D-type cyclins, *Cell* 73:487 (1993).
27. A. Murray, M.J. Solomon, and M.W. Kirschner, The role of cyclin synthesis and degradation in the control of MPF activity, *Nature (London)* 339:280 (1989).
28. M. Glotzer, A.W. Murray, and M.W. Kirschner, Cyclin is degraded by the ubiquitin pathway, *Nature (London)* 349:132 (1991).
29. R.A. Weinberg, Tumor suppressor genes, *Science* 254:1138 (1991).
30. S.F. Dowdy, P.W. Hinds, K. Louis, S.I. Reed, A. Arnold, and R.A. Weinberg, Physical interactions of the retinoblastoma protein with human cyclins, *Cell* 73:499 (1993).
31. D.W. Goodrich, N.P. Wang, Y-W. Qian, Y-H.P. Lee, and W-H. Lee, The retinoblastoma gene product regulates progression through the G1 phase of the cell cycle, *Cell* 6:953 (1991).
32. X-Q. Qin, T. Chittenden, D.M. Livingston, and W.G. Kaelin, Jr, Identification of a growth suppression domain within the retinoblastoma gene product, *Genes & Devel* 6:953 (1992).
33. P.W. Hinds, S. Mittnacht, V. Dulic, A. Arnold, S.I. Reed, and R.A. Weinberg, Regulation of retinoblastoma protein functions by ectopic expression of human cyclins, *Cell* 70:993 (1992).
34. J.S. Greenberger, M.A. Sakakeeny, R.J. Humphries, C.J. Eaves, and R.J. Eckner, Demonstration of permanent factor-dependent multipotential (erythroid/neutrophil/basophil) hematopoietic progenitor cell lines, *Proc Natl Acad Sci USA* 80:2931 (1983).
35. C.L. Rosenberg, E. Wong, E.M. Petty, A.E. Bale, Y. Tsujimoto, N.L. Harris, and A. Arnold, PRAD1, a candidate BCL1 oncogene: Mapping and expression in centrocytic lymphoma, *Proc Natl Acad Sci USA* 88:9638 (1991).
36. D.A. Withers, R.C. Harvey, J.B. Faust, O. Melnyk, K. Carey, and T.C. Meeker, Characterization of a candidate *bcl*-1 gene, *Mol Cell Biol* 11:4846 (1991).
37. G.A. Lammie, V. Fantl, R. Smith, E. Schuuring, S. Brookes, R. Michalides, C. Dickson, A. Arnold, and G. Peters, D11S287, a putative oncogene on chromosome 11q13, is amplified and expressed in squamous cell and mammary carcinomas and is linked to BCL-1, *Oncogene* 6:439 (1991).
38. W. Jiang, S.M. Kahn, N. Tomita, Y-J. Zhang, S-H. Lu, and I.B. Weinstein, Amplification and expression of the human cyclin D gene in esophageal cancer, *Cancer Res* 52:2980 (1992).
39. S. Stilgenbauer, H. Dohner, M. Bulgay-Morschel, S. Weitz, M. Bentz, and P. Lichter, High frequency of monoallelic retinoblastoma gene deletion in B-cell chronic lymphoid leukemia shown by interphase cytogenetics, *Blood* 81:2118 (1993).

3

PHOSPHORYLATION IN THE REGULATION OF PROTEIN PHOSPHATASES

David L. Brautigan, Jian Chen, Fran Pinault, Jeremy Somers, and Richard Zimmerman

Brown University
Division of Biology and Medicine
Providence, RI 02912

PROTEIN PHOSPHORYLATION CYCLES

Cell cycles are controlled by the cycling of proteins through phosphorylation and dephosphorylation. These reactions are catalyzed by kinases and phosphatases that are now appreciated as being highly specific. Dephosphorylation is not the simple reverse of phosphorylation because the introduction of a phosphate group will lead to formation of hydrogen bonds with the guanidinium side chains of arginine residues, accompanied by a refolding of the protein (1). The phosphorylated site then occupies a new position on the protein surface hydrogen-bonded to a portion of the protein distant from the original location of the unphosphorylated site. Phosphatases recognize the new conformation and hydrolyze the phosphate, abolishing the hydrogen bonds formed following phosphorylation, thereby reversing the conformational change. The cycle of phosphorylation and dephosphorylation functions as a molecular switch at appropriate times during the cell cycle and in response to extracellular signals (Figure 1A). From results of the last twenty years of research the idea

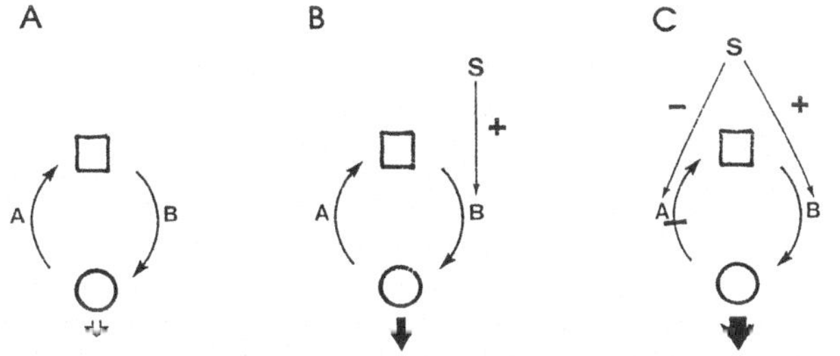

Figure 1. Regulation of protein phosphorylation cycles in signal transduction

has evolved that second messengers such as cAMP and Ca^{2+} increase in cells after stimulation to activate specific protein kinases. In contrast, the protein phosphatases were thought to be constituitively active and relatively non-specific to reverse phosphorylation events (Figure 1B). This in effect is a "tortoise and hare" model for protein phosphorylation, with kinases intermittently bursting with activity while the phosphatases plod along at a steady level of activity. Based on this thinking, phosphatases have been considered as

The Cell Cycle: Regulators, Targets, and Clinical Applications
Edited by V.W. Hu, Plenum Press, New York, 1994

"housekeeping" enzymes. What a pejorative reference to enzymes that we now beleive are important and essential regulators of a wide range of vital functions!

Here we present new data to revise this thinking about kinases and phosphatases - to replace stale thinking with a fresh viewpoint. The fresh idea is that both kinases and phosphatases are regulated in a coordinated and reciprocal manner to efficiently interconvert intracellular proteins between active and inactive states at particular times of the cell cycle or in response to extracellular growth signals (Figure 1C). Proteins can be phosphorylated reversibly either on the the phenolic side chains of Tyr residues or the alcoholic side chains Ser and Thr side chains. With a few notable exceptions there are separate kinases and separate phosphatases for these two sets of reactions (see 3). Here we will show two examples where a phosphatase in one cycle is itself a substrate for a kinase in another cycle. We propose that regulation of phosphatases involves phosphorylation of the catalytic subunits; Ser phosphorylation to activate protein Tyr phosphatases (e.g. PTP1B and *cdc25*) and Tyr phosphorylation to inactivate protein Ser/Thr phosphatase type 2A (PP2A). In fact many protein phosphatases of both the Tyr and Ser/Thr families are now known to be phosphorylated (Figure 2). Phosphorylation can be an ellusive mechanism with phosphatases. It is likely that when cells are stimulated with agents that alter phosphatase

PHOSPHATASES are PHOSPHORYLATED

Tyrosine phosphatases	Serine phosphatases
PTP1B (Ser, Tyr)	PP2A (Tyr, Thr, Ser)
cdc25 (Ser,Thr)	PP1 (Tyr, Thr)
SH-PTP (Tyr)	calcineurin [PP2B] (Ser)
CD45 (Ser, Tyr)	MPP [PP2C] (Ser)

Figure 2. Examples of phosphorylation of protein phosphatases.

activity no changes in phosphatase activity would be detected with *in vitro* assays of extracts prepared from those cells because rapid self-dephosphorylation would occur during preparation of the extracts, thereby returning the phosphatase activity to the same level in the control and experimental samples. It is our view that this is a prime reason for the apparent lack of modulation in phosphatase activity, though a recent report shows cell-cycle fluctuation in the activity of type-1 protein phosphatase (2).

REGULATION OF PROTEIN TYR PHOSPHATASE (PTP1B) BY SER PHOSPHORYLATION

The protein Tyr phosphatase (PTP) family is a diverse collection of proteins including those related to CD45 that span the plasma membrane and are composed of both extracellular and intracellular domains, as well as a collection of "intracellular" proteins related to the human placental PTP1B, the first PTP whose amino acid sequence was determined. These intracellular PTPs show sequence diversity to either the N-terminal or C-terminal of the catalytic domain, and contain a domain related to the cytoskeletal protein band 4.1, or *src*-homology 2 (SH2) domains or hydrophobic sequences responsible for membrane association (see review, 3). More distant relatives of these PTPs are found in viruses (4, 5) and the proteins that function as mitotic inducers in yeast and other organisms, encoded by the *cdc25* (see 3) and *pyp*3 (6) genes. In our laboratory cell fractionation experiments with extra precautions to limit proteolysis yields 97% of the PTP activity in the particulate fraction. This PTP activity measured with artificial substrates can be solubilized with detergents such as Triton X-100, indicating that it is associated with membranes. Recent results using indirect immunofluorescence show that PTP1B localizes to the endoplasmic reticulum (7, 8). In addition, the C-terminal 25-35 amino acids of PTP1B, including a stretch of aliphatic amino acids uninterrupted by charged side chains, are necessary and sufficient for localization of chimeric proteins to the endoplasmic reticulum (7, 8). We prepared affinity purified anti-PTP1B-peptide antibodies against residues 161-178 in the catalytic domain. This region of the catalytic domain is not conserved between various members of the PTP

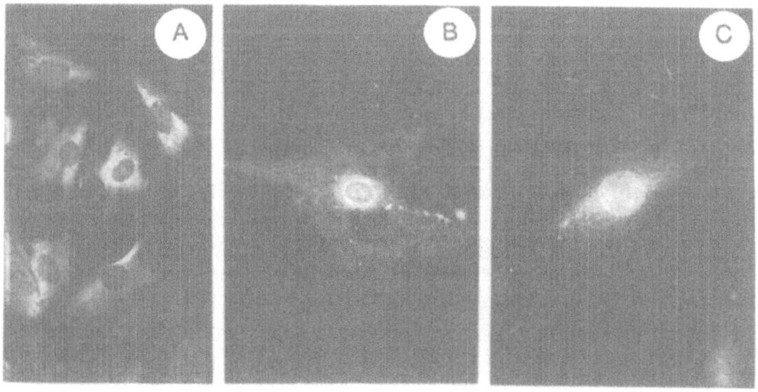

Figure 3. Localization of PTP1B in rat embryo fibroblasts during G_1, G_2, and prophase.

family, so the antibodies are specific for PTP1B and, for example, recognize only PTP1B out of the multiple forms of PTP found in liver membranes (9). Dr. Ned Lamb of Montpellier, France, has used these antibodies to stain rat embryo fibroblasts at different phases of the cell growth cycle. The cells were synchronized by serum deprivation for 24-48 hr, then refed serum for 12 hr at which point cells were in the G_1 phase of the cell cycle. After fixation with methanol, staining used the rabbit anti-PTP1B-peptide and fluorescence-conjugated goat anti-rabbit antibody. Images were examined with a Zeiss Axiophot and recorded on Kodak Tri-X film. The pattern of staining (Figure 3, panel A) showed generally uniform distribution throughout the cytoplasm with little staining in the nuclei, which appear as dark circles in the center of the cells. This pattern is consistent with the distribution in the endoplasmic reticulum as seen by others. Interestingly, following DNA synthesis (S phase) as the cells progressed through G_2 the PTP1B staining displayed a different pattern. Figure 3, Panel B shows, at higher magnification, that the PTP1B staining during G_2 was highest in the perinuclear region and within the nucleus. The staining was quite non-uniform in the rest of the cell and appeared concentrated along a fibrillar network containing several points with especially high intensity staining. This pattern is consistent with PTPase association with intracellular membranes and cytoskeletal structures. Later, as the cells entered prophase, and onward through mitosis the staining was concentrated within the nucleus and outside the nucleus staining appeared in a punctate or stippled pattern (Figure 3, Panel C). These points of staining may represent the remnants of the endoplasmic reticulum which is vesicularized in the process of cell division. This differential distribution of PTP1B during the cell growth cycle may be related to the observed changes in serine phosphorylation of PTP1B, especially the enhanced phosphorylation observed in nocodazole-blocked cells (10, 11).

The endogenous PTP1B found in monkey kidney CV-1 cells or human HeLa cells that was solubilized from membranes with Triton-X100 elutes from gel filtration columns in the absence of detergent with a M_r = 150 kDa. We have reported that this form of PTP1B can be dissociated with the chaotropic salt LiBr to a monomeric catalytic subunit of 50 kDa or can be converted to a 40 kDa truncated PTP1B by limited trypsin digestion (13). Either dissociation or truncation of the PTP1B catalytic subunit causes a shift in specificity. The activity with [^{32}P]Tyr-myelin basic protein as substrate decreased as the activity with [^{32}P]Tyr-RCM-lysozyme increased. This result implies that the other proteins in the 150 kDa form of the PTP1B, and the C-terminal portions of the catalytic subunit removed by trypsin both have a role in substrate specificity.

We investigated whether Ser/Thr phosphorylation would affect the activity of PTP1B in cells. As a start we tested whether pharmacologic agents that would increase the activity of cAMP-dependent or Ca2+/phospholipid-dependent protein kinases (PKA and PKC respectively) could alter the membrane PTP1B activity in intact cells. We found that as cells reached confluence there was a significant increase in the specific activity of membrane PTP1B (Units PTP/mg protein). This phenomenon is dependent on cell density and has been reported by another laboratory (14), but the basis for the effect is not yet understood. We found that even on top of this density dependent increase in PTP1B activity agents expected to stimulate PKA or PKC further increase the PTP1B activity 2-3 fold in a time and dose-dependent manner (13). Effects were evident within 2 min. and persisted for several minutes before returning to basal levels within 15 min. Activators of PKA including 100 nM isoproterenol, 10 μM forskolin, and 10 μM 8-Br-cAMP or 8-CPT-cAMP were all effective at increasing the activity of PTP1B. Likewise, activators of PKC [10 μM phorbol ester or 10 μM dioctanoyl glycerol (DiC8)] increased the PTP1B activity. If the extracts from stimulated cells were prepared in the presence of protein Ser/Thr phosphatase inhibitors such as fluoride or okadaic acid the observed activation of PTP1B was enhanced to 4-fold. These results are consistent with the activation of PTP1B by Ser/Thr phosphorylation and its inactivation by Ser/Thr dephosphorylation. The activated form of PTP1B was recovered as the 150 kDa complex, indicating that neither dissociation or association of other proteins could account for the activation process. Subsequently, we found an 8-fold increase in the ^{32}P-phosphorylation of the 50 kDa PTP1B catalytic subunit when living cells were stimulated with cAMP or diacylglycerol analogues (15). Phosphoamino acid analyses showed that only serine residues were phosphorylated in the PTP1B. The same ^{32}P-labeled cyanogen bromide (CNBr) peptides were recovered from each sample, indicating that the same site(s) in the PTP1B was being phosphorylated in response to different stimuli. From the sizes of the CNBr peptides recovered after partial cleavage, and the sequences surrounding the serine residues in the radiolabeled segment, we conclude that Ser352 is the likely site of phosphorylation.

To further investigate the relationship between Ser phosphorylation and activation of PTP1B we constructed a maltose binding protein (MAL) fusion with PTP1B, expressed it in bacteria and purified the fusion protein by Zn^{2+}-affinity chromatography and gel filtration chromatography (Figure 4). We found that purified protein kinases could activate the

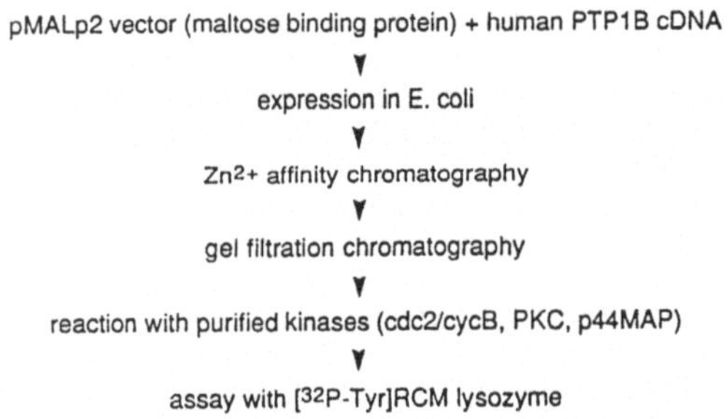

Figure 4. Experimental plan for expression of MAL-PTP1B fusion protein and *in vitro* activation by protein Ser/ Thr kinases.

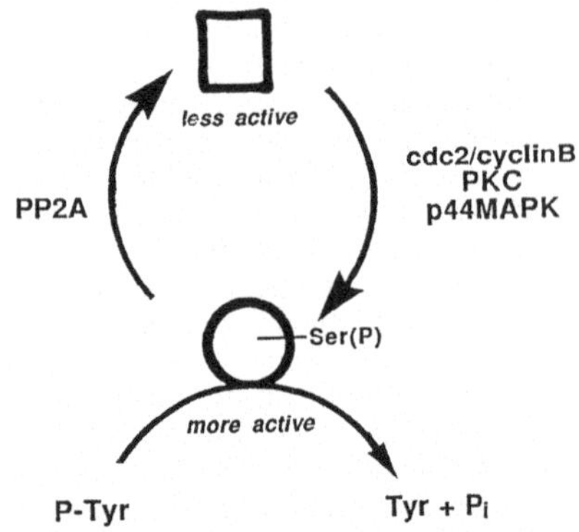

Figure 5. Regulation of PTP1B by Ser phosphorylation and dephosphorylation

PTP1B-MAL fusion protein 2.5 to 3.5-fold. These kinases included PKC, $p44^{MAPK}$, $p34^{cdc2}$:cyclin B, but interestingly not PKA (Figure 5). These results would imply that the increase in PTP1B activity in response to analogues of cAMP was not mediated by a direct phosphorylation by PKA, instead there must be an alternative mechanism involving another kinase for activation of the PTP1B. Current efforts have focused on identifying which kinase(s) in cells stimulated with analogues of cAMP are activated to cause the phosphorylation and activation of PTP1B.

REGULATION OF PROTEIN SER/THR PHOSPHATASE TYPE 2A BY TYR PHOSPHORYLATION

Regulation of protein phosphatase 2A (PP2A) also involves phosphorylation. In this case we found that phosphorylation of Tyr inactivated the phosphatase, which slowly underwent reactivation by self dephosphorylation. Our interest in this process arose from studies of various laboratories showing that in response to insulin and growth factors a series of protein Ser/Thr kinases were activated. These were called Mitogen Activated Protein (MAP) kinases or Extracellular Signal Regulated Kinases (ERK) that are sequentially phosphorylated in a "cascade" arrangement (Figure 6, see reviews, 16, 17). Activation of a MAP kinase pathway seems to be important for the G_0 to G_1 transition in the cell cycle in

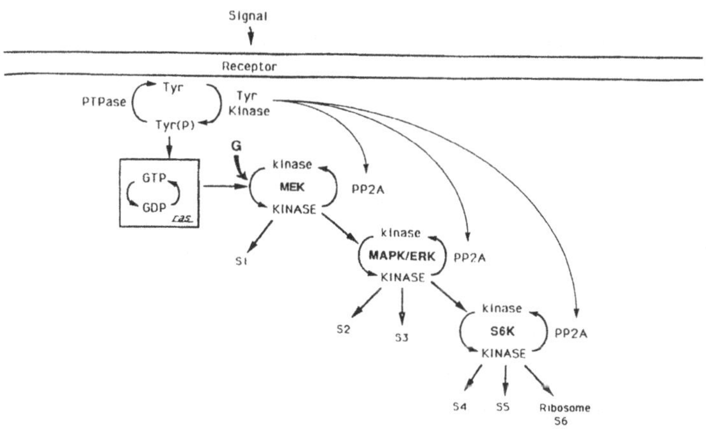

Figure 6. The "fresh" view of the MAP Kinase cascade for signal transduction.

various systems. The ribosomal S6 kinases were identified as one obvious target of the cascade. These kinases were phosphorylated and activated by MAP kinase (3 types are known to date), which itself was activated by a dual specificity kinase that phosphorylated both Tyr and Thr in MAP kinase. This enzyme called MAP kinase kinase or MEK (for MAP or ERK Kinase) is itself activated by phosphorylation by the oncogene protein *raf*. Various investigators, including John Kyriakis and Joe Avruch, with whom we have collaborated, have shown that the S6 kinases, MAP kinases and MEK could all be dephosphorylated and inactivated by PP2A. This reaction was quite specific because there was little effect even substantially higher concentrations of the protein Ser/Thr phosphatase type 1 (PP1) were used (18). From our perspective PP2A is in a critical position to be one enzyme capable of inactivating multiple steps in the cascade system. Believing that regulation of the

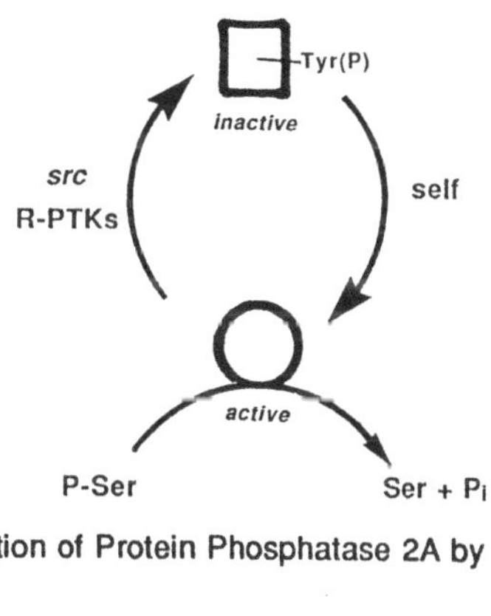

Figure 7. Regulation of Protein Phosphatase 2A by Tyr phosphorylation.

dephosphorylation reactions would be equally important as the regulation of the phosphorylation reactions we set out to investigate whether "upstream" signals, presumably involving protein phosphorylation, could regulate the activity of PP2A. A specific hypothesis was that signals could inactivate PP2A in the process of signal transduction via the MAP kinase cascade. By inactivating the PP2A the inactivation of the kinases in the cascade would be reduced, and the signal amplified. Indeed, we found that immune complexes of the $p60^{v\text{-}src}$ kinase catalyzed the ^{32}P-labeling of the PP2A catalytic subunit (Figure 7). The rate of this reaction was increased 2 to 5 fold by okadaic acid, a PP2A inhibitor and upon prolonged incubation phosphorylation reached 0.8-0.9 mol^{32}P/mol PP2A. However, the ^{32}P was rapidly lost from the PP2A catalytic subunit, precluding assay of the activity of the phosphorylated form of PP2A. To circumvent this problem we used g-thioATP to produce the thiophosphorylated PP2A catalytic subunit. It is well known that thiophosphorylated proteins are especially resistant to enzymatic dephosphorylation (20). In this way we were able to show that the PP2A catalytic subunit, either alone or complexed with a 60 kDa regulatory subunit, was inactivated up to 90% by incubation with $p60^{v\text{-}src}$. Other experiments showed that modification was entirely on Tyr residues and placed this site of phosphorylation at Tyr 307, two residues from the C-terminus of PP2A in a tail segment of the protein that is easily removed by partial trypsin digestion (19). These experiments establish that PP2A could be regulated by a cycle of Tyr phosphorylation and dephosphorylation. The *in vitro* phosphorylation could be catalyzed by both the insulin and the EGF receptor kinases as well as by the $p56^{lck}$ and $p60^{v\text{-}src}$ non-receptor protein Tyr kinases.

More recently we have addressed the question of whether this modification of PP2A occurs in intact cells. We have found that Tyr phosphorylation of PP2A occurs within one minute of surface activation of normal human lymphocytes (21). Using 2D gel electrophoresis and immunoblotting with affinity purified anti-PP2A-peptide antibodies made against residues 285-309 of the catalytic subunit we find that serum-stimulated 10T1/2 fibroblasts or *v-src*-transformed 10T1/2 fibroblasts both show 5:1 ratio of acid-shifted PP2A relative to a 0.6:1 ratio in serum-starved 10T1/2 fibroblasts. This acid-shifted form of PP2A also immunoblots with antiphosphotyrosine antibodies. The results show that serum stimulation or *src* transformation caused similar increase in the tyrosine phosphorylation of PP2A. We interpret these results to indicate that tyrosine phosphorylation and inactivation of PP2A is an important part of promoting the growth of cells. In this sense PP2A acts as a growth suppressor enzyme, which must be inactivated by diverse signals to allow progression through the cell cycle and cell division.

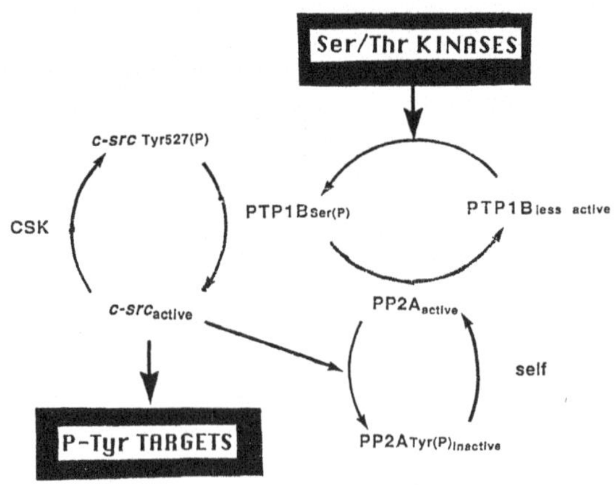

Figure 8. An interactive network of phosphatases and kinases that operates at mitosis

To bring together the two cycles of phosphatase phosphorylation we propose a "network" of PTP1B and PP2A actions (Fig. 8). This hypothetical network might be part of the program of events that occur when cells enter mitosis. Activation of the PTP1B either by MAP kinase or by cdc2/cyclinB would set off an amplification loop wherein PTP1B dephosphorylates Tyr 527 in the C-terminal region of *c-src*, which phosphorylates Tyr 307 in PP2A to inactivate it, which no longer could inactivate the PTP1B by dephosphorylation. This loop would cycle and result in accumulation of both Tyr phosphorylated and Ser/Thr phosphorylated proteins because of *src* activation and PP2A inactivation. The process would then wind down to a halt by removal of the input stimulus and the self dephosphorylation of PP2A that would dephosphorylate and inactivate PTP1B which could then no longer maintain *c-src* in its active, dephosphorylated form. Although this is but a hypothetical scheme we consider it a worthwhile exercise in thinking about how kinases and phosphatases interact with one another to modify the activity of enzymes that are vital to cells in a simultaneous and coordinated manner. We propose that understanding the molecular basis for regulation of the cell growth cycle will require definition of the physical and temporal architecture of kinase and phosphatase networks within the cell.

ACKNOWLEDGMENTS

Research in this laboratory has been supported by grants from the National Institutes of General Medical Sciences (GM35266) and from the American Cancer Society (BE130).

REFERENCES

1. Brautigan, D.L. Protein phosphatases, *Curr. Prog. Horm. Res.*, 49 (in press, 1993).

2. Walker, D.H., DePaoli-Roach, A.A., Maller, J.L. Multiple Roles for Protein Phosphatase 1 in Regulating the *Xenopus* early embryonic cell cycle, *Mol. Biol. of the Cell*, 3:687-698 (1992).

3. Brautigan, D.L. Great expectations: protein tyrosine phosphatases in cell regulation, *Biochim. et Biophys. Acta*, 1114:63-77 (1992).

4. Dixon, J.E., Guan, K. Protein tyrosine phosphatase activity of an essential virulence determinant in *Yersinia.*, *Science*, 8:553-556 (1990).

5. Sheng, Z., Charbonneau, H. The baculovirus *Autographa californica* encodes a protein tyrosine phosphatase, *J. Biol. Chem.*, 268:4728-4733 (1993).

6. Millar, J.B.A., Lenaers, G., Russell, P. *Pyp3* PTPase acts as a mitotic inducer in fission yeast, *EMBO J.*, 11:4933-4941 (1992).

7. Woodford-Thomas, T.A, Rhodes, J.D., Dixon, J.E. Expression of a protein tyrosine phosphatase in normal and v-src-transformed mouse 3T3 fibroblasts, *J. Cell Biol.*, 117:401-414 (1992).

8. Frangioni, J.V., Beahm, P.H., Shifrin, V., Jost, C.A., Neel, B.G. The nontransmembrane tyrosine phosphatase PTP 1B localizes to the endoplasmic reticulum via its 35 amino acid c-terminal sequence, *Cell*, 68:545-560 (1992).

9. Gruppusso, P.A., Boylan, J.M., Smiley, B.L., Fallon, R.J., Brautigan, D.L. Hepatic protein tyrosine phosphatase in the rat, *Biochem. J.*, 274:361-367 (1991).

10. Schievella, A.R., Paige, L.A., Johnson, K.A., Hill, D.E., Ericson, R.L. Protein tyrosine phosphatase 1B undergoes mitosis-specific phosphorylation of serine, *Cell Growth & Diff.*, 4:239-246 (1993).

11. Flint, A.J., Gebbink, M.F.G.B., Franza, B.R.Jr., Hill, D.E., Tonks, N.K. Multisite phosphorylation of the protein tyrosine phosphatase PTP 1B: identification of cell cycle regulated and phorbol ester stimulated sites of phosphorylation, *EMBO J.*, 12:1937-1946 (1993).

12. Cool, D.E., Tonks, N.K., Charbonneau, H., Fischer, E.H., Krebs, E.G. Expression of a human T-cell protein-tyrosine-phosphatase in baby hamster kidney cells, *Proc. Natl. Acad. Sci. USA*, 87:7280-7284 (1990) .

13. Brautigan, D.L., Pinault, F.M. Activation of membrane protein-tyrosine phosphatase involving cAMP and Ca^{2+} phospholipid-dependent protein kinases, *Proc. Natl. Acad. Sci. USA*, 88:6696-6700 (1991).

14. Pallen, C.J., Tong, P.H., Elevation of membrane tyrosine phosphatase activity in density-dependent growth-arrested fibroblasts, *Proc. Natl. Acad. Sci. USA*, 88:6996-7000 (1991).

15. Brautigan, D.L., Pinault, F.M., Serine phosphorylation of protein tyrosine phosphatase (PTP1B) in HeLa cells in response to analogues of cAMP or diacylglycerol plus okadaic acid, *Mol. Cell.Biochem.*, (in press, 1993).

16. Nishida, E., Gotoh, Y. The MAP kinase cascade is essential for diverse signal transduction pathways, *TIBS*, 18:128-131 (1993).

17. Lange-Carter, C.A., Pleieman, C.M., Gardner, A.M., Blumer, K.J., Johnson, G.L. A divergence in the MAP kianse regulatory network defined by MEK kinasc and *raf*, *Science*, 260:315-319 (1993).

18. Kyriakis, J.M., Brautigan, D.L., Ingebritsen, T.S., Avruch, J. pp54 Microtubule-associated protein-2 kinase requires both tyrosine and serine/threoinine phosphorylation for activity, *J. Biol. Chem.*, 226:10043-10046 (1991).

19. Chen, J., Martin, B.L., Brautigan, D.L. Regulation of protein serine-threonine phosphatase type-2A by tyrosine phosphorylation, *Science*, 257:1261-1264 (1992).

20. Li, H., Simonelli, P.F., Huan, L. Preparation of protein phosphatase-resistant substrates using adenosine 5'-O-(γ-thio)triphosphate, *Meth. Enzymol.*,159:346-356 (1988).

21. Brautigan, D.L., Chen, J., Thompson, P. Protein phosphatase 2A: specificity with physiological substrates and inactivation by tyrosine phosphorylation in transformed fibroblasts and normal lymphocytes, *Adv. Prot .Phos.* 7:49-65 (1993).

4

POSITIVE AND NEGATIVE REGULATION OF CELL CYCLE PROGRESSION BY SERINE/THREONINE PROTEIN PHOSPHATASES

Arthur S. Alberts[1] and Axel Schönthal[2,*]

[1]Departments of Pharmacology and Medicine
University of California at San Diego
La Jolla, CA 92093-0636.

[2]Department of Microbiology
University of Southern California
2011 Zonal Ave. HMR-401
Los Angeles, CA 90033-1054

INTRODUCTION

Reversible protein phosphorylation events play a key role in intracellular signal transduction pathways that regulate gene expression and cell proliferation. To analyze the involvement of protein phosphatases in these processes, we applied two different approaches. First, we used okadaic acid, a specific inhibitor of the two major serine/threonine protein phosphatases, type 1 (PP-1) and type 2A (PP-2A).[1-3] Second, we microinjected subunits of PP-2A into living cells and analyzed the effects on gene expression.

The previous use of okadaic acid has generated seemingly contradictory data (see ref. 4, for a review). On one side, it has been found that okadaic acid is a potent tumor promoter,[5,6] that it can activate histone H1 kinase,[7] a key regulator of cell cycle progression, and that it induces various mitotic events in G2 synchronized cells.[8-10] From this, it has been concluded that okadaic acid sensitive phosphatases may act as negative regulators of cellular growth.

On the other side, okadaic acid has also been shown to revert raf and retII oncogene transformed cells,[11] to block proliferation induced by platelet derived growth factor (PDGF),[12] and to induce apoptosis in normal and transformed cells.[13] In addition, two studies showed that okadaic acid is able to inhibit tumor promotion in vitro.[14,15] These results suggested that okadaic acid sensitive phosphatases can be positive regulators of growth.

What could be the basis for this obvious discrepancy? We propose that okadaic acid sensitive protein phosphatases may have positive as well as negative roles in cellular growth control. Okadaic acid inhibits PP-1 and PP-2A at different concentrations, the IC_{50} *in vitro* is 50-200 and 0.2-1.0 nM, respectively.[1,16,17] Another only recently discovered protein phosphatase, PP-3, is inhibited at intermediary concentrations (IC_{50} is 3-5 nM).[18] Moreover, there may be additional okadaic acid sensitive protein

*to whom reprint request should be sent

phosphatases yet to be discovered.[19] Thus, it is difficult to compare experiments where different concentrations of this inhibitor had been used. Moreover, different cell types may yield varying results not only because of quantitatively and qualitatively different phosphatase expression, but also because of potential variation in the abundance of phosphatase substrates. Differences in the uptake of the drug may play a role as well.

Another factor that strongly contributes to the quality of okadaic acid effects is the position of cells in the cell cycle. For example, whereas okadaic acid induces premature mitosis in G2-arrested cells,[7,8,10] it inhibits their progression through mitosis and through late G1.[20-22] In support of a cell cycle dependent role of phosphatases, it has been described recently that the activity of PP-2A, and of inhibitor-2, a PP-1 regulatory subunit, oscillate throughout the cell cycle.[23,24] Moreover, in yeast several genes with similarities to PP-1 and PP-2A have been described that appear to be involved in cell cycle regulation.[25-29]

In order to learn more about these complex processes, we studied the effects of okadaic acid on the expression of genes whose protein products are regulators of cell cycle progression. The first set of genes we analyzed was the group of immediate early genes that are induced very rapidly in response to serum stimulation of G0-arrested cells.[30,31] They are thought to be necessary for cells to enter and progress through early G1. Indeed, inhibition of expression of c-fos, the best analyzed member of this group, has been shown to prevent G0-arrested cells from reentering the cell cycle.[32,33]

The other important group of genes we analyzed is coding for two protein families, the cyclin dependent kinases (cdk's) and their regulatory subunits, the cyclins (see refs. 34,35, for a review). Members of these two families constitute the enzyme collectively referred to as histone H1 kinase. The importance of H1 kinase in cell cycle progression has been shown by experiments where defined subunits of this kinase have been inhibited in living cells. Anti-sense oligonucleotides, or microinjection of antibodies specific to either cdc2 (=cdk1) or to cyclin A, completely blocked progression through the cell cycle and arrested cells at either the G1/S or the G2/M boundary.[36-38]

In this report we provide evidence for positive and negative roles of okadaic acid sensitive protein phosphatases in the regulation of expression of growth related genes. We show that cells treated with okadaic acid respond with elevated levels of immediate early, e.g. c-fos, gene expression, whereas at the same time expression of cdc2 and cyclin A is down regulated. (Parts of this work have been published elsewhere.[20,39,40])

RESULTS

Murine fibroblasts were treated with okadaic acid and c-fos mRNA expression was analyzed at different times thereafter. Figure 1 shows that okadaic acid induces a prolonged increase in c-fos expression. This effect is very different from earlier observations where cells had been stimulated with serum growth factors or phorbol ester type tumor promoters. In those cases, induction of c-fos was maximal within the first hour after stimulation and returned to basal levels within the second hour.[41] The effects of okadaic acid appear to result from a combination of transcriptional activation of the c-fos gene as well as an increase in the stability of its otherwise very short-lived mRNA.[39] We and others have found similar effects on other immediate early genes as well.[42-45] These findings indicate that the inhibition of okadaic acid sensitive phosphatases leads to the increased expression of genes that are necessary for entry and progression through early G1. It suggests, therefore, that the respective phosphatases may play a negative role in these very early events of the cell cycle.

In an attempt to identify the specific phosphatase responsible for c-fos repression, we microinjected PP-2A into cells that harbored a reporter gene fused to the serum response element (SRE) of the c-fos promoter. The SRE is the major promoter element activating c-fos expression in response to extracellular growth stimuli. We microinjected the catalytic subunit of PP-2A, either alone or in combination with one of its regulatory subunits, into cells that had been sychronized in G0 by serum deprivation. Then the cells were stimulated with fresh growth medium and the activity of the SRE was analyzed. Since in our cells the SRE had been cloned in front of the bacterial gene for ß-galactosidase (ß-gal), we simply stained cells *in situ* for ß-gal expression with the

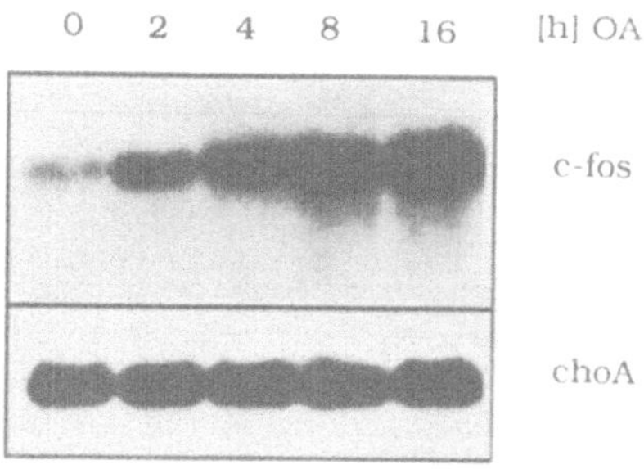

Figure 1. NIH 3T3 murine fibroblasts were grown in Dulbecco's modified Eagle's medium supplemented with 10% fetal calf serum. 50 nM okadaic acid (OA) was added into the culture medium and mRNA was harvested at the indicated time points. C-fos mRNA expression was analyzed by Northern blotting and hybridization to a specific probe as described.[39] As a control for equal amounts of mRNA loaded in each lane, the filter was stripped and rehybridized with a probe for choA, that is clone A of some highly abundant mRNA species originally isolated from chinese hamster ovary (CHO) cells.[53]

chromogenic substrate X-gal.[46] Positive cells (indicating SRE activity) turned blue (black in our black & white prints).

The results of this experiment are shown in Figure 2. Most of the cells that were injected with a neutral protein stained positive for ß-gal, confirming that the expected serum-inducibility of the SRE was not affected by microinjection *per se* (Figure 2, E and F). In contrast, the vast majority of cells that were microinjected with the catalytic subunit of PP-2A did not express ß-gal, indicating suppression of SRE activation (Figure 2, A and B). Microinjection of the catalytic subunit in combination with one of its regulatory subunits did not inhibit the serum inducibility of the SRE (Figure 2, C and D). Moreover, microinjection of the PP-1 catalytic subunit inhibited expression from the SRE as well (not shown). Thus, in combination with the results obtained with okadaic acid, this experiment indicates that in the presence of elevated PP-1 or PP-2A activity growth factor stimulated c-fos expression is inhibited.

Despite the above described positive effects of okadaic acid, we found that the cells did not proliferate as long as okadaic acid was present in the growth medium.[20] We therefore analyzed the effects of this drug on the expression of two other genes, cdc2 and cyclin A, whose protein products had been shown to be necessary for progression through later steps of the cell cycle.[36-38] Murine fibroblasts were treated with okadaic acid for various times and cdc2 and cyclin A mRNA levels were analyzed. Figure 3 shows that okadaic acid caused a down regulation of both mRNAs. Since this experiment was performed under the same conditions as the one shown in Figure 1, it clearly demonstrates that okadaic acid has differential effects on gene expression: it stimulates c-fos, but it represses cdc2 and cyclin A.

We next analyzed how repression of cdc2 and cyclin A mRNA expression by okadaic acid would correlate with the activity of their protein product, histone H1 kinase. For this purpose we immunoprecipitated cdc2 and cyclin A protein ($p34^{cdc2}$ and $p60^{cycA}$, respectively) from okadaic acid treated cells and analyzed the associated kinase activity in vitro. As shown in Figure 4, in both cases the kinase activity declines in the presence of okadaic acid. Down regulation of $p60^{cycA}$ associated kinase activity appears to be more rapid than in the case of $p34^{cdc2}$, which correlates with the differences observed at the mRNA level (compare Figure 3). Since by western blot analyses we also found a concomitant decline in $p34^{cdc2}$ and $p60^{cycA}$ protein levels,[20] we conclude that okadaic acid negatively affects histone H1 kinase activity by blocking the expression and synthesis of its subunits. Thus, the major inhibitory effect of okadaic acid on histone H1 kinase appears to be at the level of gene expression, not posttranslational at the level of the enzyme.

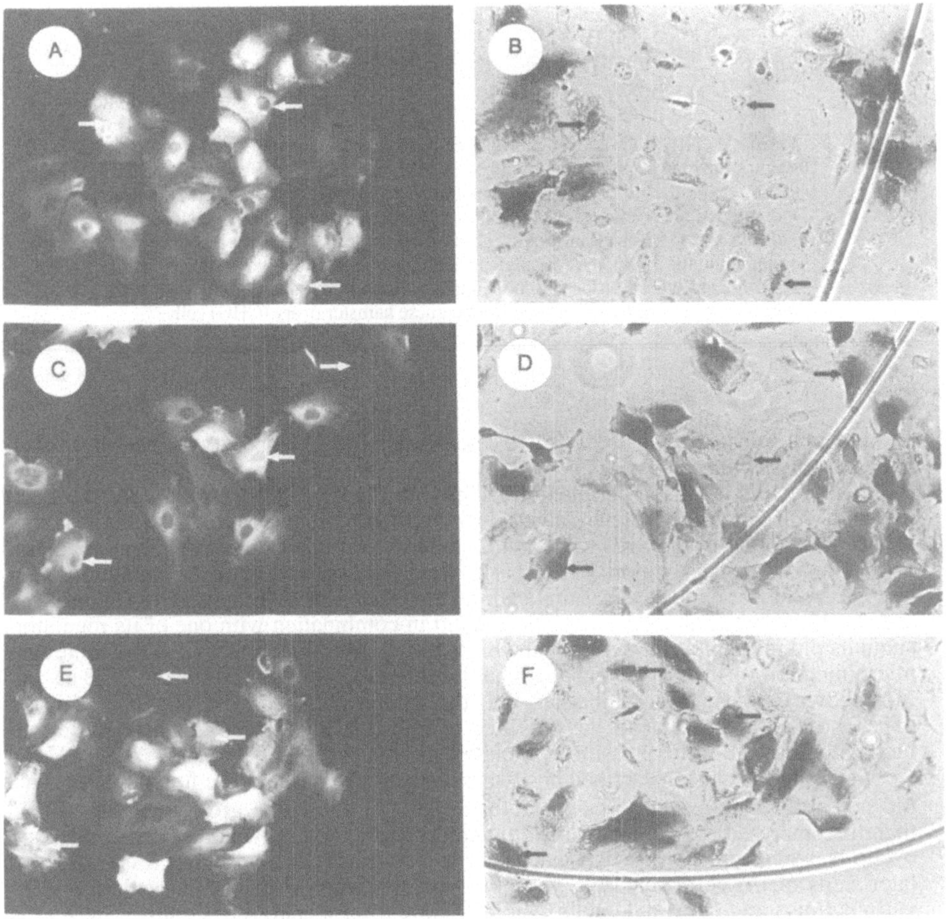

Figure 2. Balb/c 3T3 murine fibroblasts with a stably integrated SRE ß-gal construct (established by J.L. Meinkoth, UCSD) were synchronized in G0 by serum deprivation. Then cells were microinjected with the catalytic subunit of PP-2A (panel A and B), with the holoenzyme composed of the PP-2A catalytic subunit complexed with a regulatory subunit (panel C and D), or with mouse immunoglobulin (IgG) as a control (panel E and F). To enable the later identification of microinjected cells, the microinjection buffer used for all microinjections was supplemented with mouse IgG. After microinjection the cells were stimulated by the addition of 20% fetal calf serum. Four hours later the cells were fixed and stained for ß-gal expression and for the presence of the co-injected mouse IgG. The left hand photomicrographs show indirect immunoflourescence of the marker IgG. The right hand panels are phase-contrast photographs of the same fields stained with X-gal to detect ß-gal expression. Arrows are provided for orientation between panels. Note that the majority of microinjected cells in A (roughly the triangular field between the arrows) do not show ß-gal expression as indicated by black cells in B. For technical details on this experiment see ref. 40.

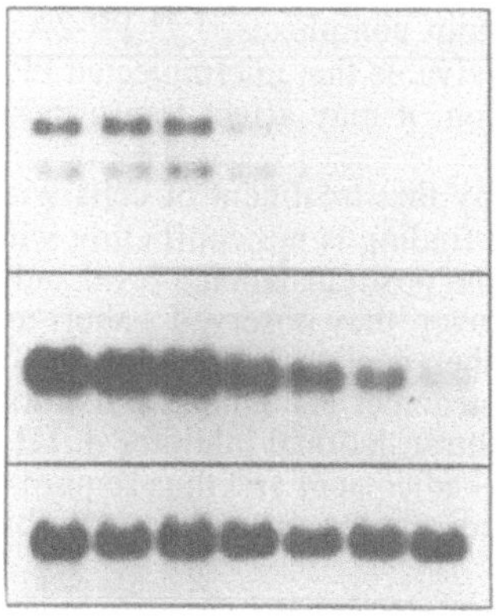

Figure 3. NIH 3T3 were treated with okadaic acid (OA) as described in the legend to Figure 1. Then cyclin A (cycA) and cdc2 mRNA levels were analyzed by Northern blot analyses and use of specific hybridization probes. ChoA was used to control for equal amounts of mRNA loaded in each lane.

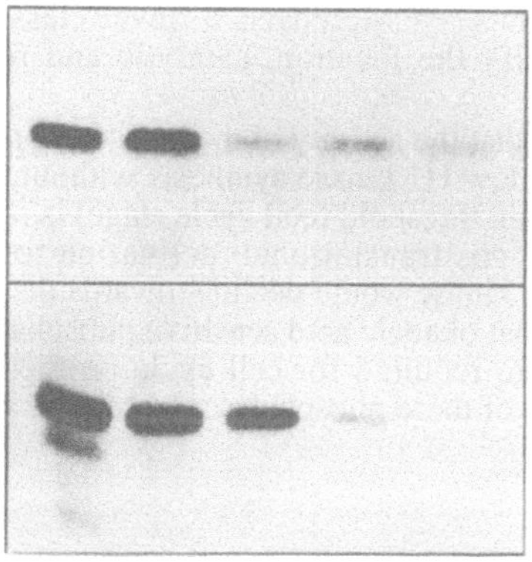

Figure 4. NIH 3T3 cells were treated with okadaic acid (OA) as described in the legend to Figure 1. Then histone H1 kinase acitivity was analyzed by immunoprecipitation of either $p60^{cycA}$ (cycA) or $p34^{cdc2}$ (cdc2) and subsequent incubation of the precipitated antigens with radiolabeled ATP and histone H1 protein as a substrate. This *in vitro* kinase reaction was separated by polyacrylamide gel electrophoresis and the phosphorylated substrate visualized by autoradiography (see ref. 20 for technical details).

DISCUSSION

In this report and others[20,39,42] we showed that treatment of cells with okadaic acid affects gene expression. While some immediate early genes, such as c-fos, are constitutively induced, cdc2 and cyclin A, two genes coding for histone H1 kinase, are repressed. This indicates that the respective okadaic acid sensitive protein phosphatases have positive as well as negative regulatory roles in the control of gene expression. Since several phosphatases may be affected by okadaic acid, it needs to be established whether the opposing effects on gene expression are mediated by the same or by different phosphatases. Our microinjection experiments indicated that both PP-1 and PP-2A are able to negatively regulate c-fos expression.

The ability of both catalytic subunits of PP-1 and PP-2A to inhibit SRE-regulated gene expression may be based on multiple phosphorylation steps which occur during signal transduction in this pathway. Okadaic acid activates MAP (mitogen activated protein) kinases which in turn have been shown to phosphorylate nuclear factors that are associated with SRE-bound protein complexes.[47-49] PP-2A inactivates several MAP kinases in vitro.[50] Thus it is conceivable that microinjected PP-2A may inactivate one or more of these kinases. In addition, it may affect transcription factor phosphorylation directly within the nucleus.

It has been reported recently that treatment of cells with okadaic acid stimulates histone H1 kinase activity.[7] That finding is not conflicting with ours since these authors describe H1 kinase activation at the posttranslational level and only at high okadaic acid concentrations (>200 nM). Moreover, they observed a short-term effect where induction is maximal within one hour and then declines. At 50 nM, the concentration that we used in our experiments, they did not detect H1 kinase activation. Although these results suggest that PP-1 may be the posttranslational inhibitor of H1 kinase activity (since this phosphatase is less sensitive to okadaic acid and thus requires higher concentrations for inhibition), findings by others indicate that it may actually by PP-2A that accounts for this function.[51]

Thus, there seems to be an interesting paradox: on one side okadaic acid sensitive phosphatases appear to negatively regulate the activity of histone H1 kinase at the posttranslational level, on the other side they appear to be required for the synthesis of this kinase at the gene expression level. However, one could envision that the respective functions of the involved phosphatases are separated spatially or temporally, for example by cellular compartmentalization or by cell cycle dependent activation.[23,24] Various mechanisms may contribute to achieve a highly regulated specificity. For example, the enzymatic activity of the catalytic subunit is controlled by the differential interaction with the respective regulatory subunits.[3,19] Moreover, a novel class of proteins, known as targetting subunits, may specify the location, catalytic and regulatory properties of protein phosphatases.[52]

Our experiments indicate that the activity of the respective phosphatase, presumably PP-2A, would be required to allow H1 kinase synthesis without immediate activation of the kinase activity. Once a certain threshold (cell cycle stage) is reached, PP-2A would be inactivated, contributing to posttranslational activation of H1 kinase activity. Concomitantly, synthesis of the kinase would decline towards the next cell cycle.

In any case, our finding that okadaic acid sensitive phosphatases positively regulate the expression of genes that are required for cell cycle progression provides a unique model to study the positive role of these phosphatases in cellular growth regulation.

ACKNOWLEDGMENTS

We are grateful to James R. Feramisco (University of California, San Diego) in whose laboratory most of the above presented work was performed.

REFERENCES

1. C. Bialojan, and A. Takai. Inhibitory effect of a marine-sponge toxin, okadaic acid, on protein phosphatases. *Biochem. J.* 256:283 (1988).
2. J. Hescheler, G. Mieskes, J.C. Rüegg, A. Takai, and W. Trautwein. Effects of a protein phosphatase inhibitor, okadaic acid, on membrane currents of isolated guinea-pig cardiac myocytes. *Eur. J. Physiol.* 412:248 (1988).
3. P. Cohen. The structure and regulation of protein phosphatases. *Ann. Rev. Biochem.* 58:453 (1989).
4. A. Schönthal. Okadaic acid: a valuable new tool for the study of signal transduction and cell cycle regulation? *New Biol.* 4:16 (1992).
5. M. Suganuma, H. Fujiki, H. Suguri, S. Yoshizawa, M. Hirota,M. Nakayasu,M. Ojika, K. Wakamatsu, and K. Yamada. Okadaic acid: an additional non-phorbol-12-O-tetradecanoate-13-acetate-type promoter. *Proc. Natl. Acad. Sci. USA* 85:1768 (1988).
6. F. Katoh, D.J. Fitzgerald, L. Giroldi, H. Fujiki, T. Sugimura, and H. Yamasaki. Okadaic acid and phorbol esters: comparative effects of these tumor promoters on cell transformation, intercellular communication and differentiation *in vitro*. *Jpn. J. Cancer Res.* 81:590 (1990).

7. K. Yamashita, H. Yasuda, J. Pines, K. Yasumoto, H. Nishitani, M. Ohsubo, T. Hunter, T. Sugimura, and T Nishimoto. Okadaic acid, a potent inhibitor of type 1 and type 2A protein phosphatases, activates cdc2/H1 kinase and transiently induces a premature mitosis-like state in BHK21 cells. *EMBO J.* 9:4331 (1990).
8. M.A. Felix, P. Cohen, and E. Karsenti. Cdc HI kinase is negatively regulated by a type 2A phosphatase in the Xenopus early embryonic cell cycle: evidence from the effect of okadaic acid. *EMBO J.* 9:675 (1990).
9. E.T. Kipreos, and J.Y.J. Wang. Differential phosphorylation of c-Abl in cell cycle determined by cdc2 kinase and phosphatase activity. *Science* 248:217 (1990).
10. B. Lüscher, L. Brizuela, D. Beach, and R.N. Eisenman. A role for the p34cdc2 kinase and phosphatases in the regulation of phosphorylation and disassembly of lamin B2 during the cell cycle. *EMBO J.* 10:865 (1991).
11. R. Sakai, I. Ikeda, H. Kitani, H. Fujiki, F. Takaku, U. Rapp, T. Sugimura, and M. Nagao. Flat reversion by okadaic acid of raf and ret-II transformants. *Proc. Natl. Acad. Sci. USA* 86:9946 (1989).
12. N.M. Dean, L.J. Mordan, K. Tse, S.L. Mooberry, and A.L. Boynton. Okadaic acid inhibits PDGF-induced proliferation and decreases PDGF receptor number in C3W10T1/2 mouse fibroblasts. *Carcinogenesis* 12:665 (1991).
13. R. Boe, B.T. Gjertsen, O.K. Vintermyr, G. Houge, M. Lanotte, and S.O. Doskeland. The protein phosphatase inhibitor okadaic acid induces morphological changes typical of apoptosis in mammalian cells. *Exp. Cell Res.* 195:237 (1991).
14. L.J. Mordan, N.M. Dean, R.E. Honkanen, and A.L. Boynton. Okadaic acid: a reversible inhibitor of neoplastic transformation of mouse fibroblasts. *Cancer Commun.* 2:237 (1990).
15. E. Rivedal, S.O. Mikalsen, and T. Sanner. The non-phorbol ester tumor promoter okadaic acid does not promote morphological transformation or inhibitjunctional communication in hamster embryo cells. *Biochem. Biophys. Res. Commun.* 167:1302 (1990).
16. P. Cohen, S. Klumpp, and D.L. Schelling. An improved procedure for identifying and quantitating protein phosphatases in mammalian tissues. *FEBS Lett.* 250:596 (1989).
17. S. Nishiwaki, H. Fujuki, M. Sugnuma, H. Furuya-Suguri, R. Matsushima, Y. Iida, M. Ojika, K. Yamada, D. Uemura, T. Yasumoto, F.J. Schmitz, and T. Sugimura. Structure-activity relationship within a series of okadaic acid derivatives. *Carcinogenesis* 11:1837 (1990).
18. R.E. Honkanen, J. Zwiller, S.L. Daily, B.S. Khatra, M. Dukelow, and A.L. Boynton. Identification, purification, and characterization of a novel serine/threonine protein phosphatase from bovine brain. *J. Biol. Chem.* 266:6614 (1991).
19. S. Shenolikar, and A.C. Narin. Protein phosphatases: Recent progress. *Adv. Sec. Messgr. and Phos.phoprot. Res.* 23:1 (1991).
20. A. Schönthal, and J.R. Feramisco. Inhibition of histone HI kinase expression, retinoblastoma protein phosphorylation, and cell proliferation by the phosphatase inhibitor okadaic acid. *Oncogene* 8:433 (1993).
21. Y. Ishida, Y. Furukawa, J.A. DeCaprio, M. Saito, and J.D. Griffin. Treatment of myeloid leukemic cells with the phosphatase inhibitor okadaic acid induces cell cycle arrest at either G1/S or G2/M depending on the dose. *J. Cell. Physiol.* 150: 484 (1992).
22. D.D. Vandre, and V.L. Wills. Inhibition of mitosis by okadaic acid: possible involvement of a protein phosphatase 2A in the transition from metaphase to anaphase. *J. Cell Sci.* 101:79 (1992).
23. J.W. Ludlow. Selective ability of S-phase cell extracts to dephosphorylate SV40 large T antigen *in vitro. Oncogene* 7:1011 (1992).
24. D.L. Brautigan, J. Sunwoo, J.-C. Labbe, A. Fernandez, and N.J.C. Lamb. Cell cycle oscillation of phosphatase inhibitor-2 in rat fibroblasts coincident with p34cdc2 restriction. *Nature* 344:74 (1990).
25. H. Okhura, N. Kinoshita, S. Miyatiani, T. Toda, and M. Yanagida. The fission yeast dis2+ gene required for chromosome disjoining encodes one of two putative type 1 protein phosphatases. *Cell* 57:997 (1989).
26. R. Booher, and D. Beach. Involvement of a type 1 protein phosphatase encoded by bws1+ in fission. *Cell* 57:1009 (1989).
27. J.H. Doonan, and N.R. Norris. The bimG gene of *Aspergillus nidulans*, required for completion of anaphase, encodes a homolog of mammalian phosphoprotein phosphatase 1. *Cell* 57:987 (1989).
28. A.M. Healy, S. Zolnierowicz, A.E. Stapleton, M. Goebl, A.A. DePaoli-Roach, and J.R. Pringle. CDC55, a *Saccharomyces cerevisiae* gene involved in cellular eisusuudsmorphogenesis: identification, characterization, and homology to the B subunit of mammalian type 2A protein phosphatase. *Mol. Cell. Biol.* 11:5767 (1991).
29. A. Sutton, D. Immanuel, and K.T. Arndt. The SIT4 protein phosphatase functions in late G1 for progression into S phase. *Mol. Cell. Biol.* 11:2133 (1991).
30. J. M. Almendral, D. Sommer, H. Macdonald-Bravo, J. Burckhardt, J. Perea, and R. Bravo. Complexity of the early genetic response to growth factors in mouse fibroblasts. *Mol. Cell. Biol.* 8:2140 (1988).

31. L.F. Lau, and D. Nathans. Identification of a set of genes expressed during the GO/GI transition of cultured mouse cells. *EMBO J.* 4:3145 (1988).
32. J.T. Holt, T. Venkat-Gopal, A.D. Moulton, and A.W. Nienhuis. Inducible production of c-fos anti-sense RNA inhibits 3T3 proliferation. *Proc. Natl. Acad. Sci. USA* 83:4794 (1986).
33. K. Nishikura, and J.M. Murray. Antisense RNA of proto-oncogene c-fos blocks renewed growth of quiescent 3T3 cells. *Mol. Cell. Biol.* 7:639 (1987).
34. T. Hunter, and J. Pines. Cyclins and cancer. *Cell* 66:1071 (1991).
35. T. Hunt. Maturation promoting factor, cyclin and the control of M-phase. *Curr. Opin. Cell Biol.* 1:268 (1989).
36. K. Riabowol, G. Draetta, L Brizuela, D. Vandre, and D. Beach. The cdc2 kinase is a nuclear protein that is essential for mitosis in mammalian cells. *Cell* 57:393 (1989).
37. F. Girard, U. Strausfeld, A. Fernandez, and N.J.C. Lamb. Cyclin A is required for the onset of DNA replication in mammalian fibroblasts. *Cell* 67:1169 (1991).
38. F. Fang, and J.W. Newport. Evidence that the G1-S and G2-M transitions are controlled by different cdc2 proteins in higher eukaryotic cells. *Cell* 66:731 (1991).
39. A. Schönthal, Y. Tsukitani, and J.R. Feramisco. Transcriptional and posttranscriptional regulation of c-fos proto-oncogene expression by the tumor promoter okadaic acid. *Oncogene* 6:423 (1991).
40. A.S. Alberts, T. Deng, A. Lin, J.L. Meinkoth, A. Schönthal, M.C. Mumby, M. Karin, and J.R. Feramisco. Protein phosphatase 2A potentiates activity of promoters containing AP-1-binding elements. *Mol. Cell. Biol.* 13:2104 (1993).
41. L.J. Ransone, and I.M. Verma. Nuclear proto-oncogenes fos and jun. *Ann. Rev. Cell Biol.* 6:539 (1990).
42. A. Schönthal, A.S. Alberts, J.A. Frost, and J.R. Feramisco. Differential regulation of jun family gene expression by the tumor promoter okadaic acid. *New Biol.* 3:977 (1991).
43. S.-J. Kim, R. Lafyatis, K.Y. Kim, P. Angel, H. Fujiki, M. Karin, M.B Sporn, and A.B. Roberts. Regulation of collagenase gene expression by okadaic acid, an inhibitor of protein phosphatases. *Cell Regulation* 1:269 (1990).
44. C. Thevenin, S.-J. Kim, and J.H. Kehrl. Inhibition of protein phosphatases by okadaic acid induces AP-1 in human T cells. *J. Biol. Chem.* 266:9363 (1991).
45. X. Cao, R. Mahendran, G.R. Guy, and Y.H. Tan. Protein phosphatase inhibitors induce the sustained expression of the egr-1 gene and hyperphosphorylation of its gene product. *J. Biol. Chem.* 267:12991 (1992).
46. J.L. Meinkoth, A. Alberts, and J.R. Feramisco. Construction of mammalian cell lines with indicator genes driven by regulated promoters. *CIBA Found. Symp.* 150:47 (1990).
47. R. Treisman. The SRE: A growth factor responsive transcriptional regulator. *Semin. Cancer Biol.* 1:47 (1990).
48. C.S. Hill, R. Marais, S. John, J. Wynne, S. Dalton, and R. Treisman. Functional analysis of growth factor-responsive transcription factor complex. *Cell* 73:395 (1993).
49. R. Marais, J. Wynne, and R. Treisman. The SRF accessory protein Elk-1 contains a growth factor-regulated transcriptional activation domain. *Cell* 73:381 (1993).
50. M.H. Cobb, T.G. Boulton, and D.J. Robbins. Extracellular signal-regulated kinases: ERKs in progress. *Cell Regulation* 2:965.
51. T.H. Lee, M.J. Solomon, M.C. Mumby, and M.W. Kirschner. INH, a negative regulator of MPF, is a form of protein phosphatase 2A. *Cell* 64:415 (1991).
52. M.J. Hubbard, and P. Cohen. On target with a new mechanism for the regulation of protein phosphorylation. *Trends Biochem. Sci.* 18:172 (1993).
53. M.M. Harpold, R. M. Evans, M. Salditt-Georgieff, and J.E. Darnell. Production of mRNA in Chinese hamster cells: Relationship of the rate of synthesis to the cytoplasmic concentration of nine specific mRNA sequences. *Cell* 17:1025 (1979).

5

EFFECTS OF PHOSPHATASE INHIBITORS ON MAMMALIAN p34cdc2 KINASE ACTIVITIES

Xiao-Wen Guo*, John P.H. Th'ng*, Richard A. Swank*, and E. Morton Bradbury*+

*Department of Biological Chemistry,
School of Medicine
University of California
Davis, CA 95616;

+Division of Life Sciences
Los Alamos National Laboratories
Los Alamos, NM 87545

ABSTRACT

The activity of cyclin-dependent protein kinase p34cdc2 is regulated by phosphorylation. In this study, we show that the presence of phosphatase inhibitors can have major effects on the levels of phosphorylation and the activities of the kinase extracted from cells. The combination of okadaic acid and sodium vanadate was most effective in protecting p34cdc2 against cellular phosphatases. In the absence of these inhibitors, p34cdc2 was dephosphorylated with an altered activity, indicating that phosphatase activities remained high during extractions. In contrast to when both inhibitors were used, lower activity of the kinase was found when only sodium vanadate was used, whereas higher activity was found in the presence of okadaic acid. Other conventional phosphatase inhibitors such as NaF, $NaHSO_3$ and glycerol 2-phosphate, were not effective in preventing dephosphorylation from p34cdc2 in whole cell lysates.

INTRODUCTION

The cyclin-dependent protein kinase, p34cdc2, plays important roles in regulating the cell cycle. The level of the protein in the cell remains relatively constant throughout the cell cycle. However, the enzymatic activity increases as the cells traverse S and G2 phases, and peaks during mitosis. This cell cycle-dependent oscillation in kinase activity results from its states of phosphorylation of $p34^{cdc2}$ and its association with the cyclins. Four cell cycle-dependent phosphorylation sites were identified in chicken $p34^{cdc2}$,[1]. During G1, phosphorylation is only found on Ser277. How this modification affects the function of the kinase remains unclear, although it is believed to be dependent on the nutritional status of the cell[2]. As cells enter S phase, Thr14 and Tyr15 become increasingly phosphorylated reaching maximum states at G2 phase. Dephosphorylation at these sites occurs abruptly at the G2/M transition. Thr161 is maximally phosphorylated during G2/M phase of the cell cycle (in *S. pombe* Thr167 is the equivalent site). The dephosphorylation of Thr14/Tyr15 and phosphorylation of Thr161 are concomitant with the activation of the kinase at the G2/M phase of the cell cycle[1,3-5]. Changes of Thr14 and Tyr15 of $p34^{cdc2}$ to non-phosphorylatable residues result in induction of premature mitotic events, confirming the importance of dephosphorylation of these residues for the activation of $p34^{cdc2}$,[2,6,7].

The phosphorylation states of $p34^{cdc2}$ are regulated in turn by other protein kinases and phosphatases. The wee1/mik1 genes code for protein tyrosine kinases that phosphorylate Tyr15 of the $p34^{cdc2}$ and inhibit the kinase[8-10]. The activity of wee1 is negatively regulated by nim1/cdr1 protein kinases[11]. The cdc25 gene codes for a tyrosine phosphatase that removes phosphate from Tyr15 to activate the $p34^{cdc2}$ kinase[12-17]. The cdc25 phosphatase itself is activated by phosphorylation on its C-terminal domains at the G2/M transition[18], and was shown to be carried out by the $p34^{cdc2}$-cyclin B kinase[19]. This phosphorylation and dephosphorylation cycle may be responsible for the rapid activation of $p34^{cdc2}$-cyclin B activation during mitosis[19]

The $p34^{cdc2}$ activating kinase (CAK/MO15), was partially purified and shown to phosphorylate Thr161 of $p34^{cdc2}$ to activate the kinase[5,20]. The dephosphorylation of Thr161 is believed to be carried out by the protein phosphatase type 2A (PP2A), or a related protein phosphatase referred to as INH[4,17,21].

There are two classes of protein phosphatases (PPase), one has specificity towards phosphotyrosine (PTPase) and the other has specificity towards phosphoserine and phosphothreonine. The activity of PTPase can be inhibited by either zinc chloride or sodium vanadate *in vitro*. The serine and threonine phosphatases are further classified into type 1 and type 2A, 2B and 2C, according to their sensitivity to specific inhibitors and their dependence on divalent cations [21,22]. Okadaic acid, a tumor promoting agent, can inhibit the activity of PP2A at concentration of 1 nM and PP1 at concentration of 1 μM *in vitro* [23]. Whereas the activity of PP2B is Ca^{2+} dependent and stimulated by calmodulin, the activity of PP2C is dependent on Mg^{2+} . Using Xenopus extracts, okadaic acid was shown to stimulate the hyperphosphorylation of cdc25 which in turn activated the protein kinase activity of $p34^{cdc2}$, [18]. These results indicate that a type 2A phosphatase may be involved in the down-regulation of cdc25 *in vivo*.

In this report, we show that the use of phosphatase inhibitors during extraction of $p34^{cdc2}$ is essential to preserve the activities of the kinase. Thus the choice of phosphatase inhibitors has profound effects on the activities of the resulting kinase.

MATERIAL AND METHODS

Tissue Culture

Mouse mammary gland carcinoma cells FM3A were routinely maintained at 32.5°C, at a density of 5 to 10×10^5 cells/ml in RPMI 1640-Hepes buffered medium supplemented with 10% calf serum.

Preparation of Whole Cell Extracts

Cells were harvested by centrifugation and washed once with cold PBS. All subsequent steps were carried out at 0-4°C. The cells were lysed in either Lysis Buffer 1 (LB1: 10 mM Tris, pH 7.4, 150 mM NaCl, 0.1% SDS, 1.0% NP-40, 1.0% sodium deoxycholate, and 2 mM EDTA) or Lysis Buffer 2 (LB2: 50 mM Tris, pH 7.4, 150 mM NaCl, 5 mM $MgCl_2$, 0.5% Triton X-100 and 0.3 M sucrose). Both lysis buffers were supplemented with micrococcal nuclease (50 units/ml) and the protease inhibitors, aprotonin (50 μg/ml), leupeptin (50 μg/ml) and phenylmethylsulfonyl fluoride (PMSF, 1 mM) just before use. Specific phosphatase inhibitors were included in the lysis buffers as indicated in the text. Cell lysis was achieved by incubating on ice for 30 minutes with vigorous pipetting. Cellular debris was removed by centrifugation at 10,000g for 10 minutes. Protein concentrations were determined with BCA reagents from Pierce, using BSA as a standard. For Western blot analysis, all samples were boiled 10 minutes in lysis buffer supplemented with 1X SDS sample buffer (50 mM Tris HCl, pH 6.8, 10% glycerol, 1% SDS, 10% β-mercaptoethanol and trace amount of bromphenyl blue) and 50 μg protein was loaded per lane on 9.5% polyacrylamide gels containing 0.1% SDS.

Protein Analysis by Western Blot

Proteins from polyacrylamide gels were transferred to PVDF immobilon membrane (Millipore) using a Bio-Rad semi-dry blotter. The membrane was then incubated with Blocking Buffer (BB: 25 mM Tris HCl, pH 7.4 and 150 mM NaCl) supplemented with 3% non-fat dry milk for 30 minutes at room temperature with rotation. The primary antibody at 1 to 2000 dilution was added to the membrane in BB containing 3% milk, 0.05% Tween 20, and 350 mM NaCl. Incubation was continued for 1 to 2 hours at room temperature with rotation. The membrane was then washed 3 times with BB containing 0.05% Triton-X 100 and 1% milk. Horse radish peroxidase conjugated anti-rabbit antibody raised in goat was then incubated with the membrane as described above for an additional hour and followed by 3 washes. To remove residual amount of milk, the membrane was washed two more times with BB. The blot was then developed with ECL (Amersham) system and exposed 10 to 60 seconds for proper exposure.

Polyclonal antibody to $p34^{cdc2}$ was prepared as described[24] and antibody to cyclin A was a generous gift from Drs. Pines and Hunter.

Immuno-precipitation and Kinase Assay

Cells were lysed as described above. Supernatants containing 1 mg of total cellular proteins was carefully removed from the lysate. To immuno-precipitate the $p34^{cdc2}$ kinase, the lysates were pre-cleared with 0.5 μl of pre-immune serum and 50 μl of a 50% protein A-Sepharose 6MB suspension on ice with rotation for 30 minutes. After centrifugation to remove the protein A-Sepharose 6MB, 1 μl anti-serum made against the C-terminal peptide (CDNQIKKM) of $p34^{cdc2}$, and 100 μl of a 50% protein A-Sepharose 6MB suspension was added to the supernatant. Following an incubation period of an hour, the protein A Sepharose 6 MB was pelletted by centrifugation and the precipitates were washed extensively 5 times with 0.5 ml lysis buffer. The proteins that were precipitated could either be analyzed by Western blot or used for kinase assays. For Western blots, 50 μl of 1XSDS sample buffer were mixed with the beads and boiled for 10 minutes before loading 25 μl on a SDS-PAGE. The protein from the gel was transferred to a membrane as described above and the presence of $p34^{cdc2}$ and cyclins were detected with specific antibodies.

To assay the activity of the kinase, immuno-precipitates were washed once with 100 μl of histone H1 kinase buffer (50 mM Tris.HCl, pH 7.4, 2 mM $MgCl_2$, 1 mM DTT, 100 mM NaCl, and 50 μM ATP). To start the reaction, 30 μl of 1 X H1 kinase buffer supplemented with 1 mM sodium vanadate, 1μM okadaic acid, 0.625 mg/ml of steer thymus histone H1s (gift from Dr. R.D. Cole) and 0.25 mCi/ml γ-^{32}P-ATP (specific activity at 1000 mCi/mmole) was added to the immuno-precipitates and incubated at 30 °C for 10 minutes. The reaction was stopped by the addition of an equal volume of 2 X SDS buffer and boiled for 10 minutes. 5 μl were loaded per lane on a 13% SDS PAGE. Phosphorylation of histone H1 was determined by autoradiography.

RESULTS AND DISCUSSION

The Presence of Protein Phosphatase Inhibitors Leads to Changes in Phosphorylation Patterns of $p34^{cdc2}$

The protein phosphatase inhibitors listed in Table I were used in this study. Whole cell lysates were prepared in Lysis Buffer 1 in the presence of specific phosphatase inhibitors as indicated in Figure 1. The proteins corresponding to $p34^{cdc2}$ and cyclin A were analyzed by Western blotting. Three bands of $p34^{cdc2}$ were detected. The faster migrating bands were shown previously to be the less phosphorylated forms of $p34^{cdc2}$ and this was later confirmed by mutational analysis of the three phosphorylation sites, namely Thr14, Tyr15 and Thr161[6,7,25]. Without any PPase inhibitors, $p34^{cdc2}$ was found in the faster migrating forms (lower and middle bands), as seen in lanes labeled (-)s. Two commonly used inhibitors of serine/threonine phosphatases, NaF, and glycerol 2-phosphate, did not prevent dephosphorylation of $p34^{cdc2}$. Sodium bisulfite ($NaHSO_3$), used to prevent the dephosphorylation of histone H1 during extraction[26], was only slightly effective in preserving the phosphorylation states of $p34^{cdc2}$, as revealed by presence of some slower migrating forms. The sulfhydryl reagent dithionitrobenzioc acid (DTNB), also used for inhibiting phosphatases during extraction of histone H1[26,27], was effective in preventing dephosphorylation on $p34^{cdc2}$. However, although DTNB prevented $p34^{cdc2}$ from dephosphorylation, its presence in whole cell lysates led to dissociation of $p34^{cdc2}$ from cyclin A and the concomitant loss of protein kinase activity (data not shown). Okadaic acid and microcystin LR are two potent inhibitors of PP1 and PP2A[23] and they showed only limited abilities in preventing dephosphorylation from $p34^{cdc2}$.

Tyrosine phosphatase inhibitors zinc chloride ($ZnCl_2$) and sodium vanadate, when used alone, did not prevent dephosphorylation of $p34^{cdc2}$. However, when zinc chloride or sodium vanadate was used in combination with okadaic acid or microcystin LR (see Figure 1), dephosphorylation of $p34^{cdc2}$ was prevented. In the presence of these inhibitors, the states of $p34^{cdc2}$ phosphorylation most likely reflect their native states in the cells. As a control, when DMSO, which was used as a solvent, was added to the lysis buffer, no protection was seen on $p34^{cdc2}$. No obvious change in mobility of cyclin A was observed in these lysates (Figure 1). We concluded from these results that when the phosphorylation states of $p34^{cdc2}$ needed to be preserved, phosphatase inhibitors to both phosphoserine/threonine phosphatases and phosphotyrosine phosphatases must be included in sample preparations.

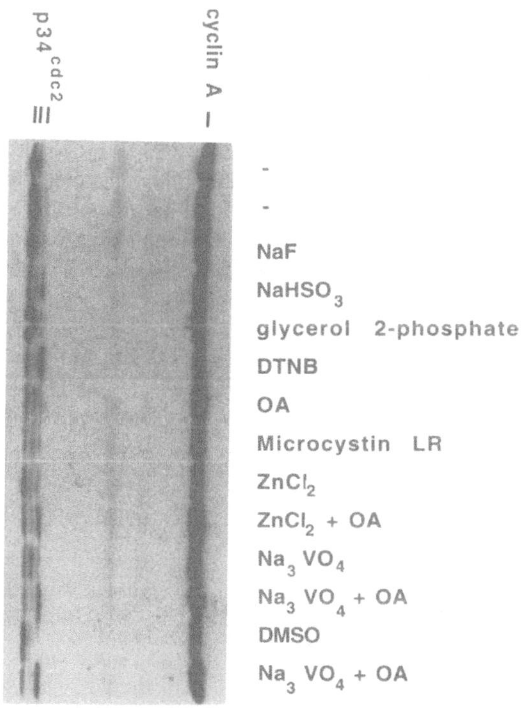

Figure 1. Western blot analysis of whole cell extracts. Whole cell extracts from FM3A cells were prepared in either LB 1 or LB 2 supplemented with PPase inhibitors as listed in Table 1. Proteins from samples 1 to 14 were loaded on a SDS-PAGE from left to right. PPase inhibitors used are marked on the top of each lane. The positions of cyclin A and $p34^{cdc2}$ are indicated to the left of the figure.

Detergents in Cell Lysis Buffer Caused Dissociation of $p34^{cdc2}$ Kinase Subunits and the Loss of Kinase Activity

We examined the conditions used for immuno-precipitation. Whole cell lysates were prepared in either LB1 or LB2, and were then immuno-precipitated using anti-$p34^{cdc2}$ antibodies. Western blot analysis of the proteins that were immuno-precipitat showed that anti-serum against C-terminal peptide of $p34^{cdc2}$ was able to specifically precipitated $p34^{cdc2}$ in both lysis buffers (Figure 2, panel A). However, cyclin A was co-precipitated only when Lysis Buffer 2 was used. When these immuno-precipitated materials were assayed for kinase activity using histone H1 as substrate, the $p34^{cdc2}$ prepared in LB1 had lost all of its enzymatic activity (panel B). It should be noted that LB1 was used in many preparations of

Table 1. The use of protein phosphatase inhibitors in the preparation of whole cell lysates. Samples 1 to 13 were prepared in LB 1 and sample 14 was prepared in LB 2. Triton X-100 (0.5%) was added to all samples except sample 1. PPase inhibitors were added to the lysis buffers just before use. DTNB*: dithionitrobenzioc acid; DMSO**: dimethyl sulfoxide; and OA***: okadaic acid.

Samples	Protein phosphatase inhibitors	Final concentration
1	-	-
2	-	-
3	NaF	50 mM
4	$NaHSO_3$	50 mM
5	glycerol 2-phosphate	10 mM
6	DTNB*	5 mM
7	OA	1 µM
8	Microcystin LR	1 µM
9	$ZnCl_2$	10 µM
10	$ZnCl_2$ + OA**	10 mM + 1 µM
11	Na_3VO_4	1 µM
12	Na_3VO_4 + OA	1 mM + 1 µM
13	DMSO***	1%
14	Na_3VO_4 + OA	1 mM + 1 µM

cell lysates and subsequently immuno-precipitations[28,29]. The consequence of protein complex dissociation and loss of enzyme activity in this buffer, therefore, must be taken into consideration.

Inhibition of PPase Activities by Okadaic Acid but not by Sodium Vanadate in The Lysis Buffer Result in the Increased Activities in $p34^{cdc2}$ Kinases

The presence various PPase inhibitors in the lysis buffer resulted in changes in the states of phosphorylation in $p34^{cdc2}$ as shown in Figure 1. To determine how these changes

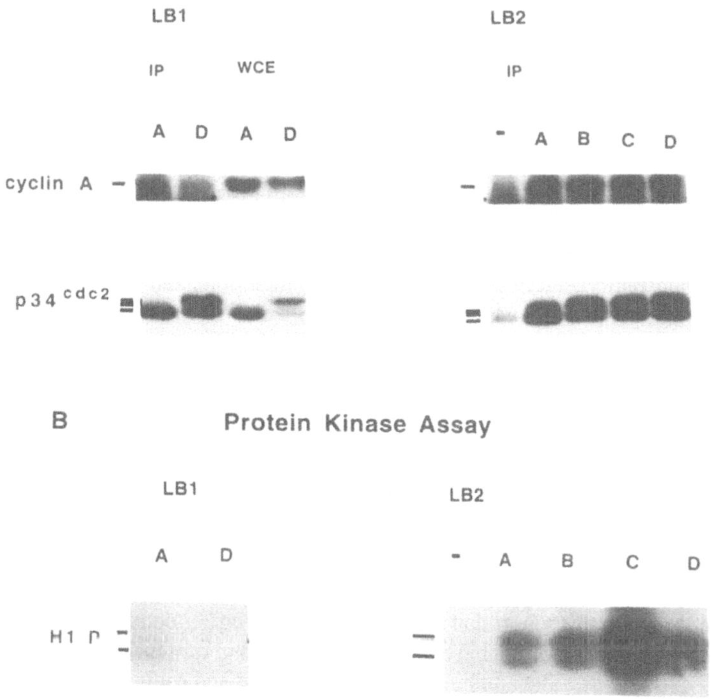

Figure 2. Immuno-precipitation and protein kinase assays. Whole cell extracts from FM3A cells were prepared in either LB1 or LB2. PPase inhibitors were supplemented as the following, sample A: no inhibitor; sample B: sodium vanadate; sample C: okadaic acid; and sample D: sodium vanadate and okadaic acid. Lanes marked WCE are proteins from whole cell extracts. The protein from all other samples were immuno-precipitated with anti-$p34^{cdc2}$ antibody except the protein from lane marked (-), which was immuno-precipitated with pre-immune serum (IP). Panel A: Western blots analysis of the immuno-precipitated material. The positions of cyclin A and $p34^{cdc2}$ are marked to the left of the panel. Panel B: Protein kinase assay of the immuno-precipitated $p34^{cdc2}$ using histone H1 as substrate. The position of histone H1 are marked at the left of the autoradiogram.

affected the enzymatic activities, $p34^{cdc2}$ was first immuno-precipitated and its kinase activities were analyzed. As shown in Figure 2, panel B, no kinase activity was detected when pre-immune serum was used (lane -). In the absence of PPase inhibitors (lane A), kinase activity was found to be either decreased or increased (data not shown) when compared to the use of both okadaic acid and sodium vanadate (lane D). When sodium vanadate was used alone, there was a slight decrease in enzymatic activity, as is seen by comparing lane B to lane D. This decrease in the kinase activity may be due to direct inhibition of the PTPase cdc25 by sodium vanadate, while other PPases present in the extracts were not inhibited. Therefore, in the presence of sodium vanadate alone, $p34^{cdc2}$ retains its phosphate on Tyr15 but not on Thr161. When okadaic acid was used alone, an increase in the kinase activity was found: compare lane C to lane D. Inhibition of PP2A, PP1 or a related kinase INH[4,17,21] by okadaic acid[23] may be responsible for protecting Thr161 of $p34^{cdc2}$ from dephosphorylation, thus activates the kinase while PTPase (for example cdc25) in the extracts remains active to remove phosphate from Tyr15. As it was shown before, in the presence of okadaic acid, cdc25 became hyperphosphorylated[18], inhibition of a PP2A or PP1 that targets cdc25 by okadaic acid may protects cdc25 from dephosphorylation thus keeps the cdc25 active. These multiple effects by okadaic acid may explain why in its presence alone, $p34^{cdc2}$ activity increased dramatically compared to when both sodium vanadate and okadaic acid were used.

When more detergents were included in the lysis buffer, as in the case of LB1, the $p34^{cdc2}$ was dephosphorylated in the absence of PPase inhibitors, and this dephosphorylation was more rapid than when LB2 was used. This suggests that the PPases that are dephosphorylating $p34^{cdc2}$ during extraction could be membrane associated, or that the presence of detergents makes the sites on $p34^{cdc2}$ more accessible to the PPases. This unfolding phenomena would be consistent with the observation that detergents breaks up the kinase complex. If it breaks up the complex, the catalytic subunit would surely be more "open" to attack by PPase.

SUMMARY

We have shown in this report that $p34^{cdc2}$ prepared in buffer conditions that do not contain the proper PPase inhibitors have lost phosphates and in turn have altered H1 kinase activity during extractions. The combination of okadaic acid and sodium vanadate in extraction buffers was most effective in preventing dephosphorylation of $p34^{cdc2}$. Furthermore, only mild detergents should be used in sample preparations when $p34^{cdc2}$ kinase activity is under investigation as the phosphatase activities seem to be stimulated by the presence of detergents.

REFERENCE

1. W. Krek, and E.A. Nigg, Differential phosphorylation of vertebrate $p34^{cdc2}$ kinase at the G1/S and G2/M transitions of the cell cycle: identification of major phosphorylation sites, EMBO J. 10:305 (1991).
2. W. Krek, J. Marks, N. Schmitz, E.A. Nigg, and V. Simanis, Vertebrate $p34^{cdc2}$ phosphorylation site mutants: effects upon cell cycle progression in the fission yeast Schizosaccharomyces pombe, J. Cell Sci., 102:43 (1992).
3. W. Krek, and E.A. Nigg, Cell cycle regulation of vertebrate $p34^{cdc2}$ activity: identification of Thr161 as an essential *in vivo* phosphorylation site, New Biologist, 4:323 (1992).
4. K.L. Gould, S. Moreno, D.J. Owen, S. Sazer, S., and P. Nurse, Phosphorylation at Thr167 is required for Schizosaccharomyces pombe $p34^{cdc2}$ function, EMBO J., 10:3297 (1991).
5. M. J. Solomon, T. Lee, and M. W. Kirschner, Role of phosphorylation in $p34^{cdc2}$ activation: identification of an activating kinase, Mol. Biol. Cell, 3:13 (1992).
6. C. Norbury, J. Blow, and P. Nurse, Regulatory phosphorylation of the $p34^{cdc2}$ protein kinase in vertebrates, EMBO J, 10:3321 (1991).
7. W. Krek, and E.A. Nigg, Mutations of $p34^{cdc2}$ phosphorylation sites induce premature mitotic events in HeLa cells: evidence for a double block to $p34^{cdc2}$ kinase activation in vertebrates, EMBO J. 10: 3331 (1991).
8. P.A. Fantes, Control of timing of cell cycle events in fission yeast by the wee 1^+ gene, Nature, 302:153 (1983).

9. P. Russell, and P. Nurse, Negative regulation of mitosis by wee1$^+$, a gene encoding a protein kinase homolog, Cell, 49:559 (1987).
10. K. Lundgren, N. Walworth, R. Booher, M. Dembski, M. Kirschner, and D. Beach, mik1 and wee1 cooperate in the inhibitory tyrosine phosphorylation of cdc2, Cell, 64:1111 (1991).
11. T.R. Coleman, Z. Tang, and W.G. Dunphy, Negative regulation of the wee1 protein kinase by direct action of the nim1/cdr1 mitotic inducer, Cell, 72:919 (1993).
12. W.G. Dunphy, and A. Kumagai, The cdc25 protein contains an intrinsic phosphatase activity, Cell, 67:189 (1991).
13. J. Gautier, M. J. Solomon, R.N. Booher, J. F. Bazan, and M. W. Kirschner, cdc25 is a specific tyrosine phosphatase that directly activates $p34^{cdc2}$, Cell, 67:197 (1991).
14. J.B. Millar, J. Blevitt, L. Gerace, K.Sadhu, C. Featherstone, and P. Russell, $p55^{CDC25}$ is a nuclear protein required for the initiation of mitosis in human cells, Proc. Natl.Acad. Sci.(U.S.A.) 88:10500 (1991)
15. U. Strausfeld, J.C. Labbe, D. Fesquet, J.C. Cavadore, A. Picard, K. Sadhu,P. Russell, and M. Doree, Dephosphorylation and activation of a $p34^{cdc2}$/cyclin B complex in vitro by human CDC25 protein, Nature, 351:242 (1991).
16. M. S. Lee, S. Ogg, M. Xu, L.L. Parker, D.J. Donoghue, J.L. Maller, H. Piwnica-Worms, cdc25+ encodes a protein phosphatase that dephosphorylates $p34^{cdc2}$, Mol.Biol.Cell, 3:73 (1992).
17. T.H. Lee, M.J. Solomon, M.C. Mumby, and M.W. Kirschner, INH, a negative regulator of MPF, is a form of protein phosphatase 2A, Cell, 64:415 (1992).
18. A. Kumagai and W. G. Dunphy, Regulation of the cdc25 protein during the cell cycle in Xenopus extracts, Cell, 70:139 (1992).
19. I. Hoffmann, P.R. Clarke, M.J. Marcote, E. Karsenti, and G. Draetta, Phosphorylation and activation of human cdc25-C by cdc2--cyclin B and its involvement in the self-amplification of MPF at mitosis, EMBO J., 12:53 (1993).
20. J. Shuttleworth, R. Godfrey, R., and A. Colman, $p40^{MO15}$, a cdc2-related protein kinase involved in negative regulation of meiotic maturation of Xenopus oocytes, EMBO J., 9:3233 (1990).
21. T. Lorca, D. Fesquet, F. Zindy, F. Le Bouffant, M. Cerruti, C. Brechot, G. Devauchelle, and M. Doree, An okadaic acid-sensitive phosphatase negatively controls the cyclin degradation pathway in amphibian eggs, Mol. Cell. Biol.,11:1171(1991).
22. P. Cohen, The structure and regulation of protein phosphatases, Ann. Rev. of Biochem., 58:453 (1989).
23. P. Cohen, S. Klumpp, and D.L. Schelling, An improved procedure for identifying and quantitating protein phosphatases in mammalian tissues, FEBS Lett.., 250:596 (1989).
24. J.R Hamaguchi,R.A. Tobey, J. Pines, H.A. Crissman, T. Hunter, and E.M. Bradbury, Requirement for $p34^{cdc2}$ kinase is restricted to mitosis in the mammalian cdc2 mutant FT210, J. Cell. Biol., 117:1041 (1992).
25. G. Draetta, and D. Beach, Activation of cdc2 protein kinase during mitosis in human cells: cell cycle-dependent phosphorylation and subunit rearrangement, Cell, 54:17 (1988).
26. D. A. Prentice, P.A. Kitos, and L.R. Gurley, Effects of phosphatase inhibitors on nuclease activity, Cell Biology International Reports, 9:1027 (1985).
27. J.R. Paulson, Sulfhydryl reagents prevent dephosphorylation and proteolysis of histones in isolated HeLa metaphase chromosomes, Eur. J. Biochem.,111:189 (1980).
28. K.L. Gould and T. Hunter, Platelet-derived growth factor induces multisite phosphorylation of $pp60^{c-src}$ and increases its protein-tyrosine kinase activity, Mol. Cell. Biol., 8:3345 (1988).
29. J. Pines, and T. Hunter, Isolation of a human cyclin cDNA: evidence for cyclin mRNA and protein regulation in the cell cycle and for interaction with $p34^{cdc2}$, Cell, 58:833 (1989).

6

THE MEIOTIC ROLE OF *twine*, A *DROSOPHILA* HOMOLOGUE OF *cdc25*

Luke Alphey, Helen White-Cooper and David Glover

Cancer Research Campaign Cell Cycle Genetics Research Group
Department of Anatomy and Physiology
The University
Dundee DD1 4HN
Scotland, UK

INTRODUCTION

cdc25 was first identified in fission yeast as a positive regulator of *cdc2* (Russell and Nurse, 1986). Homologues of *cdc25* have since been found in higher eukaryotes, including two from *Drosophila*. These are named *string* and *twine* (Edgar and O'Farrell, 1989; Jimenez *et al.*, 1990). Bacterially expressed proteins from both of these genes have been shown to have tyrosine phosphatase activity and can activate the $p34^{cdc2}$ kinase *in vitro* (Kumagai and Dunphy, 1991; Gautier *et al.*, 1991; Alphey and Clarke, unpublished). Furthermore, both homologues are functional in fission yeast in that they will rescue a temperature-sensitive mutant, *cdc25-22*. This inter-specific complementation was the basis of the original isolation of *twine* (Jimenez *et al.*, 1990), although the sequence similarity between *cdc25* homologues from different species allowed Courtot *et al.* (1992) to obtain *twine* by PCR.

Embryos homozygous for a mutation in *string* complete the first 13 mitotic cycles, but fail to enter the fourteenth mitotic division (Edgar and O'Farrell, 1989). The first 13 divisions take place in a syncytium and rely on maternally provided proteins and RNA, the fourteenth follows cellularisation and is regulated by zygotic gene expression. The introduction at cycle 14 of an extended G2 phase of variable duration leads to the complex but reproducible pattern of spatially and temporally regulated mitotic domains (Foe, 1989). *string* transcription precedes these divisions by 25 35 minutes (Edgar and O'Farrell, 1989). Ectopic expression of *string* under the control of the heat-shock promoter, will induce entry into mitosis (Edgar and O'Farrell, 1990), although heat-shock alone has been found to have dramatic effects on mitosis at this stage (Maldonado-Codina and Glover 1992). These results suggest that *string* is the primary regulator of G2-M transition in the newly-cellularised embryo. *string* expression is also seen in the proliferating cells at later stages of development (Alphey *et al.*, 1992). The pattern of expression of *string* contrasts greatly with that of *twine*. *twine* RNA is found in the growing stage of primary spermatocytes, and disappears after meiosis. Both *string* and *twine* are expressed in the ovary; these maternal transcripts persist into the syncytial embryo and are degraded on cellularisation. *twine*, unlike *string*, is not zygotically expressed in the somatic tissues during development, suggesting that it has no role in the regulation of mitosis after cellularisation.

The transcription pattern of *twine* suggests a meiotic role, and analysis of a *twine* mutation has shown a requirement for *twine* function for meiosis in both males and females (Alphey *et al.*, 1992; Courtot *et al.*, 1992). We show that *twine* is not required for all aspects of the entry into male meiosis, and that mutation in *twine* leads to abnormal spindles and defective chromosome segregation in female meiosis. We also discuss the possible existence of other *cdc25* homologues.

RESULTS AND DISCUSSION

The *twine*^{HB5} mutation

We identified the recessive female sterile mutation *mat(2)synHB5* (Schupbach and Wieschaus, 1989) as defective for *twine* and so renamed this mutation *twine*^{HB5} (Alphey *et al.*, 1992). This is the only known allele of this locus. Courtot *et al.* (1992) then showed that this allele has a point mutation which replaces a proline conserved in all *cdc25* homologues with a leucine. The phenotype of this allele is discussed in detail below. Although the experiments described use homozygous *twine*^{HB5} animals, the phenotype of *twine*^{HB5} hemizygous over a deficiency is indistinguishable. This is the classical definition of an amorphic allele and we conclude that *twine*^{HB5} has no significant residual activity. Furthermore, this mutant allele is unable to complement *cdc25-22* (Courtot *et al.*, 1992).

Meiosis is initiated but not completed in *twine*^{HB5} males

In wild-type *Drosophila*, spermatogenesis begins with the asymmetric division of one of the stem cells in the germinal proliferation centre at the apical tip of the tubular testis. The

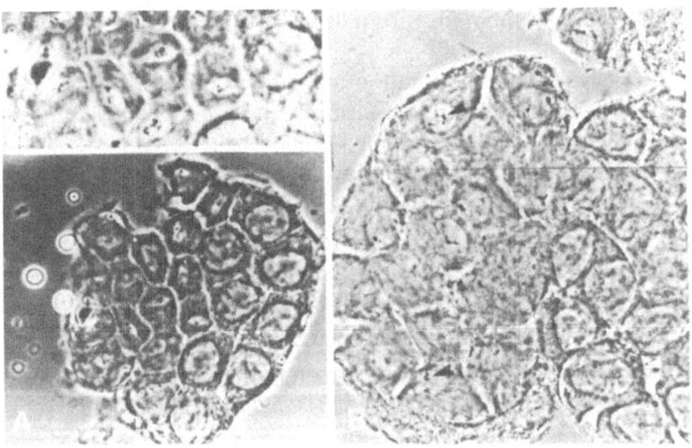

Figure 1. Chromatin condensation in spermatogenesis in wild type (A) and *twine*^{HB5} testes (B). Preparations were squashed in acetic acid as described in White-Cooper *et al.* (1993). Arrows show condensed chromosomes in *twine*^{HB5} which do not congress to a metaphase plate. Inset (C) is a part of A at higher magnification.

daughter cell undergoes four rounds of mitotic division to give a cyst of sixteen cells which remain linked by cytoplasmic bridges called ring canals. These cells grow during the next 90 hours or so and then undergo the two meiotic divisions to give a cyst of 64 cells, still linked by ring canals. The post-meiotic cells then elongate and differentiate into mature sperm. All these developmental stages are represented within a single testis, with younger cysts displacing older ones down the testis. We have previously shown that meiosis is not completed in *twine*^{HB5} males in that no 64 cell cysts are seen, but rather 16 cell cysts with a 4C DNA content, which then appear to attempt to differentiate into mature sperm (Alphey *et al.*, 1992). We have now examined the initiation of meiosis in this mutant in greater detail.

Three characteristic features of the G2-M transition are chromosome condensation, spindle formation and nuclear envelope breakdown. We looked for chromosome condensation in acetic acid squash preparations (Figure 1) and by staining with propidium iodide and were surprised to find that chromosome condensation does occur in *twine*^{HB5} males. This was unexpected as we had thought that the *twine*^{HB5} mutant males would fail to activate p34^{cdc2} kinase and so would not show any of the characteristic features of entry into M phase. Moreover, this finding is in contrast to that of Courtot *et al.* (1992) who reported that there is no such chromosome condensation in this mutant.

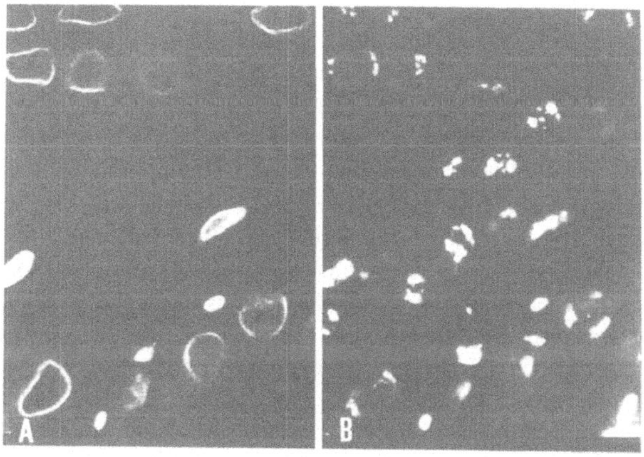

Figure 2. Nuclear envelope breakdown occurs in *twine*^{HB5} testes. Panels A and B show cysts from a *twine*^{HB5} testis stained with anti-lamin and propidium iodide respectively. The central cyst has condensed chromosomes and has broken down its nuclear envelopes (two somatic cyst cells are also present with intact nuclear envelopes and non-condensed chromatin). The surrounding cysts are at an earlier stage of development.

We have also investigated meiotic spindle assembly and nuclear lamin structure in *twine*^{HB5} males, to further characterise the meiotic defect of this mutant. Figure 2 shows that the nuclear lamina is broken down in *twine*^{HB5} males. However, despite the chromosome condensation and nuclear lamina breakdown, no spindle forms, as determined by indirect immunofluorescence with antibodies against either MPM2 or tubulin (Figure 3). Perhaps some cyclin-cdk complexes in the testis are regulated by tyrosine 15 phosphorylation while others are not. In *Xenopus* egg extracts p34^{cdc2}

/cyclin B is regulated by this inhibitory phosphorylation whereas p34^{cdc2} / cyclin A is not, so *cdc25* is only required for the activation of the cyclin B / p34^{cdc2} complex. If this is also the case in *Drosophila* then only the cyclin B complex would require activation by *twine*. The target of *twine* is clearly not required for chromosome condensation and nuclear lamina breakdown, but is required for spindle assembly.. Consistent with this, cyclin A and B associated kinases appear to have different effects on microtubule dynamics (Verde *et al.*, 1991; Buendia *et al.*, 1992), the cyclin B kinase being required for the destabilising of interphase microtubules which is essential for the establishment of the spindle. Furthermore, *Drosophila* cyclin B has been found to associate with the polar regions of the mitotic spindle (Maldonado-Codina and Glover, 1992)

twine^{HB5} oocytes fail to arrest at metaphase I

Unlike testes, *twine*^{HB5} oocytes are not blocked early in meiosis, but instead perform at least meiosis I, although in a highly aberrant manner. In the wild-type, oogenesis begins with four mitotic divisions of a precursor cell in the germarium of the ovary to produce a cyst of 16 cells connected by cytoplasmic bridges. One of these cells develops into the oocyte, the others become the highly polyploid nurse cells. The nurse cells provide the oocyte with the RNAs and proteins required for oogenesis and early embryonic development. Both *string* and *twine* are expressed in the nurse cells from stage 10 and are translocated into the oocyte by stage 13 (Alphey *et al.*, 1992; Courtot *et al.*, 1992; stages according to King, 1970). Meiotic prophase starts in the germarium and the oocyte remains in prophase I for days, as the meiotic spindle does not form until stage 13 (Therkauf and Hawley, 1992, Hatsumi and Endow, 1992). Wild-type oocytes arrest at metaphase I (stage 14) and only complete the meiotic divisions after activation of the oocyte on entry into the oviduct. Oogenesis in *twine*^{HB5} females appears normal until stage 14 when, instead of arresting at metaphase I they continue with an aberrant meiotic division. Abnormalities include splayed and kinked spindles, and spindles sharing a single pole. Metaphase II - like structures were seen, these often had chromatin masses of unequal size associated with them, suggesting gross chromosome non-disjunction. The small, non-exchange fourth chromosome often seems to escape from the mass of exchange chromosomes, as is also seen in wild-type.

Embryos from *twine*^{HB5} mothers have up to five large nuclei, which are eventually degraded and dispersed (Alphey *et al.*, 1992). These large nuclei have no spindle structures associated with them. However, a number of thin spindles associated with tiny pieces of chromatin the size of fourth chromosomes are present in some embryos. These embryos also have free asters of microtubules.

Although the *twine*^{HB5} phenotype appears to be very different in males and females, this is probably due to the significant differences between male and female meiosis. In *Drosophila*, meiotic recombination occurs only in the female, where it is important in ensuring that accurate segregation of homologous chromosomes; obviously a different mechanism must act in male meiosis I. No centrosomes are present in the oocyte and Therkauf and Hawley (1992) have suggested that the spindle microtubules are nucleated by the centromeres. This major difference may explain why meiotic spindles form in *twine*^{HB5} females but not in mutant males. The defects observed in females may indicate a requirement for *twine* function at a later stage that is never revealed in males. Similarly, it is difficult to determine whether *twine* has a role in the syncytial embryo, as embryos from *twine*^{HB5} mothers are already severely disrupted by this stage. Nevertheless, the existence of a small proportion of embryos that appear to escape from the majority phenotype, and yet still fail to reach cellularisation suggests that *twine* does indeed have an embryonic role (see Alphey *et al.*, 1992).

Does *Drosophila* have more *cdc25* homologues?

Attempts by several labs to find more *Drosophila* homologues of *cdc25* by PCR or complementation have so far failed (e.g. Courtot *et al.*, 1992; Alphey unpublished). It is impossible to be sure that all members of a given gene family have been identified, but we can address the question by asking whether there are any cell divisions in which neither of the known *cdc25* homologues are present. If so, then we may conclude that another *cdc25* or equivalent activity remains to be discovered, or else that the requirement for a *cdc25* protein tyrosine phosphatase activity is bypassed in some way. Millar *et al.* (1992) have shown that another tyrosine phosphatase, *pyp3* contributes to the tyrosyl dephosphorylation of $p34^{cdc2}$ in fission yeast. This is a member of the tyrosine phosphatase family typified by human PTP1B, and has only very weak sequence similarity to *cdc25*. *pyp3* was isolated by a complementation screen, but complements *cdc25-22* rather poorly.

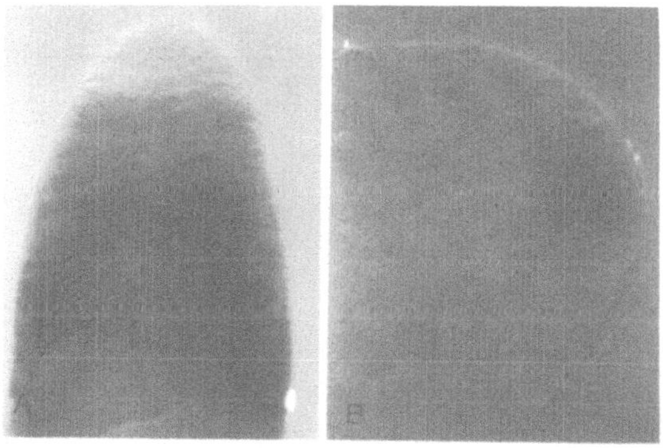

Figure 3. *string* and *twine* RNA distribution in testes. A: *twine* RNA accumulates in the growing stage. B: *string* RNA is present only at the germinal proliferation centre. Note that the gonial cells express contain neither *string* nor *twine* RNA.

Of all the proliferating tissues examined, almost all appear to express *string* or *twine* immediately before division. In the testis, there is a gap between the germinal proliferation centre, which expresses *string* and the growing stage, in which *twine* transcripts accumulate (Figure 3 and Alphey *et al.*,. 1992). In this gap, the four secondary gonial divisions occur, generating cysts of 16 primary spermatocytes (Demerec, 1950). If these divisions are regulated by *string*, as all other mitotic divisions appear to be, then the *string* protein must persist through these four divisions, whereas elsewhere it is synthesised from newly transcribed or stored RNA (see Edgar and O'Farrell, 1989, 1990; Alphey *et al.*, 1992). A similar situation may exist in the eye disc, where a single period of *string* expression drives two rounds of mitosis

The other exception is in the germarium at the apex of the ovariole. Although it contains the mitotically active oogonia and follicle cells, no expression of *string* or *twine* is observed in the germarium (Figure 4). *string* transcripts can be seen in the follicle cells from stage one until the follicle cells stop dividing at stage six. *string* transcripts are also observed in the nurse cells and oocyte at these stages, although these cells are not proliferating. *twine* transcripts do not appear until about stage 10 (Alphey *et al.*, 1992; Courtot *et al.*, 1992). As well as mitotic proliferation, entry into meiosis and the prophase establishment of the synaptonemal complexes also take place in the germarium in the

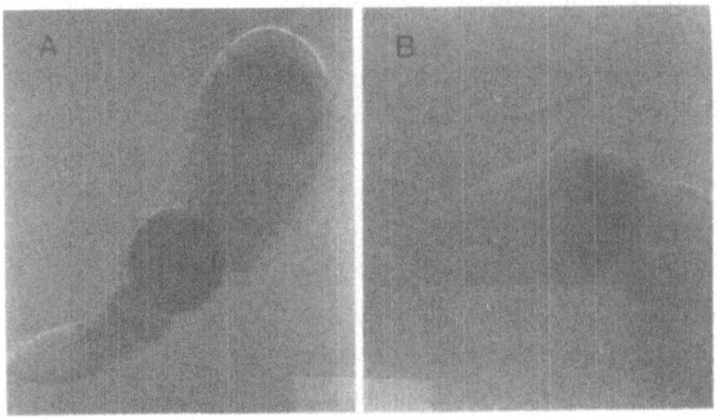

Figure 4. *string* RNA distribution in early oogenesis. *string* transcripts are present from stage one but are not detected in the germarium.

apparent absence of either *string* or *twine*. It therefore seems that the expected requirement for *string* or *twine* at this stage is replaced or avoided in some way.

ACKNOWLEDGEMENTS

We acknowledge financial support from the Cancer Research Campaign.

REFERENCES

Alphey, L., Jimenez, J., White-Cooper, H., Dawson, I., Nurse, P., and Glover, D., 1992, *twine*, a *cdc25* homolog that functions in the male and female germline of *Drosophila*, *Cell* 69:977.

Buendia, B., Draetta, G., and Karsenti, E., 1992, Regulation of the microtubule nucleating activity of centrosomes in *Xenopus* egg extracts: role of cyclin A-associated protein kinase, *J. Cell Biol.* 116:1431.

Cooper, K.W., 1950, Normal spermatogenesis in *Drosophila*, *in*: "Biology of *Drosophila*" Demerec, M., ed., John Wiley & Sons, Inc., New York.

Courtot, C., Fankhauser, C., Simanis, V., and Lehner, C.F., 1992, The *Drosophila cdc25* homologue *twine* is required for meiosis, *Development* 116:405.

Edgar, B.A., and O'Farrell, P.H., 1989, Genetic control of cell division patterns in the *Drosophila* embryo, *Cell* 57:177.

Edgar, B.A., and O'Farrell, P.H., 1990, The three postblastoderm cell cycles of *Drosophila* embryogenesis are regulated in G2 by *string*, *Cell* 62:469.

Foe, V.E., 1989, Mitotic domains reveal early commitment of cells in *Drosophila* embryos, *Development* 107:1.

Gautier, J., and Maller, J.L., 1991, Cyclin B2 and pre-MPF activation in *Xenopus* oocytes, EMBO J. 10:177.

Hatsumi, M., and Endow, S., 1992, Mutants of the microtubule protein *non claret dysjunctional*, affect spindle structure and chromosome movement in meiosis and mitosis, *J. Cell Sci.* 101:547.

Jimenez, J., Alphey, L., Nurse, P., and Glover, D., 1990, Complementation of fission yeast $cdc2^{ts}$ and $cdc25^{ts}$ mutants identifies two cell cycle genes from *Drosophila*: a *cdc2* homologue and *string*, EMBO J. 9:3565.

King, R., 1970, "Ovarian Development in *Drosophila melanogaster*," Academic Press, London.

Kumagai, A., and Dunphy, W.G., 1991, The cdc25 protein controls tyrosine dephosphorylation of the cdc2 protein in a cell-free system, *Cell* 64:903.

Maldonado-Codina, G., and Glover, D.M., 1992, Heat shock can delay and thereby synchronise the cell cycle in cellularised *Drosophila* embryos, *J. Cell Sci.* 105:711.

Maldonado-Codina, G., and Glover, D.M., 1992, Cyclins A and B associate with chromatin and the polar regions of spindles, respectively, and do not undergo complete degradation at anaphase in syncytial *Drosophila* embryos, *J. Cell Biol.* 116:967.

Millar, J.B.A., Lenears, G., and Russell, P., 1991. *pyp3* PTPase acts as a mitotic inducer in fission yeast, *EMBO J.* 11:4933.

Russell, P., and Nurse, P., 1986. Negative regulation of mitosis by *wee1* a gene encoding a protein kinase homolog, *Cell* 49:559.

Schupbach, T., and Wieschaus, E., 1989. Female sterile mutations on the second chromosome of *Drosophila melanogaster*. I. Maternal effect. *Genetics* 121:101.

Therkauf, W.E., and Hawley, R.S., 1992. Role of the nod protein, a kinesin homologue, in meiotic spindle assembly, *J. Cell Biol.* 116:1167.

Verde, F., Labbe, J.-C., Doree, M., and Karsenti, E., 1990. Regulation of microtubule dynamics by cdc2 protein kinase in cell free extracts of *Xenopus* eggs. *Nature* 343:233.

White-Cooper, H., Alphey, L., and Glover, D., 1993. The *cdc25* homologue *twine* is required for only some aspects of the entry into meiosis in *Drosophila*, *J. Cell Sci.*, submitted.

PART II

CONTROL OF CELL PROLIFERATION

7

EXTRACELLULAR SIGNAL-REGULATED PROTEIN KINASES (ERKS) 1, 2, AND 3

David J. Robbins, Erzhen Zhen, Mangeng Cheng, Colleen A. Vanderbilt, Douglas Ebert, Clark Garcia, Alphonsus Dang, and Melanie H. Cobb

University of Texas Southwestern Medical Center
Department of Pharmacology
5323 Harry Hines Blvd.
Dallas, TX 75235-9041

INTRODUCTION

Extracellular signal-regulated kinases (ERKs)[1-4] are protein kinases participating in numerous signal transduction pathways. The two best studied of these, ERK1 and ERK2, are known as MAP kinases and were first identified as 43 and 41 kDa, hormonally stimulated enzymes that used microtubule-associated protein-2 and myelin basic protein as substrates [5-8]. In addition to these two proteins they recognize a varied group of substrates many of which are known to have important regulatory functions, including protein kinases [9-12], membrane receptors [13,14], transcriptional factors [15,16], and cytoskeletal proteins [17-19], indicating a pleiotropic mode of action. It has been suggested that these enzymes cause aberrant phosphorylation of tau that occurs in Alzheimer's disease. Among the earliest studies of these enzymes were ones documenting their activation by insulin in terminally differentiated 3T3-L1 adipocytes as well as ones demonstrating that they are activated at the transition of cells from G_0 to G_1 [5,6,20-26,33].

We purified ERK1 from insulin-stimulated Rat 1 HIRc B cells and sequenced tryptic peptides. Using this sequence we then isolated cDNAs encoding the first three members of this group to be identified [3], ERK1, ERK2, and ERK3. ERK1 and ERK2 are phosphorylated and activated by MAP kinase kinase or MEK [1,28-30] (MAP kinase/ERK kinase). MEK is reported to be activated by three different protein kinases: the protooncogenes Raf [31] and Mos [26], and MEK kinase [32].

ERKs 1 and 2 are greater than 50% identical to the protein kinases KSS1 [34] and FUS3 [35] from *Saccharomyces cerevisiae*. These enzymes mediate entry into the cell cycle and growth arrest in the pheromone response pathway. The pheromone response pathway in *S. cerevisiae* [36,37] and the related ras-dependent pathway in *S. pombe* [38] are providing models for the study of the ERK phosphorylation cascade in mammals. Like the ERK pathway, these two yeast pathways are activated by ras and heterotrimeric G proteins. In *S. cerevisiae* FUS3 is activated by a receptor-G protein system and requires at least two upstream protein kinases, STE11 and STE7. The *S. pombe* pathway requires ras [39] and has functionally complementary [38] enzymes--byr2, the STE11 homolog, upstream of byr 1, the STE7 homolog; both of these are believed to be upstream of spk1, the ERK homolog. STE7 and byr1 have significant sequence identity to MEK. MEK kinase, a mammalian homolog of STE11/byr2, has recently been cloned using oligonucleotides based on STE11/byr2 sequence [32] and was found to activate MEK.

ERKs 1 and 2 were among the first dual specificity protein kinases recognized [40]; they autophosphorylate on both tyrosine and threonine. The substrates known currently, however, are phosphorylated only on the aliphatic residues. No major differences in their substrate specificity has yet been discovered. Some of their phosphorylation sites have been identified yielding a consensus of Ser or Thr followed by Pro [16,41,42]. A peptide from tyrosine hydroxylase (EAIMSPRFK) is the best peptide substrate found in vitro [42].

RESULTS AND DISCUSSION

ERKs 1 and 2 are activated by phosphorylation on tyrosine and threonine. Tyrosine phosphorylation precedes threonine phosphorylation [43], but both residues must be phosphorylated for maximum activity of the enzymes. Wildtype ERK1 and ERK2 as well as mutants of each have been synthesized as histidine-tagged proteins in bacteria. All proteins were purified by adsorption to nickel resin. The mutants include ones with defective catalytic activity and ones lacking the two activating phosphorylation sites. Wildtype ERK1 and ERK2 were phosphorylated by a partially purified preparation of MEK from rabbit muscle to a stoichiometry of 1 to 2 mol phosphate per mol ERK. MEK caused a ~1000-fold increase in activity of the two ERKs[44]. This is comparable to the specific activity of enzyme purified from insulin-stimulated cells [2]. Mutants lacking either one or both of the activating phosphorylation sites (one tyrosine and one threonine--Y185 and T183 of ERK2 [45] and Y204 and T202 of ERK1) were not activated by MEK in vitro [44]. These two phosphorylation sites fall between subdomains VII and VIII of the ERKs.

Activation of ERK1 or ERK2 by MEK

```
              MEK                  MEK        ↗ YP
ERK - - - - - - - →ERK-YP - - - - - - - →ERK
                                          ↘ TP
Inactive           Inactive              Active
```

Fig. 1

When phosphorylated by MEK to low stoichiometries, both ERK1 and ERK2 contained phosphotyrosine but little or no phosphothreonine, indicating that MEK phosphorylated tyrosine first on ERKs. These data are consistent with the time course of phosphorylation of tyrosine and threonine in ERK1 from stimulated cells [43]; the accumulation of tyrosine phosphate preceded the accumulation of threonine phosphate.

The first protein kinase found to be a substrate for ERKs 1 and 2 was one type of S6 kinase now known as Rsk [9-11]. Other protein kinases are also substrates. The phosphorylation by ERKs 1 and 2 of the proto-oncogene protein kinase Raf occurred on sites that are phosphorylated on Raf in intact cells [12]. While phosphorylation by ERKs is not believed to stimulate Raf, effects on Raf activity have been difficult to assess. ERKs 1 and 2 also phosphorylated MEK. Activity from ERKs 1 and 2 accounted for most of the MEK phosphotransferase in PC12 cells stimulated for 2 to 5 minutes with NGF. ERK phosphorylation did not appear to stimulate MEK activity. Thus, ERKs 1 and 2 phosphorylate at least three kinases in their regulatory pathway, two of which are upstream participants [31].

ERK1 and ERK2 phosphorylate Rsk, Raf and MEK.

Raf ↘ MEK – – – → ERK2 / ERK1 – – – – →Rsk
MEKK ↗

Fig. 2

At least two separate pathways activate MEK leading to activation of ERKs 1 and 2. One of these requires the small GTP-binding protein Ras [21,26,33]. Activation of ERKs by nerve growth factor, phorbol ester, and bradykinin was blocked in cells expressing S17N Ras, a dominant interfering Ras mutant. This suggested that Ras was an obligate step in NGF and phorbol ester pathways. Bradykinin is thought to work through G_q. One possible mechanism by which S17N Ras may block the effects of both phorbol ester and bradykinin is if Ras lies downstream of protein kinase C. Phorbol esters activate protein kinase C directly, while G_q probably activates protein kinase C indirectly via activation of phospholipase Cβ. Phospholipase Cβ releases diacylglycerol a cofactor for protein kinase C activation.

Table 1. Ras is involved in some but not all pathways leading to activation of MAP kinases.

Stimulus	Activation of ERKs Ras dependent?
NGF	yes
TPA	yes
Bradykinin	yes
EGF	yes
AlF_4^-	no
Okadaic acid	no

On the other hand, certain agents such as AlF_4^-, which activates heterotrimeric G proteins, and okadaic acid activated ERKs both in the absence and presence of S17N Ras. Thus, these agents act independently of Ras [21]. The Ras-independent activation of ERKs by AlF_4^- is probably mediated by one or more G_i proteins.

ERK3 is a protein of 62 kDa with a C-terminus that extends ~180 amino acids beyond the protein kinase core conserved among the ERKs. ERK3 is 50% identical to ERKs 1 and 2 within the kinase catalytic domain. To demonstrate that the protein was expressed, we generated antipeptide antibodies to four regions of ERK3. Antisera to each of these peptides recognized a protein of the same size, 62 kDa, in cell extracts; in each case the recognition of the 62 kDa protein was blocked by preincubation of the antisera with the appropriate antigenic peptides. ERK3 was found predominantly in fractions from both PC12 cells and T cells enriched in nuclei.

To facilitate biochemical analysis of ERK3 a truncated form, lacking the 180 amino acids C-terminal to the catalytic core, has been expressed in bacteria and purified on nickel-chelate resin as described for ERK2 and ERK1 [44]. This truncated form of ERK3 (ERK3ΔCt) was 45 kDa and was no longer immunoblotted by antibodies raised to peptides in the C-terminal region of the molecule. ERK3ΔCt autophosphorylated on serine residues to a small extent but did not appear to be constitutively active. This suggests that the C-terminus is not directly involved in regulating the kinase activity of ERK3. The truncated form of ERK3 was also phosphorylated by a partially purified preparation of MEK from rabbit muscle. MEK preparations from other sources also phosphorylated ERK3. Phosphoamino acid analysis of ERK3 labeled in vitro indicated that MEK phosphorylated one or more serine residues of ERK3. No phosphorylation of threonine or tyrosine was observed. Preliminary studies demonstrate that phosphorylation of EKK3 by MEK increases its protein kinase activity (Zhen, Cheng, Robbins, Boulton, and Cobb, unpublished data).

We have used a partially purified preparation of MEK from rabbit muscle for the in vitro activation studies of ERKs. Chromatography of this MEK preparation on Mono S resolves two activities that phosphorylate ERK3, one of which works poorly on ERKs 1 and 2. The finding that ERK3 is activated by a MEK-like enzyme suggests that it may be regulated in an analogous manner to ERKs 1 and 2.

In conclusion, ERKs have important regulatory functions and are controlled by remarkably complicated upstream cascades. Characterization of additional members of this family as they are identified will help to distinguish the shared from the unique features of their activation and will be important for determining how specificity in the actions of this cascade is achieved. Studies of this complex pathway will continue to provide novel information about protein kinase regulation, hormone action, and oncogenesis.

ACKNOWLEDGMENTS

We would like to thank Tom Geppert (UT Southwestern), Natalie Ahn (University of Colorado at Boulder) and Rony Seger and Jean Campbell (University of Washington) for helpful discussions, Gary Johnson (National Jewish Hospital, Denver) for recombinant MEK, and Jo Hicks for preparation of the manuscript. This work was supported by a grant from the Texas Advanced Research Program, National Institutes of Health research grant DK34128 and Research Career Development Award DK01918 (to MHC), a grant from the Juvenile Diabetes Foundation, National Institutes of Health training grant GM07062 (to DJR), and Merck training fellowship (to DE).

REFERENCES

1. Boulton TG, Yancopoulos GD, Gregory JS, Slaughter C, Moomaw C, Hsu J, et al: An Insulin-Stimulated Protein Kinase Similar to Yeast Kinases Involved in Cell Cycle Control. *Science* 1990;249:64-67
2. Boulton TG, Gregory JS, Cobb MH: Purification and Properties of Extracellular Signal-Regulated Kinase 1, an Insulin-Stimulated Microtubule-Associated Protein 2 Kinase. *Biochemistry* 1991;30:278-286
3. Boulton TG, Nye SH, Robbins DJ, Ip NY, Radziejewska E, Morgenbesser SD, et al: ERKs: A Family of Protein-Serine/Threonine Kinases That Are Activated and Tyrosine Phosphorylated in Response to Insulin and NGF. *Cell* 1991;65:663-675
4. Boulton TG, Cobb MH: Identification of Multiple Extracellular Signal-regulated Kinases (ERKs) with Antipeptide Antibodies. *Cell Regulation* 1991;2:357-371
5. Ray LB, Sturgill TW: Rapid Stimulation by Insulin of a Serine/Threonine Kinase in 3T3-L1 Adipocytes that Phosphorylates Microtubule-associated Protein 2 *in vitro*. *Proc Natl Acad Sci USA* 1987;84:1502-1506

6. Pelech SL, Tombes RM, Meijer L, Krebs EG: Activation of Myelin Basic Protein Kinases during Echinoderm Oocyte Maturation and Egg Fertilization. *Dev Biol* 1988;130:28-36
7. Ahn NG, Weiel JE, Chan CP, Krebs EG: Identification of Multiple Epidermal Growth Factor-stimulated Protein Serine/Threonine Kinases from Swiss 3T3 Cells. *J Biol Chem* 1990;265:11487-11494
8. Hoshi M, Nishida E, Sakai H: Activation of a Ca^{2+}-inhibitable Protein Kinase That Phosphorylates Microtubule-associated Protein 2 *in Vitro* by Growth Factors, Phorbol Esters, and Serum in Quiescent Cultured Human Fibroblasts. *J Biol Chem* 1988;263:5396-5401
9. Sturgill TW, Ray LB, Erikson E, Maller J: Insulin-stimulated MAP-2 Kinase Phosphorylates and Activates Ribosomal Protein S6 Kinase II. *Nature* 1988;334:715-718
10. Gregory JS, Boulton TG, Sang B-C, Cobb MH: An Insulin-stimulated Ribosomal Protein S6 Kinase from Rabbit Liver. *J Biol Chem* 1989;264:18397-18401
11. Ahn NG, Krebs EG: Evidence for an Epidermal Growth Factor-stimulated Protein Kinase Cascade in Swiss 3T3 Cells. Activation of Serine Peptide Kinase Activity by Myelin Basic Protein Kinases *In Vitro*. *J Biol Chem* 1990;265:11495-11501
12. Lee R-M, Cobb MH, Blackshear PJ: Evidence that Extracellular Signal-regulated Kinases (ERKs) are the Insulin-activated Raf-1 Kinase Kinases. *J Biol Chem* 1992;267:1088-1092
13. Takishima K, Griswold-Prenner I, Ingebritsen T, Rosner MR: Epidermal Growth Factor (EGF) Receptor T669 Peptide Kinase From 3T3-L1 Cells is an EGF-stimulated "MAP" Kinase. *Proc Natl Acad Sci USA* 1991;88:2520-2524
14. Northwood IC, Gonzalez FA, Wartmann M, Raden DL, Davis RJ: Isolation and Characterization of Two Growth Factor-stimulated Protein Kinases That Phosphorylate the Epidermal Growth Factor Receptor at Threonine 669. *J Biol Chem* 1991;266:15266-15276
15. Pulverer BJ, Kyriakis JM, Avruch J, Nikolakaki E, Woodgett JR: Phosphorylation of c-jun Mediated by MAP Kinases. *Nature* 1991;353:670-674
16. Alvarez E, Northwood IC, Gonzalez FA, Latour DA, Seth A, Abate C, et al: Pro-Leu-Ser/Thr-Pro Is a Consensus Primary Sequence for Substrate Protein Phosphorylation. Characterization of The Phosphorylation of *c-myc* and *c-jun* Proteins by an Epidermal Growth Factor Receptor Threonine 669 Protein Kinase. *J Biol Chem* 1991;266:15277-15285
17. Drechsel DN, Hyman AA, Cobb MH, Kirschner MW: Modulation of the Dynamic Instability of Tubulin Assembly by the Microtubule-associated Protein Tau. *Mol Biol Cell* 1992;3:1141-1154
18. Gotoh Y, Nishida E, Matsuda S, Shiina N, Kosako H, Shiokawa K, et al: *In Vitro* Effects on Microtubule Dynamics of Purified *Xenopus* M Phase-activated MAP Kinase. *Nature* 1991;349:251-254
19. Tsao H, Aletta JM, Greene LA: Nerve Growth Factor and Fibroblast Growth Factor Selectively Activate a Protein Kinase That Phosphorylates High Molecular Weight Microtubule-associated Proteins. Detection, Partial Purification, and Characterization in PC12 Cells. *J Biol Chem* 1990;265:15471-15480
20. Gotoh Y, Nishida E, Yamashita T, Hoshi M, Kawakami M, Sakai H:Microtubule-associated-protein (MAP) Kinase Activated by Nerve Growth Factor and Epidermal Growth Factor in PC12 Cells. Identity With the Mitogen-activated MAP Kinase of Fibroblastic Cells. *Eur J Biochem* 1990;193:661-669
21. Robbins DJ, Cheng M, Zhen E, Vanderbilt C, Feig LA, Cobb MH: Evidence for a Ras-dependent Extracellular Signal-regulated Protein Kinase (ERK) Cascade. *Proc Natl Acad Sci USA* 1992;89:6924-6928
22. Boulton TG, Gregory JS, Jong S-M, Wang L-H, Ellis L, Cobb MH: Evidence for Insulin-dependent Activation of S6 and Microtubule-associated Protein-2 Protein Kinases Via a Human Insulin Receptor/v-Ros Hybrid. *J Biol Chem* 1990;265:2713-2719
23. Levers SJ, Marshall CJ: Activation of Extracellular Signal-regulated Kinase, ERK2, by p21*ras* Oncoprotein. *EMBO J* 1992;11:569-574
24. Gupta SK, Gallego C, Johnson GL, Heasley LE: MAP Kinase Is Constitutively Activated in *gip2* and *src* Transformed Rat 1a Fibroblasts. *J Biol Chem* 1992;267:7987-7990
25. Wood KW, Sarnecki C, Roberts TM, Blenis J: ras Mediates Nerve Growth Factor Receptor Modulation of Three Signal-Transducing Protein Kinases: MAP Kinase, Raf-1, and RSK. *Cell* 1992;68:1041-1050
26. Posada J, Yew N, Ahn NG, Vande Woude GF, and Cooper JA, Mos stimulates MAP kinase in *Xenopus* oocytes and activates a MAP kinase kinase in vitro. *Molec. Cell. Biol.* 1993;13:2546-2553
27. Ahn NG, Seger R, Bratlien RL, Diltz CD, Tonks NK, Krebs EG: Multiple Components in an Epidermal Growth Factor-stimulated Protein Kinase Cascade. *In Vitro* Activation of a Myelin Basic Protein/Microtubule-associated Protein 2 Kinase. *J Biol Chem* 1991;266:4220-4227

28. Seger R, Ahn NG, Posada J, Munar ES, Jensen AM, Cooper JA, et al: Purification and Characterization of Mitogen-activated Protein Kinase Activator(s) from Epidermal Growth Factor-stimulated A431 Cells. *J Biol Chem* 1992;267:14373-14381
29. Crews C, Alessandrini A, Erikson R: The Primary Structure of MEK, a Protein Kinase that Phosphorylates the ERK Gene Product. *Science* 1992;258:478-480
30. Seger R, Seger D, Lozeman FJ, Ahn NG, Graves LM, Campbell JS, et al: Human T-cell Mitogen-activated Protein Kinase Kinases Are Related to Yeast Signal Transduction Kinases. *J Biol Chem* 1992;267:25628-25631
31. Kyriakis JM, App H, Zhang X-F, Banerjee P, Brautigan DL, Rapp UR, et al: Raf-1 Activates MAP Kinase-Knase. *Nature* 1992;358:417-42132. Lange-Carter CA, Pleiman CM, Gardner AM, Blumer J, Johnson GL: A Divergence in theMAP Kinase Regulatory Network Defined by MEK Kinase and Raf. *Science* 1993;315-319
33. Thomas SM, DeMarco M, D'Arcangelo G, Halegoua S, Brugge JS: Ras Is Essential for Nerve Growth Factor- and Phorbol Ester-Induced Tyrosine Phosphorylation of MAP Kinases. *Cell* 1992;68:1031-1040
34. Courchesne WE, Kunisawa R, Thorner J: A Putative Protein Kinase Overcomes Pheromone-induced Arrest of Cell Cycling in *S. cerevisiae*. *Cell* 1989;58:1107-1119
35. Elion EA, Grisafi PL, Fink GR: FUS3 Encodes a cdc2+/CDC28-related Kinase Required for the Transition From Mitosis into Conjugation. *Cell* 1990;60:649-664
36. Sprague Jr. GF, Thorner J: Pheromone Response and Signal Transduction during the Mating Process of *Saccharomyces cerevisiae*. *Cold Spring Harbor Laboratory Press* 1993;(In Press)
37. Marsh L, Neiman AM, Herskowitz I: Signal Transduction During Pheromone Response in Yeast. *Annu Rev Cell Biol* 1991;7:699-728
38. Neiman AM, Stevenson BJ, Xu H-P, Sprague GF,Jr., Herskowitz I, Wigler M, et al: Functional Homology of Protein Kinases Required for Sexual Differentiation in *Schizosaccharomyces pombe* and *Saccharomyces cerevisiae* Suggests a Conserved Signal Transduction Module in Eukaryotic Organisms. *Mol Biol Cell* 1993;4:107-120
39. Wang Y, Xu H-P, Riggs M, Rodgers L, Wigler M: *byr2*, a *Schizosaccharomyces pombe* Gene Encoding a Protein Kinase Capable of Partial Suppression of the *ras*1 Mutant Phenotype. *Mol Cell Biol* 1991;11:3554-3563
40. Seger R, Ahn NG, Boulton TG, Yancopoulos GD, Panayotatos N, Radziejewska E, et al: Microtubule-associated Protein 2 Kinases, ERK1 and ERK2, Undergo Autophosphorylation on Both Tyrosine and Threonine Residues: Implications for Their Mechanism of Activation. *Proc Natl Acad Sci USA* 1991;88:6142-6146
41. Clark-Lewis I, Sanghera JS, Pelech SL: Definition of a Consensus Sequence for Peptide Substrate Recognition by $p44^{mpk}$, the Meiosis-activated Myelin Basic Protein Kinase. *J Biol Chem* 1991;266:15180-15184
42. Haycock JW, Ahn NG, Cobb MH, Krebs EG: ERK1 and ERK2, Two Microtubule-associated Protein 2 Kinases, Mediate the Phosphorylation of Tyrosine Hydroxylase at Serine-31 *in situ*. *Proc Natl Acad Sci USA* 1992;89:2365-2369
43. Robbins DJ, Cobb MH: ERK2 Autophosphorylates on a Subset of Peptides Phosphorylated in Intact Cells in Response to Insulin and Nerve Growth Factor: Analysis by Peptide Mapping. *Mol Biol Cell* 1992;3:299-308
44. Robbins DJ, Zhen E, Okami H, Vanderbilt C, Ebert D, Geppert TD, et al: Regulation and Properties of Extracellular Signal-regulated Protein Kinases 1 and 2 *in vitro*. *J Biol Chem* 1993;268:5097-5106
45. Payne DM, Rossomando AJ, Martino P, Erickson AK, Her J-H, Shananowitz J, et al: Identification of the Regulatory Phosphorylation Sites in pp42/mitogen-activated Protein Kinase (MAP kinase). *EMBO J* 1991;10:885-892

8

CELL CYCLE TRAVERSE AND GROWTH ARREST CONTROL IN SENESCENT HUMAN FIBROBLASTS

Eugenia Wang and Menq-Jer Lee

The Bloomfield Center for Research on Aging
Lady Davis Institute for Medical Research
The Sir Mortimer B. Davis Jewish General Hospital
McGill University
Montréal, Québec H3T 1E2 Canada

INTRODUCTION

Since the discovery in the 1960's that cells, like organisms, are mortal and have a limited life span,[1,2] the *in vitro* aged human fibroblast has become a popular model for studying the control of the loss of replicative potential. During the 1970's, a significant body of work was devoted to analyzing what makes senescent fibroblasts different, and how one might relate the *in vitro* aging process to *in vivo* aging in organisms. Extensive studies conducted by several groups of investigators have established a parallelism between the life span of cultured skin fibroblasts and the age of their donors. In brief, an inverse relationship pertains between *in vitro* lifespan and donor age, *i.e.* a decreased replicative capacity is observed with increasing donor age,[3-6] and cellular lifespan is genetically determined with little or no significant difference within twin pairs, but a significant difference between pairs.[7]

Perhaps the most profound finding of the research of the 1980's is the observation of the inability of DNA synthesis as the dominant phenotype in senescent cells. This conclusion, by four independent groups of investigators, initiated the quest to understand the mechanism determining the permanent replicative dormancy in senescent cells.[8-12] In brief, when senescent cells are fused with growing young cells, DNA synthesis activity is turned off in the nuclei from the younger cells within the heterokaryon. Later, this inhibitory activity was shown to be dependent upon protein synthesis, and specifically traced to the poly(A)+ RNA of the senescent cells.[13] This observation of senescence-related genes has recently been localized in chromosome 1 by Sugawara, *et al.* (1990),[14] as well as in chromosome 4 by Ning, *et al.* (1991).[15] Moreover, studies on shortened telomere length in aging fibroblasts have illustrated another dimension:[16] a counting mechanism which may be important in playing the role of the clocking process for defining *in vitro* lifespan.

The Cell Cycle: Regulators, Targets, and Clinical Applications
Edited by V.W. Hu, Plenum Press, New York, 1994

Following the initial report of senescence-related inhibitory action on DNA synthesis, intensive efforts have been organized to identify unique nonproliferation-dependent gene expressions, in the hope that they may function as inhibitors. With the popular differential subtraction of cDNA libraries of different growth status, several uniquely-expressed genes have been identified: a member of the EF-1α family, fibronectin, prohibitin, and several RNAs common to senescent and Werner's syndrome fibroblasts.[17-20] With increasing sensitivity of detection, this list of unique senescent gene expressions will surely increase further in number. We have, however, used a different tactic, largely depending upon the production of unique monoclonal antibodies by the use of the strategy of immunosuppression. This approach has allowed us to characterize three genes, statin, S1 and terminin, as nonproliferation-related.[21-25] Of the three nonproliferation-specific antigens, statin is the best-characterized; the following paragraphs summarize the knowledge so far of the biology of statin relating to the physiological status of nonproliferation.

BIOLOGY OF STATIN PROPERTIES

Characterization of Statin Appearance or Disappearance during Entry or Exit from the G_0-quiescent state

Several studies have established the fact that the presence of statin in nongrowing cells is conserved and universal among many cell systems, including 5 strains of human fibroblasts,[26-27] mouse 3T3 fibroblasts,[28] and cultured lens epithelial cells.[29] In all these studies, we have examined the kinetic relationship between the rate of exit from or entry to the quiescent G_0 phase and the presence of statin, by immunofluorescence and immunoblotting techniques. Results of all these systems show that statin is present in the nuclei of all cells arrested at G_0 phase, and that this presence is rapidly lost once the exit from the quiescent state is initiated by means such as the readdition of serum to the culture. Pulse-chase experiments reveal that statin presence is rapidly lost from cells, as soon as 2-4 hours after the exit from the G_0 phase.[22] In mouse 3T3 fibroblasts, treatment of serum-starved cells with platelet-derived (PDGF) or fibroblast (FGF) growth factors induces the loss of statin presence as does the addition of serum, while epidermal growth factor (EGF) exerts no effect upon statin presence.[28] These results suggest that statin is present exclusively in those cells positioned in the G_0 phase, and not in early G_1 cells, which have already commenced the exit journey from growth arrest.

Characterization of Statin Presence in Onset of Terminal Differentiation or Cellular Senescence in Culture

As described previously, statin presence is found not only in temporarily growth-arrested quiescent cells, but also in permanently nonproliferating cells such as *in vitro* aged human fibroblasts,[26-27] as well as terminally-differentiated cultured lens epithelial cells[29] and myotubes.[30] In parallel with the irreversible loss of DNA synthesis, statin presence in these cells is permanent, and cannot be removed by addition of serum or growth factors. This fact will allow us to advance our study of clonal heterogeneity, colony formation and growth status in any cultured system; for example, the statin-positive labelling index may be used as a means to quantitate in a mass population the parameters of growth potential and response to proliferation and differentiation stimuli.

Characterization of the Inverse Relationship between Statin Presence and the Absence of PCNA, Ki67 and ^{3}H-thymidine Incorporation

During our study of statin presence in nonproliferating cells, we realized that two other proteins (PCNA and Ki67) exhibit opposite behaviour; their nuclear presence is mutually exclusive with that of statin.[30-32] Proliferating cell nuclear antigen (PCNA), also known as DNA polymerase δ auxiliary protein, is the more thoroughly studied of the two; it is found in growing cells of many tissue origins.[33] In senescent cells, its absence is post-transcriptionally controlled;[33] thus, it is rare for a cell to be positive for both PCNA and statin. To prove that statin-positive cells are not participating in DNA synthesis, we have used simultaneous labelling on the same cells with statin immunofluorescence and ^{3}H-thymidine autoradiography, in both *in vitro* cultured fibroblasts and glia[34-35] and *in vivo* intestinal epithelium.[34] In both circumstances, statin immunofluorescence labelling was observed to be restricted to those cells lacking positive ^{3}H-thymidine autoradiographic reaction; this result further supports the notion that statin is present only in nonproliferating cells.

Statin Presence as an Index for Terminal Differentiation, and Absence as a Negative Marker for Tissue Regeneration and Tumour Malignancy

To expand our initial discovery of statin presence in nongrowing cultured cells, we performed extensive immunohistochemical surveys in many human, dog, rat, mouse and chick tissues.[26] Throughout, nuclear labelling of statin presence was found in the terminally-differentiated subpopulation of cells: for example, neurons, myocardium and liver. In some tissues, however, statin presence was significantly reduced after surgical manipulation such as hepatectomy (which induces liver proliferation)[36] and rumenectomy (which induces proliferation in the duodenum).[37] In most statin-positive tissues, statin presence is detected shortly after birth and remains at a constant level throughout the lifespan; our extensive immunofluorescence and immunoblotting survey showed constant levels of nuclear p57 statin from 3 to 26 months in rats.[34-36]

Along the line of statin as a negative growth status marker, we have collaborated with 3 groups of oncologists in large-scale studies on brain, colorectal, thyroid and breast tumours;[38-44] the results so far show that the degree of malignancy is inversely related to statin positivity. For example, of the 53 brain tumours studied, statin levels in pituitary adenoma were near those observed in normal tissues, while metastatic samples had almost none.[40] In another application, we examined the increased gliogenesis associated with neurodegenerative diseases such as Alzheimer's,[45] and assayed the effectiveness of chemotherapeutic drugs (most of which induce growth arrest) in cultures.[38]

Advantages of using the statin labelling index include: (1.) avoiding labour-intensive autoradiography in cultured systems or experimental animals, and (2.) allowing the examination of the growth status of solid tumours where ^{3}H-thymidine or Budr labelling cannot be performed. We have proven the case of adaptation of statin labelling to at least 15 different tissues in routine pathological examinations.

Application of Statin Presence and c-*fos* and p53 Absence as a Tool to Test the Field Defect Theory in Colorectal Carcinoma

Recurrence of local disease after therapeutic resection is a major cause of death in colorectal carcinoma; one explanation for this is the theory of a mucosal field defect adjacent to the neoplasm.[46] We examined tissues from 10 colorectal carcinomas, along with the mucosae 1, 5 and 10 cm removed on both sides of the primary lesions; immunohistochemical and Western blotting assays with antibodies to c-*fos*, mutant p53 and statin revealed a disruption in the delicate balance between cell proliferation and non-proliferation, not only within the carcinoma as would be expected, but also in the immediately adjacent histologically normal mucosa.[42,46] The changes in statin expression are opposite to those for c-*fos* and mutant p53, and the field defect extends for at least 2 cm. We conclude that this 'transitional zone' should be removed at the time of original surgery, to obviate anastomotic recurrence.

Biochemical Properties of Statin

Immunoprecipitation and immunoblotting assays reveal statin as a protein of 57 kilodalton. Immunoprecipitation assay of ^{35}S-methionine-labelled cell extracts shows the presence of statin in both nuclear and cytoplasmic fractions; pulse-chase experiments, however, show the cytoplasmic fraction as the chased product of the nuclear form.[22] p57 statin presence in the nucleus is largely an insoluble component of the nuclear lamin and nuclear matrix;[47] in particular, immunoblotting with purified nuclear lamin fractions reveals a close association with lamin A and C components. Immunoprecipitation with ^{3}H-glucosamine-labelled cell extracts shows that p57 statin is not glycosylated; 2-D SDS-PAGE of senescent cell extracts shows that both the nuclear and cytoplasmic forms share one isomeric pI of 5.3,[22] while reduction immunoblotting 2-D gels of rat liver extracts show two isomers with pI's of 6.8 and 7.2. We have succeeded in the first stage of statin purification by step-wise column chromatography including DEAE, heparin and chromatofocussing approaches;[23,48] the purified rat liver statin contains two isomers with pI's of 6.8 and 7.2, which seem identical by peptide mapping.

Phosphorylated Nature of p57 Statin and its Associated p45 Kinase[49]

Perhaps the most intriguing characterization of statin's properties lies in its phosphorylated nature; immunoprecipitation of ^{32}P-labelled senescent cell extracts shows that p57 statin is phosphorylated, while amino acid analysis reveals that this occurs at serine and threonine sites. Coprecipitated with p57 is another phosphoprotein of 45 kDa (p45); *in situ* ^{32}P-ATP analysis of casein-impregnated gels shows that p57 is not phosphorylated autocatalytically, but rather by the kinase activity of the p45 protein. Further analysis reveals that the insoluble nature of nuclear p57 is attributed to its phosphorylation; phosphatase treatment converts it to a 56 kDa detergent-soluble form.

p45 kinase, once dissociated from p57, phosphorylates other substrates such as casein, enolase and immunoglobulin, all at serine or threonine sites. While it coprecipitates with p57 statin on Western blot, it does not react with the statin antibody. Cell fractionation studies show p45 kinase present only in the nucleus; unlike nonproliferation-specific nuclear statin, however, it is found in both growing and non-growing cells. Characterization reveals that it is dependent upon ATP, and not GTP. Neither Ca^{++}, EGTA, nor both affects its relative phosphorylation efficiency; this negative result rules out the possibility that p45 might be a member of the calmodulin-protein kinase C family. Similarly, adding either heparin or spermine fails to affect the enzymatic activity, which excludes p45 from the casein kinase II family. We are now testing whether p45 kinase may pertain to the mitogen-activating promoter (MAPII) kinase family, and its dependency upon cyclic AMP presence.

Biochemical Relationship among p57 Statin, p45 Statin-associated Kinase, and p110 RB Protein in Senescent Cells

Comparative analysis of the statin antibody-immunoprecipitated profile from radioisotope-labelled senescent cells shows that, in addition to the p57 and p45 proteins, a band migrating at 110 kDa is observed in the ^{35}S-methionine-labelled, but is not as readily detectable in ^{32}P-labelled, extracts; this suggests that the p110 protein is not phosphorylated. (On Western blots, neither the p45 kinase nor the p110 protein reacts with the S44 antibody, which suggests that their presence in the immunoprecipitates is due to biochemical complex interaction with the statin polypeptide, rather than shared antigenicity with the antibody.) Along with its electrophoretic mobility, this feature suggests the possibility that p110, coprecipitated with p57 statin and its p45 kinase by S44 antibody, is the RB protein; preliminary results indicate that the coprecipitated p110 reacts with antibody to RB on immunoblot. Also, immunoprecipitation of nongrowing young fibroblasts with RB antibody shows two major proteins at p110 and p57, while the same antibody applied to immunoprecipitates of growing young fibroblasts shows no p57 band, but several faint bands around 112 to 116 kDa; this group of bands in the 112 to 116 kDa region may prove to be RB in different states of phosphorylation. This p57 comigrates with exactly the same electrophoretic mobility as statin; and the presence of p57 in quiescent cell extracts, along with its absence in growing cell extracts, reflects both that this protein is indeed statin, and its nonproliferation-specific nature. These results suggest that p57 and p110 can both be coprecipitated from nongrowing cell extracts by antibodies to either RB or statin, and preliminary peptide maps of both proteins precipitated by either antibody are similar.

Further support of the close relationship between p110 RB and p57 statin is their nuclear colocalization, observed by confocal image analysis. Laser scanning of a composite of many defined planes in the nucleus of senescent cells reveals that the two antibodies stain in exactly the same pattern, with the exception that the RB antibody reaction is seen as the predominant one in the nucleoli. This colocalization is especially prominent in the nuclear envelope region.

REPRESSION OF c-fos EXPRESSION AND RB PHOSPHORYLATION AND "REPLICATIVE SENESCENCE"

Equally important to the activation of nonproliferation-dependent gene expression is the deactivation or loss of proliferation-dependent gene expressions in senescent cells. The finding by Seshadri and Campisi that c-*fos* is not transcribed in senescent fibroblasts suggests the involvement of oncogene loss in the acquisition of the growth-arrest phenotype;[50] moreover, this result leads us to ponder whether fibroblast senescence is the manifest destiny of "anti-oncogenesis".[51] Other related work on posttranscriptional blockade in the expression of PCNA (a DNA polymerase δ) in senescent cells further supports the notion of a lack or repression of growth-promoting factors as part of the mechanism blocking replication in senescent cells.[51-52] Based upon this logic, providing a strong oncogene such as the SV-40 T-antigen to senescent cells should allow them to overcome the blockage of DNA synthesis and extend their lifespan. However, this provision cannot indefinitely support proliferation; after a few rounds of cell cycle traverse, the permanent loss of replication is again observed.[53-54]

The finding that retinoblastoma (RB) gene is not phosphorylated in senescent cells provides another clue to the mystery of their nonproliferative nature.[55] Since the abundance of RB protein does not appear to be regulated, Northern analysis and pulse-chase experiments show that RB messages and p110 RB protein are stable for at least 6 hours;[56] therefore, modulation of RB activity may prove to be dependent upon the phosphorylation state of the RB proteins.[55,57-64] Unphosphorylated RB has been shown to bind the transforming product of SV-40 T-antigen,[58,65] and to associate with the transcription-activating factor E2F.[64-69] Some workers have suggested that RB binding results in sequestration or inactivation of these genes; if this is so, the cascade effect is then the consequent result of the failed expression of growth-promoting genes.

The importance of the RB phosphorylation state in monitoring cell cycle traverse, and the fact that RB gene products are not phosphorylated at all in senescent cells, lead us to ask, "Why is the RB gene product not phosphorylated; what happened to the kinases needed for this process?" An obvious answer is that the necessary kinases are either absent or unavailable. Recent findings of Stein, *et al.*[70] on the absence of cdc2 kinase, as well as cyclin A and cyclin B, in senescent cells confirms one of these possibilities. Other reports, however, have shown that the first stage of RB phosphorylation occurs in early G_1 phase, before the activation of $p34^{cdc2}$ expression at the G_1/S transition;[71] antisense oligodeoxynucleotides to $p34^{cdc2}$ can reduce this expression, although RB is still phosphorylated.[72] Thus $p34^{cdc2}$ cannot be the only kinase responsible for the pre-S-phase phosphorylation of RB, and its absence cannot be the sole explanation for the lack of phosphorylation of RB in senescent cells.

An answer to the lack of complete correlation between the absence of $p34^{cdc2}$ and RB phosphorylation may lie in the following explanation. Several laboratories have reported that RB phosphorylation is cell cycle-dependent;[57-64] it turns out that RB becomes phosphorylated in at least three stages during the entry and traverse of the cell cycle.[71] The initial phosphorylation is noted in early G_1, after the exit from G_0; the second occasion for RB to become phosphorylated occurs at G_1/S or after the beginning of the S phase, and the final phosphorylation occurs during G_2 phase. These results of Livingston and coworkers[71] show that in synchronized cultures, there are at least 4 species of RB proteins; the lowest molecular weight, unphosphorylated one is found during the G_0 phase, while sequential addition of three other species with progressively higher molecular weights occurs in early G_1, G_1/S and G_2. The corresponding examination of cdc2 presence shows that $p34^{cdc2}$ probably participates in the S and G_2 phases of RB phosphorylation.[71] This leaves the suggestion that another enzyme(s) must be responsible for RB phosphorylation during early G_1 phase.

STATIN-ASSOCIATED p45 KINASE AS THE MISSING KINASE CONTRIBUTING TO THE EARLY G1 STEP OF RB PHOSPHORYLATION

Here, we propose the statin-associated p45 kinase as a candidate for the missing enzyme required for the critical first-step phosphorylation; during the nongrowing G_0 stage, the p57 sequestration of this enzyme prevents RB from being phosphorylated. Results so far show that both proteins are present in the nucleus; p57 statin only in nongrowing, quiescent or senescent cells, and the p45 kinase in both proliferating and nonproliferating cells. In addition, a p110 protein (which our results so far identify with RB) forms a complex with both p57 statin and its associated p45 kinase. Since statin presence disappears rapidly at about 2-4 hours following the entrance into G_1 from the G_0 phase,[22] this absence may quickly releases the blockade of the p45 kinase, thus allowing it to perform the first (early G_1) stage of RB phosphorylation.

The expression of statin in senescent cells is permanent and independent of serum concentrations. This permanent presence is also observed in terminally-differentiated cells in nervous, myocardial and skeletal muscle tissues, where regeneration or reinitiation of replication is almost impossible, and may be detrimental to the normal functioning of these tissues. According to our hypothesis, the permanent presence of statin may serve as a lock, preventing p45 kinase from functioning in the early phosphorylation of RB; this, along with the absence of later RB phosphorylation associated with the $p34^{cdc2}$ kinase, leaves RB in the unphosphorylated state to bind to E2F, SV-40 T-antigen, *etc.* In temporarily growth-arrested cells, this locking mechanism may be transiently removed, while in senescent or terminally-differentiated cells it is permanent.

POSSIBLE MODE FOR STATIN FUNCTIONING AS THE SEQUESTER FOR THE p45 KINASE WHICH IN TURN BLOCKS THE FIRST STAGE OF RB PHOSPHORYLATION IN QUIESCENT AND SENESCENT CELLS

In quiescent or senescent cells, p57 statin, p45 statin-associated kinase and p110 RB may be complexed. The fact that RB is not phosphorylated in the presence of the potent p45 kinase suggests that the steric topography of the 3 proteins may be as illustrated in **Figure 1**, with statin positioned between the kinase and RB, thus functioning as a barrier or sequester preventing phosphorylation. The insolubility associated with phosphorylated p57 may provide the biochemical means to anchor the whole complex to the nuclear envelope or the matrix, thus optimizing the sequestering function of statin.

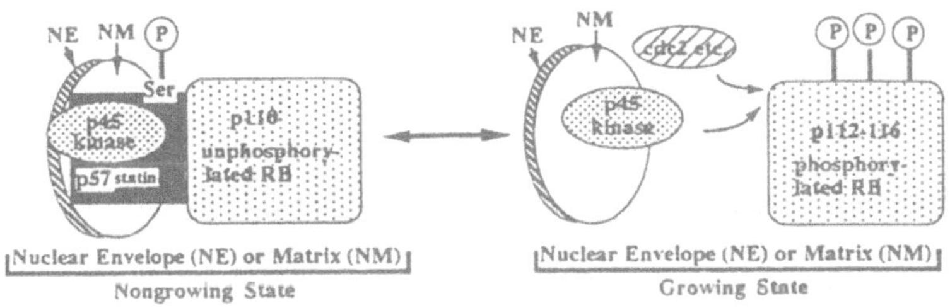

Figure 1. Proposed steric topography of p57 statin, p45 kinase and RB in nongrowing and growing states.

Our understanding so far leads us to suggest that during the exit from growth arrest, statin first disappears, thus releasing p45 kinase and allowing the first step of RB phosphorylation (**Fig. 2**). As the cells traverse to the remaining cell cycle phases (G_1/S/G_2), the expressions of p34^{cdc2} kinase and other associated members of its family are activated, thus accomplishing the subsequent steps of RB phosphorylation. In senescent cells, the lack of expression of p34^{cdc2} and the associated cyclins A and B, along with the permanent presence of statin as a kinase sequester, results in a double disadvantage militating against the success of RB phosphorylation, a prerequisite for DNA synthesis. We therefore propose this "two-hit antiphosphorylation" theory as a possible model for replicative senescence in *in vitro* aged fibroblasts; the experiments in our next funding cycle are designed to test this hypothesis.

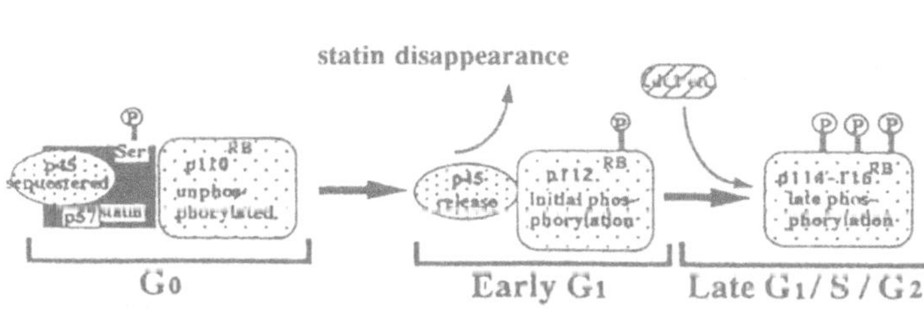

Figure 2. Proposed model of two possible stages of RB phosphorylation during cell cycle traverse.

Biochemically, multiple phosphorylation phases requiring separate kinases is not an uncommon event. We propose that the multiple stages of RB phosphorylation may require at least two separate classes of kinase, the initial step in early G_1 phase calling for the p45 statin-associated kinase, while the G_1/S and G_2 events are mediated by $p34^{cdc2}$. In senescent cells, where statin functions as a sequester for the p45 kinase and the gene expression of the $p34^{cdc2}$ kinase is absent, the "two-hit antiphosphorylation" notion for the RB protein becomes a testable model. Questions such as (1) can one induce RB phosphorylation in senescent fibroblasts by the molecular manipulation of removing statin and adding cdc2 kinase, or *vice versa*, and (2) can one block RB phosphorylation in growing cells by adding statin and removing cdc2 kinase, are experimentally verifiable. While other possibilities may exist to complicate the present hypothesis, this notion that statin may function as a sequester of the kinase needed for initial RB phosphorylation provides the first handle to begin our investigation, and offers an intriguing and workable model in our continuing research quest for the understanding of the molecular mechanism determining the cessation of replication in *in vitro* aged human fibroblasts. The answer to this question may add to our knowledge of cell cycle control, a fundamental mechanism pivotal to the wellbeing of significant processes such as terminal differentiation, development and aging.

ACKNOWLEDGEMENT

The authors wish to thank Mr. Alan N. Bloch for the preparation and proofreading of this manuscript. This work was supported by a grant (AG07444) to E. W. from the National Institute on Aging of the National Institutes of Health of the United States of America.

REFERENCES

1. L. Hayflick and P.S. Moorhead, The serial cultivation of human diploid cell strains, *Exp. Cell Res.* **25**:585 (1961).
2. H.E. Swim and R.I. Parker, Culture characteristics of human fibroblasts propagated serially, *Amer. J. Hyg.* **66**:235 (1957).
3. L. Hayflick, The limited *in vitro* lifetime of human diploid cell strains, *Exp. Cell Res.* **37**:614 (1965).
4. S. Goldstein, Cellular senescence, *in*: "Endocrinology", Vol. 3, L.J. Degroot, *et al.*, ed., pp. 2525-2599, W.B. Saunders Co., Philadelphia, Pa. (1989).
5. G.M. Martin, C.A. Sprague, and C.J. Epstein, Replicative life-span of cultivated human cells. Effects of donor's age, tissue and genotype, *Lab. Invest.* **23**:86 (1970).
6. E.L. Schneider and Y. Mitsui, The relationship between *in vitro* cellular aging and *in vivo* human age, *Proc. Nat'l Acad. Sci. USA* **73**:3584 (1976).
7. J.M. Ryan, D.G. Ostrou, X.O. Brookefield, E.G. Gershon, and L. Upchurch, A comparison of the proliferative and replicative life span kinetics of cell cultures derived from monozygotic twins. *In Vitro* **17**:20 (1981).
8. T.H. Norwood, W.R. Pendergrass, C.A. Sprague, and G.M. Martin, Dominance of the senescent phenotype in heterokaryons between replicative and post-replicative human fibroblast-like cells, *Proc. Nat'l Acad. Sci. USA* **71**:2231 (1974).
9. O.M. Pereira-Smith and J.R. Smith, Expression of SV40 T antigen in finite life-span hybrids of normal and SV40-transformed fibroblasts, *Somatic Cell Genetics* **7**:411 (1981).
10. G.C. Burmer, C.J. Ziegler, and T.H. Norwood, Evidence for endogenous polypep-

10. G.C. Burmer, C.J. Ziegler, and T.H. Norwood, Evidence for endogenous polypeptide-mediated inhibition of cell-cycle transit in human diploid cells, *J. Cell Biol.* **84**:187 (1982).
11. O.M. Pereira-Smith and J.R. Smith, Genetic analysis of indefinite division in human cells: Identification of four complementation groups, *Proc. Nat'l Acad. Sci. USA* **85**:6042 (1988).
12. G.H. Stein and R. Yanishewsky, Entry into S phase is inhibited in two immortal cell lines fused to senescent human diploid cells, *Exp. Cell Res.* **120**:155 (1979).
13. C.K. Lumpkin, J.K. McClung, O.M. Pereira-Smith, and J.R. Smith, Existence of high abundance antiproliferative mRNA's in senescent human diploid fibroblasts, *Science* **232**:393 (1986).
14. O. Sugawara, M. Oshimura, M. Koi, L.A. Annalo, and J.G. Barrett, Induction of cellular senescence in immortalized cells by human chromosome 1, *Science* **247**: 707 (1990).
15. Y. Ning, J.L. Weber, A.M. Killary, D.H. Ledbetter, J.R. Smith, and O.M. Pereira-Smith, Genetic analysis of indefinite division in human cells. Evidence for a cell senescence-related gene(s) on human chromosome 4. *Proc. Nat'l Acad. Sci. USA* **88**:5635 (1991).
16. C.B. Harley, A.B. Futcher, and C.W. Greider, Telomerases shorten during aging of human fibroblasts, *Nature* **345**:458 (1990).
17. T. Giordano and D.N. Foster, Identification of a highly abundant cDNA isolated from senescent WI-38 cells, *Exp. Cell Res.* **185**:399 (1989).
18. J.K. McClung, D.B. Danner, D.A. Stewart, J.R. Smith, E.L. Schneider, C.K. Lumpkin, R.T. Dell'Orco, and M.J. Nuell, Isolation of a cDNA that hybrid selects antiproliferative mRNA from rat liver. *Biochem. Biophys. Res. Comm.* **164**: 1346 (1989).
19. S. Goldstein, S. Murano, H. Benes, E. J. Moerman, R. A. Jone, R. Thweatt, R. J. Shmooker-Reis, and B. H. Howard. Studies on the molecular genetic basis of replicative senescence in Werner Syndrome and normal fibroblasts. *Exp. Gerontol.* **24**:461 (1989).
20. S. Murano, R. Thweatt, R.J. Shmookler-Reis, R.A. Jones, E.J. Moerman, and S. Goldstein, Diverse gene sequences are overexpressed in Werner syndrome fibroblasts undergoing premature replicative senescence. *Mol. Cell Biol.* **11**:3905 (1991).
21. E. Wang, A 57,000-mol-wt protein uniquely present in nonproliferating cells and senescent human fibroblasts, *J. Cell Biol.* **100**:545 (1985).
22. G. Ching and E. Wang, Characterization of two populations of statin and the relationship of their syntheses to the state of cell proliferation, *J. Cell Biol.* **110**: 255 (1990).
23. U. Sester, M. Sawada, and E. Wang, Purification and biochemical characterization of statin, a nonproliferation-specific protein from rat liver, *J. Biol. Chem.* **265**: 19966 (1990).
24. E. Wang and G. Tomaszewski, The granular presence of terminin is the marker to distinguish between senescent and quiescent states, *J. Cell Physiol.* **147**:514 (1991).
25. D.K. Ann, I.K. Moutsatsos, T. Nakamura, P.-L. Mao, M.J. Lee, S. Chin, R.K. Liem, and E. Wang, Isolation and characterization of the rat chromosomal gene for a polypeptide (pS1) antigenically related to statin, *J. Biol. Chem.* **266**:10429 (1991).
26. E. Wang and J.G. Krueger, Application of a unique monoclonal antibody as a marker for nonproliferating subpopulations of cells and some tissues, *J.*

27. E. Wang, Rapid disappearance of statin, a nonproliferating and senescent cell-specific protein, upon reentering the process of cell cycling, *J. Cell Biol.* **101**:1695 (1985).
28. E. Wang and S.L. Lin, Disappearance of statin, a protein marker for nonproliferating and senescent cells, following serum-stimulated cell cycle entry, *Exp. Cell Res.* **167**:135 (1986).
29. A.L. Muggleton-Harris and E. Wang, Statin expression associated with terminally differentiating and postreplicative lens epithelial cells, *Exp. Cell Res.* **182**:152 (1989).
30. J.A. Connolly, V.E. Sarabia, D.J. Kelvin, and E. Wang, The disappearance of cyclin-like protein and the appearance of statin are correlated with the onset of differentiation during myogenesis *in vitro*, *Exp. Cell Res.* **174**:461 (1988).
31. C. Pellicciari, M. Danova, M. Giordano, A.M.F. Conti, G. Mazzinni, E. Wang, E. Ronchetti, A. Riccardi, and M.G.M. Romanini, Expression of cell cycle-related proteins (PCNA and Statin) during adaptation and deadaptation of EUE cells to a hypertonic medium, *Cell Proliferation* **24**:469 (1991).
32. S. Fedoroff, I. Ahmed, and E. Wang, The relationship of expression of statin, the nuclear protein of non-proliferating cells, to the differentiation and cell cycle of astroglia in cultures and *in situ*, *J. Neurosciences Research* **26**:1 (1990).
33. C.-D. Chang, P.D. Phillips, K.E. Lipson, V.J. Cristofalo, and R. Baserga, Senescent human fibroblasts have a posttranscriptional block in the expression of the proliferating cell nuclear antigen gene, *J. Biol. Chem.* **266**:8663 (1991).
34. R. Bissonnette, M.J. Lee, and E. Wang, The differentiation process of intestinal epithelial cells is associated with the appearance of statin, a non-proliferation-specific nuclear protein, *J. Cell Science* **95**:247 (1990).
35. H.M. Schipper, R. Mauricette, J.J. Liang, M.J. Lee, and E. Wang, Expression of the non-proliferation-specific protein, statin, in grey matter neuroglia of the aging rat brain, *Brain Res.* **591**:129 (1992).
36. M. Sandig, G. Tomaszewski, C. Liu, and E. Wang, Characterization of isoforms of statin in rat hepatocytes *in situ*. *J. Cell Biol.* **115**:429a (1991).
37. S. Kyzer, B. Mitmaker, P.H. Gordon, H.M. Schipper, and E. Wang, Rumenectomy-induced proliferation in duodenal villus epithelium is mechanistically related to the disappearance of statin, a nonproliferation specific nuclear protein, *J. Histochem. Cytochem.* **39**:1611 (1991).
38. H.M. Schipper, V. Skalski, L.C. Panasci, and E. Wang, Statin expression in the untreated and SarCNU-exposed human glioma cell line, SK-MG-1, *Cancer Chemotherapy and Pharmacology* **26**:383 (1990).
39. H.M. Schipper and E. Wang, Expression of statin, a nonproliferation-dependent nuclear protein, in the postnatal rat brain: evidence for substantial retention of neuroglial proliferative capacity with aging, *Brain Research* **528**:250 (1990).
40. A.M.C. Tsanaclis, S. Brenn, S. Gately, H.M. Schipper, and E. Wang, Statin immunolocalization in human brain tumors, *Cancer* **68**:786 (1991).
41. I. Bayer, B. Mitmaker, P.H. Gordon, and E. Wang, Modulation of statin expression in nucleus of rat thyroid follicle cells following administration of thyroid stimulating hormone, *J. Cell Physiol.* **150**:276 (1992).
42. M. Chung, B. Mitmaker, P.H. Gordon, and E. Wang, Expression of the c-*fos* protooncogene, the p53 anti-oncogene and statin in colorectal carcinoma and adjacent mucosa, *Surgical Forum* 427 (1991).
43. M. Trudel, L. Oligny, S. Caplan, C. Caplan, H.M. Schipper, and E. Wang, Expression and quantitation of statin, a novel G_O marker, in normal and neoplastic hematolymphoid cells, *Am. J. Clin. Path.* in press (1993).
44. B. Mitmaker, S. Kyzer, P.H. Gordon, and E. Wang, Histochemical localization of

statin - a non-proliferation-specific nuclear protein - in nuclei of normal and abnormal human thyroid tissue, *Euro. J. Histochem.* **36**:123 (1992).
45. H.M. Schipper, J.J. Liang, and E. Wang, Quiescent and cycling cell compartments in the senescent and Alzheimer-diseased brain, *Neurology* **43**:87 (1993).
46. S. Kyzer, B. Mitmaker, P.H. Gordon, H.M. Schipper, and E. Wang, Proliferative activity of colonic mucosa at different distances from primary adenocarcinoma as determined by the presence of statin: a nonproliferation-specific nuclear protein, *Dis. Colon Rectum* **35**:879 (1992).
47. E. Wang, Statin, a nonproliferation-specific protein, is associated with the nuclear envelope and is heterogeneously distributed in cells leaving quiescent state, *J. Cell Physiol.* **140**:418 (1989).
48. U. Sester, I.K. Moutsatsos, and E. Wang, A rat liver 57-kDa protein is identified to share antigenic determinants with statin, a marker for nonproliferating cells, *Exp. Cell Res.* **182**:550 (1989).
49. M.J. Lee, M. Sandig, and E. Wang, Statin, a protein specifically present in nonproliferating cells, is a phosphoprotein and forms a complex with a 45 kilodalton serine/threonine kinase, *J. Biol. Chem.* **267**:21773 (1992).
50. T. Seshadri and J. Campisi, Repression of c-*fos* transcription and an altered genetic program in senescent human fibroblasts, *Science* **247**:205 (1990).
51. S. Goldstein, Replicative senescence. The human fibroblast comes of age. *Science* **249**:1129 (1990).
52. S.R. Rittling, K.M. Brooks, V.J. Cristofalo, and R. Baserga, Expression of cell cycle-dependent genes in young and senescent WI-38 fibroblasts, *Proc. Nat'l Acad. Sci. USA* **83**:3316 (1986).
53. W.E. Wright, O.M. Pereira-Smith, and J.W. Shay, Reversible cellular senescence: implications for immortalization of normal human diploid fibroblasts, *Mol. Cell Biol.* **9**:3088 (1989).
54. J.W. Shay, W.E. Wright, and H. Werbin, Defining the molecular mechanisms of human cell immortalization, *Biochim. Biophys. Acta* **1072**:1 (1991).
55. G.H. Stein, M. Beeson, and L. Gordon, Failure to phosphorylate the retinoblastoma gene product in senescent human fibroblasts, *Science* **249**:666 (1990).
56. J.M. Horowitz, D.W. Yandell, S.-H. Park, S. Canning, P. Whyte, K. Buchkovich, E. Harlow, R.A. Weinberg, and J.P. Dryja, Point mutational inactivation of the retinoblatoma antioncogene, *Science* **243**:937 (1989).
57. K. Buchkovich, L.A. Duffy, and E. Harlow, The retinoblastoma protein is phosphorylated during specific phases of the cell cycle, *Cell* **58**:1097 (1989).
58. J.A. De Caprio, J.W. Ludlow, D. Lynch, Y. Furukawa, J. Griffin, H. Piwnica-Worms, C.-M. Huang, and D.M. Livingston, The product of retinoblastoma susceptibility, *Cell* **58**:1085 (1989).
59. J.W. Ludlow, J.A. De Caprio, C.M. Huang, W. Lee, E. Paucha, and D.H. Livingston, SV-40 large T antigen binds preferentially to an underphosphorylated member of the retinoblastoma susceptibility gene product family, *Cell* **56**:57 (1989).
60. P. Chen, P. Scully, J. Shew, J.Y.Y.J. Wang, and W.H. Lee, Phosphorylation of the retinoblastoma gene product is modulated during the cell cycle and cellular differentiation, *Cell* **58**:1193 (1989).
61. M. Laiho, J. A. DeCaprio, J. W. Ludlow, D. M. Livingston, and J. Massague, Growth inhibition by TGF-β linked to suppression of retinoblastoma protein phosphorylation, *Cell* 62: 175 (1990).
62. Y. Furukawa, J.A. De Caprio, A. Freedman, Y. Kanakura, M. Nakamura, T.J. Ernst, D.M. Livingston, and J.D. Griffin, Expression and state of phosphorylation of the retinoblastoma susceptibility gene product in cycling and noncycling

human hematopoietic cells, *Proc. Nat'l Acad. Sci. USA* **87**:2770 (1990).

63. K. Mihara, X. Cao, A. Yen, S. Chandler, B. Driscoll, A.L. Murphree, A. T'ang, and Y. Fung, Cell cycle-dependent regulation of phosphorylation of the human retinoblastoma gene product, *Science* **246**:1300 (1989).
64. H. Xu, S. Hu, T. Hashimoto, R. Takahashi, and W.F. Benedict, The retinoblastoma susceptibility gene product: a characteristic pattern in normal cells and abnormal expression in malignant cells, *Oncogene* **4**:807 (1989).
65. J.W. Ludlow, J. Shon, J.M. Pipas, D.M. Livingston, and J.A. De Caprio, The retinoblastoma susceptibility gene product undergoes cell-cycle dependent dephosphorylation binding to and release from SV-40 large T, *Cell* **60**:387 (1990).
66. S. Bagchi, R. Weinmann, and P. Raychaudhuri, The retinoblastoma protein copurifies with E2F-1, an E1A-regulated inhibitor of the transcription factor E2F. *Cell* **65**:1063 (1991).
67. L.R. Bandara, J. P. Adamczewski, T. Hunt, and N.B. La Thangue, Cyclin A and the retinoblastoma gene product complex with a common transcription factor. *Nature* **352**:249 (1991).
68. S.P. Chellappan, S. Hiebert, M. Mudryi, J.M. Horowitz, and J.R. Nevins, The E2F transcription factor is a cellular target for the RB protein, *Cell* **65**:1053 (1991).
69. M.C. Blake and J.C. Azizkhan, Transcription factor E2F is required for efficient expression of the hamster dihydrofolate reductase gene *in vitro* and *in vivo*, *Mol. Cell Biol.* **9**:4994 (1989).
70. G.H. Stein, L.F. Drullinger, R.S. Robetorye, O.M. Pereira-Smith, and J.R. Smith, Senescent cells fail to express *cdc2*, cycA, and cycB in response to mitogen stimulation, *Proc. Nat'l Acad. Sci. USA* **88**:11012 (1991).
71. J.A. De Caprio, Y. Furakawa, F. Ajchenbaum, J.D. Griffin, and D.M. Livingston, The retinoblastoma-susceptibilty gene product becomes phosphorylated in multiple stages during cell cycle entry and progression, *Proc. Nat'l Acad. Sci. USA* **89**:1795 (1992).
72. Y. Furakawa, H. Piwnica-Worms, T.J. Ernst, Y. Kanakura, and J.D. Griffin, Cdc2 gene expression at the G_1 to S transition in human T lymphocytes, *Science* **250**: 805 (1990).

9

CELLULAR SENESCENCE AND THE CELL CYCLE

J. Carl Barrett[1] and Cynthia A. Afshari[2]

[1]Laboratory of Molecular Carcinogenesis
Environmental Carcinogenesis Program
National Institute of Environmental Health Sciences
National Institutes of Health
Research Triangle Park, North Carolina 27709

[2]Duke University Medical School
Center for the Study of Aging and Human Development
Durham, North Carolina 27705

INTRODUCTION

Normal cells in culture can be grown for only a limited number of cell divisions after which they exhibit morphological changes and cease proliferation, a process termed cellular senescence or cellular aging (1). Hayflick and Moorhead (2) reported this finding with human fibroblasts over 30 years ago, and it has been subsequently confirmed by many investigators using cells from different tissues and species. The failure of cells to grow beyond this limit is an inherent property of the cells that cannot be explained simply by inadequate media components or growth conditions (1,2). The key determinant in the life span of cells in culture is the number of cell doublings, not the length of time in culture (1). Normal cells transplanted serially in vivo also exhibit a finite life span, suggesting that cellular senescence is not a cell culture artifact (3). Several lines of evidence suggest that the aging of cells in culture may be related to the aging of the organism (1,4). These lines of evidence, although not conclusive, provide provocative support for the hypothesis that aging of cells is related to the aging process of the organism.

Escape from cellular senescence is an important step in neoplastic progression of human and rodent cancers (5). Many, but not all, tumor cells can be grown indefinitely in culture and therefore have escaped senescence and are termed immortal. It is not clear whether the failure of some tumor cells to grow in culture is a technical artifact or an indication that escape from senescence is not required for these cancers. A model to explain these observations is discussed later in this review.

Normal cells escape senescence following treatment with diverse carcinogenic agents, including chemicals, radiation, viruses, and oncogenes (5). This observation suggests that immortalization is important in carcinogenesis. While immortality is not sufficient for neoplastic transformation, most immortal cells have an increased propensity for spontaneous, carcinogen-induced or oncogene-induced neoplastic progression (5). Therefore, escape from senescence is a preneoplastic change that predisposes a cell to neoplastic conversion. It is clear that immortal cells are further along the multistep pathway to neoplasia than normal cells. Since cellular senescence limits the growth of cells, it is reasonable that senescence might be one mechanism by which tumor suppressor genes operate (5,6).

Genetic Basis for Cellular Senescence

Two major theories of cellular senescence have been debated for many years (7). One is the error catastrophe or damage model, which proposes that the random accumulation of damage or mutations in DNA, RNA, or proteins leads to the loss of proliferative capacity. The experimental evidence supporting the error accumulation hypothesis has been criticized (7). A second hypothesis is that senescence is a genetically programmed process. Strong experimental support for a genetic basis for senescence is provided by studies of Pereira-Smith and Smith (8) and by Sugawara et al. (9), which are discussed below.

It is possible to fuse cells of different origins and to selectively grow the hybrid cells using biochemical markers for drug sensitivity or resistance that differ in the parental cells. When cells with a finite life span are fused to immortal cells with an indefinite life span, the majority of these hybrids senesce (8,10,11). Even hybridization of two different immortal human cell lines with each other can result in senescence, indicating that different complementation groups exist for the senescence function lost in these cells (8). By fusing different immortal human cell lines with each other, Pereira-Smith and Smith (8) established four complementation groups, suggesting that loss or inactivation of one of multiple genes allows cells to escape from senescence. If this hypothesis is correct, it should be possible to map the genes involved in cellular senescence, and recent findings with hamster and human interspecies hybrids and microcell-mediated chromosome transfer experiments have mapped putative senescence genes to specific human chromosomes (9,12,13).

The initial mapping by Sugawara et al. (9) of a senescence gene to chromosome 1 was demonstrated by three, independent experimental approaches: 1) interspecies hybrids of normal human cells with immortal hamster cells that showed nonrandom losses of human chromosome 1 in hybrids that escaped senescence, 2) interspecies cell hybrids with human cells carrying a t(1;X) chromosome that allowed selective pressure for the long arm of chromosome 1 and a corresponding increased frequency of senescent hybrids, and 3) microcell-mediated chromosome transfer which demonstrated that introduction of a single copy of human chromosome 1, but not other chromosomes, restored the program of senescence in certain immortal cell lines.

Using the technique of microcell-mediated chromosome transfer, Klein et al. (12) mapped another senescence gene to the X chromosome, and Ning et al. (13) mapped a senescence gene for HeLa cells to chromosome 4. In addition, several additional studies indicate the presence of senescence genes on other chromosomes (Table 1).

These results have led us to propose the following hypothesis: Cellular senescence is controlled by genes that are activated or whose functions become manifested at the end of the life span of the cell. Defects in the function of these gene products allow cells to escape the program of senescence and become immortal. Immortalization relieves one constraint on tumor cell growth, allowing malignant progression.

According to this hypothesis, a family of senescence genes exists. Immortalization occurs due to defects in these genes. This explains the complementation studies of Pereira-Smith and Smith (8), which show that different immortal cell lines when fused together can complement each other and senesce. This also explains our results that introduction of a specific human chromosome causes senescence in some cell lines but not others. It appears that senescence gene mutations in a specific immortal cell line are not related to tumor histology or activated oncogenes in the cell. Current efforts are focused on the cloning of these genes and on understanding how they operate to arrest cell growth.

We have included in the Rb and p53 tumor suppressor genes in Table 1 as putative senescence genes. Operationally, these genes fit the criteria for senescence genes in certain contexts. Down-regulation of their expression by antisense methods results in extension of the life span of normal human cells (14), and reintroduction of the genes into certain immortal cells causes cessation of growth and morphological changes similar to senescent cells (15,16). According to our hypothesis, specific genes are involved in one or more pathways leading to a program of senescence. In normal cells, Rb and p53 proteins, which are regulated by other proteins, are negative regulators of the cell cycle, allowing cell cycle progression. In senescent cells, a program is activated that blocks entry into the DNA synthesis phase of the cell cycle, and the cell becomes irreversibly growth arrested. Rb and p53 may participate in one or more pathways that activate or affect the senescence program. Defects in the senescence program can result from inactivation or mutations of different genes. In some immortal cells, p53 and Rb can be normal and genes that control their phosphorylation or other post-translational modifications may be defective. In other cell lines, deletions or mutations of p53 or Rb genes could result in the inability to activate the senescence program.

Recently, the gene for Werner syndrome has been mapped to chromosome 8 (17). Mutations of this gene result in premature senescence of the Werner cells and premature aging of individuals with this syndrome. The Werner gene differs from the senescence genes discussed above. In contrast to putative senescence genes, the inactivation of which causes cells to become immortal, the Werner gene should be considered an anti-senescence or anti-aging gene because its loss results in accelerated senescence, not immortalization.

Recently, we observed that introduction of different chromosomes can induce senescence in the same immortal cell line (18). Using a human endometrial carcinoma cell line, senescence genes for these cells were mapped to chromosomes 1 and 18 (18). This finding implies that multiple pathways of senescence exist and that immortal cells arise due to defects in each of these pathways. As illustrated in Figure 1, the senescence program could be activated by a single pathway and immortal cells would arise due to mutations in any of the genes that encode proteins involved in this pathway. An alternative hypothesis is that the senescence program is activated by multiple, independent pathways. Immortal cells would require at least one mutation in each pathway. Mutations that affect only a single pathway would not result in cells that were immortal, but the cells might have an extended life span. For example, SV40 virus infection (which inactivates Rb and p53 proteins) results in extended life span but not immortalization of infected cells (19). Additional genetic changes, possibly loss of chromosome 6, are required for immortalization of SV40-infected human cells (20). Reintroduction of a normal chromosome 6 results in senescence of SV40 immortal cells (53,54). Antisense downregulation of Rb and p53 mRNAs also results in extension of the life

Table 1. Mapping Putative Senescence Genes

Chromosome	Gene	Affected Cell Line	Reference
1	unknown	Syrian hamster fibrosarcoma (10W)	9
		endometrial carcinoma (HHUA)	51
		endometrial carcinoma (Ishikawa)	18
		osteosarcoma (TE85)	52
4	unknown	cervical carcinoma (HeLa)	13
		bladder carcinoma (J82)	
6	unknown	SV40 immortalized	53,54
9	unknown	melanoma (H32941)	55
		leukemia (K562)	
11	unknown	bladder carcinoma (H2)	56,57
13	retinoblastoma	many	15
17	P53	many	16
18	unknown	endometrial carcinoma (HHUA)	18
X	unknown	Chinese hamster (N:2)	12
		ovarian carcinoma (Hoc8)	
		breast carcinoma (ELCO)	

Mapping Putative Anti-Senescence Gene

Chromosome	Gene	Syndrome	Reference
8	unknown	Werner Syndrome	17

span of human cells without immortalization (14). The multiple pathways to senescence hypothesis is also consistent with the multistep nature of immortalization observed in chemically induced immortalization (5). Furthermore, the inability to assign certain immortal cells to a single complementation group is also explained by this hypothesis (21,22).

An important aspect of the hypothesis of multiple pathways to cellular senescence is that it explains why many tumor-derived cells are not immortal. Hayflick has shown that cells from adults can be grown in culture for 14 to 29 population doublings (1). If all the changes necessary for tumorigenic conversion were to accumulate in an adult cell without loss or gain of lifespan potential (which may be unlikely), then this cell could grow to form a tumor of only 16,354 cells (14 doublings or 2^{14} cells) to 5.4 x 10^8 cells (29 doublings or 2^{29} cells). It is estimated that a tumor formed after 30 cell doublings would be approximately 1 cm^2 in size (Fig. 2). Interestingly, Paraskeva and coworkers have shown that colon adenomas of < 1 cm^2 in size are rarely capable of indefinite growth in vitro whereas cells from adenomas of > 1 cm^2 are often immortal (23-25), which supports the hypothesis that escape from senescence

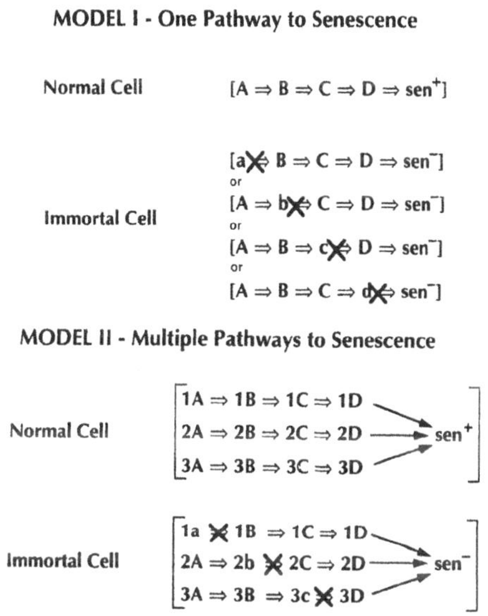

Figure 1

is a requirement for tumor growth beyond a certain size or cell number. A tumor of < 1 cm may be lethal in some cases but not generally. For the tumor to expand, an extension of the lifespan would be necessary. Extension of the life span to 35 population doublings would yield a tumor of 1 foot in diameter whereas an extension to 50 population doublings would yield a tumor of 1000 kg, which would very certainly be sufficient to kill the host. Thus, immortalization is not as important as extension of life span for neoplastic progression within an organism. Mutation of a senescence gene in one pathway may result in an extended life span without immortalization according to the multiple pathways model. This hypothesis may explain why tumor cells are not always immortal.

Cell Cycle Controls in Senescent Cells

Many studies of senescent cells are beginning to focus on the proteins involved in cell cycle progression in order to discern how the irreversible block to S-phase entry occurs in senescent cells (Table 2). Senescent fibroblasts exhibit downregulation of c-fos mRNA and protein and hence downregulation of AP-1 activity (26-28). In addition, it has been shown that the RB protein, which becomes phosphorylated as cells enter into S-phase remains underphosphorylated in senescent cells (29,30). These data indicate that the block in senescence is in G1 and that alterations in signalling in this part of the cell cycle must be instrumental in the arrest. In pursuit of this mechanism, we have analyzed the $p34^{cdc2}$ protein.

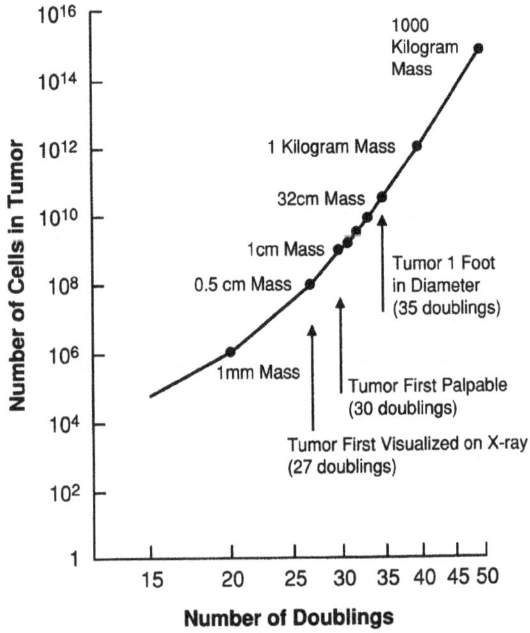

Figure 2

Adapted from DeVita (ref. 59).

We have shown that cdc2 is down-regulated in senescent cells and that expression of transfected cdc2 protein in senescent cells is insufficient for rescue from arrest (31). Cyclin A is also downregulated in senescent cells. Similar results have been reported by Stein and coworkers (32). While the expressions of cdc2, cyclin A, and cyclin B mRNAs are downregulated in senescent cells, mRNAs for cyclin C, D1, and E are still abundant (31-34). Further studies may reveal whether these proteins or their associations with other proteins are altered in senescent cells.

While downregulation of positive signals may be important for cellular senescence, it is clear that activation of negative signals is also important. We have been interested in candidate genes that fall into this second category. Protein phosphorylation is a signal elicited during transduction of mitogenic pathways and is also important in neoplastic transformation, as indicated by the many oncogene products that code for protein kinases (35). Therefore, we decided to investigate whether there may be a role for protein phosphatases in the negative growth regulation of normal cells during quiescence and senescence.

For this purpose, we used the protein phosphatase inhibitors, sodium orthovanadate (NaV) and okadaic acid (OA), to determine if treatment of cells with these agents and inhibition of protein phosphatases would alter different growth arrest states. In vitro studies using these inhibitors show that NaV inhibits tyrosine phosphatases (IC_{50}, 0.17 nM), while OA inhibits type I serine-threonine protein phosphatases at high concentrations (IC_{50}, 3.4 nM), type IIA phosphatases at low levels (IC_{50}, 0.07nM), and type III phosphatase at intermediate levels (36-38). When we treated Syrian hamster embryo (SHE) cells in the presence of the inhibitors and measured growth effects by colony forming assays, we observed a slight increase in colony forming efficiency (CFE) with levels of 0.16 nM OA and 1 μM NaV and a marked decrease in CFE or toxicity with high doses (2nM OA or 5 μM NaV) in this cell growth assay (39). Further investigation showed that apoptosis was induced in SHE cells by OA levels that were greater than 100 nM and that induction of cell death was accompanied by induction of c-fos but not by induction of DNA synthesis (28). The characteristic rounding of apoptotic cells, as well as the fragmentation of nuclei and DNA ladders, were visible as early as four hours after treatment with high levels of OA (28).

In contrast to induction of apoptosis at high doses of OA, we observed that treatment of low serum-arrested quiescent SHE or MRC5 human fibroblasts with very low levels of NaV (1 μM) or OA (0.16 nM) resulted in induction of DNA synthesis in low serum (39). Hence, it appears that inhibition of protein phosphatases that may be active in quiescent cells allows cells to exit from G_0 into S-phase. This transition was accompanied by tyrosine

Table 2. Cell Cycle Proteins in Senescent Cells

Protein	State in senescent cells	Reference
c-fos; AP-1	downregulated	26-28
$p34^{cdc2}$	downregulated	31,32
cyclin A, cyclin B	downregulated	32,34
cdk2	mRNA downregulated	32,34
Rb	unphosphorylated	29-30
p53	no observable change	34,50
cyclins C, D1, E	mRNA expressed	33,34,58
c-myc, H-ras	mRNA expressed	50
ornithine decarboxylase	mRNA expressed	50
thymidine kinase	mRNA expressed	50
cdk4	mRNA expressed	58
cdk5	mRNA expressed	58

phosphorylation of MAP-kinase, phosphorylation of the retinoblastoma (RB) protein, and induction of the $p34^{cdc2}$ kinase. Interestingly, we also observed the ability of the same levels of OA and NaV to induce DNA synthesis in senescent cells (39).

It is interesting to note that we observed a pleiotropic effect of OA on cells that is dose dependent (Table 3). Low levels of OA, below 1 nM, result in increases in cell growth as measured by CFE, induction of DNA synthesis in quiescent cells (28,39), and progression in SHE cells (40). Intermediate levels of OA (10-100nM) have been reported to inhibit DNA synthesis in serum stimulated cells (41), while higher levels of OA (> 100nM) caused an induction of apoptosis that was not accompanied by an induction of DNA synthesis in serum starved cells (28). This data suggests that protein phosphatases may be affected in vivo by low doses of NaV and OA and these phosphatases act as growth inhibitors or suppressors of DNA synthesis. The observation that both NaV and OA override this inhibition suggests that there is a network of proteins regulated by both serine and threonine phosphorylation, as well as tyrosine phosphorylation.

Table 3. Multiple Effects of OA Treatment on Cellular Response

Dose	Effect
below 1 nM	increase in colony forming efficiency induction of DNA synthesis in quiescent cells induction of DNA synthesis in senescent cells induction of progression in SHE cell transformation model
10-100 nM	decreased colony forming efficiency inhibition of DNA synthesis
above 100 nM	induction of apoptosis no induction of DNA synthesis induction of c-fos mRNA

One possible candidate for the phosphatase involved in growth suppression is phosphatase 2A (PP2A) or another protein with similar properties. First, PP2A activity is inhibited in vitro by low levels of OA (42). In addition, its activity is regulated by phosphorylation on tyrosine (43). It also has been shown that PP2A is targeted for binding by the transforming virus, SV40, and that binding of SV40 proteins to the enzyme may inhibit its activity (44-47). Interestingly, it has been shown that treatment of senescent human diploid fibroblasts with T antigen also results in induction of DNA synthesis at levels similar to what we observed following treatment with OA and NaV (48). This suggests that a common negative growth regulator may be inactivated by both SV40 and protein phosphatase inhibitors. In further support of PP2A as a candidate for this activity, it has been shown that microinjection of the catalytic subunit of PP2A caused an increased affinity of the RB protein for the nucleus as well as inhibition of DNA synthesis (49). We are currently pursuing studies of the activity of protein phosphatases in senescent cells and testing the hypothesis that one or more of these enzymes may be responsible for this growth arrest state.

CONCLUSIONS

Cell senescence is characterized by the activation of a program at the end of the life span of a cell that results in irreversible growth arrest and a failure of the cell to enter into S-phase. Immortal cells arise due to mutations or inactivation of senescence genes that are involved in the activation pathways of the senescence program. Nearly ten senescence genes and one putative anti-senescence gene have been mapped. Two known genes, Rb and p53, are involved in senescence, but other senescence genes remain to be cloned. Irreversible growth arrest in senescence cells involves alterations in cell cycle control. Downregulation of cell cycle proteins (e.g., cdc2, cyclin A, and phosphorylation of Rb) are observed in senescent cells. Expression of these genes in senescent cells, however, does not override the senescence program. This observation, as well as the finding that multiple senescence genes are altered in immortal cells, support the hypothesis that multiple cellular pathways regulate the senescence program. Negative controls, in particular protein phosphatases, are likely involved in cellular senescence.

REFERENCES

1. Hayflick, L. The cell biology of human aging. *N. Engl. J. Med.* 295:1302-1308 (1976).
2. Hayflick, L. and Moorhead, P.S. The serial cultivation of human diploid cell strains. *Exp. Cell Res.* 5:585-621 (1961).
3. Daniel, C.W., DeOme, K.B., Young, J.T., Blair, P.B., and Faulkin, L.J., Jr. The *in vivo* span of normal and preneoplastic mouse mammary glands: a serial transplantation study. *Proc. Natl. Acad. Sci. USA* 61:53-60 (1968).
4. Barrett, J.C. Cell Senescence and Apoptosis. Molecular Genetics of Nervous System Tumors. 61-72 (1993).
5. Barrett, J.C. and Fletcher, W.F. Cellular and molecular mechanisms of multistep carcinogenesis in cell culture models, *in*: "Mechanisms of Environmental Carcinogenesis: Multistep Models of Carcinogenesis," Volume II, J. C. Barrett, ed., CRC Press, Boca Raton, 1987.
6. Sager, R. Genetic suppression of tumor formation: a new frontier in cancer research. *Cancer Res.* 46:1573-1580 (1986).
7. Maciera-Coelho, A. Biology of normal proliferating cells *in vitro*. Relevance for *in vivo* aging, *in*: "Interdisciplinary Topics in Gerontology," Volume 23, H.P. von Hang, ed. Karger, Basel (1988).
8. Pereira-Smith, O.M. and Smith, J.R. Genetic analysis of indefinite division in human cells: Identification of four complementation groups. *Proc. Natl. Acad. Sci. USA* 85:6042-6046 (1988).
9. Sugawara, O.M., Oshimura, M., Koi, M., Annab, L., and Barrett, J.C. Induction of cellular senescence in immortalized cells by human chromosome 1. *Science* 247:707-710 (1990).
10. Koi, M. and Barrett, J.C. Loss of tumor-suppressive function during chemically induced neoplastic progression of Syrian hamster embryo cells. *Proc. Natl. Acad. Sci. USA* 83:5992-5996 (1986).
11. Bunn, C.L., and Tarrant, G.M. Limited lifespan in somatic cell hybrids and cybrids *Exp. Cell Res.* 127:385-396 (1980).
12. Klein, C.B., Conway, K., Wang, X.W., Bhamra, R.K., Lin, X., Cohen, M.D., Annab, L., Barrett, J.C., and Costa, M. Senescence of nickel-transformed cells by a mammalian X chromosome: possible epigenetic control. *Science* 251:796-799 (1991).
13. Ning, Y., Weber, J.L., Killary, A.M., Ledbetter, D.H., Smith, J.R., and Pereira-Smith, O.M. Genetic analysis of indefinite division in human cells: Evidence for a senescence-related gene(s) on human chromosome 4. *Proc. Natl. Acad. Sci. USA* 88:5635-5639 (1991).
14. Hara, E., Tsurui, H., Shinozaki, A., Nakada, S., and Oda, K. Cooperative effect of antisense-Rb and antisense-p53 oligomers on the extension of life span in human diploid fibroblasts, TIG-1. *Biochem. Biophys. Res. Commun.* 179:528-534 (1991.
15. Hinds, P.W., Mittnacht, S., Dulic, V., Arnold, A. Reed, S.I., and Weinberg, R.A. Regulation of retinoblastoma protein functions by ectopic expression of human cyclins. *Cell* 70:993-1006 (1992).
16. Levine, A.J., Momand, J., and Finlay, C.A. The p53 tumor suppressor gene. *Nature* 351:453-456 (1991).
17. Goto, M., Rubenstein, M., Weber, J., Woods, K., and Drayna, D. Genetic linkage of Werner's syndrome to five markers on chromosome 8. *Nature* 355:735-738 (1992).
18. Sasaki, M., Honda, T., Yamada, H., Wake, N., Barrett, J.C., and Oshimura, M. Evidence for multiple pathways to cellular senescence. Submitted.
19. Wright, W.E., and Shay, J.W. The two-stage mechanism controlling cellular senescence and immortalization. *Exp. Gerontol.* 27:383-389 (1992).

20. Hubbard-Smith, K., Patsalis, P., Pardinas, J.R., Jha, K.K., Henderson, A.S., and Ozer, H.L. Altered chromosome 6 in immortal human fibroblasts. *Mol. Cell. Biol.* 12:2273- (1992).

21. Duncan, E.L., Whitaker, N.J., Moy, E.L., and Reddel, R.R. Assignment of SV40-immortalized cells to more than one complementation group for immortalization. *Exp. Cell Res.* 205:337-344 (1993).

22. Berry, I.J., Burns, J.E., and Parkinson, E.K. Two human epidermal squamous cell carcinoma cell lines assign to more than one complementation group for the immortal phenotype. *Mol. Carcinog.*, in press.

23. Paraskeva, C., Finarty, S., and Powell, S. Immortalization of a human colorectal adenoma cell line by continuous in vitro passage: possible involvement of chromosome 1 in tumour progression. *Int. J. Cancer* 41:908-912 (1988).

24. Paraskeva, C., Finarty, S., Mountford, R.A., and Powell, S.C. Specific cytogenetic abnormalities in two new human colorectal adenoma-derived epithelial cell lines. *Cancer Res.* 49:1282-1286 (1989).

25. Paraskeva, C., Harvey, A., Finarty, S., and Powell, S. Possible involvement of chromosome 1 in in vitro immortalization: evidence from progression of a human adenoma-derived cell line in vitro. *Int. J. Cancer* 43:743-746 (1989).

26. Seshadri,T. and Campisi, J. 1990. Repression of c-fos transcription and an altered genetic program in senescent human fibroblasts. *Science* 247:205-209 (1990).

27. Riabowol, K., Schiff, J., and Gilman, M.Z. Transcription factor AP-1 activity is required for initiation of DNA synthesis and is lost during cellular aging. *Proc. Natl. Acad. Sci. USA* 89:157-161 (1992).

28. Afshari, C.A., Bivins, H.B., and Barrett, J.C. Utilization of a fos-lacZ plasmid to investigate the activation of c-*fos* during cellular senescence and okadaic acid-induced apoptosis. Submitted.

29. Stein, G.H., Beeson, M., and Gordon, L. Failure to phosphorylate the retinoblastoma gene product in senescent human fibroblasts. *Science* 249:666-669 (1990).

30. Futreal, P.A., and Barrett, J.C. Failure of senescent cells to phosphorylate the RB protein. *Oncogene* 6:1109-1113 (1991).

31. Richter, K.H., Afshari, C.A., Annab, L.A., Burkhart, B.A., Owen, R.D., Boyd, J., and Barrett, J.C. Down-regulation of cdc2 in senescent human and hamster cells. *Cancer Res.* 51:6010-6013 (1991).

32. Stein, G.H., Drullinger, L.F., Robetorye, R.S., Pereira-Smith, O.M., and Smith, J.R. Senescent cells fail to express cdc2, cycA, and cycB in response to mitogen stimulation. *Proc. Natl. Acad. Sci. USA* 88:11012-11016 (1991).

33. Won, K-A., Xiong, Y., Beach, D., and Gilman, M.Z. Growth regulated expression of D-type cyclin genes in human diploid fibroblasts. *Proc. Natl. Acad. Sci. USA* 89:9910-9914 (1992).

34. Afshari, C.A., Vojta, P.J, Bivins, H.B., Annab, L.A., Willard, T.B., Futreal, A.F., and Barrett, J.C. Investigation of the role of G_1/S cell cycle mediators in cellular senescence. Submitted.

35. Hunter, T. A thousand and one protein kinases. *Cell* 50:823-829 (1987).

36. Gordon, J.A. Use of vanadate as protein-phosphotyrosine phosphatase inhibitor. *Meth. Enzymol.* 201:477-482 (1991).

37. Cohen, P., Holmes, C.F.B., and Tsukitani, Y. Okadaic acid: a new probe for the study of cellular regulation. *Trends Biochem. Sci.* 15:98-102 (1990).

38. Honkanen, R.E., Zwiller, J., Daily, S.L., Khatra, B.S., Dukelow, M., and Boynton, A.L. Identification, purification, and characterization of a novel serine/threonine protein phosphatase from bovine brain. *J. Biol. Chem.* 266:6614-6619 (1991).

39. Afshari, C.A. and Barrett, J.C. Disruption of G_0G_1 arrest in quiescent and senescent cells treated with phosphatase inhibitors. Submitted.

40. Afshari, C.A., Kodama, S., Bivins, H.M., Willard, T.B., Fujiki, H., and Barrett, J.C. Induction of neoplastic progression to Syrian hamster embryo cells treated with phosphatase inhibitors. *Cancer Res.* 53:1777-1782 (1992).

41. Schönthal, A., and Feramisco, J.R. Inhibition of histone H1 kinase expression, retinoblastoma protein phosphorylation, and cell proliferation by the phosphatase inhibitor okadaic acid. *Oncogene* 8:433-441 (1993).

42. Hardie, D.G., Haystead, T.A.J., and Sim, A.T.R. Use of okadaic acid to inhibit protein phosphatases in intact cells. *Meth. Enzymol.* 201:469-476 (1991).

43. Chen, J., Martin, B.L., and Brautigan, D.L. Regulation of protein serine-threonine phosphatase type 2A by tyrosine phosphorylation. *Science* 257:1261-1264 (1992).

44. Pallas, D.C., Shahrik, L.K., Martin, B.L., Jaspers, S., Miller, T.B., Brautigan, D.L., and Roberts, T. Polyoma small and middle T antigens and SV40 small t antigen form stable complexes with protein phosphatase 2A. *Cell* 60:167-176 (1990).

45. Walter, G., Ruediger, R., Slaughter, C., and Mumby, M. 1990. Association of protein phosphatase 2A with polyoma virus medium tumor antigen. *Proc. Natl. Acad. Sci. USA* 87:2521-2525 (1990).

46. Yang, S.-I., Lickteig, R.L., Estes, R., Rundell, K., Walter, G., and Mumby, M. Control of protein phosphatase 2A by Simian virus 40 small-t antigen. *Mol. Cell. Biol.* 11:1988-1995 (1991).

47. Scheidtmann, K.H., Virshup, D.M., and Kelly, T.J. Protein phosphatase 2A dephosphorylates simian virus 40 large T antigen specifically at residues involved in regulation of DNA-binding activity. *J. Virol.* 65:2098-2101 (1991).
48. Ide, T., Tsuji, Y., Ishibashi, S., and Mitsui, Y. Reinitiation of host DNA synthesis in senescent human diploid cells by infection with Simian virus 40. *Exp. Cell Res.* 143:343-349 (1983).
49. Alberts, A.S., Thorburn, A.M., Shenolikar, S., Mumby, M.C., and Feramisco, J.R. Regulation of cell cycle progression and nuclear affinity of the retinoblastoma protein by protein phosphatases. *Proc. Natl. Acad. Sci. USA* 90:388-392 (1993).
50. Rittling, S.R., Brooks, K.M., Cristofalo, V.J., and Baserga, R. Expression of cell cycle-dependent genes in young and senescent WI-38 fibroblasts. *Proc. Natl. Acad. Sci. USA* 83:3316-3320 (1986).
51. Yamada, H., Wake, N., Fujimoto, S., Barrett, J.C., and Oshimura, M. Multiple chromosomes carrying tumor suppressor activity for a uterine endometrial carcinoma cell line identified by microcell-mediated chromosome transfer. *Oncogene* 5:1141-1447 (1990).
52. Annab, L.A., Barrett, J.C., Hensler, P., and Smith, O., unpublished.
53. Kodama, S., Oshimura, M., and Barrett, J.C. Introduction of a normal human chromosome 8 into SV40-transformed Werner syndrome cells by microcell fusion. Submitted.
54. Ozer, H. Personal communication.
55. Diaz, M. Personal communication.
56. Oshimura, M. Personal communication.
57. Koi, M., Johnson, L.A., Kalikin, L.M., Little, P.F.R., Nakamura, Y., and Feinberg, A.P. Tumor cell growth arrest caused by subchromosomal transferable DNA fragments from chromosome 11. *Science* 260:361-364.
58. Lucibello, F.C., Sewing, A., Brüsselbach, S., Bürger, C., and Müller, R. Deregulation of cyclins D1 and E and suppression of cdk2 and cdk4 in senescent human fibroblasts. *J. Cell Sci.* 105:123-133 (1993).
59. DeVita, V.T. Jr., Young, R.C., and Canellos, G.P. Combination versus single agent chemotherapy: A review of the basis for selection of drug treatment of cancer. *Cancer* 35:98-110, 1975.

10

CELL CYCLE TARGETS OF VIRAL ONCOPROTEINS

Joseph R. Nevins

Section of Genetics
Howard Hughes Medical Institute
Duke University Medical Center
Durham, N. C. 27710

INTRODUCTION

Although the viruses classified as the small DNA tumor viruses, which includes the adenoviruses, the polyomaviruses (SV40), and the human papillomaviruses possess the capacity to transform cells in culture to an oncogenic state, and indeed the papillomaviruses represent an important causative agent for human cervical carcinoma, this activity is generally the result of a non-productive viral infection. Rather, these viruses are lytic viruses that have evolved mechanisms to efficiently replicate in a variety of cells. Indeed, it would appear that it is the strategy that these viruses have developed to achieve a replicative cycle in a quiescent cell that likely also results in oncogenic transformation when a lytic infection cannot proceed. Recent events have revealed details of these mechanisms, including the ability of these viruses to inactivate key regulatory proteins involved in cell growth control. An understanding of these events in the context of the lytic infection has shed light upon the manner by which these viruses can cause oncogenic transformation and cancer.

A COMMON ACTIVITY OF THE DNA TUMOR VIRUSES

Experiments aimed at identifying targets for the action of the adenovirus E1A protein revealed a series of cellular proteins in association with E1A (Harlow et al., 1986; Yee and Branton, 1985). One of these E1A-associated proteins turned out to be the product of the retinoblastoma tumor suppressor gene (Whyte et al., 1988) in addition to other key cellular regulatory proteins such as cyclin A and its associated kinase, cdk2 (Pines and Hunter, 1990, Tsai et al., 1991, Faha et al., 1991, Ewen et al., 1991). Subsequent experiments demonstrated that both the SV40 T antigen as well as the human papillomavirus E7 product also could be found to associate with Rb as well as many of the other E1A-associated proteins (DeCaprio et al., 1988; Dyson et al., 1989). Clearly, these DNA tumor virus oncogenic products possessed a common ability to bind to key cellular regulatory proteins.

The Cell Cycle: Regulators, Targets, and Clinical Applications
Edited by V.W. Hu, Plenum Press, New York, 1994

Other experiments, directed at the action of E1A as a transcriptional regulatory protein, provided a context in which to view the action of E1A, T antigen, and E7. An investigation of the ability of the transcription of the early adenovirus E2 gene identified a cellular transcription factor termed E2F that was essential for the E1A-mediated activation of E2 transcription (Kovesdi et al., 1986). Later experiments demonstrated an interaction of E2F with other cellular proteins and the capacity of the E1A protein to release E2F from these complexes. Through the use of specific antibodies, it became clear that one of the proteins in association with E2F was the retinoblastoma gene product (Bandara and LaThangue, 1991; Bagchi et al., 1991; Chellappan et al., 1991; Chittenden et al., 1991). Moreover, many of the other proteins identified as E1A-associated proteins were also found in complexes with E2F. Indeed, at least four distinct E2F complexes, schematically depicted in Figure 1, have now been identified in cell extracts.

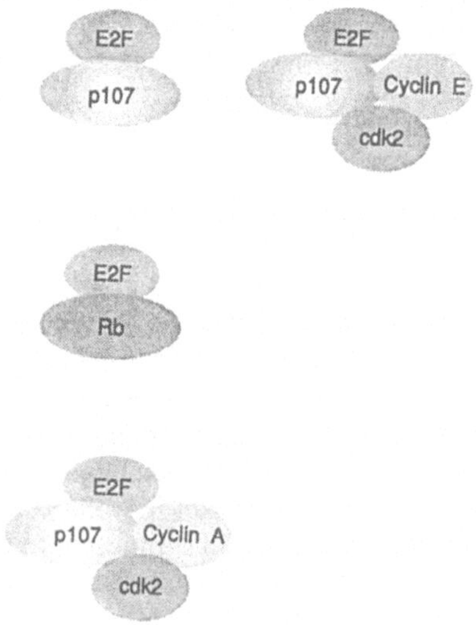

Figure 1. E2F containing complexes.

As described above, one complex contains the Rb protein while a second complex, that accumulates during S phase of the cell cycle contains the Rb related p107 protein together with cyclin A and the cdk2 kinase (Mudryj et al., 1991; Bandara et al., 1991; Devoto et al., 1992; Cao et al., 1992; Shirodkar et al., 1992; Pagano et al., 1992). Finally, at least two complexes have been detected in G1 cell extracts, one containing E2F in association with p107 (Schwarz et al., 1993) and another containing the same components as well as the cyclin E protein and the cdk2 kinase (Lees et al., 1993).

As shown in Figure 2, it is now clear that E1A, as well as T antigen and E7, can dissociate the E2F complexes, releasing free E2F and leaving E1A bound to the Rb protein (Chellappan et al., 1992).

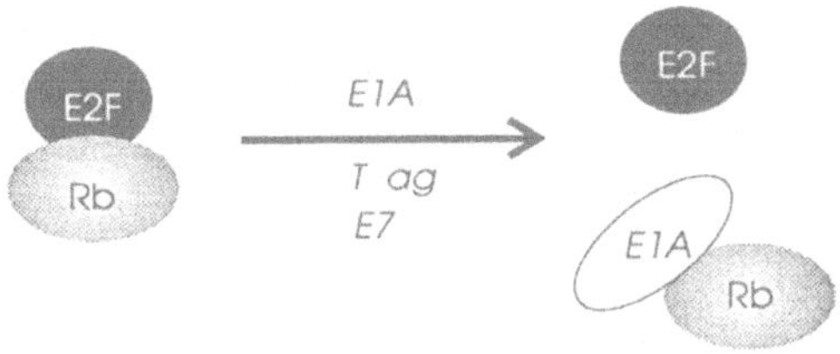

Figure 2. Dissociation of E2F complexes by adenovirus E1A, SV40 T antigen, and human papillomavirus E7.

Thus, the common activity of E1A, T antigen, and E7 that allows binding to the Rb protein may reflect the ability of these viral proteins to release E2F from complexes containing the Rb protein.

FUNCTIONAL CONSEQUENCE OF E2F COMPLEX DISSOCIATION

Transfection assays have investigated the functional significance of the release of E2F from complexes containing Rb or other cellular regulatory proteins. Introduction of a reporter gene under the control of the adenovirus E2 promoter into an Rb-deficient cell line provides an assay for E2F dependent transcription. Using this assay, it has been shown that the Rb protein inhibits the capacity of E2F to activate transcription, dependent on the ability of Rb to interact with E2F (Hiebert et al., 1992; Hamel et al., 1992; Zamanian and LaThangue, 1992). Moreover, recent experiments have shown that the Rb-related p107 protein also inhibits E2F-dependent transcription (Schwarz et al., 1993). Whether this latter result is the consequence of the interaction of the p107 protein alone or in combination with the cyclin A/cdk2 kinase complex is not yet clear.

The ability of Rb to inhibit E2F transcriptional ability, dependent on the capacity to interact with E2F, is consistent with the initial structure-function analyses of the recently cloned E2F1 gene (Helin et al., 1992; Kaelin et al., 1992; Shan et al., 1992). In particular, it has been shown that sequences required for transcriptional activation, that include an acidic domain at the C terminus of the protein, include sequences that are also required for Rb binding. Thus, it is possible that the interaction of Rb with E2F, and possibly p107 with E2F, physically obscures the ability of the activation domain to make appropriate contacts with other components of the transcriptional machinery.

A COMMON STRATEGY FOR E2F ACTIVATION BY VIRAL ONCOPROTEINS

Whereas the activation of E2F by E1A makes sense for adenovirus transcription, given the role of E2F in transcription of the viral E2 gene (Kovesdi et al., 1986; Yee et al., 1989; Loeken and Brady, 1989), the activation by T antigen and E7 cannot affect SV40 or HPV transcription since there are no E2F sites in the genomes of either of these viruses (Kraus et al., 1993). Thus, why would E2F be a target for these viral oncoproteins? To state the question differently - what do these three DNA tumor viruses share in common that would necessitate the activation of E2F?

A possible answer lies in the cellular genes that may be subject to E2F control. As shown in Figure 3, identification of cellular genes that appear to utilize E2F reveals a set of genes with common regulatory properties and that encode proteins with related activities (Nevins, 1992).

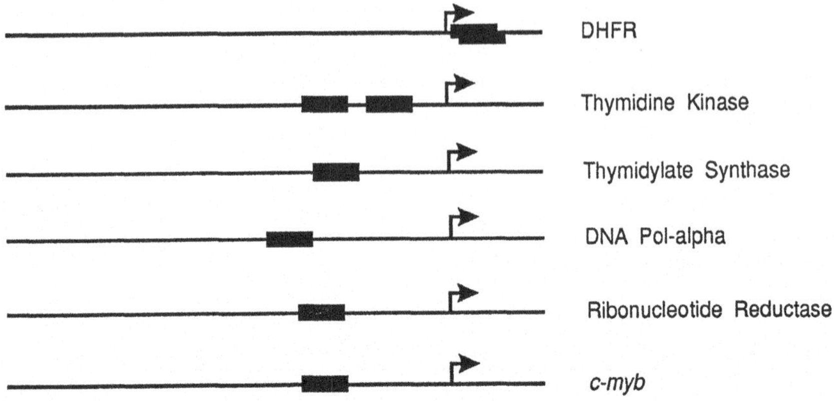

Figure 3. E2F sites in promoters of cellular genes.

Each of these genes is activated in late G1 as a quiescent cell is stimulated to re-enter the growth cycle. Moreover, with the possible exception of c-myb for which a function is yet to be determined, each gene encodes a protein that plays a role in producing the appropriate environment for DNA replication to occur. In this context, an activation of E2F leading to an activation of these S phase genes, may represent the common ground shared by these three DNA tumor viruses. It has been known for over twenty years that adenovirus, SV40, and polyomavirus can induce a quiescent cell to enter S phase and that this is dependent on E1A or T antigen function (Tooze, 1984). Together with the entry into S phase is an activation of various activities that encode S phase functions. Given the fact that many of these genes contain E2F binding sites, together with the fact that direct experiments have shown a role for E2F in the activation of the DHFR gene, leads to the speculation that the ability of E1A, T antigen, and E7 to activate E2F may be responsible for the ability of these viruses to induce a quiescent cell to enter S phase and thus promote the replication of the viral DNA.

CONCLUSIONS

The experiments outlined above suggest that the E2F transcription factor may well be a significant target for the DNA tumor virus oncoproteins in driving a quiescent cell into S phase. Since these viruses are genetically distinct, and undoubtedly have evolved independently, the fact that they share a common mechanism for inducing entry to S phase would suggest that this represents a key regulatory step in the normal transition of cells from G1 to S phase. This possibility is further strengthened by the recent observations that overexpression of the E2F1 cDNA in quiescent cells, as a result of microinjection of an E2F1-encoding plasmid, results in the stimulation of DNA synthesis in these otherwise quiescent cells (D.G. Johnson, W. D. Cress, and J. R. Nevins, submitted). These results, together with the previous observations of the ability of E1A and T antigen to induce cellular proliferation, suggests the possibility that E2F is indeed a critical regulatory activating in mediating cellular growth control.

REFERENCES

Bagchi, S., Weinmann, R., and Raychaudhuri, P., 1991, The retinoblastoma protein copurifies with E2F-1, an E1A-regulated inhibitor of the transcription factor E2F, *Cell* 65:1063-1072.

Bandara, L.R., and LaThangue, N.B., 1991, Adenovirus E1a prevents the retinoblastoma gene product from complexing with a cellular transcription factor, *Nature* 351:494-497.

Bandara, L.R., Adamczewski, J.P., Hunt, T. and LaThanque, N.B., 1991, Cyclin A and the retinoblastoma gene product complex with a common transcription factor, *Nature* 352:249-251.

Cao, L.M., Faha, B., Dembski, M., Tsai, L-H., Harlow, E., and Dyson, N., 1992, Independent binding of the retinoblastoma protein and p107 to the transcription factor E2F, *Nature* 355:176-179.

Chellappan, S.P., Hiebert, S., Mudryj, M., Horowitz, J.M., and Nevins, J. R., 1991, The E2F transcription factor is a cellular target for the Rb protein, *Cell* 65:1053-1061.

Chellappan, S., Kraus, V.B., Kroger, B., Munger, K., Howley, P. M., Phelps, W.C., and Nevins, J.R., 1992, Adenovirus E1A, simian virus 40 tumor antigen, and human papillomavirus E7 protein share the capacity to disrupt the interaction between transcription factor E2F and the retinoblastoma gene product. *Proc. Natl. Acad. Sci. USA* 89:4549-4553.

Chittenden, T., Livingston, D.M., & Kaelin, W.G., 1991, The TE1A- binding domain of the retinoblastoma product can interact selectively with a sequence-specific DNA-binding protein, *Cell* 65:1073-1082.

DeCaprio, J.A., Ludlow, J.W., Figge, J., Shew, J.Y., Huang, C.M., Lee, W.H., Marsilio, E., Paucha, E., and Livingston, D.M., 1988, SV40 large tumor antigen forms a specific complex with the product of the retinoblastoma susceptibility gene. *Cell* 54:275-283.

Devoto, S.H., Mudryj, M., Pines, J., Hunter, T., and Nevins, J.R., 1992, A cyclin A-cdc2 kinase complex possesses sequence-specific DNA binding activity: p33 cdk2 is a component of the E2F-cyclin A complex, *Cell* 68:167-176.

Dyson, N., Howley, P.M., Munger, K., and Harlow, E., 1989, The human papilloma virus 16 E7 oncoprotein is able to bind to the retinoblastoma gene product, *Science* 243:934-937.

Ewen, M.E., Xing, Y., Lawrence, J.B., and Livingston, D.M., 1991, Molecular cloning, chromosomal mapping, and expression of the cDNA for p107, a retinoblastoma gene product-related protein, Cell 66:1155-1164.

Faha, B., Ewen, M., Tsai, L-H., Livingston, D., and Harlow, E., 1992, Interaction between human cyclin A and adenovirus E1A-associated p107 protein, *Science* 255:87-90.

Hamel, P.A., Gill, R.M., Phillips, R.A., and Gallie, B.L., 1992, Transcriptional repression of the E2-containing promoters EIIaE, c-myc and RB1 by the product of the RB1 gene, *Mol. Cell. Biol.* 12:3431-3438.

Harlow, E., Whyte, P., Franza, B.R., and Schley, C., 1986, Association of adenovirus early region 1A proteins with cellular polypeptides, *Mol. Cell. Biol.* 6:1579-1589.

Hiebert, S.W., Chellappan, S.P., Horowitz, J.M., and Nevins, J.R., 1992, The interaction of RB with E2F coincides with an inhibition of the transcriptional activity of E2F, *Genes & Develop.* 6:177-185.

Helin, K., Lees, J.A., Vidal, M., Dyson, N., Harlow, E., and Fattaey, A., 1992, A cDNA encoding a pRB-binding protein with properties of the transcription factor E2F, *Cell* 70:337-350.

Kaelin, W.G., Krek, W., Sellers, W.R., DeCaprio, J.A., Ajchenbaum, F., Fuchs, C.S., Chittenden, T., Li, Y., Farnham, P. J., Blanar, M.A., Livingston, D.M., Flemington, E. K., 1992, Expression cloning of a cDNA encoding a retinoblastoma-binding protein with E2F-like properties, *Cell* 70:351-364.

Kovesdi, I., Reichel, R., & Nevins, J.R., 1986. Identification of a cellular transcription factor involved in E1A trans-activation, *Cell* 45:219-228.

Kraus, V.B., and Nevins, J.R. unpublished data.

Lees, E., Faha, B., Dulic, V., Reed, S. I., and Harlow, E., 1993, Cyclin E/cdk2 and cylcin A/cdk2 kinases associate with p107 and E2F in a temporally distinct manner, *Genes & Develop.* 6:1874-1885.

Loeken, M.R., and Brady, J., 1989, The adenovirus EIIA enhancer. Analysis of regulatory sequences and changes in binding activity of ATF and EIIF following adenovirus infection, *J. Biol. Chem.* 264:6572-6579.

Mudryj, M., Devoto, S.H., Hiebert, S.W., Hunter, T., Pines, J., and Nevins, J.R., 1991, Cell cycle regulation of the E2F transcription factor involves an interaction with cyclin A, *Cell* 65:1243-1253.

Nevins, J.R., 1992. E2F: A link between the Rb tumor suppressor protein and viral oncoproteins, *Science* 258:424-429.

Pagano, M., Draetta, G., and Jansen-Durr, P., 1992, Association of cdk2 kinase with the transcription factor E2F during S phase, *Science* 255:1144-1147.

Pines, J., and Hunter, T., 1990, Human cyclin A is adenovirus E1A- associated protein p60 and behaves differently from cyclin B, *Nature* 346:760-763.

Schwarz, J., Devoto, S.H., Smith, E.J., Chellappan, S., Jakoi, L., and Nevins, J.R., 1993, Interactions of the p107 and Rb proteins with E2F during the cell proliferative response, *The EMBO J.,* 12:1013-1020.

Shan, B., Zhu, X., Chen, P.L., Durfee, T., Yang, Y., and Sharp, D., and Lee, W., 1992, Molecular cloning of cellular genes encoding retinoblastoma-associated proteins: Identification of a gene with properties of the transcription factor E2F, *Mol. Cell. Biol.* 12:5620-5631.

Shirodkar, S., Ewen, M., DeCaprio, J.A., Morgan, J., Livingston, D.M., and Chittenden, T., 1992, The transcription factor E2F interacts with the retinoblastoma product and a p107-cyclin A complex in a cell cycle- regulated manner, *Cell* 68:157-166.

Tooze, J., 1984, "DNA Tumor Viruses," Cold spring Harbor Press, Cold Spring Harbor, New York.

Tsai, L.H., Harlow, E., and Meyerson, M., 1991, Isolation of the human cdk2 gene that encodes the cyclin A- and adenovirus E1A-associated p33 kinase. *Nature* 353:174-177.

Whyte, P., Buchkovich, J., Horowitz, J.M., Friend, S.H., Raybuck, M., Weinberg, R. A. and Harlow, E., 1988, Association between an oncogene and an anti-oncogene: the adenovirus E1A proteins bind to the retinoblastoma gene product, *Nature* 334:124-129.

Yee, A.S., Raychaudhuri, P., Jakoi, L., and Nevins, J.R., 1989, The adenovirus-inducible factor E2F stimulates transcription after specific DNA binding, *Mol. Cell. Biol.*, 9:578-585.

Yee, S., & Branton, P.E., 1985, Detection of cellular proteins associated with human adenovirus type 5 early region 1A polypeptides, *Virology* 147:142-153.

Zamanian, M. and LaThangue, N.B., 1992, Adenovirus E1A prevents the retinoblastoma gene product from repressing the activity of a cellular transcription factor, *The EMBO J.* 11:2603-2610.

11

A GTPase CYCLE COUPLED TO THE CELL CYCLE

Elias Coutavas,[1] Mindong Ren,[2] Joel D. Oppenheim,[3] Vijay Yajnik,[4]
Peter D'Eustachio,[1,5] and Mark G. Rush[1,5]

[1]Department of Biochemistry
[2]Department of Cell Biology
[3]Department of Microbiology
[4]Department of Pharmacology
[5]Kaplan Cancer Center
NYU Medical Center
New York, NY 10016

ABSTRACT

Ran/TC4, a Ras-related nuclear protein first identified in a human teratocarcinoma cDNA library, is postulated to function in conjunction with another protein, RCC1 (Regulator of Chromosomal Condensation-1), to prevent the initiation of mitosis before the completion of DNA synthesis. RCC1 is a chromatin-associated DNA-binding protein as well as a Ran/TC4-specific guanine nucleotide exchange factor. One model for the function of Ran/TC4 and RCC1 is that RCC1 monitors DNA synthesis and maintains Ran/TC4 in a GTP-bound form (Ran/TC4·GTP) until the completion of DNA synthesis and that Ran/TC4·GTP in turn inhibits the activation of MPF (Mitosis Promoting Factor). We are testing this model, in which a GTPase cycle is coupled to the cell cycle via MPF, by expressing mutant Ran/TC4 proteins in human cells, and by using a Ran/TC4·[^{32}P]GTP complex as a probe to isolate Ran-interacting proteins.

INTRODUCTION

Ran/TC4 is a RAS-related protein

In eukaryotes, several kinds of guanine nucleotide-binding and hydrolyzing proteins have been identified, including the initiation and elongation factors of protein synthesis, subunits of the signal recognition particle and its receptor, ADP-ribosylation factors, the α subunits of heterotrimeric G proteins involved in signal transduction, and the RAS protein family (Bourne et al., 1990, 1991). The RAS protein family consists of small proteins with sequence similarity, especially in their guanine nucleotide-binding domains, to the products of *RAS* proto-oncogenes. Before the discovery of Ran/TC4 and its relatives, the RAS protein family was divided into three main subfamilies, RAS and RAS-like, Rho, and

Rab, on the basis of protein sequence comparisons (Downward, 1992; Hall, 1990).

Ran/TC4 is the prototype of a fourth major subfamily of RAS-related genes (Figure 1). It was first identifed in a human cDNA library (Drivas et al., 1990, 1991b); homologues have since been identified in other mammals (dog, mouse), *Drosophila*, *Caenorhabditis*, *Dictyostelium*, and in yeasts (*Saccharomyces cerevisiae*, *Schizosaccharomyces pombe*). The Ran subfamily is very well conserved: the predicted sequences of all known Ran-like proteins are at least 80% identical. However, the subfamily shows only limited resemblance to other RAS genes. Outside of the four stretches of sequence that specify the RAS guanine nucleotide binding site, there are no obvious homologies between Ran sequences and those of other RAS family members. One striking difference is that, while other RAS-related proteins share a carboxyterminal cysteine motif that mediates isoprenylation and membrane association, Ran proteins have a stretch of acidic amino acid residues at this position and are localized in the nuclei of interphase cells (Bischoff and Ponstingl, 1991a; Ren et al., 1993).

```
                           *                                                        *
TC4  MAAQ-G-EPQ V-QFKLVLVG DGGTGKTTFV KRHLTGEFEK KYVATLGVEV HPLVFHTNRG PIKFNVWDTA GQEKFGGLRD
M1   ....-.-... .-........ .......... .......... .......... .......... .......... ..........
M2   ....-.-... .-...V.... .........M .......... E......... .......... .......... ..........
SPI1 ..-.PQN--- .PT....... .......... .......... ..I....... ...H....F. E.C....... ....L.....
GSP1 .S.PA-ANGE .PT....... .......... .......... ..I..I.... ...S.Y..F. E...D..... ..........
GSP2 .S.PAQNNAE .PT....... .......... .......... ..I..I.... ...S.Y..F. E...D..... ..........

TC4  GYYIQAQCAI IMFDVTSRVT YKNVPNWHRD LVRVCENIPI VLCGNKVDIK DRKVKAKSIV FHRKKNLQYY DISAKSNYNF
M1   .......... .......... .......... .......... .......... .......... .......... ..........
M2   .......... .......... .....S..K. .......... ........V. .M.....P.L .......... ....R.....
SPI1 .....G..G. ........I. .....H.W.. .......... ........V. E......A.T .......... ..........
GSP1 ....N..... ........I. .......... .......... ........V. E......T.T .......... ..........
GSP2 ....N..... ........I. .......... .......... ........V. E......T.T .......... ..........

                                                         **** **
TC4  EKPFLWLARK LIGDPNLEFV AMPALAPPEV VMDPALAAQY EHDLEVAQTT ALPDEDD-DL
M1   .......... .......... .......... .......... .......... .......-..
M2   ....F..... .......... .......... .......... .......... .....E.-..
SPI1 .......... .V.N...... .S........ QV.QQ.L... QQEMNE.AAM P......A..
GSP1 .......... .A.N.Q.... .S........ QV.EQ.MQ.. QQEM.Q.TAL P......A..
GSP2 .......... .A.N.Q.... .S........ QV.EQ.MH.. QQEMDQ.TAL P......A..
```

Figure 1. Comparison of five Ran protein sequences with Ran/TC4. Dots indicate identities and dashes indicate gaps. The four elements of the guanine nucleotide binding site are underlined. Asterisks above the Ran/TC4 sequence indicate residues altered in the expression studies described here. Sources of the sequences are: human Ran/TC4 - Ren et al., 1993; mouse Ran/M1 and Ran/M2 - Coutavas, E., et al., manuscript in preparation; *S. pombe SPI1* - Matsumoto and Beach, 1991; *S. cerevisiae GSP1* and *GSP2* - Belhumeur et al., 1993.

GTPases are associated with diverse cellular functions, but all share the same cycle of biochemical reactions (Bourne et al., 1990, 1991). Binding and hydrolysis of GTP drive transitions between two conformational states, GDP-bound and GTP-bound. The function of each GTPase in the cell is determined by the differing abilities of its conformational states to interact with specific macromolecular effectors and regulators. The ratio of GTP-bound to GDP-bound forms depends on the relative rates of two reactions: hydrolysis of bound GTP, and dissociation of bound GDP. Intrinsic rates of both reactions are generally slow. The effective rates in the cell are determined by interactions of the GTPase with accessory proteins including (Bollag and McCormick, 1991; Bourne et al., 1990; Takai et al., 1992) GTPase activating proteins (GAPs), which stimulate the hydrolysis of GTP and therefore promote the GDP-bound form of the protein; and guanine nucleotide dissociation stimulators (GDSs), which stimulate the exchange of bound for free guanine nucleotide. In the high-GTP milieu of the cell, they promote the GTP-bound form of the protein.

Ran/TC4 and RCC1 are components of an intrinsic checkpoint control

The tsBN2 cell line, a temperature-sensitive mutant of baby hamster kidney cells, grows normally at 33.5° but ceases to grow at 39.5°. In cells shifted to nonpermissive temperature early in S phase, chromosome condensation and other signs of entry into mitosis occur prematurely, indicating that a feedback-checkpoint control coupling mitosis to the completion of DNA replication is defective (Nishimoto et al., 1978). Through DNA-mediated gene transfer, the gene complementing the tsBN2 mutation has been cloned. It is named RCC1 (Regulator of Chromosome Condensation-1) (Ohtsubo et al., 1987). The properties of tsBN2 cells are due to a point mutation in RCC1 that renders the protein unstable at the nonpermissive temperature (Uchida et al., 1990). Premature entry into mitosis upon loss of RCC1 requires MPF activation and the presence of cdc25 (the phosphatase that activates preMPF) (Nishitani et al., 1991; Seino et al., 1991; Seki et al., 1992). Genes homologous to RCC1 have been isolated from organisms including humans (Bischoff and Ponstingl, 1991a), *Xenopus* (Nishitani et al., 1990), *Drosophila* (Frasch, 1991), *S. cerevisiae* (Clark and Sprague, 1989; Aebi et al., 1990) and *S. pombe* (Matsumoto and Beach, 1991).

RCC1 is a nuclear protein that associates with chromatin and can bind DNA (Ohtsubo et al., 1989; Frasch, 1991). It is present at about one million copies per nucleus, equivalent to one copy per nucleosome (Seino et al., 1991). Although not a component of the basal complex required for initiation of eukaryotic DNA replication in vitro (Hurwitz et al., 1990), RCC1 is absolutely required for the replication of added sperm chromatin DNA in *Xenopus* egg extracts (Dasso et al., 1992).

A fission yeast temperature-sensitive mutant, defective in coupling mitosis to the completion of DNA replication, is caused by mutation in the RCC1-homologous gene *PIM1* (premature initiation of mitosis-1) (Matsumoto and Beach, 1991). *pim1* mutants are rescued by overexpression of the wild-type allele of *SPI1* (suppressor of *pim1*), the fission yeast homologue of Ran/TC4 (Matsumoto and Beach, 1991, 1993). Human RCC1 protein can be purified from HeLa cells in the form of a complex with Ran/TC4 (Bischoff and Ponstingl, 1991a), and RCC1 protein specifically catalyzes exchange of guanine nucleotides on Ran/TC4 protein (Bischoff and Ponstingl, 1991b). Collectively, the genetic and biochemical evidence suggest that Ran/TC4 and RCC1 are key components of a feedback-checkpoint control that monitors the progress of DNA replication and transduces signals to regulate the activity of MPF.

MPF is a complex of two proteins, cyclin B and a cyclin-dependent serine-threonine protein kinase $p34^{cdc2}$ (Freeman & Donoghue 1991; Hunt 1991; Norbury & Nurse 1992). MPF normally plays a key role in eukaryotic cell cycle control; its regulation involves interactions among tyrosine and serine-threonine protein kinases, protein phosphatases, a cyclin protease, and other activators and inhibitors. It is formed late in G1 or in S, depending on the cell type, by the association of newly synthesized cyclin B with $p34^{cdc2}$, and is immediately inactivated by phosphorylation of specific threonine and tyrosine residues. Conversion of inactive pre-MPF to active MPF in prophase requires removal of esterified phosphate from these $p34^{cdc2}$ tyrosine and threonine residues, while passage through anaphase requires destruction of cyclin B.

The cell cycle may be coupled to a Ran GTPase cycle as shown in Figure 2 (Bischoff and Ponstingl, 1991a, 1991b; Ren et al. 1993; Roberge, 1992). Ran may switch between an active GTP-bound form and an inactive GDP-bound form, and this switch may regulate the MPF-driven cell cycle. In this model, START, the commitment of the cell to DNA replication, stimulates the guanine nucleotide exchange function of RCC1, and this in turn

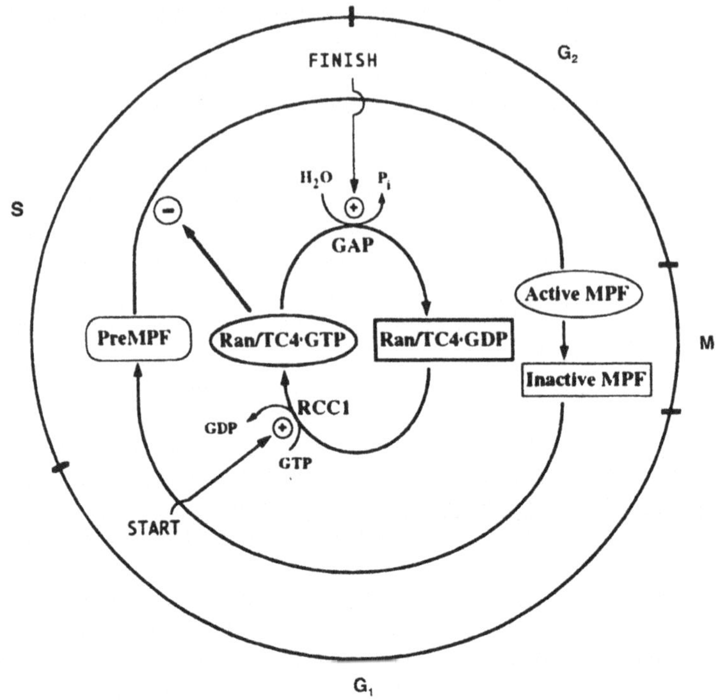

Figure 2. A model coupling the cell cycle and a GTPase cycle via the nuclear proteins Ran/TC4 and RCC1. The inner circle represents the Ran/TC4 GTPase cycle, the second circle represents the MPF cycle, and the outer circle represents the cell cycle. START is the point of commitment to DNA synthesis, as distinguished from the point at which DNA replication actually begins. FINISH is the end of DNA replication. Key links between the cycles in the model are 1) the stimulation, at START, of the guanine nucleotide exchange activity of RCC1, and 2) the role of Ran/TC4 · GTP as an inhibitor of the conversion of preMPF to MPF.

converts Ran to its active GTP-bound form. Ran · GTP transduces a signal that inhibits the conversion of preMPF to MPF. FINISH, the completion of DNA replication, is associated with the inactivation of RCC1 and the activation of a Ran-specific GAP, restoring Ran to its inactive GDP-bound form.

According to this model, Ran function would not be required during the pre-START G1 phase of the cell cycle since preMPF is absent during this period. This model predicts that expression and accumulation of GTPase-deficient Ran · GTP in a cell will result in cell cycle arrest at the G2/M boundary.

RESULTS AND DISCUSSION

Expression of a GTPase-deficient Ran/TC4 protein inhibits cell cycle progression

To test the proposed role for Ran · GTP in cell cycle regulation, we constructed a mutated recombinant Ran/TC4 cDNA expected to encode a protein deficient in GTP hydrolysis, and determined the effects of transient expression of mutant protein on cell cycle progression in mammalian cells.

That the mutant protein was indeed GTPase-defective was shown as follows. A wild-type Ran/TC4 cDNA and a cDNA mutated at codons 19 and 69 (corresponding to GTPase defective mutations [dm] of *HRAS* codons 12 and 61 - Ren et al., 1993) were cloned into the bacterial expression vector pET-9C and transfected into *E. coli* BL21(DE3). Ran/TC4 proteins were purified from the bacterial lysates. Wild-type and mutant proteins both bound GTP and neither protein catalyzed appreciable hydrolysis of the bound nucleotide over a 60 min incubation period.

These GTP-charged proteins were then used to search for a GAP activity. When wild-type Ran/TC4 · GTP was incubated in the presence of a HeLa cell S-100 extract, hydrolysis of the bound GTP proceeded briskly, reaching a plateau within 20 min of the start of the incubation (Figure 3). When this HeLa S-100 extract was fractionated on a Superose 12HR 10/30 column (Pharmacia), essentially all GAP activity was recovered as a single broad peak of >100,000 Kd. The GAP activity defined in this assay may be Ran-specific. Incubation of wild-type Ran/TC4 · GTP with purified RAS GAPs (p120 and NF-1), a Rap-1 GAP, and Rho GAP p190 (gifts from F. McCormick and R. Weinberg) stimulated no GTP hydrolysis above background levels. Bischoff and Ponstingl (this meeting) have recently purified a 65 Kd polypeptide, present in HeLa cell extracts as a 130 Kd homodimer, with similar GAP properties.

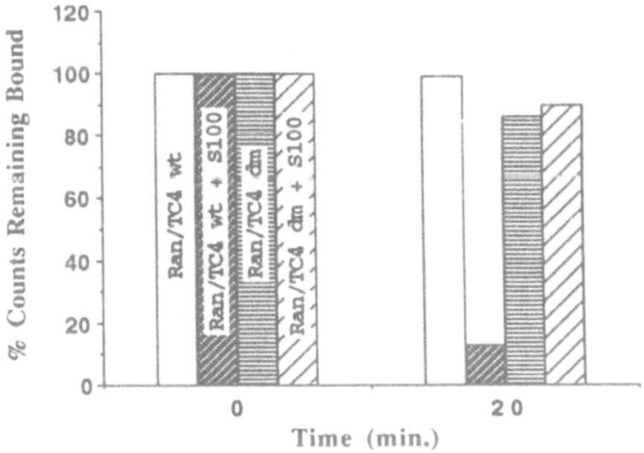

Figure 3. Hydrolysis of GTP bound to wild-type, but not to mutant, Ran/TC4 in the presence of HeLa S-100 extract. Recombinant wild-type or mutant Ran/TC4 proteins charged with $\gamma[^{32}P]GTP$ were incubated for the indicated times in the presence or absence of S-100, and assayed for GTP hydrolysis by filter binding. Ran/TC4 proteins (1 μM) charged with $\gamma[^{32}P]GTP$ (0.5 μM, 1000 Ci/mmole, NEN) by incubation for 20 min at 25° in 30-35 μl 50 mM Tris pH 7.4, 40 mM NaCl, 0.5 mM EDTA, 1 mM DTT, 1 mg/ml BSA, were adjusted to 10 mM $MgCl_2$, 5 mM NaCl, and incubated in the presence or absence of 1 μg of HeLa S-100 protein in a total volume of 20 μl at 25°. Reactions were terminated by addition of 1 ml ice-cold 50 mM Tris pH 7.4, 20 mM $MgCl_2$, 1 mM DTT, 2 mg/ml BSA, and the amount of radioactivity bound to Ran/TC4 determined by nitrocellulose binding and scintillation counting. The HeLa S-100 extract was prepared by sonicating cells on ice in 25 mM HEPES pH 7, 150 mM NaCl, 5 mM $MgCl_2$, 3 mM DTT, 100 U/ml aprotinin, 10 μg/ml leupeptin, followed by centrifugation at 100,000Xg for 1 hr at 4°.

Mutant Ran/TC4 · GTP showed no GTPase activity in the presence of HeLa S-100 extract (Figure 3), confirming the prediction that mutations in Ran/TC4 codons 19 and 69 would block GTPase activity. The intrinsic GTP hydrolysis rate of wild-type Ran/TC4 is very low, so the mutant defect is manifested in vitro as its insensitivity to stimulation by GAP.

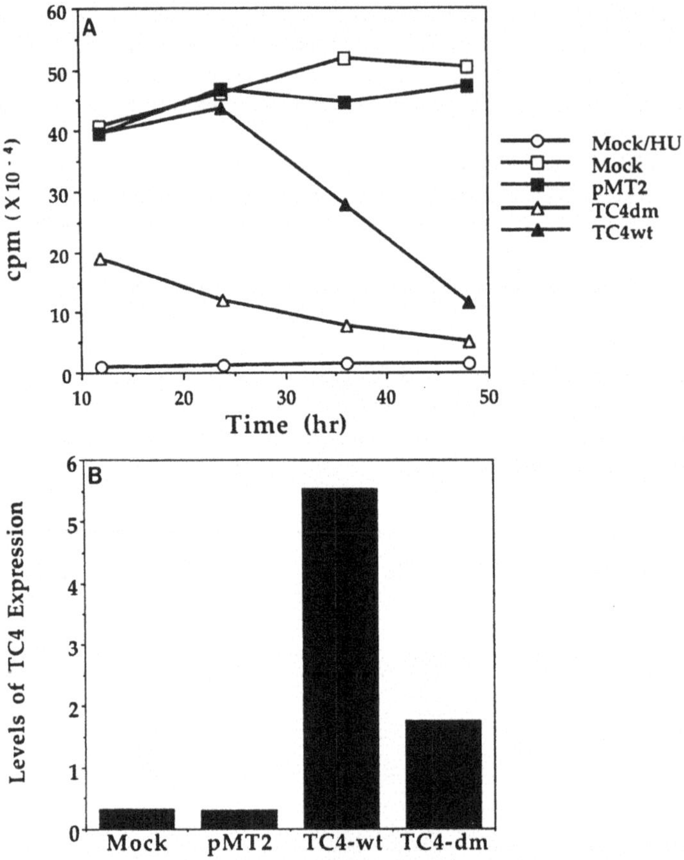

Figure 4. Inhibition of DNA synthesis in 293/Tag cells transfected with pMT2-Ran/TC4 constructs.

A. Kinetics of inhibition. Subconfluent cultures of 293/Tag cells in 10-cm dishes were transfected with nothing (Mock), 20 μg of pMT2 vector, or 20 μg of pMT2 constructs encoding either wild-type (wt) or GTPase-deficient mutant (dm) Ran/TC4 by calcium phosphate co-precipitation. At the indicated times after transfection, [^{3}H]Thymidine was added for an additional three hr. DNA synthesis was assayed as acid-precipitable thymidine incorporation. To measure background [^{3}H]Thymidine incorporation, mock-transfected cultures were treated with Hydroxyurea (10 mM) 3 hr before addition of [^{3}H]Thymidine (Mock / HU).

B. Quantitiation of Ran/TC4 levels in 293/Tag cells transfected with pMT2-Ran/TC4 constructs. Untransfected (Mock) 293/Tag cells, or cells transfected with the indicated amounts of pMT2 vector, or with pMT2-Ran/TC4 wt or dm constructs, were incubated for 48 hr and extracts containing 50 μg protein were analyzed for Ran/TC4 levels by immunoblotting and phosphorimaging (Ren et al., 1993). Ran/TC4 protein levels are given in arbitrary units.

We showed previously that in individual monkey (COS) cells transiently expressing this mutant protein (but not wild-type Ran/TC4), both chromosomal and extrachromosomal DNA replication were inhibited, as would be expected (Figure 2) if Ran/TC4 · GTP blocked cell cycle progression in G2 (Ren et al., 1993). In order to quantitate this inhibition, and to determine the points in the cell cycle at which cells expressing mutant Ran/TC4 protein are blocked, we carried out transient expression assays using human embryonal kidney 293/Tag cells under conditions where populations of transfected cells could readily be studied. Like COS cells, 293/Tag cells express SV40 large T antigen, allowing extensive replication of the pMT2 expression vector, which contains an SV40 origin of replication. In addition, the transfection efficiency of 293/Tag cells is very high; at least 80% of 293/Tag cells are transfected per experiment (M. Ren, unpublished observations). Low level expression of the GTPase-defective, but not the wild-type, Ran/TC4 protein in 293/Tag cells inhibited DNA synthesis (Figure 4). Very high level expression of wild-type Ran/TC4 also caused some inhibition. Immunoblotting of cells transfected with 20 μg DNA/culture of Ran/TC4 wild-type or mutant cDNAs revealed less than doubled amounts of Ran/TC4 protein relative to endogenous levels 12 hr after transfection (data not shown), but substantial increases 48 hr after transfection (Figure 4). As observed previously, transfection of Ran/TC4 wild-type construct led to substantially higher levels of Ran/TC4 protein synthesis than did transfection of the GTPase defective construct. Transfection of 293/Tag cells with a lower dose (10 μg/culture) of Ran/TC4 wild-type construct failed to inhibit DNA replication significantly at any time up to 48 hr after transfection (data not shown). This observation, and the kinetics of inhibition after transfection of high doses of wild-type construct, suggest that expression of very high levels of wild-type Ran/TC4 protein mimics some of the effects observed with low doses of GTPase-deficient mutant protein. This result parallels ones obtained in studies of the effects of true Ras proteins on cell proliferation.

As expected from the DNA replication data (Figure 4), cell proliferation, measured by cell counting, was inhibited in cultures transfected with GTPase defective Ran/TC4 constructs or with high doses of wild-type constructs. While mock and pMT2 transfected cells underwent two to three doublings in 48 hr, Ran/TC4 wild-type transfected cells underwent one to two doublings, and mutant transfected cells underwent less than one.

To determine the point(s) in the cell cycle at which cells expressing mutant Ran/TC4 protein are blocked, cellular DNA content was analyzed by flow cytometry. If Ran/TC4 functions as a GTPase switch whose GTP-bound form generates a signal that blocks premature initiation of mitosis, cells expressing the Ran/TC4 GTPase-deficient mutant should arrest late in G2 (Figure 2). Flow cytometry of cell populations arrested by expression of the Ran/TC4 mutant construct revealed a major accumulation of cells with a fully replicated DNA complement, and visual analysis of these cells indicated that most were in G2 rather than in M. Flow cytometry of these cells also revealed a minor accumulation of cells with unreplicated DNA (M. Ren et al., manuscript in preparation).

There are several explanations for the G1 block that are plausible and consistent with the GTPase switch model shown in Figure 2. One is that progression through G1 may be regulated by a second GTPase switch whose normal operation is overwhelmed by the presence of high levels of Ran/TC4·GTP. Although Ran proteins show little homology to other members of the RAS family, they show close homology to one another. The two transcriptionally active Ran genes identified to date in the mouse, for example, specify proteins whose amino acid sequences are 95% identical (Drivas et al., 1991b; Coutavas, E., et al., manuscript in preparation). It is thus plausible to imagine that two or more Ran GTPase switches monitor cell cycle progression, and that overexpression of any one Ran protein could disrupt some or all of these switches.

In summary, expression of a GTPase defective Ran/TC4 mutant in proliferating human (293/Tag) cells blocked chromosomal DNA replication by inducing arrest predominantly in the G2 phase of the cell cycle. This G2 arrest is presumably due to inhibition of MPF activation, although this has not yet been investigated.

Deletion of the carboxyterminal hexapaptide sequence, DEDDDL, from the mutant protein blocked its inhibitory activites (Ren, M., et al., manuscript in preparation). The same deletion in an otherwise wild-type protein had no detectable effect on cell proliferation. The silencing of the cell cycle arrest phenotype of the GTPase deficient mutant by deletion of six carboxyterminal amino acids, confirms the importance of these Ran subfamily-specific residues. One explanation for the silencing is that the carboxyterminal region is required for Ran nuclear localization. Another is that it is required for interaction with an effector. In either case, the carboxyterminal-deleted Ran/TC4 · GTP complex would be unable to exert its otherwise dominant inhibitory effect.

Together, these results provide direct evidence for the role of a small GTPase in mammalian cell cycle control. The results can be accomodated by a straightforward model of a GTPase switch. Equally important, they will help to focus the search for the effector molecules through which Ran/TC4 exerts its effects on cell cycle progression.

Ran/TC4 · GTP, but not Ran/TC4 · GDP, binds to specific proteins on renatured blots of mammalian cell extracts

To search for Ran/TC4 interacting proteins, we probed renatured protein blots of human (HeLa) and hamster (tsBN2) cell soluble extracts with Ran/TC4 · $\gamma[^{32}P]$GTP or Ran/TC4 · $\alpha[^{32}P]$GTP, and reproducibly detected a polypeptide of apparent molecular weight 27,000. No proteins were detected in blots probed with Ran/TC4 · $\alpha[^{32}P]$GDP, and binding by Ran/TC4 · $\gamma[^{32}P]$GTP was inhibited in the presence of excess unlabelled

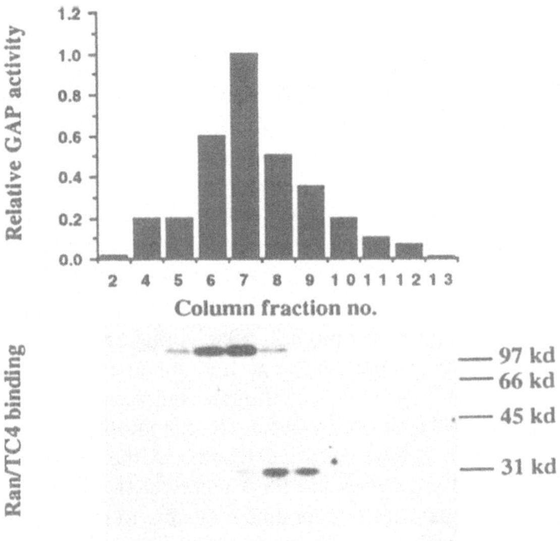

Figure 5. Ran/TC4 · GTP binds to two proteins in HeLa S-100 extracts. Equal volumes of fractions from a Superose 12 HR 10/30 column (Pharmacia) were analyzed for Ran/TC4-specific GAP activity using the assay described in the legend to Figure 4. To assay fractions for Ran/TC4 binding, they were subjected to electrophoresis on SDS - 12% polyacrylamide gels, transferred to nitrocellulose membranes in 25 mM Tris, 192 mM glycine, 15% methanol, pH 8.3, and renatured in situ by a 48 hr incubation at 4° in 50 mM HEPES pH 7, 100 mM KOAc, 5 mM $Mg(OAc)_2$, 3 mM DTT, 0.3% Tween20, 1% BSA (Connolly and Gilmore, 1989). Blots were preincubated in 40 ml 25 mM HEPES pH 7, 100 mM KOAc, 5 mM $Mg(OAc)_2$, 3 mM DTT, 0.1% Tween20, 1% BSA, 250 mM GTP for 30 min at 25°, and then in the same buffer containing 0.5 µg/ml Ran/TC4 · $\alpha[^{32}P]$GTP (30 Ci/mMole) for 45 min at 4°. Filters were washed five times in preincubation buffer without GTP and autoradiographed. The Ran/TC4 · $\alpha[^{32}P]$GTP probe was an equimolar mixture of wild-type and GTPase-deficient recombinant proteins (8 µM total).

Ran/TC4 · GTP but not in the presence of excess unlabelled Ran/TC4 · GDP (Coutavas, E., et al., manuscript in preparation). A polypeptide of apparent molecular weight 110,000 with the same binding specificity was also detected in some experiments. Immunoblotting with Ran/TC4 antipeptide antibody (Ren et al., 1993) demonstrated that the protein corresponding to the 27 Kd band was distinct from Ran/TC4 itself, and Superose 12 HR 10/30 column fractionation of HeLa cell extracts resolved it from both Ran/TC4 GAP activity and the 110 Kd Ran/TC4 · GTP-binding protein (Figure 5). It is not known whether the 110 Kd protein has GAP activity. We named the 27 Kd protein Ran Binding Protein-1 (RanBP1).

Ran/TC4 · GTP can be used as a probe for the cloning of RanBP1 cDNAs

To clone and further characterize RanBP1, we probed a λEXlox(+) expression library of 16-day mouse embryo cDNAs (Novagen, Madison WI) with Ran/TC4 · $\alpha[^{32}P]$GTP, and identified and plaque-purified five different strongly positive clones that all contained the same 609-bp open reading frame. Four contained the full open reading frame and one lacked the 60 amino terminal codons. A computer search revealed that the nucleotide sequence of the open reading frame was 98% identical to that of the mouse HTF9A gene (Bressan et al., 1991; Stapleton et al., 1993). HTF9A is one of the open reading frames flanking a bidirectional promoter; its function is unknown. The protein is highly charged (40% glu + asp + lys + arg), acidic, and has a predicted molecular weight of 23,600. It contains no clear diagnostic catalytic domains, but does contain two regions close to the consensus sequence for leucine zippers, one close to that for a helix-turn-helix, and one close to that for an RNA binding motif (Coutavas, E., et al., manuscript in preparation).

The 609-bp open reading frame from one of the λEXlox(+) cDNAs was subcloned into the pET-9C bacterial expression vector and used to generate recombinant protein. When analyzed by SDS polyacrylamide gel electrophoresis, purified recombinant RanBP1 migrated at the same rate as RanBP1 from HeLa cell extracts. The anomalous slow mobility of RanBP1 on SDS gels may be due to its low isoelectric point.

The facts that Ran/TC4 · GTP, but not Ran/TC4 · GDP, interacts with RanBP1 and that RanBP1 possesses no detectable GAP activity, support the hypothesis that RanBP1 may be a Ran/TC4 effector protein. Ran proteins are highly conserved evolutionarily (Figure 1), leading us to expect that effector domains interacting with them might likewise be well conserved. To determine whether cells of two distantly related eukaryotes contained RanBP-like activities, a cytosolic extract of *Xenopus* oocytes and a total extract of *S. pombe* were subjected to SDS polyacrylamide gel electrophoresis, blotted, renatured, and probed with Ran/TC4 · $\alpha[^{32}P]$GTP. Single Ran/TC4-interacting proteins were identified in each extract (Coutavas, E., et al., manuscript in preparation).

How does Ran/TC4 regulate MPF?

In regard to the mechanism of MPF activation, our model predicts that the unreplicated DNA present between START and FINISH alters the levels of Ran/TC4 · GTP. We suggest that this change is an essential step of a pathway that leads to MPF inactivation, although there may be multiple intermediate steps. For example, Ran/TC4 · GTP could stimulate, either directly or through regulatory kinases and phosphatases, the activation of the kinase that phosphorylates $p34^{cdc2}$ on tyrosine to inactivate MPF (Smythe and Newport, 1992). Ran/TC4 · GTP could also inhibit, directly or indirectly, the tyrosine phosphatase that dephosphorylates $p34^{cdc2}$ and that is required for MPF activation (Millar and Russell, 1992).

Ran and RCC1 proteins have been found in all eukaryotes examined (Dasso, 1993), and are implicated in diverse cellular functions. As previously mentioned, in fission yeast and mammals alterations in either protein perturb a checkpoint between the completion of DNA synthesis and the onset of mitosis (Dasso, 1993; Matsumoto and Beach, 1991, 1993; Nishitani et al., 1991; Ren et al., 1993; Roberge, 1992). In addition, a mutant allele of the *S. cerevisiae* RCC1 homologue, *prp20*, causes accumulation of precursor mRNAs in the nucleus, aberrant transcription termination, and disruption of nuclear and nucleolar morphologies (Aebi et al., 1990; Forrester et al., 1992). Two *S. cerevisiae* Ran GTPases, *GSP1* and *GSP2*, have been identified. Each has over 80% amino acid sequence identity with Ran/TC4. Overexpression of either *GSP1* or *GSP2* in the *prp20-1* mutant strain restores its viability and restores normal nucleoplasmic morphology (Belhumeur et al., 1993). These results suggest that Ran GTPases might play a role in nuclear organization and nuclear transport, a suggestion supported by the observation that the hamster tsBN2 RCC1 mutant cell line, which exhibits premature entry into mitosis, is also defective in mRNA transport at nonpermissive temperature (Amberg et al., 1993). Finally, in one in vitro model for protein import into nuclei, purified recombinant Ran/TC4 protein substitutes for the 25 Kd "B1" *Xenopus* protein fraction (Moore and Blobel, 1992, and personal communication).

All of these functions may be interconnected. The induction of premature chromosome condensation in tsBN2 cells at nonpermissive temperature requires new protein synthesis (Uchida et al., 1990). Wild-type Ran/TC4 and RCC1 might modulate the production of cell cycle control proteins, as part of a role in the general regulation of nuclear transport, by affecting traffic between the nucleus and cytosol of the relevant mRNAs or newly synthesized proteins. RanBP1, which interacts specifically with GTP-charged Ran and is an acidic protein with a putative RNA binding site, is a plausible intermediary in this process.

ACKNOWLEDGEMENTS

We thank D. Sabatini, M. Adesnik, and G. Teebor for their advice and encouragement throughout this project, and D. Frendewey, B. Margolis, F. McCormick, M. Moore, and R. Weinberg for gifts of reagents and helpful discussions. This research was supported in part by a PHS Basic Research Support Grant to NYU Medical Center, and E. Coutavas was supported by a PHS Training Grant. Nucleotide sequence analyses were performed at the NYU Medical Center Computer Facility (supported by NSF), and utilized the GCG Program Package v7 (Madison WI).

REFERENCES

Aebi, M., Clark, M.W., Vijayraghavan, U., and Abelson, J., 1990, A yeast mutant, *PRP20*, altered in mRNA metabolism and maintenance of the nuclear structure, is defective in a gene homologous to the human gene *RCC1* which is involved in the control of chromosome condensation, *Mol. Gen. Genet 224*:72.

Amberg, D.C., Fleischmann, M., Stagljar, I., Cole, C.N., and Aebi, M., 1993, Nuclear PRP20 protein is required for mRNA export, *EMBO J. 12*:233.

Barbacid, M., 1987, *ras* genes, *Annu. Rev. Biochem. 56*:779.

Belhumeur, P., Lee, A., Tam, R., DiPaolo, T., Fortin, N., and Clark, M.W., *GSP1* and *GSP2, genetic suppressors of the prp20-1* mutant in *Saccharomyces cerevisiae*: GTP-binding proteins involved in the maintenance of nuclear organization, *Mol. Cell. Biol. 13*:2152.

Bischoff, F.R., and Ponstingl, H., 1991a, Mitotic regulator protein RCC1 is complexed with a nuclear *ras*-related polypeptide, *Proc. Natl. Acad. Sci. USA 88*:10830.

Bischoff, F.R., and Ponstingl, H., 1991b, Catalysis of guanine nucleotide exchange on Ran by the mitotic regulator RCC1, *Nature 354*:80.

Bollag, G., and McCormick, F., 1991, Regulators and effectors of Ras proteins, *Annu. Rev. Cell Biol. 7*:601.

Bourne, H.R., Sanders, D.A., and McCormick, F., 1990, The GTPase superfamily: a conserved switch for diverse cell functions, *Nature 348*:125.

Bourne, H.R., Sanders, D.A., and McCormick, F., 1991, The GTPase superfamily: conserved structure and molecular mechanism. *Nature 349*:117.

Bressan, A., Somma, M.P., Lewis, J., Santolamazza, C., Copeland, N.G., Gilbert, D.J., Jenkins, N.A., and Lavia, P., 1991, Characterization of the opposite-strand genes from the mouse bidirectionally transcribed *HTF9* locus, *Gene 103*:201.

Clark, K., and Sprague, G., 1989, Yeast pheromone response pathway: characterization of a suppressor that restores mating to receptorless mutants, *Mol. Cell. Biol.* 9:2682.

Connolly, T., and Gilmore, R., 1989, The signal recognition particle receptor mediates the GTP-dependent displacement of SRP from the signal sequence of the nascent polypeptide, *Cell 57*:599.

Dasso, M., 1993, RCC1 in the cell cycle: the regulator of chromosome condensation takes on new roles, *TIBS 18*:96.

Dasso, M., Nishitani, H., Kornbluth, S., Nishimoto, T., and Newport, J.W., 1992, RCC1, a regulator of mitosis, is essential for DNA replication, *Mol. Cell. Biol. 12*:3337.

Downward, J., 1992, Rac and Rho in tune, *Nature 359*:273.

Drivas, G.T., Shih, A., Coutavas, E.E., Rush, M.G., and D'Eustachio, P., 1990, Characterization of four novel RAS-related genes expressed in a human teratocarcinoma cell line, *Mol. Cell. Biol. 10*:1793.

Drivas, G.T., Palmieri, S., D'Eustachio, P., and Rush, M.G., 1991a, Evolutionary grouping of the RAS-protein family, *Biochem. Biophys. Res. Comm. 176*:1130.

Drivas, G.T., Massey, R., Chang, H.-Y., Rush, M.G., and D'Eustachio, P., 1991b, *Ras*-like genes and gene families in the mouse, *Mamm. Genome* 1:112.

Forrester, W., Stutz, F., Rosbasch, M., and Wickens, M., 1992, Defects in mRNA 3'-end formation, transcription initiation, and in RNA transport associated with the yeast mutation *prp20*: possible coupling of mRNAprocessing and chromatin structure, *Genes Devel. 6*:1914.

Frasch, M., 1991, The maternally expressed *Drosophila* gene encoding the chromatin-binding protein BJ1 is a homolog of the vertebrate gene regulator of chromatin condensation, *RCC1*, *EMBO J. 10*:1225.

Freeman, R.S., and Donoghue, D.J., 1991, Protein kinases and protooncogenes: biochemical regulators of the eukaryotic cell cycle, *Biochemistry 30*:2293.

Hall, A., 1990, The cellular functions of small GTP-binding proteins, *Science 249*:635.

Hunt, T., 1991, Summary: put out more flags, *CSH Symp. Quant. Biol. 56*:757.

Hurwitz, J., Dean, F.B., Kwong, A.D., and Lee, S.-H., 1990, The in vitro replication of DNA containing the SV40 origin, *J. Biol. Chem. 265*:18043.

Matsumoto, T., and Beach, D., 1991, Premature initiation of mitosis in yeast lacking RCC1 or an interacting GTPase, *Cell 66*:347.

Matsumoto, T., and Beach, D., 1993, Interaction of the pim1/spi1 mitotic checkpoint with a protein phosphatase, *Mol. Biol. Cell 4*:337.

McCormick, F., 1992, Coupling of ras p21 signalling and GTP hydrolysis by GTPase activating proteins, *Phil. Trans. R. Soc. Lond. B 336*:43.

Millar, J.B.A., and Russell, P., 1992, The cdc25 M-phase inducer: an unconventional protein phosphatase, *Cell 68*: 279.

Moore, M.S., and Blobel, G., 1992, The two steps of nuclear import, targeting to the nuclear envelope and translocation through the nuclear pore, require different cytosolic factors, *Cell 69*:939.

Nishimoto, T., Eilen, E., and Basilico, C., 1978, Premature chromosome condensation in a ts DNA- mutant of BHK cells, *Cell 15*:475.

Nishitani, H., Kobayashi, H., Ohtsubo, M., and Nishimoto, T., 1990, Cloning of *Xenopus* RCC1 cDNA, a homolog of the human *RCC1* gene: complementation of tsBN2 mutation and identification of the product, *J. Biochem. 107*:228.

Nishitani, H., Ohtsubo, M., Yamashita, K., Iida, H., Pines, J., Yasudo, H., Shibata, Y., Hunter, T., and Nishimoto, T., 1991, Loss of RCC1, a nuclear DNA binding protein, uncouples the completion of DNA replication from the activation of cdc2 protein kinase and mitosis, *EMBO J. 10*:1555.

Norbury, C., and Nurse, P., 1992, Animal cell cycles and their control, *Annu. Rev. Biochem. 61*:441.

Ohtsubo, M., Kai, R., Furuno, T., Sekiguchi, T., Sekiguchi, M., Hayashida, H., Kuma, K., Miyata, T., Fukushige, S., Murotu, T., Matsubara, K., and Nishimoto, T., 1987, Isolation and characterization of the active cDNA of the human cell cycle gene (*RCC1*) involved in the regulation of the onset of chromosome condensation, *Genes Devel. 1*:585.

Ohtsubo, M., Okazaki, H., and Nishimoto, T., 1989, The RCC1 protein, a regulator for the onset of chromosome condensation locates in the nucleus and binds to DNA, *J. Cell. Biol. 109*:1389.

Ren, M., Drivas, G., D'Eustachio, P., and Rush, M.G., 1993, Ran/TC4: a small nuclear GTP-binding protein that regulates DNA synthesis, *J. Cell. Biol.* 120:313.

Roberge, M., 1992, Checkpoint controls that couple mitosis to completion of DNA replication, *Trends Cell Bio.* 2:177.

Seino, H., Nishitani, H., Seki, T., Hisamoto, N., Tazunoki, T., Shiraki, N., Ohtsubo, M., Yamashita, K.,

Sekiguchi, T., and Nishimoto, T., 1991, RCC1 is a nuclear protein required for coupling activation of cdc2 kinase with DNA synthesis and for Start of the cell cycle, *CSH Symp. Quant. Biol. 56*:367.
Seki, T., Yamashita, K., Nishitani, H., Takagi, T., Russell, P., and Nishimoto, T., 1992, Chromosome condensation caused by loss of RCC1 function requires the cdc25C protein that is located in the cytoplasm, *Mol. Biol. Cell 3*:1373.
Smythe, C., and Newport, J.W., 1992, Coupling of mitosis to the completion of S phase in Xenopus occurs via modulation of the tyrosine kinase that phosphorylates $p34^{cdc2}$, *Cell 68*:787.
Stapleton, G., Somma, M.P., and Lavia, P., 1993, Cell type-specific interactions of transcription factors with a housekeeping promoter *in vivo*, *Nucleic Acids Res. 21*:2465.
Takai, Y., Kaibuchi, K., Kikuchi, A., and Kawata, M., 1992, Small GTP-binding proteins, *Int. Rev. Cytol. 133*:187.
Uchida, S., Sekiguchi, T., Nishitani, H., Miyauchi, K., Ohtsubo, M., and Nishimoto, T., 1990, Premature chromosome condensation is induced by a point mutation in the hamster RCC1 gene, *Mol. Cell. Biol. 10*:577.

12

SPHINGOLIPIDS METABOLITES: A NEW CLASS OF SECOND MESSENGERS IN THE REGULATION OF CELL GROWTH

Sarah Spiegel

Department of Biochemistry and Molecular Biology
Georgetown University Medical Center
Washington, DC 20007

INTRODUCTION

The interaction of growth factors with specific cell surface receptors triggers multiple intracellular signaling pathways that culminate in DNA synthesis and cell division.[1,2] Growth signaling networks in which glycerophospholipid metabolites, such as diacylglycerol, inositol 1,4,5-trisphosphate ($InsP_3$), phosphatidic acid, and arachidonic acid, serve as second messengers have been well characterized.[3-5] Much less is known of the second messengers derived from another major class of membrane lipids, the sphingolipids. All sphingolipids, including ceramide, sphingomyelin, cerebrosides, gangliosides, and sulfatides, contain (1) a long-chain sphingoid base as their backbone, of which sphingosine is the most prominent, (2) an amide-linked fatty acid, and (3) a polar head group (hydroxyl for ceramide, phosphorylcholine for sphingomyelin, and carbohydrate residues of varying complexity for glycosphingolipids). These ubiquitous cellular components have long been known to play an important, yet undefined, role in cell growth regulation.[6-8]

A tantalizing link between sphingolipids and signal transduction has emerged from the discovery that sphingosine, a metabolite of cellular sphingolipids, inhibits protein kinase C (PKC) *in vitro* and diverse PKC-dependent processes *in vivo*.[9-11] Thus, sphingosine has been proposed to function as an endogenous inhibitor of PKC, a pivotal regulatory enzyme in cell growth and transformation, and, consequently, to oppose the action of the PKC activator diacylglycerol.[9-11] Although a wealth of information exists that sphingosine has diverse biological functions that

The Cell Cycle: Regulators, Targets, and Clinical Applications
Edited by V.W. Hu, Plenum Press, New York, 1994

are related to its effect on PKC, my collaborators and I, as well as others, have discovered a more versatile role of sphingosine, which does not appear to rely soley on its effect on PKC.[11-15]

SPHINGOLIPID METABOLITES AND CELLULAR PROLIFERATION

We have shown that low concentrations of sphingosine stimulate the proliferation of quiescent Swiss 3T3 fibroblasts as well as potentiate the mitogenic responses to other growth factors via a PKC-independent pathway.[13] In contrast to its effect on PKC, the mitogenic effect of sphingosine is very specific: Only the naturally occurring D-*erthyro* isomer of sphingosine is active; structurally related analogs of sphingosine or L-*threo* stereoisomers do not mimic the mitogenic action of sphingosine. These results suggest that this breakdown product of cellular sphingolipids may play an important role as a positive modulator of cell growth in a PKC-independent pathway and that additional targets of sphingosine action remain to be uncovered.

In recent years, an extensive effort has been targeted to identifying the biochemical pathways affected by sphingosine. This search has led us to explore the effects of sphingosine on many of the intracellular signaling molecules thought to play a role in proliferation, including phosphatidic acid, Ca^{2+}, $InsP_3$, and adenosine 3',5'-monophosphate (cyclic AMP), and to investigate sphingosine metabolism.[8,12,14-16] There has been a recent surge of information regarding sphingosine-mediated regulation of phosphatidic acid metabolism. We have shown that the mitogenic effect of sphingosine is accompanied by an increase in the amount of phosphatidic acid.[12] Sphingosine stimulates phospholipase D (PLD) in several cell types,[12,17,18] including Swiss 3T3 fibroblasts, activates the 80-kD form of diacylglycerol kinase *in vitro*,[19] and inhibits phosphatidic acid phosphohydrolase.[20-23] Thus, in a wide variety of cells with different pathways for formation of phosphatidic acid, sphingosine uniformly increases phosphatidic acid concentrations. Furthermore, similar to the actions of phosphatidic acid on signal transduction in Swiss 3T3 cells, mitogenic concentrations of sphingosine also inhibit cyclic AMP accumulation and trigger the hydrolysis of inositol phospholipids, resulting in an increase in $InsP_3$ concentrations.[12,24,25]

Phosphatidic acid has been implicated in growth regulation in a variety of cell types, including Swiss 3T3 fibroblasts.[26,27] Recently, it has been shown that certain growth factors stimulate the production of phosphatidic acid by receptor-coupled activation of PLD, and it has been suggested that an increase in the concentration of phosphatidic acid is essential in mitogenic signal transduction cascades.[5,28] Perhaps most importantly, phosphatidic acid has been shown to regulate the biological action of cellular Ras activity.[29,30]

Cellular Ras protein plays a central role as a molecular switch in the early steps of signal transduction pathways associated with cell growth, differentiation, and neoplasia.[31] When the protein is in the guanosine triphosphate (GTP)-complexed form, it is active in signal transduction pathways, whereas it is inactive in its guanosine diphosphate (GDP)-complexed form. Activation of Ras is an initial and key

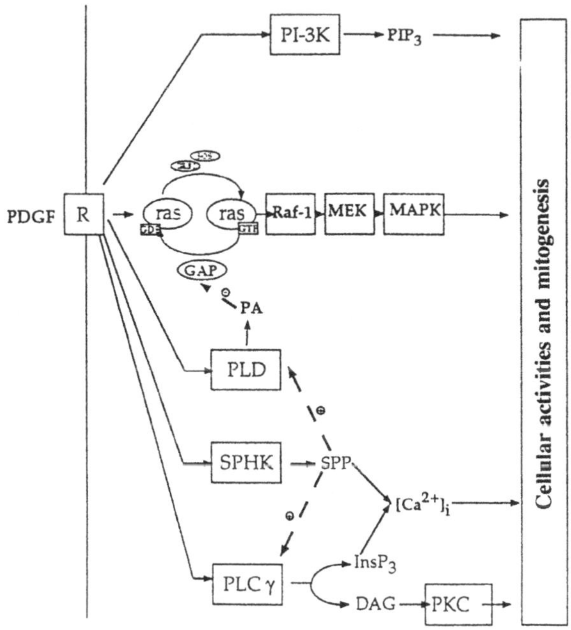

Figure 1. The major growth factor signaling pathways and possible sites of action of sphingolipid metabolites. Tyrosine kinase-linked receptors (R) for growth factors (GF) activate many effector systems, including: (1) phosphatidylinositol 3-hydroxy kinase (PI3K), which catalyzes the phosphorylation of phosphatidylinositol to phosphatidylinositol 3-phosphate (PI3P); (2)PLD, which catalyzes the hydrolysis of membrane phospholipids to phospatidic acid (PA); (3) Phospholipase C-γ (PLC), which catalyzes the hydrolysis of phosphatidylinositol 4,5-biphosphate to InsP3 and diacylglycerol (DAG); and (4) Sphingosine kinase (SPHK), which catalyzes the phosphorylation of sphingosine to SPP. Tyrosine-kinase receptors also recruit the key effector systems of GAP and the adaptor complex GRB2-SOS, which regulate ras. Ras activation triggers a kinase cascade that includes Raf-1, MEK, and MAP kinase (MAPK), finally culminating in the phosphorylation of transcription factors (such as Jun and Fos) and other factors that bring about the cellular response. For more detailed information, see references[51-53]. Dashed arrows indicate influences on pathways. Stimulatory and inhibitory effects are indicated by (+) and (-), respectively.

Raf-1 kinase, and the various components of the mitogen-activated protein (MAP) kinase cascade and culminates in the phosphorylation and activation of S6 kinase and transcription factors implicated in cell growth (Figure 1).[32]

The function of Ras in mammalian cells depends on its interaction with a cytoplasmic GTPase activating protein (GAP), which stimulates the GTPase activity of Ras and thereby accumulation of the inactive form.[33] Recently, phosphatidic acid has been shown to bind and inhibit Ras GAP and homologous proteins.[29,30,34] In addition, phosphatidic acid stimulates a cytoplasmic Ras GTPase inhibitory protein.[30] The combined effect of these actions of phosphatidic acid is an inhibition of the intrinsic GTPase activity of Ras and thereby stabilization of the active form. Thus, phosphatidic acid has been proposed to positively modulate Ras activity during cellular proliferation.[29,30] Regulation of the GTPase activity of small GTP-binding proteins such as Ras could represent a convergence point where signals carried by lipid-derived messengers are integrated. In view of the prominent role of phosphatidic acid in signal transduction, activation of Ras, and cellular proliferation, the connection between sphingosine and phosphatidic acid may have important implications for the biological actions of sphingolipid metabolites.

SPHINGOSINE 1-PHOSPHATE MAY MEDIATE THE ACTIONS OF SPHINGOSINE

Recent studies from our lab,[14] have provided new insights into the formation and function of a metabolite of sphingosine, sphingosine 1-phosphate (SPP). We showed that mitogenic concentrations of sphingosine induced an increase in SPP concentration that preceded the increase in phosphatidic acid. SPP itself induced a more rapid increase in phosphatidic acid, suggesting that it may mediate the effects of sphingosine on phosphatidic acid accumulation.[12] Indeed, SPP also stimulated the activity of PLD more potently than sphingosine.[15] Furthermore, we have accumulated substantial evidence suggesting that SPP mediates the mitogenic effect of sphingosine. SPP stimulated DNA synthesis to a greater extent than sphingosine and required much lower concentrations for the maximum response.[14] Although both sphingosine and SPP acted synergistically with a wide variety of growth factors, there was no additive or synergistic response to a combination of sphingosine and SPP, indicating that both compounds modulate cellular proliferation through a common pathway. Furthermore, inhibition of SPP production markedly inhibited the sphingosine-induced increase in DNA synthesis.

Sphingosine has also been shown to induce a rapid and marked translocation of Ca^{2+} from intracellular stores in permeabilized smooth muscle cells, and it was proposed that this effect is mediated via the conversion of sphingosine to SPP.[35] However, direct experimental evidence that SPP itself can induce release of Ca^{2+} from internal sources was not presented. With a digital imaging fluorescence system for measurement of cytosolic free Ca^{2+}, we observed that SPP is a very potent Ca^{2+}-mobilizing agonist in viable 3T3 fibroblasts.[14] The temperature dependence of the effect of sphingosine on cytosolic Ca^{2+} concentration, the longer lag time for its action relative to that for SPP, the fact that SPP is more potent than sphingosine, and the observation that inhibition of the conversion of sphingosine to SPP prevents sphingosine-induced Ca^{2+} release, collectively support the suggestion that sphingosine requires enzymatic conversion to SPP for its function.[14,35] Although SPP, like sphingosine, increased $InsP_3$ concentrations in Swiss 3T3 Fibroblasts, complete inhibition of inositol phosphate formation by prior treatment of cells with 12-O-tetradecanoyl phorbol-13-acetate did not block SPP-mediated Ca^{2+} responses indicating that SPP can induce the release of Ca^{2+} without formation of $InsP_3$.[36] Our recent results suggest that SPP mobilizes Ca^{2+} primarily via a previously undescribed mechanism that is independent of Ca^{2+} influx, inositol phospholipid hydrolysis, and arachidonic acid release.[36] Because cells can regulate their concentrations of sphingosine[11,37] and SPP,[38] our data suggest that sphingolipid turnover may be important in the regulation of Ca^{2+} homeostasis.

SPP is produced from sphingosine by the action of a specific kinase, which catalyzes the phosphorylation of the primary hydroxyl group of sphingosine to produce SPP in the major sphingolipid degradation pathway.[39,40] SPP is rapidly degraded to *trans*-2-hexadecanal and phosphoethanolamine by the action of a microsomal lyase,[41] reported to be located in the endoplasmic reticulum.[42] Thus, we propose that SPP is a suitable candidate for an intracellular second messenger because it can be rapidly synthesized and degraded. As mentioned above, our recent results are consistent with such an hypothesis. SPP is rapidly produced in the presence of mitogenic concentrations of sphingosine, is a more potent mitogen than sphingosine for 3T3 fibroblasts, and triggers the dual signal transduction pathways of Ca^{2+} mobilization and PLD activation, which are prominent events in the control of cellular proliferation. Thus, SPP may be an important component of the intracellular second messenger system that contributes to Ca^{2+} release and the regulation of cell growth.

If SPP functions *in vivo* as a mediator of cell growth, then its concentrations should be low in quiescent cells, tightly regulated, and increased by growth-promoting agents. Indeed, in quiescent Swiss 3T3 fibroblasts labeled with [^{3}H]serine, a precursor of cellular sphingolipids, concentrations of SPP are very low, and are rapidly and transiently increased in response to the potent mitogens, platelet-derived growth factor (PDGF) and fetal bovine serum (FBS).[38] The SPP response is specific for certain mitogens, because epidermal growth factor (EGF), which also stimulates proliferation of 3T3 cells, did not induce significant changes in SPP concentration. [^{3}H]Sphingosine also increased in the same time period in response to FBS and PDGF. In contrast, the concentration of [^{3}H]sphingomyelin, which contains a much greater proportion of the total [^{3}H]serine incorporated into lipids (15.8%) than either free sphingoid bases (0.32%) or SPP (0.13%), was not significantly altered. The rapid and transient increases in concentrationss of intracellular sphingosine and SPP in response to PDGF and FBS may have important implications for their biological roles in growth factor-activated signal transduction pathways.

To determine whether a growth factor-induced increase in intracellular sphingosine is the sole regulatory factor influencing SPP concentrations we examined the effects of various growth factors on the activity of sphingosine kinase, which has been suggested to catalyze the rate-limiting step in the metabolism of sphingosine. FBS and PDGF induced a transient increase in sphingosine kinase activity, as measured by ^{32}P incorporation into sphingosine, that reached a maximum within 15 min and returned to basal values after 1 hour. In contrast, other mitogens, including bombesin, bradykinin, insulin, and EGF, had little or no effect. Even a combination of insulin and EGF, which induced marked proliferation of quiescent 3T3 fibroblasts, did not activate sphingosine kinase, indicating that the effect is restricted to specific growth-promoting agents.

To substantiate further a potential role for SPP in cellular proliferation, we examined the effects of a competitive inhibitor of sphingosine kinase, DL-*threo*-dihydrosphingosine, that has been shown to inhibit the production of SPP in isolated platelets.[40] In intact Swiss 3T3 fibroblasts, DL-*threo*-dihydrosphingosine inhibited by 60% the production of SPP induced by a mitogenic concentration of sphingosine and also inhibited the initiation of DNA synthesis in response to sphingosine by more than 55%, supporting our notion that SPP mediates the mitogenic effects of sphingosine. DNA synthesis induced by PDGF and FBS was also markedly inhibited.

The mechanisms by which eukaryotic cells regulate proliferation remain poorly understood. As knowledge of these mechanisms has increased, it has become apparent that not all of the relevant intracellular second messengers have been identified. Numerous studies have demonstrated that inositol phospholipid hydrolysis is not required for growth factor-induced mitogenesis[43-45] and that activation of PLD rather than phospholipase C correlates with mitogenesis induced by PDGF.[46] Furthermore, PDGF may stimulate the formation of an intracellular second messenger other than $InsP_3$ that is capable of releasing Ca^{2+} from intracellular sources, because this growth factor has virtually no effect on $InsP_3$ concentration in Swiss 3T3 fibroblasts.[47]

Our recent data provide the first clues to the identity of the possible missing link between the plasma membrane (where the growth factor receptors lie), the intracellular Ca^{2+} stores, and cell growth. Thus, we propose that SPP has appropriate properties to function as an intracellular second messenger: (1) It elicits diverse cellular responses;[14,15,48] (2) it is rapidly produced from sphingosine by the action of a specific kinase;[39,49,50] (3) its turnover is rapid, mediated by an endoplasmic reticulum lyase;[41,42] (4) its concentration is low in quiescent cells but increases rapidly and transiently in response to FBS and PDGF; (5) it releases Ca^{2+} from internal sources in an $InsP_3$-independent manner;[14,35] and (6) SPP has the potential to act in a positive feedback loop, to amplify the cascade of events that follow receptor stimulation, via its effect on phosphatidic acid concentrations,[15] Thus linking growth factor signaling to cellular Ras activity.[30]

Although the existence of a sphingolipid cycle that can act as a signal-transducing element, similar to phospholipid cycles, is an intriguing hypothesis, many questions remain open. Are sphingosine and SPP the actual bioactive molecules for this cycle, or are other sphingolipid metabolites involved? Does the cycle function similarly in all tissues and cells? Are different routes activated within the cycle, and, if so, are different messengers generated depending on the cell type or stimulus? Increasing knowledge of the activity of the products of sphingolipid turnover could add new outlooks toward the understanding of the regulation of cell growth.

ACKNOWLEDGMENTS

The author gratefully acknowledged the contributions of his collaborators especially Drs. Ana Olivera and Hong Zhang. Work by the author's laboratory cited in this review was supported by research grants 1RO1 GM43880 from the National Institutes of Health, and 3018M from the Council for Tobacco Research.

REFERENCES

1. S.A. Aaronson, Growth factors and cancer, *Science* 254:1146 (1991).
2. A. Ullrich and J. Schlessinger, Signal transduction by receptors with tyrosine kinase activity, *Cell* 61:203 (1990).
3. M.J. Berridge, Inositol trisphosphates and calcium signalling, *Nature* 361:315 (1993).
4. Y. Nishizuka, Intracellular signaling by hydrolysis of phospholipids and activation of protein kinase C, *Science* 258:607 (1992).
5. J.H. Exton, Signaling through phosphatidylcholine breakdown, *J. Biol. Chem.* 265:1 (1990).
6. S. Hakomori, Bifunctional role of glycosphingolipids. Modulators for transmembrane signaling and mediators for cellular interactions, *J. Biol. Chem.* 265:18713 (1990).
7. R.K. Yu, Gangliosides: structure and function, in: "Ganglioside Structure, Function, and Biomedical Potential," R. Ledeen, R.K. Yu, M.M. Rapport, and K. Suzuki, Eds. Plenum, New York, pp. 39-45 (1984).
8. A. Olivera and S. Spiegel, Ganglioside GM1 and sphingolipid breakdown products in cellular proliferation and signal transduction pathways, *Glycoconjugate J.* 9:109 (1992).
9. Y.A. Hannun and R.M. Bell, Lysosphingolipids inhibit protein kinase C: Implcations for the sphingolipidoses, *Science* 235:670 (1987).
10. Y.A. Hannun and R.M. Bell, Functions of sphingolipids and sphingolipid breakdown products in cellular regulation, *Science* 243:500 (1989).
11. A.H. Merrill, Cell regulation by sphingosine and more complex sphingolipids, J. Bioenerg. Biomembr. 23:83 (1991).
12. H. Zhang, N.N. Desai, J.M. Murphey and S. Spiegel, Increases in phosphatidic acid levels accompany sphingosine-stimulated proliferation of quiescent Swiss 3T3 cells, *J. Biol. Chem.* 265:21309 (1990).

13. H. Zhang, N.E. Buckley, K. Gibson and S. Spiegel, Sphingosine stimulates cellular proliferation via a protein kinase C-independent pathway, *J. Biol. Chem.* 265:76 (1990).
14. H. Zhang, N.N. Desai, A. Olivera, T. Seki, G. Brooker and S. Spiegel, Sphingosine-1-phosphate, a novel lipid, involved in cellular proliferation, *J. Cell Biol.* 114:155 (1991).
15. N.N. Desai, H. Zhang, A. Olivera, M.E. Mattie and S. Spiegel, Sphingosine-1-phosphate, a metabolite of sphingosine, increases phosphatidic acid levels by phospholipase D activation, *J. Biol. Chem.* 267:23122 (1992).
16. S. Spiegel, A. Olivera and R.O. Carlson, The role of sphingosine in cell growth and transmembrane signaling. in "Advances in Lipid Research: Sphingolipids in Signaling, Part A", R. M. Bell, Y. A. Hannun and A. H. Merrill. Eds., Academic Press Inc., Orlando. p105 (1993).
17. Z. Kiss and W.B. Anderson, ATP stimulates the hydrolysis of phosphatidylethanolamine in NIH 3T3 cells. Potentiating effects of guanosine triphosphates and sphingosine, *J. Biol. Chem.* 265:7188 (1990).
18. Y. Lavie and M. Liscovitch, Activation of phospholipase D by sphingoid bases in NG108-15 neural-derived cells, Biochem. Biophys. Res. Commun. 167:607 (1990).
19. F. Sakane, K. Yamada and H. Kanoh, Different effects of sphingosine, R59022 and anionic amphiphiles on two diacylglycerol kinase isozymes purified from porcine thymus cytosol, *Eur. J. Pediatr.* 149:31 (1989).
20. Y. Lavie, O. Piterman and M. Liscovitch, Inhibition of phosphatidic acid phosphohydrolase activity by sphingosine: Dual action of sphingosine in diacylglycerol signal termination, *FEBS Lett.* 277:7 (1990).
21. T.J. Mullmann, M.I. Siegel, R.W. Egan and M.M. Billah, Sphingosine inhibits phosphatidate phosphohydrolase in human neutrophils by a protein kinase C-independent mechanism, *J. Biol. Chem.* 266:2013 (1991).
22. Z. Jamal, A. Martin, A.G. Munoz and D.N. Brindley, Plasma membrane fractions from rat liver contain a phosphatidate phosphohydrolase distinct from that in the endoplasmic reticulum and cytosol, *J. Biol. Chem.* 266:2988 (1991).
23. D.K. Perry, W.L. Hand, D.E. Edondson and J.D. Lambeth, Role of phospholipase D-derived diradylglycerol in activation of the human neutrophil respiratory burst oxidase, Inhibition by phosphatidic acid phosphoydrolase inhibitors. *J. Immunol.* 149:2749 (1992).
24. J.A. Johnson and R.B. Clark, Multiple non-specific effects of sphingosine on adenylate cyclase and cyclic AMP accumulation in S49 lymphoma cells preclude its use as a specific inhibitor of protein kinase C, *J. Biol. Chem* 265:9333 (1990).
25. M. Wahl and G. Carpenter, Regulation of the epidermal growth factor-stimulated formation of inositol phosphates in A-431 cells by calcium and protein kinase C, *J. Biol. Chem* 263:7581 (1988).
26. W.H. Moolenaar, W. Krujer, B.C. Tilly, I. Verlaan, A.J. Bierman and S.W. deLaat, Growth factor-like action of phospatidic acid, *Nature* 323:171 (1986).
27. C. Yu, M. Tsai and D.W. Stacey, Cellular ras activity and phospholipid metabolism, *Cell* 52:63 (1988).
28. P. Ben-Av and M. Liscovitch. Phospholipase D activation by the mitogens platelet-derived growth factor and 12-O-tetradecanoyl phorbol 13-acetate in NIH-3T3 cells. *FEBS Lett.* 259:64-66 (1989).
29. M.H. Tsai, C.L. Yu, F.S. Wei and D.W. Stacey, The effect of GTPase activating protein upon ras is inhibited by mitogenically responsive lipids, *Science* 243:522 (1989).
30. M. Tsai, C. Yu and D.W. Stacey, A cytoplasmic protein Inhibits the GTPase activity of H-Ras in a phospholipid-dependent manner, *Science* 250:982 (1990).
31. M. Barbacid, Ras genes, *Annu. Rev. Biochem.* 56:779 (1987).
32. S.L. Pelech and J.S. Sanghera, MAP kinases: charting the regulatory pathways, *Science* 257:1355 (1992).
33. M. Trahey and F. McCormick, A cytoplasmic protein stimulates normal N-ras p21GTPase, but does not affect oncogenic mutants, *Science* 238:524(1987).

34. M. Tsai, M. Roudebush, S. Dobrowolski, C. Yu, J. B. Gibbs and D. W. Stacey, Ras GTPase-activating protein physically associates with mitogenically active phospholipids, *Mol. Cell. Biol.* 11:2785(1991).

35. T.K. Ghosh, J. Bian and D.L. Gill. Intracellular calcium release mediated by sphingosine derivatives generated in cells, *Science* 248:1653(1990).

36. M. E. Mattie, G. Brooker and S. Spiegel Sphingosine-1-phosphate, a putative second messenger, mobilizes calcium from internal stores via an inositol trisphosphate-independent pathway. *J. Biol. Chem.*, Submitted (1993).

37. E. Wilson, E. Wang, R.E. Mullins, D.J. Uhlinger, D.C. Liotta, J.D. Lambeth and A.H. Merrill, Modulation of free sphingosine levels in human neutrophils by phorbol esters and other factors *J. Biol. Chem.* 263:9304 (1988).

38. A. Olivera and S. Spiegel, Sphingosine-1-phosphate, a novel second messenger involved in PDGF- and serum-induced cellular proliferation, *Nature,* in press, (1993)

39. W. Stoffel, B. Hellenbroich and G. Heimann, Properties and specificities of sphingosine kinase from blood platelets, *Hoppe-Seyler's Z. Physiol. Chem.* 354:1311(1973).

40. B.M. Buehrer and R.M. Bell, Inhibition of sphingosine kinase in vitro and in platelets. Implications for signal transduction pathways, *J. Biol. Chem.* 267:3154 (1992).

41. W. Stoffel and G. Assmann, Metabolism of sphingoid bases, XV. Enzymatic degradation of 4t-sphingenine 1-phosphate (sphingosine-1-phosphate) to 2t-hexadecan-1-al and ethanolamine phosphate, *Hoppe-Seyler's Z. Physiol. Chem.* 351:1041 (1970).

42. P.P. Van Veldhoven and G.P. Mannaerts, Subcellular localization and membrane topology of sphingosine-1-phosphate lyase in rat liver, *J. Biol. Chem.* 266:12502 (1991).

43. L.T. Williams, Signal transduction by the platelet-derived growth factor receptor. *Science* 243:1564 (1989).

44. K.G. Peters, J. Marie, E. Wilson, H.E. Ives, J. Escobedo, M.D. Rosario, D. Mirda and L.T. Williams, Point mutation of an FGF receptor abolishes phosphatidylinositol turnover and Ca^{2+} flux but not mitogenesis, *Nature* 358:678 (1992).

45. B. Margolis, A. Zilberstein, C. Franks, S. Felder, S. Kremer, A. Ullrich, S.G. Rhee, K. Skorecki and J. Schlessinger, Effect of phospholipase C-γ overexpression on PDGF-induced second messengers and mitogenesis, *Science* 247:607 (1990).

46. K. Fukami and T. Takenawa, Phosphatidic acid that accumulates in platelet-derived growth factor Balb/c 3T3 cells is a potential mitogenic signal, *J. Biol. Chem.* 267:10988 (1992).

47. A. Lopez-Rivas, S.A. Mendoza, E. Nanherg, J. Sinnett-Smith and E. Rozengurt, Ca^{2+}-mobilizing actions of platelet-derived growth factor differ from those of bombesin and vasopressin in Swiss 3T3 mouse cells,*Proc. Natl. Acad. Sci. USA* 84:5768 (1987).

48. Y. Sadahira, F. Ruan, S. Hakomori and Y. Igarashi, Sphingosine 1-phosphate, a specific endogenous signaling molecule controlling motility and tumor cell invasivenes, *Proc. Natl. Acad. Sci. USA* 89:9686 (1992).

49. D.D. Louie, A. Kisic and G.J. Schroepfer, Sphingolipid base metabolism. Partial purification and properties of sphinganine kinase of brain, *J. Biol. Chem.* 251:4557 (1976).

50. C.B. Hirschberg, A. Kisic and G.J. Schroepfer, Enzymatic formation of dihydrosphingosine 1-phosphate, *J. Biol. Chem.* 245:3084 (1970).

51. N.W. Gale, S. Kaplan, E.J. Lowenstein, J. Schlessinger and D. Bar-Sagi, Grb2 mediates the EGF-dependent activation of guanine nucleotide exchange on Ras, *Nature* 363:88 (1993).

52. S.E. Egan, D.W. Giddings, M.W. Brooks, L. Buday, A.M. Sizeland and R.A. Weinberg, Association of Sos Ras exchange protein with Grb2 is implicated in tyrosine kinase signal transduction and transformation, *Nature* 363:45 (1993).

53. M. Rozakis-Adcock, R. Fernley, J. Wade, T. Pawson and D. Bowtell, The SH2 and SH3 domains of mammalian Grb2 couple the EGF receptor to the Ras activator mSos1, *Nature* 363:83 (1993).

PART III

CELL CYCLE CONTROL OF GENE EXPRESSION AND REPAIR

13

HISTONE GENE TRANSCRIPTION DURING THE CELL CYCLE

Franca LaBella, Rosanna Martinelli, Neil Segil and Nathaniel Heintz

Howard Hughes Medical Institute
Laboratory of Molecular Biology
The Rockefeller University
New York, New York

INTRODUCTION

It has become evident during the last decade that periodic transcription of a wide variety of mRNAs is a critical component of the eucaryotic cell cycle. The first genes to be identified whose transcription is modulated during cell cycle progression were the histone genes (1). In vertebrates, the histone genes comprise a small gene family in which there are approximately a dozen genes encoding each of the four core histones (H4, H3, H2a, H2b), and approximately half that number encoding the linker histone H1 (2). Since histones are essential structural proteins, they are transcribed in most eucaryotic cells. However, in actively dividing cells their transcription is very tightly coupled to S phase of the cell cycle. Thus, as cells enter S phase, histone gene transcription is increased approximately ten fold, and as the cell exits S phase the rate of histone gene transcription returns to its pre-S phase levels. This tight coupling of histone gene transcription with DNA replication provides an opportunity to dissect molecular mechanisms of transcription that are responsive to specific cues during traverse through the cell cycle, and that may be essential for orderly progression toward division. The interest of our laboratory has been to define these transcriptional regulatory mechanisms, and to relate them to other critical events that are important for cell cycle regulation.

Analysis of DNA sequences controlling histone gene transcription in animal cells has established that genes encoding a given histone are controlled by subtype specific consensus elements (3,4,5). For example, Figure 1 shows the structures of the histone H4, H2b and H1 promoters and the sequences that are critical for S phase specific transcription. In each case, there are specific sequences that are essential for S phase specific transcription and are shared between the dozen genes encoding that subtype, although they are not found in the other histone genes. These sequences each interact with distinct transcription factors that show no obvious biochemical similarities other than their binding to S phase specific regulatory elements in histone gene promoters. These observations prompt two key questions concerning histone gene cell cycle regulation: what is the mechanism that couples the activity of these transcription factors to S phase control in the context of histone gene promoters and what is the mechanism responsible for coordinate regulation of these factors as cell traverse the cell cycle?

The Cell Cycle: Regulators, Targets, and Clinical Applications
Edited by V.W. Hu, Plenum Press, New York, 1994

To address these issues, we have begun an in depth characterization of the transcription factors Oct1 and H1TF2 and their regulation during the cell cycle. Our recent studies have demonstrated that Oct1 is regulated by a complex program of phosphorylation during the HeLa cell cycle (6) that culminates in its inactivation by phosphorylation of the POU-homeodomain during mitosis (7). Although these direct posttranslational modifications to Oct1 late during the cell cycle could explain histone H2b transcriptional repression as cells exit S phase and complete the cell cycle, we have not yet identified such modifications as candidate regulatory steps in activation of histone gene transcription as cells enter S-phase. The lack of a posttranslational modification to Oct1 that could explain histone H2b activation has led to the search for an Oct1 interacting protein that might modulate its function during the cell cycle as a possible candidate for a coordinate regulator of histone gene transcription (5). Similar studies of the histone H1 transcription factor H1TF2 are also being pursued to identify such a molecule. We describe here our progress toward that goal.

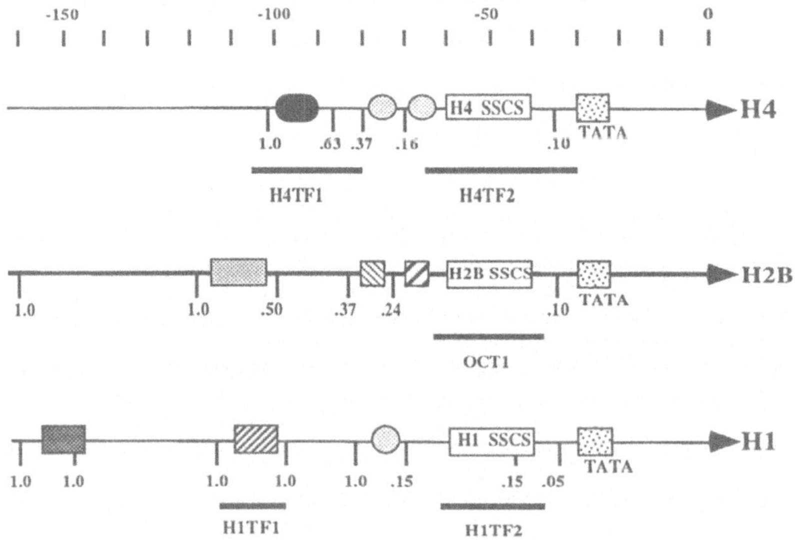

RESULTS AND DISCUSSION

Oct1. To assess the status of Oct1 within the cell, a series of biochemical and immunochemical analyses have been performed. The major point to be made from this work is that there are multiple forms of Oct1 within the cell, and that these forms are found in different macromolecular complexes. For example, gel filtration experiments on crude HeLa cell nuclear extracts demonstrate that Oct1 elutes in two positions: the first corresponding to monomeric Oct1 at a native molecular weight of approximately 90kd, and a faster migrating complex of approximately 200kd native molecular weight. Electrophoretic mobility shift assays reveal that the high molecular weight peak of Oct1 activity can specifically bind DNA, and that this complex can be dissociated into free Oct1 and as yet unknown Oct1 associated proteins. Detection of the Oct1 species present in these different complexes using monoclonal antibodies suggests that the form of Oct1 present in these high molecular weight complexes is different from that found in the monomeric Oct1 peak because they migrate at slightly different positions in denaturing polyacrylamide gels. The nature of this difference is not clear, although obvious possibilities for heterogeneity include differences in the state of phosphorylation or glycosylation and, possibly, alternative splicing.

Although these experiments have not yet fully matured, they raise the interesting possibility that distinct forms of Oct1 are functionally segregated by interaction with important cofactors. It seems likely that these factors could be important in regulating the activity of Oct1 during the cell cycle, or in targeting specific Oct1 species to their sites of action. This type of diversity in Oct1 species in the cell might, for example, explain the distinct roles of Oct1 as an important component of histone gene transcriptional regulation versus its role in snRNA transcription (8), which is not known to be cell cycle regulated. Identification of the molecules complexed with Oct1, analysis of their functional consequences for Oct1 transcriptional activation and identification of the Oct1 species found in these complexes are our current goals.

H1TF2. S-phase specific transcription of histone H1 genes involves two highly conserved sequences within the promoter (Figure 1). The proximal H1 subtype specific consensus element contains a CCAAT box and binds to a novel heterodimeric CCAAT binding protein. We have recently cloned the 43kd subunit of this factor (H1TF2A) and shown it to be a glutamine rich polypeptide with very limited sequence similarity to other CCAAT binding proteins (9). To assess whether H1TF2 is coregulated with Oct1 late during the cell cycle, a series of immunoprecipitation studies were performed that demonstrate that H1TF2A is a phosphoprotein *in vivo*. In contrast to Oct1, no obvious increase in H1TF2A phosphorylation during the cell cycle was observed. However, several proteins that coimmunoprecipitate with H1TF2A are heavily phosphorylated during mitosis, suggesting that they might contribute to H1TF2 regulation. Our present efforts are to continue detailed investigations into the regulation of H1TF2 and its associated proteins during the cell cycle. It is hoped that these efforts will lead both to an understanding of histone H1 transcriptional control, and possibly to the identification of common factors regulating histone H2b and H1 transcription during the cell cycle.

REFERENCES

1. Osley, M.A. (1991). Ann. Rev. Biochem. **60**:827-861.
2. Wells, J. R. E. (1986). Compilation analysis of histones and histone genes. Nucleic Acids Res. **14**:119-149.
3. LaBella, F., Sive, H.L., Roeder, R.G. and N. Heintz (1988). Cell-cycle regulation of a human histone gene H2B gene is mediated by the H2B subtype-specific DNA consensus element. Genes and Devel. **2**, 32.
4. Dalton, S. and J. R. E. Wells (1985). A gene specific promoter element is required for optimal expression of the histone H1 gene in S-phase. EMBO J. **7**:49-56.
5. LaBella, F., Gallinari, P., McKinney, J., and N. Heintz (1990). Histone H1 subtype-specific consensus elements mediate cell cycle-regulated transcription *in vitro* . Genes and Development **3**, 1982.
6. Roberts, S. B., Segil, N. and N. Heintz (1991). Oct1 (OTF1) is differentially phosphorylated during the cell cycle. Science **253**:1022-1026.
7. Segil, N., Roberts, S. B. and N. Heintz (1991). Mitotic phosphorylation of the Oct1 POU-homeodomain and regulation of DNA binding. Science **254**, 1814.
8. Carbon, P., Murgo, S., Ebel, J.-P., Krol, A., Tewbb, G. and I. W. Mattaj (1987). A common octamer motif binding protein is involved in the transcription of U6 snRNA by RNA polymerase III and U2 snRNA by RNA polymerase II. Cell **51**:71-79.
9. Martinelli, R. and N. Heintz (1993). H1TF2A: the large subunit of a heteromeric, glutamine rich CCAAT-binding transcription factor involved in histone H1 cell cycle regulation. In preparation.

14

G1-S REGULATORY PROMOTER ELEMENTS AND THEIR INTERACTING TRANSCRIPTION FACTORS

Gregory S. Naeve, Li-jing Li, Lin Guo, Ajay Sharma, and Amy S. Lee

Department of Biochemistry and Molecular Biology
University of Southern California School of Medicine, and the Norris Cancer Center
1441 Eastlake Avenue
Los Angeles, CA 90033

INTRODUCTION

Modulation of S-phase linked transcription in cell cycle regulated genes can be attributed to cis regulatory elements found in their promoters (Naeve et al., 1991; Johnson, 1992). Characterization of the factors that act through these elements provides the basic principle for understanding the mechanism that regulates the periodic transcription changes of cell cycle regulated genes. The G1/S border is a key regulatory point in the cell cycle that directs the fate of a cell to duplicate itself. As such, the events leading to this point must provide the appropriate signals for the replication machinery to be duplicated in anticipation for the division of a daughter cell. The signals for the transcriptional regulation of the replication-dependent genes are generated by what are proving to be multi-functional transcription factors. Our laboratory is interested in the target regulatory elements these factors act on in cell cycle regulated genes as studied in growth stimulated cells. The hamster histone H3.2 and human thymidine kinase genes contain characterized promoter elements that are known to be involved in their replication-dependent transcriptional increase. Current work is focusing on the identification and characterization of the factors that interact with these elements and to understand how they elicit their regulation.

The Hamster Histone H3.2 Gene

Previous deletion analysis of the promoter from the hamster histone H3.2 gene identified a 32 base pair (bp) region that is 160 bp from the TATA element which is important for the transcriptional regulation during the cell cycle (Artishevsky et al., 1987).

Within this region is an element similar to that of a Jun/AP-1 binding site (site X) found in the coding strand spanning from base pairs -238 to -231 (Sharma et al., 1989). The sequence, 5' CGAGTCA 3', is similar to the canonical AP-1 site, TGAG/cTCA, as well as the cyclic AMP response element (CRE), TGACGTCA. Linker insertion mutation of this element abolishes appropriate transcriptional increase at the G1/S border of messages under control of the H3.2 promoter (Naeve et al., 1992). Figure 1 demonstrates the elimination of the S-phase associated increase of CAT message driven by a mutated H3.2 promoter (Xm) in serum synchronized cells harboring stably integrated H3.2/CAT

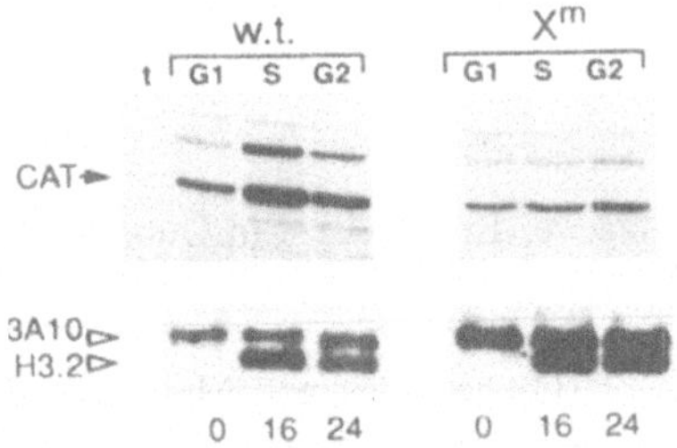

Figure 1. RNase protection assay of synchronized, stable CAT transfectants during the cell cycle. Plasmid pHCAT426 (w.t.) or pHCAT426Xm (Xm) was transfected into K12 hamster fibroblasts with pNEO3. Cytoplasmic RNA was extracted from pooled, stable transfectants resistant to G418 at indicated time points after serum release and analyzed for CAT mRNA levels by RNase protection. The position of the major CAT mRNA band (153 nucleotides) is shown. The 0 time point refers to cells arrested at the G0/G1 border. In Lane t, 10 μg of tRNA were used. Endogenous levels of the H3.2 mRNA and an invariant mRNA, p3A10, in these RNA samples are shown below as measured by RNA blot hybridizations.

constructs. The functional significance of this site may extend to the regulation of other histone H3 genes as the promoters from other species demonstrate similar sites at comparable distances from their TATA boxes (Table I).

Table I. Comparison between histone promoter elements and Jun/CRE binding sites.

	Promoter sequence	Distance from the TATA element (bp)	Reference
Hamster H3.2	C G A - G T C A	160	Sharma, A., et al.
Human H3	T G A C G T C A	130	Pauli, U., et al.
Mouse H3.2	T G G C G T C A	70	Taylor, J.D., et al.
Wheat H3	C C A C G T C A	140	Tabatta, T., et al.
Xenopus H3	T T A C G T C A	150	Old, R., et al.
AP-1 consensus	T G A - G/C T C A		Angel, P., et al.
CRE consensus	T G A C G T C A		Yamamoto, K.K., et al.

[a]Bases found to be identical with the indicated consensus sequences are connected by dots.

Gel retardation assays, which use a 19 bp oligomer containing site X as a probe, suggest that the element behaves as a poor AP-1 binding site; oligonucleotides containing characterized AP-1 binding sites compete efficiently for the X bound complex and antibodies specific for the Jun and Fos family members can disrupt the formation of an X complex. The AP-1 characteristic of site X can be removed by the addition of extended promoter sequence flanking the core element (Figure 2). The X probe (X* panel) shows the X complex efficiently competed by AP-1 containing DNA whereas, the X probe containing additional flanking sequence (XY* panel) is only efficiently competed by the wild type cold competitor. These results demonstrate that although the AP-1 like element is necessary for binding and appropriate cell cycle regulation, the flanking region directs the promoter specificity of the complex.

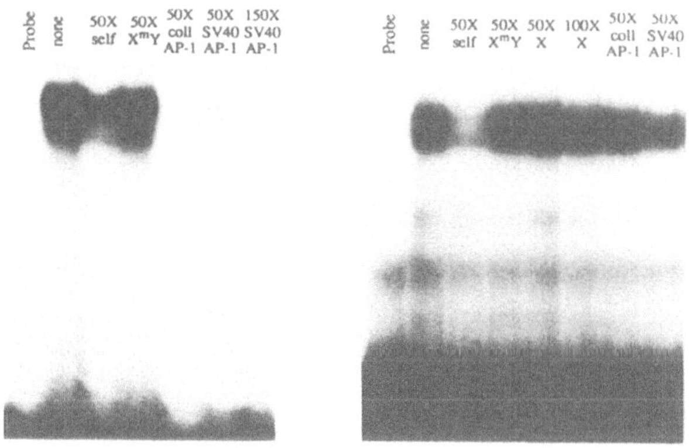

Figure 2. Binding specificity of nuclear factors. Gel mobility shift assays using either X or XY as probe and nuclear extracts from the hamster K12 cells. The fold molar excess of the unlabeled competitors are indicated on top. The DNA sequences of the competitors used are: X, tcaTGGCG AGTCAGCCtca (19mer); XY, ggcaagcttGCAGAACCTTGGCGAGTCAGCCAGCCGCGGGGCTG GACAA (49 mer); XmY, CCCTCGCAGAACCTTttaagcttctCCAGCCGCGGGGCTG (40 mer); Collagenase AP-1, agcttAAAGGCATGAGTCAGACACgtagc (29 mer); SV-40 AP-1, AGCTTGAT TAGTCAGCCG[5x(18mer)]. Underlined sequence indicates putative AP-1 element. Capital letters represent the top strand sequence of the oligomer, lower case indicates polylinker, and lower case italics shows the location of the mutation in the H3abp site.

Although the promoter specific complex bound to the extended site X, the H3abp site, behaves distinctly from an AP-1 complex, it is not clear if a member of the AP-1 family of transcription factors is involved in binding. Methylation interference analysis of the H3abp complex reveals that the guanine residues involved in binding are different from those used in the collagenase AP-1 site known to bind the Jun family members (Figure 3). Interestingly, the residues unique to the H3abp site appear to be the most important for the H3.2 complex. Figure 4 depicts a model which describes how the factors binding to the H3abp may interact distinctly from the Juns bound to the collagenase AP-1a site as described by König et al., 1992. This model suggests that if a Jun/Fos family member is interacting at site X it is doing so in conjunction with accessory proteins. Alternatively, they may not be required at all. Specific antibodies to

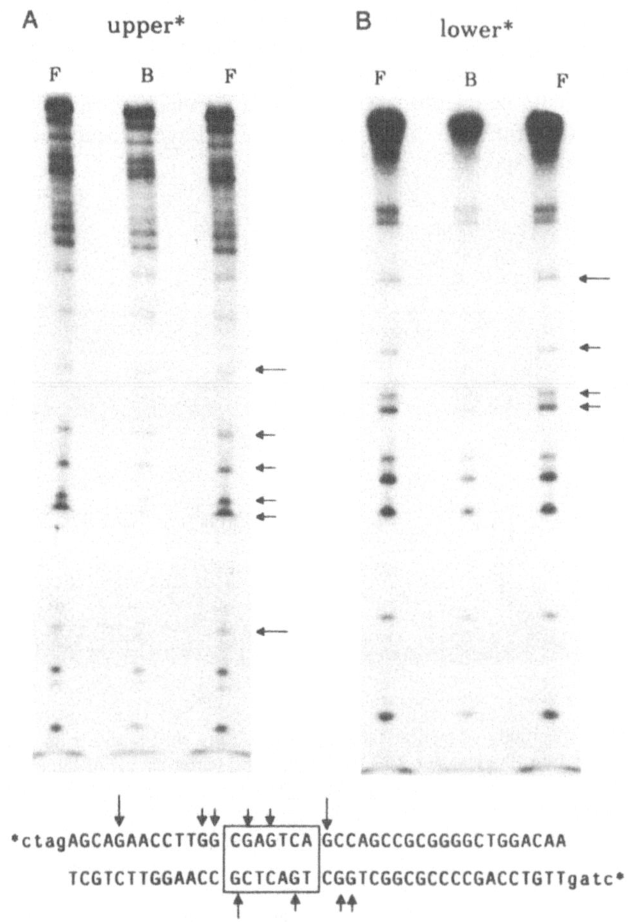

Figure 3. Methylation interference assay of the H3abp site. Forty micrograms of K12 nuclear extract was bound to methylated oligomers labeled on either the top or bottom strand. Remaining free DNA was separated from bound DNA by gel mobility shift assay. Free and bound DNA was electroeluted, piperidine digested, and separated on a 12% polyacrylamide gel with 8M Urea. A, Interference of the labeled upper strand. F: Free, uncomplexed DNA. B: Bound, protein complexed DNA. Relative protection of guanine residues is indicated by arrows. B, Interference of the labeled lower strand. H3abp oligomers used in this assay are shown below.

the various Jun/Fos family members can disrupt nuclear extract complex formation in gel retardation assays when either the minimal site X or a collagenase AP-1 site is used as a probe. However, when the H3abp site and nuclear extracts are used, the antibodies have no significant effect on the formation of the H3abp complex (Table II). Although not listed in this table, an antibody which recognizes members of the CREB family was also assayed and did not have an effect on complex formation. Only an antibody that cross-reacts with all the characterized Jun species shows any reactivity. This suggests that if the H3.2 complex contains a member of the characterized Jun family the epitopes are masked in complex by additional accessory proteins, or that a novel member of this immediate early gene family is involved in binding which shares some antigenic similarity to the Jun transcription factor family.

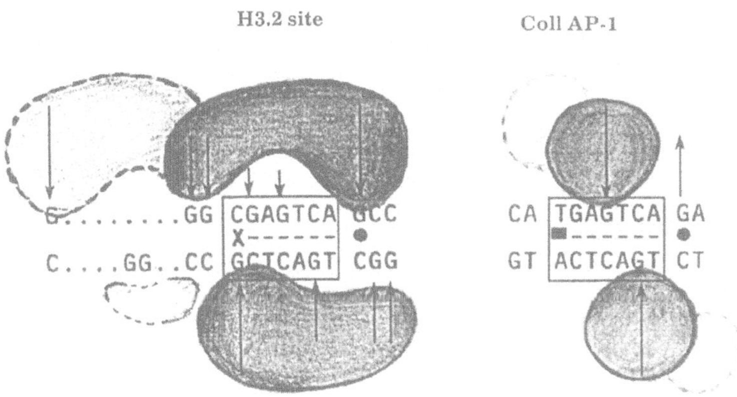

Figure 4. Protein binding models for the H3abp and AP-1 (König et al., 1992) based on methylation interference and protection data.

positive fractions, enriched for DNA binding proteins, are again pooled and dialyzed. An affinity column containing concatamers of the H3abp site is used in the next three steps. The first two steps processively remove non-specific DNA-binding proteins while efficiently retaining the H3abp activity. In the third affinity step, the activity is loaded at a higher salt concentration to more stringently remove non-specific proteins that bind at lower salt concentrations and are slow to elute. These proteins characteristically cross

To identify the factor(s) involved in binding to the H3abp, we are currently attempting to purify this complex from a hamster fibroblast nuclear extract. The scheme we have developed is outlined in Figure 5. Nuclear extract is partially purified by passing crude nuclear extract over a weak cation exchanger, BioRex 70. Positive fractions are then pooled, dialyzed, and passed over a non-specific DNA (calf thymus) column. The contaminate the enriched H3abp activity. In order to determine if a classical AP-1 activity is involved in H3abp binding a comparative analysis of the enriched fractions was done using a radiolabeled collagenase AP-1 site and an H3abp site in a side by side gel retardation assay (Figure 6). As the amount of non-specific competitor is reduced, using

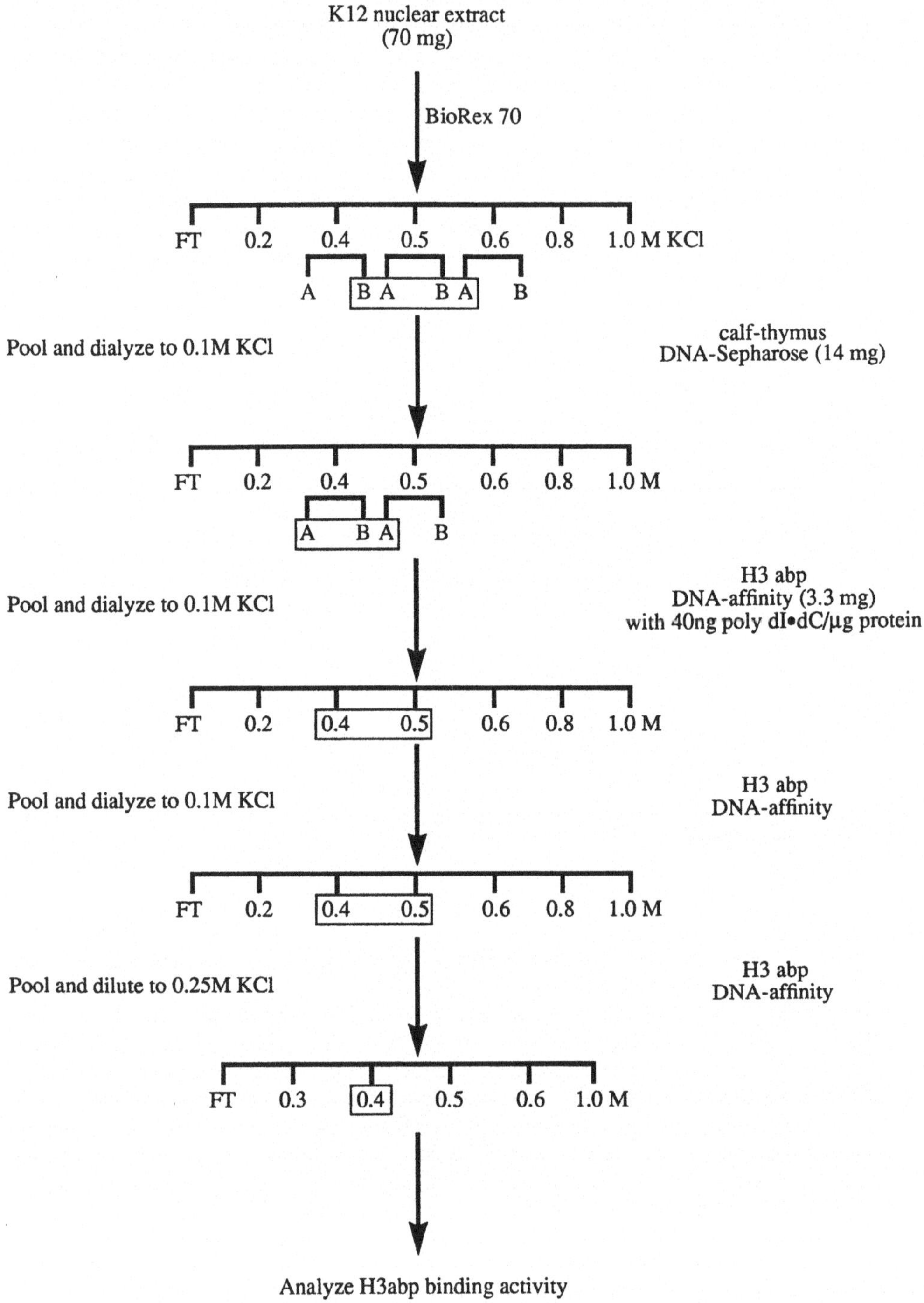

Figure 5. Flow chart for affinity purification of H3abp activity.

Table II. Summary of the degree of inhibition of DNA-protein complex formation by preincubation with the various antisera indicated.*

Probe	Protein source	Antisera				
		α c-Jun	α JunB	α JunD	α c-Fos	α Fra-1
Coll AP-1	IVT c-Jun	+++	-	-	-	
	IVT c-Jun/c-Fos	+++	-	-	+++	
	IVT JunB/c-Fos	++	+++	-	+++	
	IVT JunD/c-Fos	+	-	+++	+++	
	Hamster NE	++	+++	-	+++	++
XY	Hamster NE	+	-	-	-	-

*(+), relative ability of the antiserum to inhibit the formation of the protein complex; (-), inability of the antiserum to interfere with the complex formation. The probes used were either the collagenase AP-1 (Coll AP-1) or XY probe as described in Figure 2. The protein sources were either rabbit reticulocyte lysates containing in vitro translated (IVT) Jun and Fos proteins or nuclear extract (NE) from exponentially growing hamster cells.

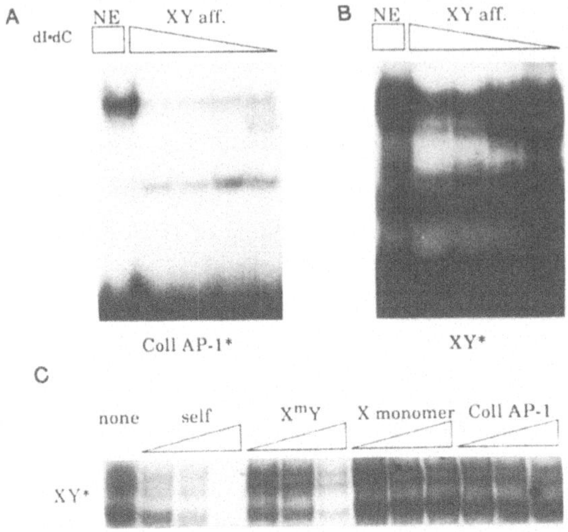

Figure 6. Comparative gel mobility shift assay with affinity purified fraction containing H3abp activity. Using 1 μl of the affinity fraction and either the Collagenase AP-1 or the XY oligomers described in Figure 2 as probes, the relative binding activity with decreasing amounts of poly dI·dC was compared with 500 ng of K12 nuclear extract containing 100 ng dI·dC. A, Binding activity of the H3abp fraction to the Collagenase AP-1 probe. B, Binding activity to the XY probe using the same fraction. C, Competition analysis of affinity purified H3abp activity. Using 1 μl of the affinity fraction and the radiolabeled XY oligomer, the complex was competed with 5, 10, and 50 fold molar excess of the indicated unlabeled competitors.

a constant amount of protein and probe, the amount of complex bound to the H3abp site steadily increases (Panel B), whereas the activity bound to the AP-1 element remains relatively constant (Panel A). This result indicates that we are enriching for an H3abp specific binding activity that is not directly analogous to the AP-1 activity found in the same nuclear extracts. However, this does not completely rule out the presence of an AP-1 family member in the H3abp fraction. The competition analysis shown in panel C demonstrates that the complex maintains its specificity for the H3abp site throughout purification. Current studies with the affinity purified fractions are being done to further characterize the enriched proteins visible on silver stained gels and determine which ones are specific for generating an H3.2 specific complex.

The Human Thymidine Kinase Gene (htk)

Experiments in our lab using the htk promoter fused to heterologous reporter genes identified a 70 base pair region that can confer an S-phase linked transcriptional increase of reporter message (Kim and Lee, 1991). Within this 70 bp region we identified a 25 bp protein binding site that is necessary for proper cell cycle regulation (Kim and Lee, 1992). As shown in Figure 7, the 25 bp binding site has multiple elements known to bind proteins in other replication dependent promoters and may bind factors involved in regulation of the htk gene. Most notably there are three E2-like binding sites. These sites are different from the prototypical adenovirus E2 site but do share identity with E2 sites in the N-myc and epidermal growth factor receptor promoters. The E2 binding site is known to bind a cellular transcription factor, E2F, which is thought to be important for regulation of cellular proliferation. More recently, E2F has been shown to form complexes with factors known to be critical in the modulation of cell cycle progression. These factors include the retinoblastoma protein (pRb), p107, cyclin A, p34cdc2 and p33cdk2 kinases (Shirodkar et al., 1992; Devoto et al., 1992; Cao et al., 1992). In addition to the E2 like binding sites, there are two sites that are similar to the Yi elements found in the promoter of the mouse tk gene. The mouse tk Yi elements have been shown to undergo changes in the complex composition as cells pass through the G1/S transition and these complexes have recently been shown to contain pRb and p34cdc2 (Dou et al., 1992). The in vivo functional significance of the E2 and Yi binding sites has not yet been established in their respective systems and the relationship between pRb and cdc binding to E2 and Yi in vitro is not understood.

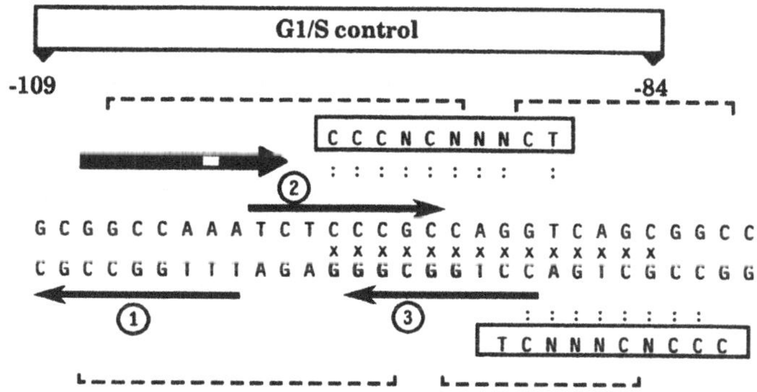

Figure 7. Sequence organization of the G1/S regulatory domain of the human tk promoter. The G1-S control domain spans -109 to -84. The footprinted regions are bracketed. The three thin arrows indicate E2-like binding sites. The identical bases between the human tk sequence and the mouse Yi protein binding site (box) are connected by dots. The thick arrow represents a CCAAT-like motif. A G-rich domain resembling the Sp1 motif is highlighted. The bases of the htk promoter mutated in the LS(-97/-84) construction are indicated by the crosses.

The entire 25 bp binding site is protected in DNaseI footprinting and yields four distinct complexes in a gel retardation assay. Previous gel retardation assays done in our lab using the 70 bp fragment, which contains the 25 bp binding site, as probe and whole-cell synchronized extracts from G1, S, and G2 phases of the cell cycle showed constitutive binding of two major complexes. When a 32 bp subfragment of the 70 bp region is used as probe four separate bands can be resolved, as seen in Figure 8. Using nuclear extracts from serum synchronized cells, harvested every two to four hours post stimulation, the relative binding activities of complexes I, II, and IV remain constant as cells progress through the cell cycle. Interestingly, complex III undergoes a significant increase in binding as cells enter S phase and this increase is maintained until DNA synthesis subsides. This change in binding activity of complex III correlates with the increase in transcription of the tk gene and provides a possible mechanism for its replication-dependent transcription (Li et al., 1993).

The presence of putative control elements within the 32 bp probe suggests that the factors known to interact at these sites or with the factors themselves may be involved in regulating the htk gene. As mentioned above, other studies have demonstrated that if E2 or Yi is part of the complex it is likely that cell cycle regulatory molecules like cyclins, pRb, or p34cdc2 may also be found in the complexes (Dou et al., 1992). Table III shows a summary of data from a panel of antibodies used to address this question in the htk

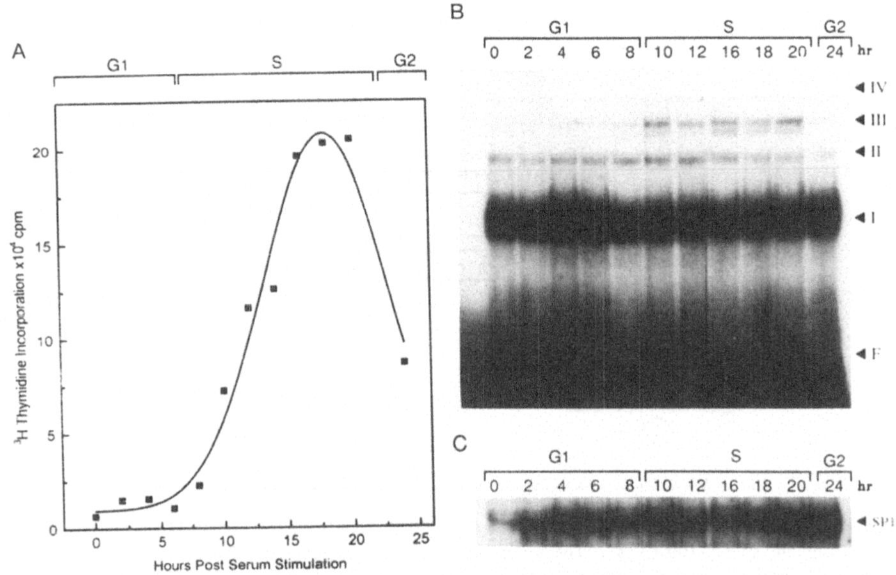

Figure 8. Protein binding profile of the htk complexes in synchronized cells. A, Synchrony of the cell cultures used for nuclear protein extraction as monitored by [3H]thymidine incorporation. B, Gel mobility shift assays performed with nuclear extracts prepared from the synchronized cells and a synthetic oligomer probe (w.t.) that spans -110 to -84 of the htk promoter. The positions of the free probe (F) and the four complexes (I-IV) are indicated. C, Identical to B with the exception that the probe used was the G8 oligomer derived from the human cardiac α-actin promoter. The Sp1 complex formed between the G8 probe and the nuclear extract at various times after serum release is shown.

system. The individual antibodies were used in gel retardation assays with the 32 bp tk probe in attempt to disrupt complex formation from S-phase nuclear extracts. These studies demonstrate that complex III contains cyclin A and p107, while complex II contains the G1/S phase specific kinase p33cdk2. How the composition of these complexes might regulate S-phase dependent transcription is schematically outlined in the model shown in Figure 9. The transition from G1 to S is presumably accompanied by the addition of complex III which may "complete" a complex that is capable of acting on the basal transcriptional machinery. This model suggests phosphorylation as the modification, but many scenarios may be invoked including the recruitment of additional factors. The persistence of the cyclin A/p107 complex throughout S phase indicates that there is a requirement for constant maintenance of the active state during DNA synthesis and may be established by the cooperation of cdk and cyclin A acting in concert. In order to test the validity of this model, we are currently characterizing the 25 bp binding site by mutational analysis and methylation interference to establish the elements required for the binding of the various complexes. We can show that a linker scanning mutation spanning -97 to -84 (Figure 7) disrupts the S-phase dependent increase in transcription of a heterologous fusion gene (Li et al., 1993) and are therefore focusing our analysis on this DNA sequence. Using affinity chromatography we are also trying to purify the various components of the individual complexes in order to more accurately understand the mechanism through which the htk gene is transcriptionally activated during S phase.

CONCLUSION

A large number of proteins have recently been implicated in controlling key events in the cell cycle. These proteins are often discovered by genetic lesion or biochemical purification. Their cellular targets however, remain unclear. We have been able to identify promoter elements in a histone (Naeve et al., 1992) and thymidine kinase (Li et al., 1993) gene that are necessary for the replication-dependent increase of their transcription. The discovery that cyclin A, p107, and p33cdk2 are involved in a transcriptional unit which directs the S-phase dependent transcription of the human thymidine kinase gene in serum stimulated cells provides a downstream link to a master pathway directing a cell to replicate its DNA. Further characterization of the complexes involved in regulating cell cycle transcriptional increases may lead to the convergence of additional pathways and factor interactions. Perhaps novel associations of cell cycle control with previously thought to be unrelated pathways will develop. The observation that a Jun-like factor may be involved in the regulation of events at the G1/S border and in particular with the histone H3.2 gene has, until recently, been unexplored. There are now reports that the Jun family members are required for cell cycle progression (Kovary and Bravo, 1991) and also that c-Jun has a biphasic expression profile (Carter et al., 1991; Trejo et al., 1992); the first peak is classically associated with growth response and the second appears to correlate with the onset of DNA synthesis. Definitive identification of the protein(s) regulating the S-phase transcriptional increase of the H3.2 gene will provide insight into the mechanism of coordinate regulation of the replication-dependent histones and other G1/S regulated genes.

ACKNOWLEDGEMENT

The authors would like to thank the American Association for Cancer Research, Inc. for permission to reprint figures 1 and 2 and Tables I and II (Naeve et al., 1992).

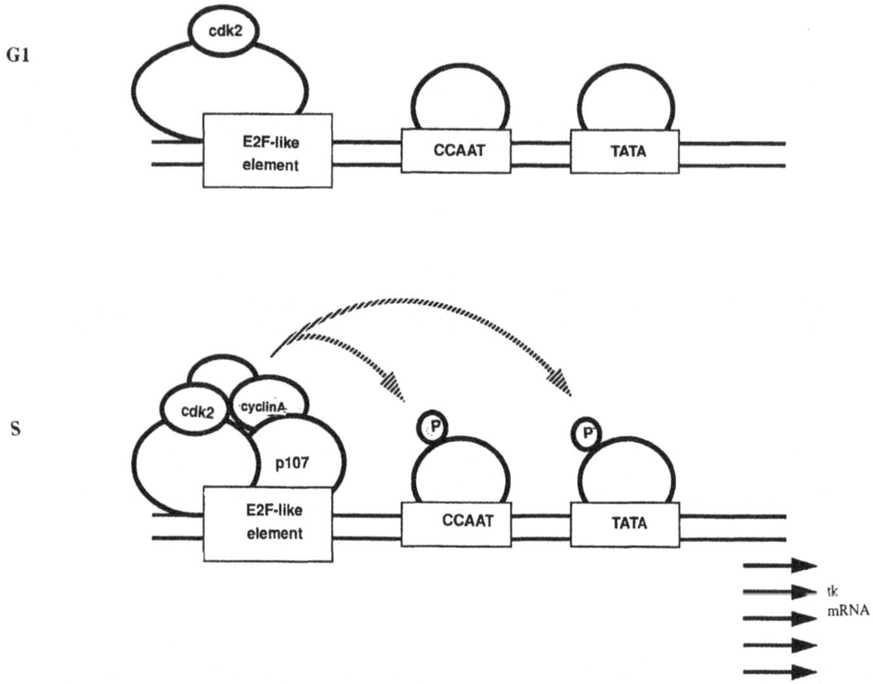

Figure 9. Proposed model of complex formation resulting in activation of the htk during S phase.

Table III. Summary of the ability to inhibit the formation of the DNA-protein complexes by pre-incubation with the various antisera indicated.*

	Antisera					
Complex	α Rb	α p107	α cycA	AC-3**	α cdc2	α cdk2
I	-	-	-	-	-	-
II	-	-	-	-	-	+
III	-	+	+	+	-	-
IV	-	-	-	-	-	-

* (+), ability of the antiserum to inhibit the formation of the protein complex; (-), inability of the antiserum to interfere with the complex formation. The probe used is the oligomer spanning -110 to -84 of the htk promoter. The nuclear extract is prepared from serum synchronized cells harvested 18 hr after stimulation (S-phase).

** AC-3 is directed against the C-terminal 12 amino acid peptide, VSLLNPPETLNL, of cyclin A. Reactivity against complex III can be eliminated by preincubation of the antisera with the peptide antigen.

REFERENCES

Angel, P., Imagawa, M., Chiu, R., Stein, B., Imbra, R.J., Rahmsdorf, H.J., Jomat, C., Herrlich, P., and Karin, M., 1987, Phorbol ester-inducible genes contain a common cis element recognized by a TPA-modulated trans-acting factor, Cell 49:729-739.

Artishevsky, A., Wooden, S., Sharma, A., Resendez, Jr., E., and Lee, A.S., 1987, Cell-cycle regulatory sequences in a hamster promoter and their interactions with cellular factors, Nature (London) 328:823-827.

Cao, L., Faha, B., Dembski, M., Tsai, L.H., Harlow, E., and Dyson, N., 1992, Independent binding of the retinoblastoma protein and p107 to the transcription factor E2F, Nature 355:176-179.

Carter, R., Cosenza, S.C., Pena, A., Lipson, K., Soprano, D.R., and Soprano, K.J., 1991, A potential role for c-jun in cell cycle progression through late G1 and S, Oncogene 6:229-235.

Devoto, S.H., Mudryj, M., Pines, J., Hunter, T., and Nevins, J.R, 1992, A cyclin A-protein kinase complex possesses sequence-specific DNA binding activity: p33cdk2 is a component of the E2F-cyclin A complex, Cell 68:167-176.

Dou, Q.P., Markill, P.J., and Pardee, A.B., 1992, Thymidine kinase transcription is regulated at G1/S phase by a complex that contains retinoblastoma-like protein and a cdc2 kinase, Proc. Natl. Acad. Sci. USA 89:3256-3260.

Johnson, L.F., 1992, G1 events and the regulation of genes for S-phase enzymes, Curr. Opinion Cell Biol. 4:149-154.

Kim, Y.K., and Lee, A.S., 1991, The identification of a 70 base pair cell-cycle regulatory unit within the promoter of the human thymidine kinase gene and its interaction with cellular factors, Mol. Cell Biol. 11:2296-2302.

Kim, Y.K., and Lee, A.S., 1992, Identification of a protein-binding site in the promoter of the human thymidine kinase gene required for the G1-S-regulated transcription, J. Biol. Chem. 267:2723-2727.

König, H., Ponta, H., Rahmsdorf, H., and Herrlich P., 1992, Interference between pathway-specific transcription factors: glucocorticoids antagonize phorbol ester-induced AP-1 activity without altering AP-1 site occupation in vivo, EMBO J. 11:2241-2246.

Kovary, K., and Bravo, R., 1991, The Jun and Fos protein families are both required for cell cycle progression in fibroblasts, Mol. Cell. Biol. 11:4466-4472.

Li, L., Naeve, G.S., and Lee, A.S., 1993, Temporal regulation of cyclin A-p107 and p33cdk2 complexes binding to a human thymidine kinase promoter element important for G1-S phase transcriptional regulation, Proc. Natl. Acad. Sci. USA 90:3554-3558.

Old, R.W., Sheikh, S.A., Chambers, A., Newton, C.A., Mohammed, A., and Aldridge, T.C., 1985, Individual Xenopus histone genes are replication-independent in oocytes and replication-dependent in Xenopus or mouse somatic cells, Nucl. Acids Res. 13:7341-7358.

Pauli, U., Chrysogelos, S., Nick, H., Stein, G., and Stein J.,1989, In vivo protein binding sites and nuclease hypersensitivity in the promoter of a cell cycle regulated human H3 gene, Nucl. Acids Res. 17:2333-2350.

Naeve, G.S., Sharma, A., and Lee, A.S., 1991, Temporal events regulating the early phases of the mammalian cell cycle, Curr. Opinion Cell Biol. 3:261-268.

Naeve, G.S., Sharma, A., and Lee, A.S., 1992, Identification of a 10-base pair protein binding site in the promoter of the hamster H3.2 gene required for the S phase dependent increase in transcription and its interaction with a Jun-like nuclear factor, Cell Growth Differ. 3:919-928.

Sharma, A., Bos, T.J., Pekkala-Flagan, A., Vogt, P.K., and Lee, A.S., 1989, Interaction of hamster cellular factors related to Jun oncoprotein with the promoter of a replication-dependent hamster histone H3.2 gene, Proc. Natl. Acad. Sci. USA 86:491-495.

Shirodkar, S., Ewen, M., DeCaprio, J.A., Morgan, J., Livingston, D.M., and Chittenden, T., 1992, The transcription factor E2F interacts with the retinoblastoma product and a p107-cyclin A complex in a cell cycle-regulated manner, Cell 68:157-166.

Tabatta, T., Takase, H.H., Takayama, S., Mikami, K., Nakatsuka, A., Kawata, T., Nakayama, T., and Iwabuchi, M., 1989, A protein that binds to a cis-acting element of wheat histone genes has a leucine zipper, Science (Washington, DC) 245:965-967.

Taylor, J.D., Wellman, S.E., and Marzluff, W.F., 1986, Sequences of four mouse histone H3 genes: implications for evolution of mouse histone genes, J. Mol. Evol. 23:242-249.

Trejo, J., Chambard, J.C., Karin, M., and Brown, H., 1992, Biphasic increase in c-jun mRNA is required for induction of AP-1-mediated gene transcription: differential effects of muscarinic and thrombin receptor activation, Mol. Cell Biol. 12:4742-4750.

Yamamoto, K.K., Gonzalez, G.A., Menzel, P., and Montminy, M.R., 1990, Characterization of a bipartite activator domain in transcription factor CREB, Cell 60:611-617.

15

REGULATION OF THYMIDYLATE SYNTHASE GENE EXPRESSION IN GROWTH-STIMULATED CELLS

Lee F. Johnson

Departments of Molecular Genetics and Biochemistry
The Ohio State University
Columbus, OH, 43210

INTRODUCTION

Thymidylate synthase (TS) catalyzes the reductive methylation of dUMP to form TMP in the *de novo* biosynthetic pathway. The enzyme is essential in proliferating cells and is an important target for a variety of cancer chemotherapeutic drugs (Danenberg, 1977). The TS gene is expressed at a very low level in non-proliferating cells. When the quiescent cells are stimulated to proliferate, the amount of TS mRNA and enzyme remain low during G1 phase, then increase 10-20 fold as the cells progress through S phase (Navalgund et al., 1980).

The increase in TS gene expression can be viewed as one of the final aspects of the signal transduction cascade that is initiated by the exposure of cells to a growth factor and that culminates in commitment of the cell to initiate DNA replication and complete the cell cycle. Much progress has been made recently in identifying some of the key regulatory factors that play important roles in G1 progression. Information is just beginning to emerge on the biochemical mechanisms by which these factors govern gene expression and cell cycle progression (Johnson, 1992).

The overall goal of our research is to understand the mechanisms for controlling the expression of the mouse TS gene in growth-stimulated cells. Our initial focus in on identifying the sequences and trans-acting factors that are important for regulating expression. We then hope to determine how these control processes are integrated into the G1 signal transduction cascade.

To permit a detailed analysis of the regulatory signals, we have cloned and analyzed the structure of the mouse TS gene and cDNA. The gene is 12 kb in length and has a 1 kb coding region that is interrupted by 6 introns (Perryman et al., 1986; Deng et al., 1986). The human TS gene has a similar structure (Kaneda et al., 1990). A number of unusual features of the TS gene have been uncovered. For example, mouse TS mRNA lacks a 3' untranslated region. The poly(A) tail extends directly from the UAA termination codon. The human TS mRNA does not share this unusual property (Takeishi et al., 1985). The mouse TS promoter is unusual in that it is very G/C rich and lacks a TATAA box (Fig. 1) (Deng et al., 1986). Primer extension and S1 nuclease protection assays revealed that transcription is initiated at many sites in an 80 nucleotide initiation window immediately upstream of the AUG codon (Fig. 2A). The human promoter has the similar properties, although the 5' untranslated region has several repeated sequences that are not present in the mouse TS gene (Kandeda et al., 1987).

The Cell Cycle: Regulators, Targets, and Clinical Applications
Edited by V.W. Hu, Plenum Press, New York, 1994

```
                                   [- - essential promoter elements - -]

                                                   [- - - - initiation window - - -
                                      (G/A)          (G/A) (G/C )
CACGTGGGGGCGGGGTCTGCCACGGATTCTGGCGGCCGGAAGTTTCCCAGCAGGAAGAGGCGGGCTGGTGTTGGAGGAA
                                                     +1                  +20


 - - - - initiation window - - - - - - - - - - -]
         ##               ###         #
AAGAGCGCCAGGAAGGTCCTGGTTTTGTCGCTGACTACACTGCTGCCAGACTGCTCCGTTATGCTGGTGGTTGGCT
        +40                  +60                  +80
```

Figure 1. Sequence of the TS 5' flanking region. The locations of the G/A and G/C-rich promoter elements in the essential region are indicated. Transcriptional initiation occurs primarily within the indicated window. The first initiation site is designated as +1. The locations of several predominant initiation sites (#) are shown. The AUG start codon is underlined.

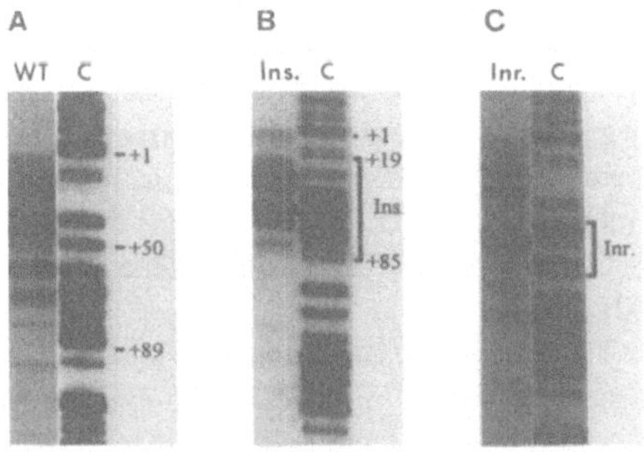

Figure 2. Initiation patterns for the wild-type TS promoter or TS promoters with insertions in the initiation window. TS minigenes were transiently transfected into TS(-) hamster V79 cells. Total cytoplasmic RNA was isolated and the pattern of 5' ends of mRNA derived from the wild-type (WT) and mutated minigenes was determined by S1 nuclease protection assays. Maxam and Gilbert sequencing reactions were performed on the probe to permit identification of the C residues (C) for each mutated promoter. The distance from the first initiation site (+1) is indicated.

Panel A: Pattern of start sites of wild-type TS mRNA.

Panel B: *E. coli* insertion mutations. The location of the insertion (immediately downstream of the essential promoter elements) is bracketed. Note that the initiation window is shifted from the normal TS promoter sequences to the *E. coli* sequences.

Panel C: TdT initiator insertion. The location of the inserted TdT initiator element is indicated. Note that the predominant transcriptional start site is now defined by the strong initiator element that was inserted downstream of the essential TS promoter elements.

ANALYSIS OF TS PROMOTER ELEMENTS

Deletion analyses of the mouse TS 5' flanking region revealed that wild-type promoter activity and the normal spectrum of transcriptional start sites are maintained with a promoter that extends only 13 nucleotides upstream of the first start site. Deletion of an additional 20 nucleotides almost completely eliminated promoter activity as well as the complete spectrum of start sites (Deng et al., 1989). G/A-rich and G/C-rich promoter elements and trans-acting factors were identified within the essential promoter region (Jolliff et al., 1991).

To determine if there are additional downstream promoter elements in the initiation window and to determine why transcription initiates across such a broad region, we have analyzed the effects of a variety of substitution, deletion, and insertion mutations in the initiation window (Geng, and Johnson, 1993). Substitution of the promoter sequences between -72 and -6 with three different *E. coli* sequences of approximately the same length had little effect on the production of TS mRNA, indicating that there are no essential promoter elements downstream of the G/C box. Analysis of the transcriptional start sites for the promoters that contained *E. coli* sequences within the initiation window revealed that the pattern for each substitution was complex but unique (Geng, and Johnson, 1993).

The boundaries of the initiation window are established by the essential promoter elements at the 5' end of the window. This was best illustrated by a promoter alteration that inserted 62 nucleotides of *E. coli* DNA immediately downstream of the essential promoter region. In this altered promoter, virtually all of the initiation sites are within the inserted region. The TS sequences which are normally in the initiation window (but which are moved farther downstream of the essential elements as a result of the insertion) are no longer used for transcriptional initiation (Fig. 2B).

The complexity of the pattern of start sites may be due to the absence of an initiator element. When the strong initiator element from the terminal deoxynucleotidyl transferase (TdT) gene was inserted within the initiation window, transcription initiated predominantly at the position directed by the initiator (Fig. 2C). However, when the initiator element was inserted immediately upstream of the window, it had no effect on the normal pattern of start sites (not shown) except for the introduction of an additional upstream start site that was directed by the initiator (Geng, and Johnson, 1993).

In conclusion, the essential promoter elements of the mouse TS promoter are located within a 20-30 nucleotide region that overlaps the upstream boundary of the transcriptional initiation window. The complex pattern of transcriptional start sites is the result of the absence of an element that specifies a unique start site (such as a TATAA box or an initiator element). The upstream essential promoter elements determine the boundaries of the initiation window. However, the actual start sites that are used are determined by the sequences that are present within the initiation window and probably reflect the thermodynamic preferences of the transcription complex for specific nucleotide sequences.

SEQUENCES IMPORTANT FOR GROWTH-REGULATED EXPRESSION

TS mRNA content increases 10-20 fold as growth-stimulated cells progress from G0 through S phase. However, nuclear run-on transcription assays showed that the rate of TS gene transcription increases only about 2-3-fold during this interval (Jenh et al.,

1985). Therefore it appears that TS mRNA content is controlled primarily at the posttranscriptional level in growth-stimulated mouse fibroblasts. Similar observations and conclusions have been made for the human TS gene (Ayusawa et al., 1986).

To identify the sequences that are important for regulating TS gene expression, we have constructed a series of minigenes that contain different 5' flanking regions, the TS coding region (with or without introns) and various polyadenylation signals. The coding region was "tagged" with a small deletion to allow for simultaneous detection of mRNA derived from the endogenous TS gene and the minigene by S1 nuclease protection assays. The minigenes were stably transfected into wild-type 3T6 cells along with a neo gene as a selectable marker. The expression of the TS minigene was then compared with that of the endogenous TS gene following growth-stimulation (Li et al., 1991; Ash et al., 1993).

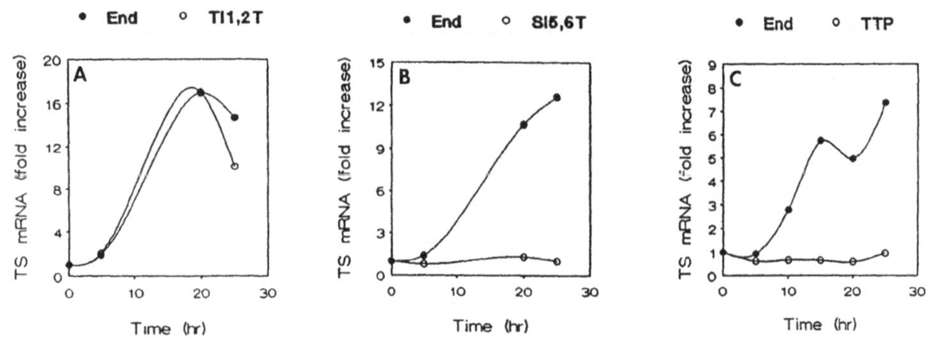

Figure 3. Regulation of stably transfected TS minigenes. The indicated minigenes were stably transfected into 3T6 cells. The cells were serum-stimulated for the indicated times and the amount of mRNA corresponding to the endogenous TS gene (closed circle) or to the minigene (open circle) was determined by S1 nuclease protection assays. Radioactivity was quantitated and normalized to the value at T=0 to permit direct comparisons.

Panel A: $TI_{12}T$, a wild-type minigene consisting of 1 kb of TS 5' flanking DNA, the TS coding region with introns 1&2 at their normal locations in the coding region and 0.25 kb of 3' flanking DNA.

Panel B: $SI_{56}T$, a TS minigene in which the SV40 promoter was substituted for the TS promoter. The minigene contains TS introns 5&6 at their normal locations.

Panel C: TTP, intronless version of the TS minigene shown in panel A.

Fig. 3A shows that a TS minigene ($TI_{12}T$) consisting of the TS promoter (extending to -1 kb), the TS coding region (including introns 1&2 at their normal positions) and the TS polyadenylation signal were regulated normally in response to growth-stimulation. The amount of mRNA derived from the TS minigene increased at the same time and to the same extent as that of the endogenous TS gene. Similar results were obtained when the minigene contained introns 5&6 (not shown) instead of 1&2 although the extent of the increase is not as great as that observed with the endogenous gene. These observations demonstrate that sequences sufficient for normal regulation are included in the minigenes, and that they are not uniquely located within a single intron of the TS gene (Li et al., 1991).

As discussed above, the TS polyadenylation signal has a highly unusual structure. To determine if the polyadenylation signal or sequences downstream of the signal are important for normal regulation, the TS polyadenylation signal was inactivated by site-directed mutagenesis and several different polyadenylation signals were inserted immediately downstream of the termination codon. When the TS polyadenylation signal of $TI_{12}T$ was replaced with the polyadenylation signal of the human beta globin gene or the bovine growth hormone gene, the minigene was regulated in the same manner as that of the minigenes that contained the normal TS polyadenylation signal (data not shown). This indicates that the region downstream of the termination codon does not contain sequences that are important for normal regulation (Ash et al., 1993).

When the TS promoter was replaced with the SV40 early promoter to form $SI_{56}T$ (Fig. 3B), there was almost no change in the level of expression of the minigene as the cells progressed from G1 through S phase (Li et al., 1991). This shows that the promoter/5' flanking region of the TS gene contains sequences that are necessary for normal regulation. To localize these sequences, we analyzed the expression of minigenes with deletions in the 5' flanking region of the gene. We found that minigenes that extend 250 nucleotides upstream of the AUG codon are regulated normally. However, minigenes that extend only 110 nucleotides upstream of the AUG codon, which is only about 5 nucleotides upstream of the essential promoter region, are not regulated (not shown). This indicates that at least some of the upstream regulatory elements are located upstream of the essential promoter elements.

Although sequences in the 5' flanking region are necessary for growth-regulated expression they are not sufficient for normal regulation. This was shown by a minigene that consisted of the TS 5' flanking region (from the AUG codon to -1kb) linked to the CAT coding region to form Tcat. The Tcat gene was expressed at a constant level in growth-stimulated cells whereas the endogenous TS gene was regulated normally (Li et al., 1991). We have recently made similar observations with a minigene that consists of the same TS 5' flanking region linked to the luciferase reporter gene, indicating that the loss of growth-regulated expression is not a peculiarity of the CAT indicator gene. Therefore, sequences that are downstream of the AUG codon are also necessary (but not sufficient) for normal regulation.

INTRONS ARE REQUIRED FOR NORMAL REGULATION

The fact that regulation was observed when introns 1&2 or 5&6 were present suggested that regulatory sequences are not located within the TS introns. However, it was possible that simply the presence of an intron was essential for proper regulation. To test this idea, intronless derivatives of various TS minigenes were constructed and stably transfected into 3T6 cells. We found that when all of the introns were removed, regulation was abolished (Fig. 3C) (Ash et al., 1993).

These observations suggest that introns must be present in the minigene for proper regulation to be observed. However, it was not clear if TS introns were required or of introns from other genes would suffice. Our studies showed that the small T intron of the SV40 gene (which is included in the CAT and luciferase indicator genes) will not suffice, but this could be due to unusual features of this intron. To further explore this question, we have recently constructed a minigene that consists of the TS 5' flanking region (immediately upstream of the AUG codon to -1 kb) linked to the human beta globin gene (from the AUG codon through the 3' flanking region but not including the 3' enhancer sequence). When the globin gene included both introns, this gene was regulated in a growth-dependent manner. However, when the introns were removed by substituting globin cDNA for the coding region, the minigene was not regulated (data not shown).

These observations strongly support our conclusion that introns are required for growth-regulated expression and that the TS 5' flanking sequences are essential for this regulation. Since it is unlikely that S-phase specific regulatory elements are contained within the globin introns, these observations are consistent with our hypothesis that the splicing reaction itself is somehow involved in proper growth-regulated expression.

How do the upstream regulatory sequences "communicate" with the introns to bring about regulation? One possibility is that the upstream elements are somehow able to affect RNA splicing (rather than transcription) in a cell cycle specific manner. A possible mechanism for communication between upstream regulatory sequences and TS RNA processing reactions is summarized in Fig. 4. In G0/G1, the TS gene is transcribed but the RNA products are very inefficiently processed into cytoplasmic TS mRNA. Most of the transcripts are degraded within the nucleus. The regulatory site upstream of the essential promoter elements is unoccupied (or perhaps occupied by an inactive factor). As the cells approach the G1/S boundary, the upstream regulatory element is occupied by a trans-acting factor(s). This factor has little if any effect on the rate of transcription of the

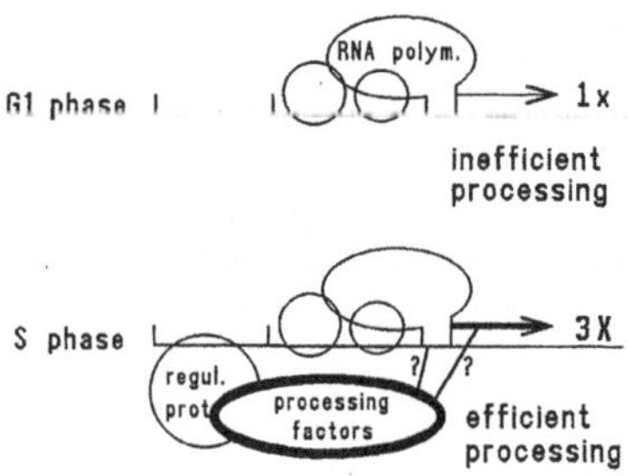

Figure 4. Model for regulation of processing by upstream elements of the TS promoter. The model is explained in the text.

gene. Instead, this regulatory factor interacts with a component of the RNA processing (splicing or polyadenylation) machinery. The splicing component could then be transferred to the nascent transcript, either directly or via the RNA polymerase. The "tagged" RNA molecule would then be processed efficiently into cytoplasmic TS mRNA.

Studies are currently underway to identify the TS upstream regulatory sequences and binding factors and to determine if the regulatory sequences will function in cooperation with introns and coding regions from other foreign genes that are not cell cycle regulated. It will also be interesting to determine if similar sequences and factors are responsible for controlling the expression of other genes for S-phase enzymes.

ACKNOWLEDGMENTS

These studies were supported by grants from the National Institute for General Medical Sciences (GM29356) and the National Cancer Institute (CA16058).

REFERENCES

Ash, J., Ke, Y., Korb, M., and Johnson, L.F., 1993, Introns are essential for growth-regulated expression of the mouse thymidylate synthase gene. Mol. Cell. Biol. 13:1565-15710.

Ayusawa, D., Shimizu, K., Koyama, H., Kaneda, S., Takeishi, K., and Seno, T., 1986, Cell-cycle-directed regulation of thymidylate synthase messenger RNA in human diploid fibroblasts stimulated to proliferate. J. Mol. Biol. 190:559-567.

Danenberg, P.V., 1977, Thymidylate synthetase - a target enzyme in cancer chemotherapy. Biochim. Biophys. Acta 473:73-92.

Deng, T., Li, D., Jenh, C.-H., and Johnson, L.F., 1986, Structure of the gene for mouse thymidylate synthase: Locations of introns and multiple transcriptional start sites. J. Biol. Chem. 261:16000-16005.

Deng, T., Li, Y., Jolliff, K., and Johnson, L.F., 1989, The mouse thymidylate synthase promoter: essential elements are in close proximity to the transcriptional inititation sites. Mol. Cell. Biol. 9:4079-4082.

Geng, Y., and Johnson, L.F., 1993, Lack of an initiator element is responsible for multiple transcriptional initiation sites of the TATA-less mouse thymidylate synthase promoter. Mol. Cell. Biol. 13:4894-4903.

Jenh, C.-H., Geyer, P.K., and Johnson, L.F., 1985, Control of thymidylate synthase mRNA content and gene transcription in an overproducing mouse cell line. Mol. Cell. Biol. 5:2527-2532.

Johnson, L.F., 1992, G1 events and the regulation of genes for S-phase enzymes. Curr. Opin. Cell Biol. 4:149-154.

Jolliff, K., Li, Y., and Johnson, L.F., 1991, Multiple protein-DNA interactions in the TATAA-less mouse thymidylate synthase promoter. Nucleic Acids Res. 19:2267-2274.

Kaneda, S., Takeishi, K., Ayusawa, D., Shimizu, K., Seno, T., and Altman, S., 1987, Role in translation of a triple tandemly repeated sequence in the 5'-untranslated region of human thymidylate synthase mRNA. Nucleic Acids Res. 15:1259-1270.

Kaneda, S., Nalbantoglu, J., Takeishi, K., Shimizu, K., Gotoh, O., Seno, T., and Ayusawa, D., 1990, Structural and functional analysis of the human thymidylate synthase gene. J. Biol. Chem. 265:20277-20284.

Li, Y., Li, D., Osborn, K., and Johnson, L.F., 1991, The 5' flanking region of the mouse thymidylate synthase gene is necessary but not sufficient for normal regulation in growth-stimulated cells. Mol. Cell. Biol. 11:1023-1029.

Navalgund, L.G., Rossana, C., Muench, A.J., and Johnson, L.F., 1980, Cell cycle regulation of thymidylate synthetase gene expression in cultured mouse fibroblasts. J. Biol. Chem. 255:7386-7390.

Perryman, S.M., Rossana, C., Deng, T., Vanin, E.F., and Johnson, L.F., 1986, Sequence of a cDNA for mouse thymidylate synthase reveals striking similarity with the prokaryotic enzyme. Mol. Biol. Evol. 3:313-321.

Takeishi, K., Kaneda, S., Ayusawa, D., Shimizu, K., Gotoh, O., and Seno, T., 1985, Nucleotide sequence of a functional cDNA for human thymidylate synthase. Nucleic Acids Res. 13:2035-2043.

16

THE ROLE OF THE TRANSCRIPTION FACTOR E2F IN THE GROWTH REGULATION OF DHFR

Jill E. Slansky and Peggy J. Farnham

McArdle Laboratory for Cancer Research
Department of Oncology
University of Wisconsin, Madison
Madison, WI 53706

INTRODUCTION

As cells enter S phase of the growth cycle there is a transient increase in the transcription and translation of a number of genes that prepare the cell for DNA synthesis. Increased levels of enzymes such as dihydrofolate reductase (DHFR), thymidine kinase, thymidylate synthase, ribonucleotide reductase, CAD, and DNA polymerase alpha help to create an environment favorable for DNA synthesis. As is true for most genes critical for cell growth, expression of these genes is controlled on many levels. Our interests are to understand the control mechanisms that lead to increased transcription at the G1/S-phase boundary. Using the DHFR promoter as a model system, we mapped a cis-acting element necessary and sufficient for a transcriptional increase late in G1[1, 2]. We propose a simple model in which the transcription factor E2F1 binds to and activates transcription from this element in a growth-regulated manner. However, this model is not complete and additional factors may play a role in the growth regulation of the DHFR gene.

RESULTS

The E2F sites in the DHFR promoter are necessary and sufficient to confer growth regulation on a heterologous gene

Endogenous DHFR mRNA and protein levels increase late in G1[3-6]. To define the region of the DHFR gene responsible for growth regulation, we first inserted the DHFR promoter region (-270 to +20 relative to the transcription initiation site) upstream of a luciferase reporter gene. The construct was transiently transfected into growing NIH 3T3 cells which were subsequently driven to quiescence by removal of serum (Figure 1). Luciferase activity was then monitored after the cells were stimulated with serum to enter the growth cycle. Luciferase activity driven by the DHFR promoter increased 10 hours after

The Cell Cycle: Regulators, Targets, and Clinical Applications
Edited by V.W. Hu, Plenum Press, New York, 1994

serum stimulation, just as the cells entered S phase. Twelve hours after stimulation, activity peaked approximately 14-fold higher than in serum-starved cells (Figure 2A, filled in boxes). Because this increase in luciferase activity mirrors the increase of the endogenous DHFR message, we conclude that this serum starvation and stimulation assay accurately reflects the regulation of DHFR transcription and that promoter sequences from -270 to +20 contain the sequences that are at least in part responsible for the serum regulation of DHFR.

Plate $2X10^5$ NIH 3T3 cells/60 mm dish in 5% BCS

↓ 12-16 h

Calcium phosphate transfect

↓ 1 h recovery

Starve cells in 0.5% BCS

↓ 48-60 h

Stimulate cells in 10% BCS

↓

Harvest cells and analyze:
•Promoter activity (luciferase assays),
•RNA levels (Northern blot assays),
•DNA content (flow cytometric analysis).

Figure 1. The serum starvation and stimulation assay. The protocol outlined above describes the method employed to synchronize cells in the experiments described in this paper. For Northern blot analysis, cells were grown on a larger scale and were not transfected. BCS, bovine calf serum.

Biochemical analysis of the -270 to +20 region of the DHFR promoter revealed four protein binding sites for the transcription factor Sp1 between -200 and -50[7] and two overlapping, inverted sites for E2F in the transcription initiation region of DHFR (-8 to +1)[8]. The Sp1 sites seemed unlikely candidates for the growth regulation elements since Sp1 activates other constitutive promoters such as the SV40 early promoter. In contrast, E2F is found transiently complexed with positive and negative regulators of cell growth and was therefore a likely candidate. To determine if the E2F elements were responsible for DHFR growth regulation, we mutated the sites. We then compared luciferase activity driven by the mutant versus wild type DHFR promoters in the serum starvation and stimulation assay. The activity of the mutant construct increased approximately 4-fold 12 hours after serum stimulation (Figure 2A, open boxes), whereas the wild type promoter construct increased approximately 14-fold. These results indicate that the E2F sites are required for DHFR growth regulation.

To determine if the E2F sites could confer regulation to a heterologous promoter, we inserted two oligonucleotides containing DHFR sequences from -20 to +9 upstream of the SV40 early promoter driving the luciferase cDNA to form E2F-SV40e-luc. The DHFR sequences include the E2F sites and the flanking region found in the footprinted region of the DHFR promoter[7]. Although the E2F sites did not significantly increase the activity of SV40e-luc in quiescent cells, they did confer growth-regulated activity to the SV40 early promoter (Figure 2B). When assayed in the serum starvation and stimulation assay, activity from E2F-SV40e-luc increased approximately 8-fold during S phase, whereas activity from SV40e-luc increased only two- to threefold during S phase. Thus, we conclude that E2F sites are necessary for the transcriptional increase of DHFR at the G1/S-phase boundary and can confer growth regulated expression on a heterolougous promoter.

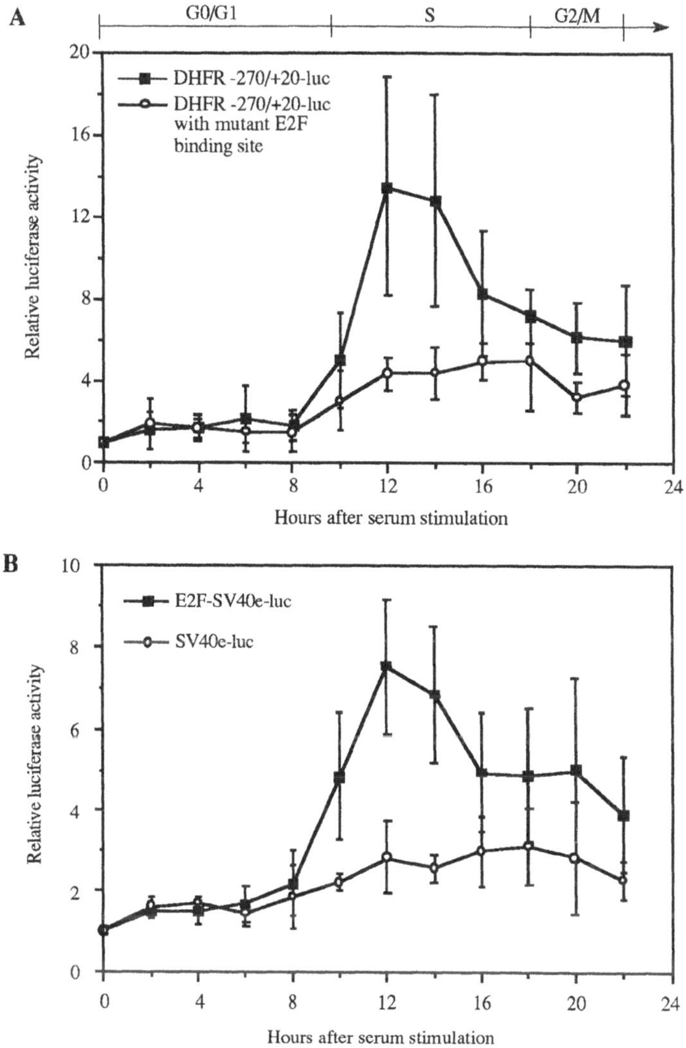

Figure 2. The E2F sites (-20 to +9) are necessary for DHFR serum regulation (A) and can confer serum regulation to a heterologous reporter gene (B). These experiments were performed 3 times as outlined in Figure 1. The error bars represent the standard error of the mean. These results are reprinted with the permission of the original publisher: A[1] and B[2].

E2F1 may regulate transcription from the DHFR promoter at the G1/S-phase boundary

The next step toward understanding the growth regulation of DHFR was to determine what proteins bind to the -20 to +9 region of the DHFR promoter. A number of genes have been cloned which bind to E2F sites including E2F1[9-11], E2F2[12], E2F3[12] and DP1[13]. Of these, only E2F1 has been shown to bind directly to the DHFR sites[9, 14] and to activate transcription through these sequences in serum-starved cells[2]. To further investigate the role of E2F1 in growth regulation, we monitored the levels of E2F1 mRNA by performing Northern blot analysis on RNA from cells that had been serum starved and stimulated. As depicted in Figure 3, we observed an increase in E2F1 mRNA that parallels the increase in the activity of the DHFR promoter. Because 1) E2F1 can bind specifically to the E2F sites in the DHFR promoter, 2) the DHFR promoter can be activated by transfected E2F1, and 3) levels of E2F1 mRNA parallel the DHFR promoter activity, we propose that E2F1 may be involved in the activation of transcription of DHFR at the G1/S-phase boundary.

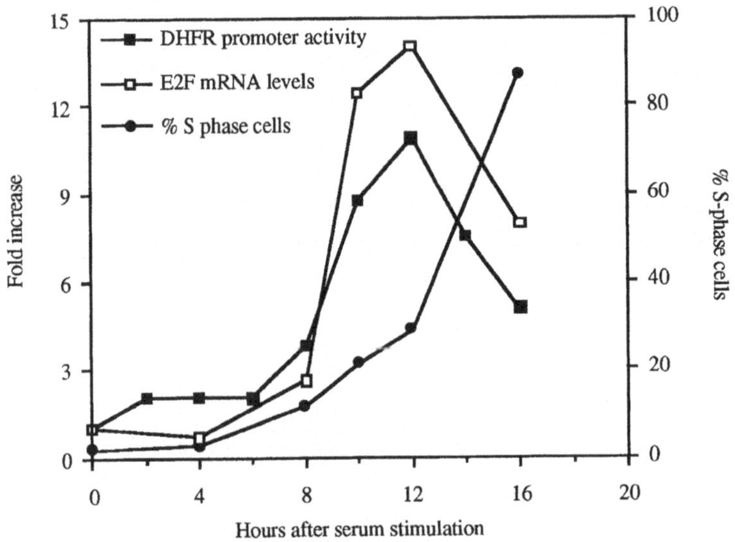

Figure 3. E2F1 mRNA levels parallel the activation of the DHFR promoter after serum stimulation. The DHFR promoter activity was determined by transient transfection of the -270 to +20 construct, followed by the serum starvation and stimulation assay (Figure 1). The E2F1 mRNA levels were determined by Northern blot analysis of cytoplasmic RNA from serum-starved and -stimulated cells. The percentage of cells entering S phase was determined by flow cytometric analysis. These results are reprinted with the permission of the original publisher[2].

Other factors may be involved in the serum regulation of the DHFR promoter

To determine if E2F alone or in cooperation with other DNA binding proteins was responsible for regulation at the G1/S-phase boundary, we tested additional DHFR promoter constructs. Although the constructs all contained the -20/+9 regions of the DHFR promoter, we found that as the 5' end of the promoter was deleted, serum-stimulated expression was lost (see Figure 4). The promoter also contains binding sites for the transcription factor Sp1 as well as unknown binding sites [15]. These results suggest that additional cis-acting sequences may influence the serum activation of the DHFR promoter. Others have observed similar results with other growth-regulated promoters. As the 5' ends of the human DNA polymerase α [16] and mouse Rep-3B [17] genes are deleted in similar assays, the activity also decreases at the G1/S-phase boundary. Experiments are on going in the lab to understand this phenomena.

CONCLUSIONS

In summary, we have presented data that supports the involvement of E2F in the transcriptional regulation of the DHFR promoter. Our simple model is that E2F1 levels increase at the G1/S-phase boundary, and this increase is responsible for activation of DHFR transcription. However, as described, it appears as if other cis-acting elements in the DHFR promoter can modulate regulation by E2F at the G1/S-phase boundary. Others have demonstrated that E2F activity can also be regulated via protein-protein interactions. Complexes including E2F and various combinations of the retinoblastoma protein, p107,

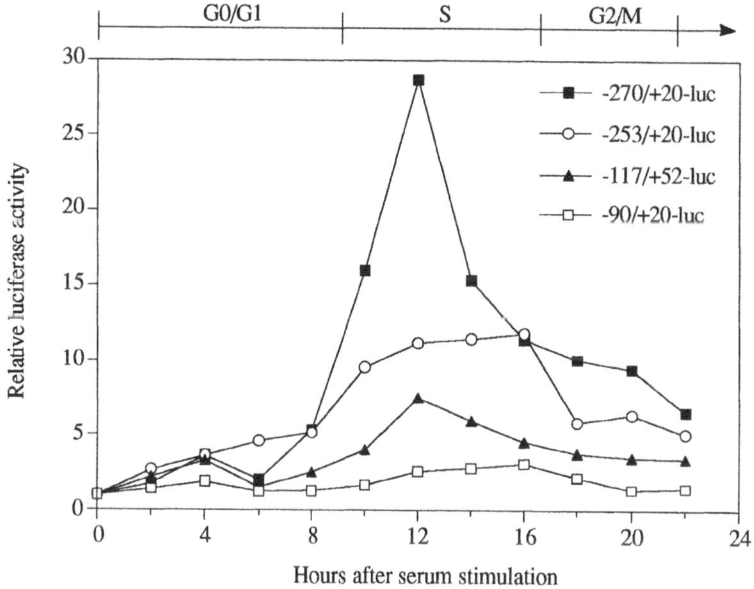

Figure 4. Serum-stimulated expression from the DHFR promoter is lost at the G1/S-phase boundary as the 5' end is deleted. These experiments were performed 3 times as outlined in Figure 1. "These results are reprinted with the permission of the original publisher (BBA 1155:125-131, 1993)".

cyclins A and E, and cdk2[18-23] have been reported. To date, all such complexes appear to be inhibitory to E2F activity[18, 24-26]. E2F may also be regulated by heterodimer formation[27] and by phosphorylation[28]. Thus, regulation of DHFR and of other genes that are involved in preparing cells for S phase is complex. This may reflect the importance of the ordered sequence of events required for normal cellular proliferation. Although understanding these cellular control mechanisms requires additional research, it is hoped that these studies may eventually lead to better understanding of cell growth and suggest clinical applications.

REFERENCES

1. A. L. Means, J. E. Slansky, S. L. McMahon, M. W. Knuth, and P. J. Farnham, The HIP1 binding site is required for growth regulation of the dihydrofolate reductase gene promoter, *Mol. Cell. Biol.* 12:1054-1063 (1992).
2. J. E. Slansky, Y. Li, W. G. Kaelin, and P. J. Farnham, A protein synthesis-dependent increase in E2F1 mRNA correlates with growth regulation of the DHFR promoter, *Mol. Cell. Biol.* 13:1610-1618 (1993).
3. P. J. Farnham and R. T. Schimke, Transcriptional regulation of mouse dihydrofolate reductase in the cell cycle, *J. Biol. Chem.* 260:7675-7680 (1985).
4. P. J. Farnham and R. T. Schimke, Murine dihydrofolate reductase transcripts through the cell cycle, *Mol. Cell. Biol.* 6:365-371 (1986).
5. L. F. Johnson, C. L. Fuhrman, and L. M. Wiedemann, Regulation of dihydrofolate reductase gene expression in mouse fibroblasts during the transition from the resting to growing state, *J. Cell. Physiol.* 97:397-406 (1978).
6. C. Santiago, M. Collins, and L. F. Johnson, In vitro and in vivo analysis of the control of dihydrofolate reductase gene transcription in serum-stimulated mouse fibroblasts, *J. Cell. Physiol.* 118:79-86 (1984).
7. A. L. Means and P. J. Farnham, Transcription initiation from the dihydrofolate reductase promoter is positioned by HIP1 binding at the initiation site, *Mol. Cell. Biol.* 10:653-661 (1990).
8. M. C. Blake and J. C. Azizkhan, Transcription factor E2F is required for efficient expression of the hamster dihydrofolate reductase gene in vitro and in vivo, *Mol. Cell. Biol.* 9:4994-5002 (1989).
9. W. G. Kaelin, W. Krek, W. R. Sellers, J. A. DeCaprio, F. Ajchenbaum, C. S. Fuchs, T. Chittenden, Y. Li, P. J. Farnham, M. A. Blanar, D. M. Livingston, and E. K. Flemington, Expression cloning of a cDNA encoding a retinoblastoma-binding protein with E2F-like properties, *Cell* 70:351-364 (1992).
10. K. Helin, J. A. Lees, M. Vidal, N. Dyson, E. Harlow, and A. Fattaey, A cDNA encoding a pRB-binding protein with properties of the transcription factor E2F, *Cell* 70:337-350 (1992).
11. B. Shan, X. Zhu, P.-L. Chen, T. Durfee, Y. Yang, D. Sharp, and W.-H. Lee, Molecular cloning of cellular genes encoding retinoblastoma-associated proteins: identification of a gene with properties of the transcription factor E2F, *Mol. Cell. Biol.* 12:5620-5631 (1992).
12. E. Harlow. Personal communication.
13. R. Girling, J. F. Partridge, L. R. Bandara, N. Burden, N. F. Totty, J. J. Hsuan, and N. B. La Thangue, A new component of the transcription factor DRTF1/E2F, *Nature* 362:83-87 (1993).
14. Y. Li and P. J. Farnham. Unpublished data.
15. P. J. Farnham and A. L. Means, Sequences downstream of the transcription initiation site modulate the activity of the murine dihydrofolate reductase promoter., *Mol. Cell. Biol.* 10:1390-1398 (1990).
16. B. E. Pearson, H.-P. Nasheuer, and T. S.-F. Wang, Human DNA polymerase α gene: sequences controlling expression in cycling and serum-stimulated cells., *Mol. Cell. Biol.* 11:2081-2095 (1991).
17. R. J. Miltenberger and P. J. Farnham. Unpublished data.
18. S. Bagchi, R. Weinmann, and P. Raychaudhuri, The retinoblastoma protein copurifies with E2F-I, an E1A-regulated inhibitor of the transcription factor E2F, *Cell* 65:1063-1072 (1991).
19. S. P. Chellappan, S. W. Hiebert, M. Mudryj, J. M. Horowitz, and J. R. Nevins, The E2F transcription factor is a cellular target for the RB protein, *Cell* 65:1053-1061 (1991).
20. S. H. Devoto, M. Mudryj, J. Pines, T. Hunter, and J. R. Nevins, A cyclin A-protein kinase complex possesses sequence-specific DNA binding activity: p33cdk2 is a component of the E2F-cyclin A complex, *Cell* 68:167-176 (1992).
21. M. Pagano, G. Draetta, and P. Jansen-Durr, Association of cdk2 kinase with the transcription factor E2F during S phase, *Science* 255:1144-1147 (1992).
22. M. Mudryj, S. H. Devoto, S. W. Hiebert, T. Hunter, J. Pines, and J. R. Nevins, Cell cycle regulation of the E2F transcription factor involves an interaction with cyclin A, *Cell* 65:1243-1253 (1991).
23. E. Lees, B. Faha, V. Dulic, S. I. Reed, and E. Harlow, Cyclin E/cdk2 and cyclin A/cdk2 kinases associate with p107 and E2F in a temporally distinct manner, *Genes and Dev.* 6:1874-1885 (1992).
24. S. J. Weintraub, C. A. Prater, and D. C. Dean, Retinoblastoma protein switches the E2F site from positive to negative element, *Nature* 358:259-261 (1992).
25. J. K. Schwarz, S. H. Devoto, E. J. Smith, S. P. Chellappan, L. Jakoi, and J. R. Nevins, Interactions of the p107 and Rb proteins with E2F during the cell proliferation response, *EMBO J.* 12:1013-1020 (1993).
26. S. W. Hiebert, S. P. Chellappan, J. M. Horowitz, and J. R. Nevins, The interaction of RB with E2F coincides with an inhibition of the transcriptional activity of E2F, *Genes Dev.* 6:177-185 (1992).
27. H. E. Huber, G. Edwards, P. J. Goodhart, D. R. Patrick, P. S. Huang, M. Ivey-Hoyle, S. F. Barnett, A. Oliff, and D. C. Heimbrook, Transcription factor E2F binds as a heterodimer, *Proc. Natl. Acad. Sci. USA* 90:3525-3529 (1993).
28. J. R. Nevins, E2F: A link between the Rb tumor suppressor protein and viral oncoproteins, *Science* 258:424-429 (1992).

17

ACTIVATION OF THE HEAT SHOCK TRANSCRIPTION FACTOR DURING G_1

Jacqueline L. Bruce, Brendan D. Price and Stuart K. Calderwood
Stress Protein Group
Dana Farber Cancer Institute and Joint Center for Radiation Therapy
Harvard Medical School
44 Binney Street
Boston, Ma 02115

INTRODUCTION

The regulation of the division cycles of eukaryotic cells has been intensively investigated in recent years and appears to involve both changes in gene expression and posttranslational modification of preexisting proteins[1-3]. Cell-cycle dependent transcription has been shown to be regulated by a number of transcription factors. Most of these factors are associated with the progression of cells from G_0 to G_1. The nature of the factors involved in cell cycle stage dependent transcription is more obscure although factors involved in the expression of the histone genes have been described[4] and the recent discovery of the E_2F transcription factor which may regulate events during G_1 and S has supplied the link between cell cycle dependent transcription and the cyclin-cdc2 system that drives the cell cycle[5-7]. The regulation of G_1 appears to be highly complex with at least three families of cyclins acting at different stages and with apparently discrete functions in the cell [2,8]. In the present report, we have examined the potential role of members of the heat shock transcription factor (HSF) family in cell cycle-dependent transcription. Cells from higher eukaryotic organisms express 2-3 distinct HSF genes[9,10]. The prototypical family member, HSF-1 regulates transcription of genes encoding heat shock proteins (hsp) during stress while the functions of the other family members remain largely unknown[9,11]. The rationale for investigating HSF activity during the cell cycle is that the expression of heat shock proteins, the downstream products of HSF activation is observed during the division cycles of many cell types[12-14]. We observed HSF activation during the early stages of G_1 which was associated with the transcription of heat shock genes and the synthesis of heat shock mRNA and proteins. Using antibodies specific for HSF-1 and HSF-2, we identified the species active during G_1 as HSF-1. HSF-1 activation possessed many of the properties associated with a key point in the cell cycle known as the restriction point including a transient lifetime (3hr) and a marked sensitivity to cycloheximide treatment.

RESULTS

We examined the binding of cellular proteins to the heat shock element at various times during the cell cycle of HeLa cells synchronized by serum stimulation (Fig 1).

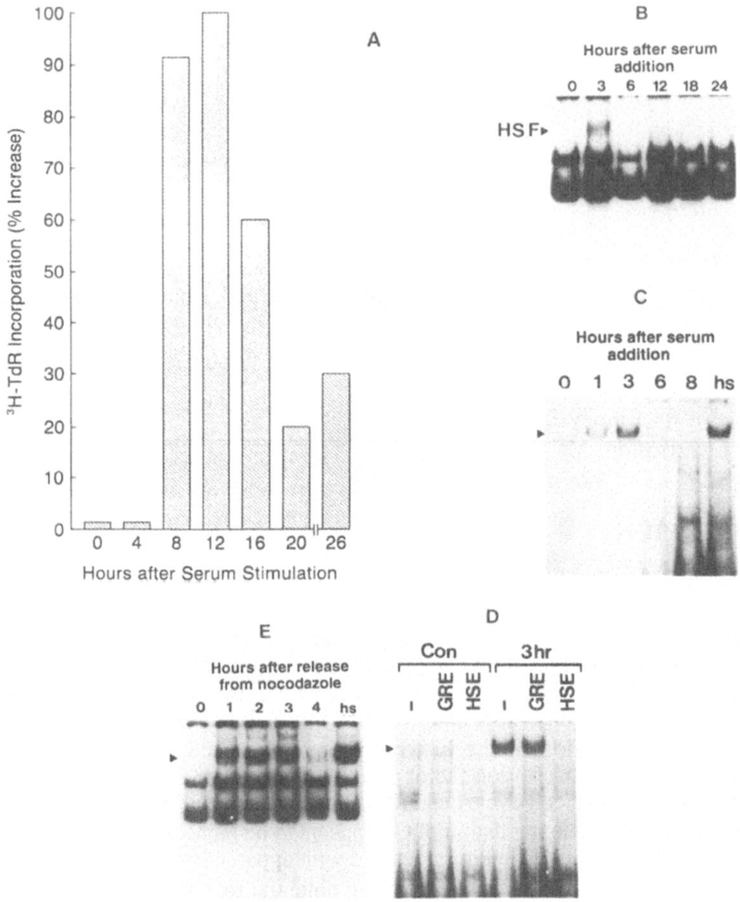

Figure 1 (A) ^{3}H-thymidine incorporation into HeLa cells induced to enter the cell cycle by serum refeeding. Cells were arrested in G_0 by 3 days culture in low serum (0.1%) and then induced to enter the cell cycle with 10% fetal calf serum (time 0). Cultures were exposed to ^{3}HTdR for 4hr periods in controls (labeled as 0) or 0-4 hr (4), 4-8 (8), 8-12 (12), 12-16 (16), 16-20 (20) and 22-26 hr after serum addition. (B) Cell cycle stage dependence of HSF-HSE binding in HeLa cells. NaCl-extracted nuclear proteins were tested for HSF binding activity after 3, 6, 12, 18, and 24 hr serum exposure. HSF extraction and binding to HSE were carried out as described in ref. 15. (C) Comparison of the magnitude of serum-induced HSF-HSE binding activation with the effects of heat shock. HeLa cells were synchronized in G_0 as described in (Fig 1A) and either untreated (C), exposed to serum for 1, 3, 6 Or 8 hr or heat shocked for 12 min at 45°C. (D) Competition of the HSF-HSE binding activity by unlabeled HSE. NaCl extracts from control cultures or after 3 hr in serum were incubated with ^{32}P labeled HSE either with no additional oligonucleotide or with a tenfold molar excess of an unrelated oligonucleotide (GRE) or HSE. (E) HSF-HSE binding in cells released from mitosis. Hela cells were arrested in M by presynchrony in thymidine, release and treatment with nocodazole. Cells were released from nocodazole arrest by washing three times in medium.

As may be observed in Fig 1A, HeLa cells remain in the G_0/G_1 phase of the cell cycle for at least four hours after serum addition and enter S phase with an approximate cell cycle time of 24 hr. We observed the appearance of a single molecular species associated with HSE using a gel shift assay and this species corresponded, in terms of electrophoretic mobility to the heat shock transcription factor (Fig1 B,C). HSF binding was observed early in the cell cycle (1 and 3 hours after serum addition; Fig1 B,C) and rapidly decayed thereafter (Fig 1B). The binding was specific by the criteria of being competed by a tenfold molar excess of unlabeled HSE but not affected by an oligonucleotide (GRE) of different sequence (Fig 1D). The data obtained using EMSA were confirmed using the methylation interference assay (not shown). In order to test for a role for HSF as a cell cycle regulated transcription factor, we examined HSF-HSE binding activity in cells arrested in M and allowed to progress into G1 phase (Fig 1E). Cells in mitosis showed no evidence of HSF activation. HSF-HSE binding was however rapidly activated by 1 hr after release from nocodazole block and persisted for 3hr before decaying to undetectable levels by 4 hr after release from the block (Fig 1E).

In order to determine which of the isoforms of HSF was activated in G_1, we carried out supershift experiments using antibodies specific for HSF-1 and HSF-2 (Fig 2). HSF activity induced 3hr after serum addition underwent a major shift in electrophoretic mobility when incubated with anti-HSF-1 antibody Ab68-3 indicating a large increase in molecular weight consistent with antibody binding (lane 8).This antibody induced a similar shift in mobility when incubated with HSF-HSE complexes from heat shocked cells (lane 2). The mobility shift was completely reversed on addition of the peptide whose sequence corresponds to that of the epitope recognized by Ab68-3 (lanes 2, 8). No significant electrophoretic mobility shift was observed in HSF-HSE complexes from heat shocked or G_1 HeLa cells using Ab3105 specific for HSF-2 (lanes 5, 11.

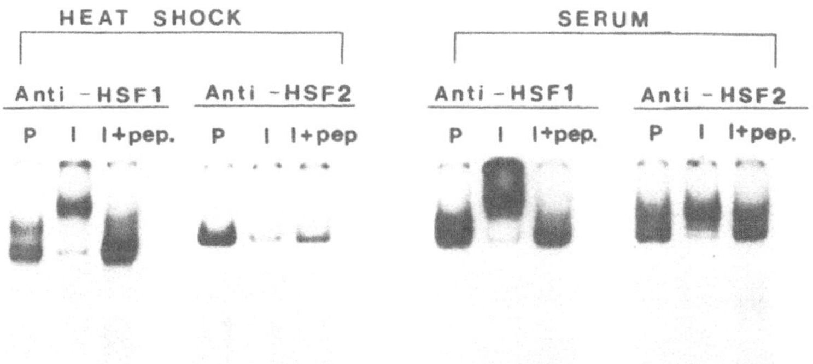

Figure 2 Identification of the HSF isoform activated during G_1 using specific anti-HSF antibodies. Extracts from either heat shocked or G_1 phase HeLa cells were incubated with ^{32}P-labelled HSE and either anti-HSF-1 Ab68-3 or anti HSF-2 Ab-3105. Extracts were incubated in the presence of either preimmune serum (P), immune serum (I), or a mixture of immune serum and blocking peptide (300uM; I+pep).

The time at which peak activity of HSF-1 occurs (5-6 hr before the onset of S phase) corresponds approximately to the restriction point in G_1 [16]. The restriction point is characterized as a time point in G_1 at which cells exposed to low concentrations of the translational inhibitor cycloheximide (CHX) become arrested presumably because they require the expression of a labile protein for progression into S phase. HSF-1 activation was similarly sensitive to CHX and was significantly reduced at 0.1 ug/ml CHX and completely inhibited at 0.5 ug/ml (Fig 3A). This circumstantial evidence indicates a potential role for HSF-1 at the restriction point. Other inhibitors which act at different stages of mRNA translation were also examined; anisomycin another inhibitor of the translocation step also effectively blocked HSF activation while puromycin, an inhibitor of peptide bond formation was entirely without effect . The data seem to rule out a role for the synthesis of specific labile protein species in HSF-1 activation. The requirement seems rather for ongoing translation, even if the products are prematurely-terminated polypeptide chains as would result from puromycin treatment.

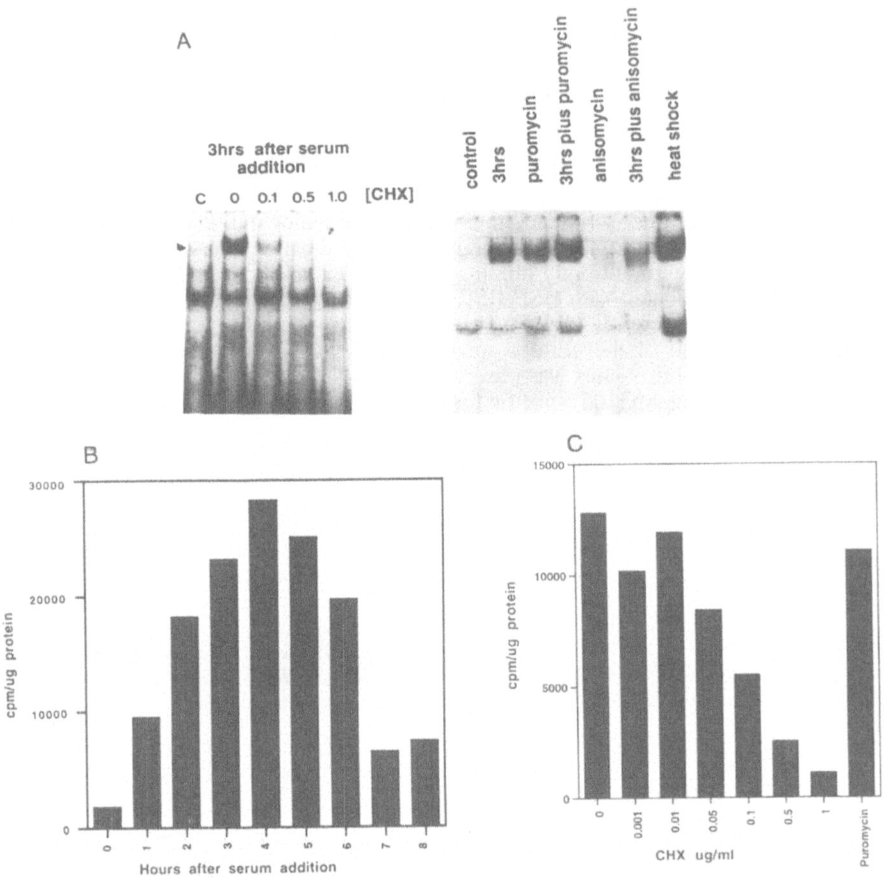

Figure 3 Effects of translational inhibitors on HSF-1 activation and protein synthesis during the cell cycle of HeLa cells. (A) Effects on HSF-HSE binding of inhibitors that act on different components of the pathway of translational elongation (B) Relative rates of protein synthesis in HeLa cells in G_0 or in the first 8 hr after serum stimulation. Protein synthesis was assayed by 1 hr pulse labelling of cell proteins with 25 uCi of ^{35}S-methionine followed by TCA precipitation of proteins and radioassay by LSC. (C) Dose-response curve for inhibition of protein synthesis by CHX.

The period of the cell cycle at which HSF-1 becomes activated corresponds to a major burst in protein synthesis (0-5 hr after entering G_1; Fig 3B indicating that HSF-1 activation and elevated protein synthesis may be causally related. When we examined the concentration dependence of CHX induced protein synthesis inhibition (Fig 3C), we found that 0.1 ug/ml CHX caused an approximate 50% decrease in ^{35}S methionine incorporation into cellular proteins and that puromycin did not significantly reduce ^{35}S-methionine incorporation at the dose used in (A).

DISCUSSION

The experiments indicate rapid activation of HSF-1 as cells, either in G_0 or in mitosis enter the G_1 phase of the cell cycle (Fig 1). The HSF species activated during this cell cycle stage is HSF-1 (Fig 2), a transcription factor previously shown to be activated by stress[11,17,18]. Activation of HSF-1 was associated with the transcription of heat shock genes and the expression of heat shock proteins (data not shown). These findings are consistent with a number of earlier studies indicating the expression of hsp70 and hsp90 å and ß after mitogenic stimulation of human cells[12-14] but this is the first demonstration of HSF-1 activation in the cell cycle. The mechanisms involved in the activation of HSF-1 by heat shock are not fully understood, although a number of steps, including HSF oligomerization, nuclear localization, complex formation with DNA and phosphorylation appear to be involved[11,17,18]. Much evidence suggests that HSF activity is repressed during normal metabolism due to binding to its downstream products, the heat shock proteins[19,20,21]. The hsps could form a chaperonin complex that binds HSF-1 in an inactive state and during stress the chaperonin complex may be competed for by denatured and aggregated proteins, releasing free HSF-1 to direct the transcription of heat shock genes[20,21]. The mechanism by which HSF-1 activation occurs in G_1 is not clear, although its antagonism by translational inhibitors CHX, anisomycin and emitine indicates a distinct pathway compared to heat shock activation which appears to occur posttranslationally[18,20]. Treatment with puromycin, an agent that causes the premature release of polypeptide chains from ribosomes without interrupting translation *per se*, is permissive for HSF-1 activation, indicating a requirement for the process of mRNA translation but not for mature protein species in HSF activation (and cell cycle progression). Such a requirement is in accord with some current models for the regulation of both HSF-1 activity and the heat shock response, which indicate that the level of free hsp70 may be the important parameter[20]. Loss of free hsp70 during heat shock by sequestration in protein aggregates [20,21] or by binding to nascent or partially completed polypeptide chains in G_1 [22] may activate the response.

The G_1 period of the cell cycle is required for cells to synthesize proteins and other products needed for S phase and subsequent cell division and is prolonged in undersized cells such as daughter cells formed during the cell division in *S.cerevisiae* and abbreviated in large cells with an abundance of stored macromolecules such as *Xenopus* oocytes[1,8]. In addition, a critical rate of protein synthesis is required in order for cells to enter the cell cycle[24]. Experiments in *S.cerevisiae* have indicated the involvement of a putative size sensor or "critical threshold requirement" in order for cells to activate the CLN1 and CLN2 genes required to be the functional equivalent in *S.cerevisiae* of the restriction point in mammalian cells. Activation of HSF may play a role in regulating the duration of G_1 by monitoring the tranlsational burst that occurs early in the cell cycle (Fig 3C). It has long been known that the restriction point in the G_1 stage of mammalian cells is exquisitely sensitive to CHX[16,23]. HSF-1 activation, an early event in the G_1 is inhibited by doses of CHX which block cell cycle progression at the restriction point(Fig 3). The studies of Pardee *et al.* indicate a CHX-tempting sensitive process with a half life similar to that of HSF-1 activation (ref16;Fig1) and it is thus tempting to speculate that HSF-1 activation is required to traverse the restriction point. It may be significant that heat shock blocks the onset of START in *S.cerevisiae* and inhibits the expression of the CLN1 and CLN2 genes[27]. Although there are no exact homologs to the CLN genes in mammalian cells, they appear to be related to the G_1 cyclins C, D and E[8]. One might thus predict that the activation of HSF-1 involved in suppressing expression of G_1

cyclins and delaying the commitment to cell division until adequate protein synthesis/cell size is attained. Recent evidence indeed indicates a role for cyclins E and A at the restriction point of mammalian cells[28].

REFERENCES

1. Murray, A.W., and Kirschner, M.C. (1989) Cyclin synthesis drives the early embryonic cell cycle. Nature 339, 275-286.
2. Hunter T. and Pines, J. (1991) Cyclins and Cancer Cell 66: 1071-1074.
3. Meek, D.W., and Street, A.J. (1992) Nuclear protein phosphorylation and growth. Biochem. J. 287, 1-15.
4. La Bella, F., Gallinari P., McKinney, J., and Heintz N. (1989) Histone H1, subtype specific consensus elements mediate cell cycle regulated transcription *in vitro* Genes Dev 3, 1982-1990.
5. Mudryj. M., Devoto, S., Hiebert, S.W., Hunter, T., Pines, J., and Nevins, J.R. (1991) Cell cycle regulation of the E2F transcription factor involves interaction with Cyclin A. Cell 65, 1243-1253.
6. Lees, E., Faha, B., Dulic V., Reed, S.I., and Harlow E. (1992) Cyclin E/ CDK2 and cyclin A/ cdk2 kinases associate with p107 and E2F in a temporally distinct manner. Genes Dev. 6, 1874-1883.
7. Nevins, J.R. (1992) E_2F; a link between the Rb tumor suppressor protein and viral oncoproteins. Science 259, 424-427.
8. Lew, D.J., Marini, N., and Reed, S.I. (1992) Different G1 cyclins control the timing of cell cycle commitment in mother and daughter cells of the budding yeast *S. cerevisiae*. Cell 69, 317-327.
9. Schuetz, T.J., Gallo, G.J., Sheldon, L., Tempst, P., and Kingston, R., E. (1991) Isolation of a cDNA for HSF-2; evidence for two heat shock factors in humans. Procs. Natl. Acad. Sci. (USA) 88, 6910-6915.
10. Nakai, A., and Morimoto, R.I. (1993) Characterization of a novel chicken heat shock transcription factor, HSF-3 suggests a new regulatory pathway. Mol. and Cell. Biol. 13, 1983-1987.
11. Baler, R., Dahl, G. and Voellmy, R.(1993) Activation of human heat shock genes is accompanied by oligomerization, modification and rapid translocation of heat shock transcription factor HSF-1. Mol and Cell. Biol. 13: 2486-2496.
12. Haire, R.N., and O'Leary J.J. (1988) Mitogen-induced preferential synthesis of proteins in the Go to S transition in human lymphocytes. Exp. Cell Res. 179: 65-78.
13. Hansen L.K., Houchins, J.P. and O'Leary, J.J. (1991) Differential regulation of hsc70, hsp70, hsp90 å and hsp90B mRNA expression by mitogen expression and heat shock in human lymphocytes. Exp. Cell Res. 192: 587-596.
14. Kao, H.T., Capasso O., Heintz N. and Nevins J. (1985) Cell cycle control of the human hsp70 gene: implications for the role of acellular E1A-like function. Mol. and Cell. Biol. 5: 628-633.
15. Price B.D. and Calderwood, S.K. (1991) Ca^{++} is essential for multistep activation of the heat shock transcription factor in permeabilized cells. Mol. and Cell. Biol. 11, 3365-3368.
16. Pardee, A.B. (1989) G_1 events and regulation of cell proliferation. Science 246, 603-608.
17. Zimarino, U., Tsai, C. and Wu, C. (1990) Complex modes of heat shock factor activation. Mol. and Cell. Biol. 10, 752-759.
18. Sarge K.D., Murphy, S.P., and Morimoto, R.I. (1993) Activation of heat shock transcription factor 1 involves oligomerization, acquisition of DNA binding activity and nuclear localization and can occur in the absence of stress. Mol. and Cell. Biol. 13, 1392-1407.
19. Sorger, P.K. (1991) The heat shock transcription factor and the heat shock response. Cell 65, 363-366.
20. Abravaya K.,Phillips, B. and Morimoto, R. 1991 Attenuation of the heat shock response in HeLa cells is mediated by release of bound HSF and is modulated by changes in growth and in heat shock temperature. Genes and Development 5: 2117-2127
21. Craig, E.A. and Gross, C.A. (1991) Is hsp70 the cellular thermometer? TIBS 16:135-140.

22. Beckman, R.P., Mizzen, L.A. and Welch, W.J. (1990) Interaction of hsp70 with newly-synthesized proteins; implications for protein folding and assembly. Science 248: 850-854.
23. Hartwell, L.H. and Unger M.W. (1977) Unequal cell division in *S. cerivisiae* and its implications for the control of cell division. J. Cell Biol. 75: 422-435.
24. Moore, S.A. (1988) Kinetic evidence for a critical rate of protein synthesis in the *S. cerivisiae* cell cycle. J. Biol. Chem. 263, 9674-9681.
25. Johnston G.C., Pringle J.R. and Hartwell, L.H. (1977) Coordination of growth with cell division in the yeast *S. cerevisiae*. Exp. Cell Res. 105: 79-98.
26. Dirick, L. and Nasmyth, K. (1991) Positive feedback control in the activation of G1 cyclins in Yeast. Nature 351: 754-757.
27. Rowley, A., Johnstone, G.C., Butler, B., Werner-Washburne, M., and Singer, R.A. (1993) Heat shock-mediated cell blockage and G_1 cyclin expression in the yeast *saccharomyces cerevisiae*. Mol. and Cell. Biol. 13, 1034-104.
28. Duo, Q.P., Levin, A., Zhao C. and Pardee, A.B. (1993) Cyclin E and cyclin A as candidates for the restriction point protein. Cancer Res. 53: 1493-1497.

18

DNA DAMAGE AND CELL CYCLE REGULATION IN *S. CEREVISIAE*

Stephen J. Elledge, Zheng Zhou, James B. Allen, Tony A. Navas, and William J. Jones

Department of Biochemistry
Baylor College of Medicine
One Baylor Plaza
Houston, TX 77030

INTRODUCTION

The capacity to efficiently sense and respond to environmental stress is central to an organism's ability to undergo complex developmental transformations and to successfully adapt to changing environmental conditions. Two fundamental types of sensory networks appear to be ubiquitous among organisms; the ability to recognize and respond to thermal shock (1-3), and the ability to recognize and respond to DNA damage (4-14). In addition, several genes have been identified which are transcriptionally activated in response to both types of stress (15, 16), suggesting a potential interaction between both sensory networks. In every organism which has been examined, cells respond to DNA damage in two distinct ways. They arrest the progression of the cell cycle at G1 or G2 and they induce the transcription of genes involved in repair of DNA damage. Damage induced cell cycle arrest, often referred to as the DNA damage checkpoint, is thought to provide additional time for DNA repair, allowing the cell to remove damaged DNA prior to entry into S-phase or mitosis. DNA damage present in S-phase may lead to increased frequencies of mutations or double strand breaks. Post-replicative recombinational repair in G2 can result in physical linkage of sister chromatids due to crossing over and if present during mitosis can produce chromosome breakage, translocations, rearrangements, aneuploidy and abnormal chromosome segregation. Therefore a mechanism that ensures DNA damage is absent prior to entry into S or M phase ensures the fidelity of inheritance of genetic information. A large number of genes are inducible in response to DNA damage. These include genes involved in excision repair, recombinational repair, DNA synthesis, and deoxyribonucleotide synthesis. Induction of these genes facilitates DNA repair and in theory, reduces the amount of time cells spend in the arrested state.

REGULATION OF DNA DAMAGE INDUCIBLE GENES

Both procaryotes and eucaryotes are able to sense and respond to DNA damage. However, this sensory network is well understood only in the procaryote *Escherichia coli*. Treatment of *E. coli* with agents that damage DNA or block replication causes the appearance of a set of physiological responses that include the induction of DNA repair processes, mutagenesis, and induction of lysogenic bacteriophage (for reviews see refs. 13

The Cell Cycle: Regulators, Targets, and Clinical Applications
Edited by V.W. Hu, Plenum Press, New York, 1994

and 14). These processes have collectively been called the SOS response because at least some of them appear to promote cell survival. In all, over 20 genes have been identified which are transcriptionally activated in response to DNA damage. The molecular mechanism of this coordinately regulated response involves the proteolytic inactivation of a common repressor, the LexA protein, by an activated form of the RecA protein. The RecA protein can become activated to facilitate proteolysis by binding to single-stranded DNA, a possible damage signal.

Unlike procaryotes, little is known about how eucaryotes respond to DNA damage. To approach this problem, we have studied the regulation of a family of DNA damage-inducible genes in yeast, the *RNR* family encoding the enzyme ribonucleotide reductase. Ribonucleotide reductase catalyzes the first step in the pathway for the production of the deoxyribonucleotides needed for DNA synthesis. The ribonucleotide reductase genes are the only genes known to be both cell cycle regulated and DNA damage inducible in all organisms examined from procaryotes to eucaryotes. It is an enzyme of structure $\alpha_2\beta_2$. In yeast, the small subunit is encoded by *RNR2* (17-20), and the large regulatory subunit is encoded by two homologous genes, *RNR1* and *RNR3* (23) *RNR3* is an alternative regulatory subunit that is only expressed in the presence of DNA damage, nucleotide depletion, or blocks in DNA replication. The properties of the *RNR* genes are shown in Table 1 (for recent review see ref. 22). *RNR3* is the most inducible of the *RNR* genes and was chosen for examination of the pathway involved in transducing the damage signal.

Table 1. Properties of the *S. cerevisiae* genes encoding ribonucleotide reductase.

Gene	Subunit	Null Phenotype	Cell Cycle Regulation	DNA Damage Inducibility
RNR1	Large	Lethal (cdc)	Yes (Strong)	5-Fold
RNR2	Small	Lethal (cdc)	Yes (Weak)	25-fold
RNR3	Large	Viable	Unknown	>100-fold

To identify genes in the signalling pathway mediating this DNA damage response, we have designed a selection for isolating spontaneous *trans*-acting mutations that alter *RNR3* expression using a chromosomal *RNR3-URA3* transcriptional fusion and an *RNR3-lacZ* reporter plasmid(23). Using this system, we isolated 202 independent *trans*-acting *crt* (constitutive RNR3 transcription) mutants that express high levels of *RNR3* in the absence of DNA damaging agents which fall into 9 complementation groups(23). In some *crt* groups, the expression of *RNR1* and *RNR2* are also elevated, suggesting that all three *RNR* genes share a common regulatory pathway. Mutations in most *CRT* genes confer additional phenotypes, among these are hydroxyurea sensitivity, temperature sensitivity, and slow growth. Five of the *CRT* genes have been identified as previously cloned genes: *CRT4* is *TUP1*, *CRT5* is *POL1/CDC17, CRT6* is *RNR2*, *CRT7* is *RNR1*, and *CRT8* is *SSN6* . *crt6-68* and *crt7-240* are the first ts alleles of *RNR2* and *RNR1*, respectively, and arrest with a large budded, *cdc* terminal phenotype at the nonpermissive temperature. The isolation of *crt5-262*, an additional *cdc* allele of *POL1/CDC17*, suggests for the first time that directly blocking DNA replication can provide a signal to induce the DNA damage response(23).

Using the same selection but in the opposite direction, we have identified 16 mutants that were uninducible for transcription of the *RNR3* gene in cells treated with DNA damaging agents or hydroxyurea (HU) an inhibitor of ribonucleotide reductase. These fell into five complementation groups. We call these genes *DUN* for DNA damage uninducible. In addition to affecting *RNR3* regulation, *dun* mutants also block inducibility of other DNA damage inducible genes such as *RNR1* and *RNR2*. This suggests that the *RNR* family forms a regulon. *DUN1* was cloned and was shown to encode a protein kinase (Zhou and Elledge, unpublished). This confirms previous data that demonstrated that the induction of the *RNR2* transcript in response to DNA damage was not dependent on protein synthesis. Thus, we think that protein phosphorylation is involved in transducing the DNA damage signal. We are currently trying to understand how this signal is transduced. Is the activity of *DUN1* altered in response to DNA damage? If so, how? We are looking for substrates using the two hybrid system to identify other proteins that associate with *DUN1*.

COORDINATION OF S-PHASE AND MITOSIS

A second avenue of research has been the coordination of S-phase with mitosis. Cells in S-phase actively inhibit entry into mitosis and we are interested in finding mutants defective in this pathway to learn how cells sense S-phase. We have looked for *sad* mutants (S-phase arrest defective) that enter mitosis in the presence of unreplicated DNA. We screened for hydroxyurea sensitive mutants that died rapidly in the presence of HU. This lethality could be suppressed by prior arrest in G1 with α-factor. The fourth phenotype we tested was the presence of abnormal mitosis in the presence of HU. Seven mutants were identified that fit these criteria and fell into five complementation groups, *sad1* to *sad5*. *sad1* mutants have the additional phenotype of being uninducible for *RNR3* transcription with DNA damage (Allen and Elledge, unpublished). This suggests the DNA damage sensing and S-phase sensing pathways overlap. *SAD1* has been cloned and encodes a protein kinase (Allen and Elledge, unpublished). It is an essential gene. We are currently in the process of making ts mutants to identify other components of the pathway. The connection between the DNA damage and S-phase signalling pathways suggests that *SAD1* may also control the DNA damage cell cycle arrest pathway controlled by *RAD9*(24), although *rad9* mutants are proficient for the S-phase arrest pathway (21). We are testing this hypothesis by looking at various *sad1cdc* mutant combinations to determine if any of the *cdc* mutants that depend on *RAD9* for their cell cycle arrest, also depend upon *SAD1*. A summary of our working model for the interaction of the DNA damage and S-phase sensory pathways in the regulation of transcriptional induction and cell cycle arrest is shown in Fig. 1. In summary, the identification of the *DUN1* and *SAD1* kinases provides an entry point from which we may begin to unravel the DNA damage and S-phase sensory networks.

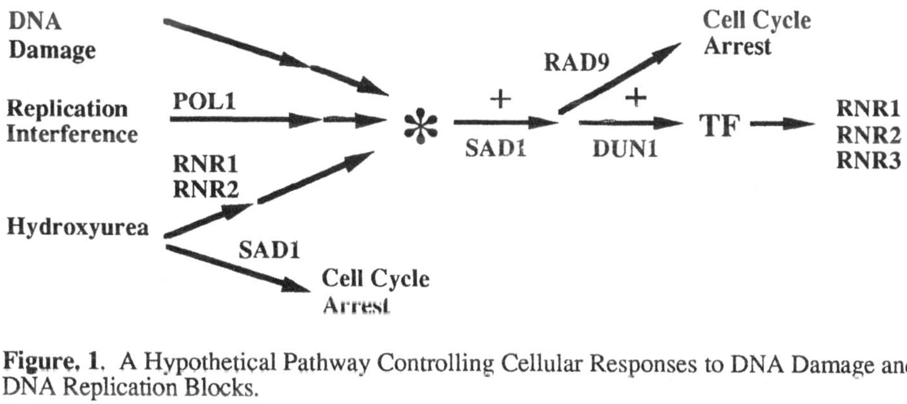

Figure. 1. A Hypothetical Pathway Controlling Cellular Responses to DNA Damage and DNA Replication Blocks.

The names of genes involved are in capital letters. TF stands for transcription factor. * represents a hypothetical DNA damage signal that may or may not be shared among the different stimuli. The *RAD9* gene is placed after *SAD1* because it is not needed for transcription of *RNR*, however this is arbitrary at this point in time because mechanisms can be imagined that could allow for a role upstream of *SAD1*. Epistasis analysis will be required to establish the true order.

This work was supported by a National Institutes of Health grant, GM44664, and a grant, Q1186, from the Robert A. Welch Foundation to S.J.E. S.J.E. is a PEW Scholar in the Biomedical Sciences and Z.Z. is a Robert A. Welch Predoctoral Fellow.

REFERENCES

1. S. Lindquist, The heat shock response. *Ann. Rev. Biochem.* 55:1151-1191 (1986).
2. E. A. Craig, The heat shock response. *Crit. Rev. Biochem.* 18:239-280 (1985).
3. M. Bienz and H.R.B. Pelham, Mechanisms of heat-shock activation in higher eukaryotes. *Adv. Genet.* 24:31-72 (1987).
4. P. Angel, A. Poting, U. Mallick, H.J. Rahmsdorf, M. Schorpp, and P. Herrlich, Induction of metallothionein and other mRNA species by carcinogens and tumor promoters in primary human skin fibroblasts. *Mol. Cell. Biol. 6*:1760-1766 (1987).
5. M.J. Defais and P.C. Hanawalt, *Adv. Rad. Biol.* 10:1-37(1983).
6. T. Kartasova, B.J.C. Cornelissen, P. Belt, and P. van de Putte, Effects of UV, 4-NQO and TPA on gene expression in cultured human epidermal keratinocytes. *Nucl.Acid Res.* 15:5945-5962 (1987).
7. M.W. Liberman, L.R. Beach, and R.D. Palmiter, Ultraviolet radiation inducedmetallothionein-I gene activation is associated with extensive DNA demethylation. *Cell* 35:207-214 (1983).
8. T. McClanahan and K. McEntee. Specific transcripts are elevated in *Saccharomyces cerevisiae* in response to DNA damage. *Mol. Cell. Biol.* 4:2356-2363 (1984).
9. S.W. Ruby and J.W. Szostak, Specific *Saccharomyces cerevisiae* genes are expressed in response to DNA-damaging agents. *Mol. Cell Biol.* 5:75-84 (1985).
10. S.W. Ruby , J.W. Szostak, and A.W. Murray, Cloning regulated yeast genes from a pool of *lacZ* fusions. *Methods Enzymol.* 101:253-269 (1983).
11. L. Samson and J.L. Schwartz, Evidence for an adaptive DNA repair pathway in CHO and human skin fibroblast cell lines. *Nature* 287:861-863 (1980).
12. K. Valerie, A. Delers, C. Bruck, C. Thiriart, H. Rosenberg, C. Debouck, and M. Rosenberg, Activation of human immunodeficiency virus type 1 by DNA damage in human cells. *Nature* 333:78-80 (1988).
13. G.C. Walker, . Mutagenesis and inducible responses to deoxyribonucleic acid damage in *Escherichia coli. Microbiol. Rev.* 48:60-93 (1984).
14. G.C. Walker, Inducible DNA repair systems. *Ann. Rev. Biochem.* 54:425-457 (1985).
15. T. McClanahan and K. McEntee, . DNA damage and heat shock dually regulate genes in *Saccharomyces cerevisiae. Mol. Cell Biol.* 6:90-96 (1986).
16. J.H. ,Krueger and G.C. Walker, *groEL* and *dnaK* genes of *Escherichia coli* are induced by UV irradiation and nalidixic acid in an *htpR*+-dependent fashion. *Proc. Natl. Acad. Sci. USA* 81:1499-1503 (1984).
17. S.J. Elledge and R. W. Davis, Identification and isolation of the gene encoding the small subunit of ribonucleotide reductase from *Saccharomyces cerevisiae*: A DNA damage-inducible gene required for mitotic viabilty. *Mol. Cell. Biol.* 7: 2783-2793 (1987).
18. H.K. Hurd, C. W. Roberts and J. W. Roberts, Identification of the gene for the yeast ribonucleotide reductase small subunit and its inducibility by methyl methanesulfonate. *Mol. Cell. Biol.* 7: 3673-3677 (1987).
19. S.J. Elledge and R.W. Davis, DNA-damage induction of ribonucleotide reductase. *Mol. Cell. Biol.* 9: 4932-4940 (1989).
20. S.J. Elledge and R.W. Davis, Identification of a damage regulatory element of *RNR2*, and evidence that four distint proteins bind to it. *Mol. Cell. Biol.* 9: 5373-5386 (1989).
21. S.J. Elledge and R.W. Davis, Two genes, differentially regulated by DNA damage and the cell cycle, encode alternate regulatory subunits of ribonucleotide reductase. *Genes Dev.* 4: 740-751 (1990).
22. S.J. Elledge, Zhou, Z., and J.B. Allen, Ribonucleotide reductase: regulation, regulation, regulation. *Trends Biochem. Sci.* 17: 119-123 (1992).
23. Zhou, Z. and S.J. Elledge, Analysis of the DNA damage sensory pathway through the isolation and characterization of the *CRT* mutants, constitutive for *RNR3* transcription. *Genetics* 131:851-866 (1992) .
24. T.A., Weinert and L.H. Hartwell, The *RAD9* gene controls the cell cycle response to DNA damage in *Saccharomyces cerevisiae. Science.* 241: 317-322 (1988).

19

PREFERENTIAL REPAIR OF CISPLATIN ADDUCTS IN THE HUMAN DHFR GENE DURING G_1 PHASE ASSAYED WITH T4 DNA POLYMERASE

Nicholas J. Rampino* and Vilhelm A. Bohr

Laboratory of Molecular Genetics, NIA, NIH
4940 Eastern Ave, Baltimore MD 21224

SUMMARY

We use a novel assay to detect strand specific repair in genes after cellular exposure to cisplatin at IC_{50} levels. Single stranded DNA capable of hybridizing to gene specific probes is generated enzymatically by the 3'-5' exonuclease activity of T4 DNA polymerase. In the presence of cisplatin adducts, the exonuclease activity of this enzyme is blocked, preventing the formation of single stranded DNA, and thus lowering the amount of complementary sequence available for probe hybridization.[1, 2]

Human ovarian carcinoma cells synchronized in M phase were treated at the beginning of G_1 phase with 5 uM cisplatin for 1 hr, and then allowed to repair in drug free media. Repair during early G_1 phase in the unexpressed δ-globin gene was compared to that in a late growth-regulated gene, dihydrofolate reductase, that is involved in nucleotide biosynthesis, and normally shows induction at the G_1-S boundary. Extensive cisplatin adduct removal was measured in the dihydrofolate reductase gene after four hours of cellular repair, whereas there was no evidence of repair in the δ-globin gene.

Finally, we show that this assay utilizing T4 DNA polymerase can detect low levels of DNA oxidation, brought about by cellular exposure to hydrogen peroxide.

INTRODUCTION

Cisplatin is a chemotherapeutic agent used in the treatment of a variety of human cancers, where its cytotoxicity involves the formation of DNA adducts.[3] One method aimed at measuring gene specific repair after DNA damage by cytotoxic agents has utilized endonuclease enzyme treatment and alkaline gel electrophoresis.[4] Using

* Corresponding author

The Cell Cycle: Regulators, Targets, and Clinical Applications
Edited by V.W. Hu, Plenum Press, New York, 1994

such a method it has been shown, for A2780 human ovarian carcinoma cells treated with cisplatin, that the DHFR gene is repaired more efficiently than a non-transcribed region of the genome.[5] However, it is noteworthy that the cisplatin dose used was 300 uM, 100 times the IC_{50} point value, and that the experiments were done in asynchronous cells. DNA repair in the late S and G_2 phase of the mammalian cell cycle has recently been investigated, and it was reported that repair rates of the transcribed and non-transcribed genes are similar during these stages of the cell cycle.[6] In the yeast *Saccharomyces cerevisiae*, there has been a tentative report that DNA repair discriminates between transcriptionally active and inactive DNA during the G_1 and S phases but not in the G_2 phase.[7, 8]

Inhibition of the synthetic activity of *Thermus aquaticus* (TAQ) polymerase by cisplatin has been exploited in Polymerase Chain Reaction (PCR) assays of gene specific repair.[9, 10] While demonstrating the advantage of not requiring specific enzymes to recognize and cleave at DNA adducts, such PCR-based assays exploit an exponential amplification process making quantitation difficult. Moreover, it appears that the PCR-based assays are not appropriate for determining strand biased repair. In a defined 43 base pair DNA sequence, it has been shown using a Maxam and Gilbert sequencing approach that the 3'-5' exonuclease activity of T4 DNA polymerase cannot hydrolyze the cyclobutane dimers or (6-4) photoproducts produced by ultraviolet light irradiation.[11] Here, using a novel approach (figure 1) to distinguish levels of platination, where single stranded DNA capable of hybridizing to gene specific probes is generated enzymatically by T4 DNA polymerase, rates of DNA repair, measured from residual platination levels, are assessed from hybridization intensities. This assay is inherently strand specific, and is sensitive to small differences in levels of DNA damage.

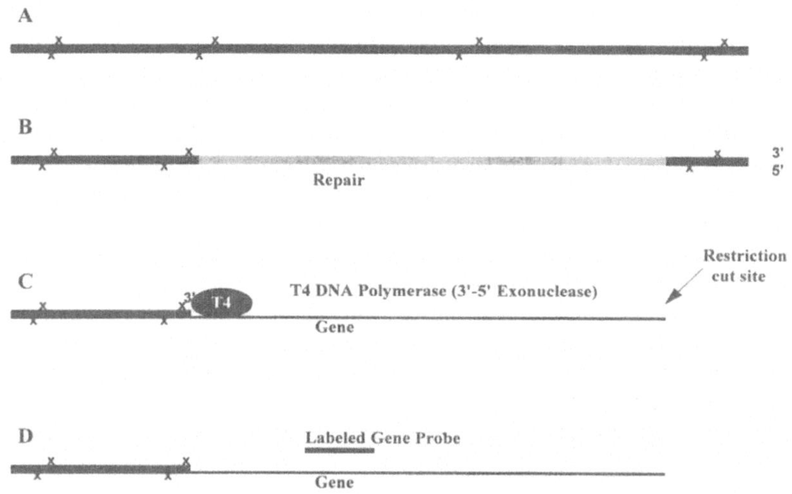

Figure 1. Schematic representation of the T4 DNA polymerase assay used to distinguish the degree of gene specific repair after low levels of DNA damage. (A) Initial distribution of cisplatin-DNA adducts, each represented as an "x", (B) Preferential repair removes adducts from specific regions in the genome, (C) To assay for gene specific repair the genomic DNA is restricted with an appropriate endonuclease, and incubated with T4 DNA polymerase to produce single stranded DNA in the regions that were free of lesions. In the absence of dNTPs the enzyme functions as a 3'-5' exonuclease that, starting at a free 3' end, will hydrolyze away one strand until it encounters a blocking lesions, (D) This DNA, single stranded in the region that was free of lesions, is slot blotted onto a nylon membrane, and hybridized against a labeled gene probe. The degree of probe binding is a measure of the proportion of molecules that were free of cisplatin adducts in the region between the restriction endonuclease cut site, and the distal end of the gene probe sequence.

RESULTS

Human A2780 ovarian carcinoma cells (cisplatin IC_{50} = 3 uM),[12] synchronized at M phase by shaking and collection, were plated and allowed 2 hrs to divide and enter G_1 phase before cisplatin treatment. The degree of synchrony.for this cell type, using the mitotic shake method, was measured by FACS analysis. Typically, by 1 hr after plating, less than 10% of the M phase cells had failed to complete division, and enter G_1 phase.

To measure the isolated effect of cisplatin dose on hybridization intensity, the synchronized cells were treated with either 2.5 or 5 uM cisplatin for 1 hr. The DNA was immediately isolated, restricted, treated with T4 DNA polymerase, and hybridized against a digoxigenin labeled DHFR gene probe. The filter hybridization intensities were measured densitometrically. A significant ($p<0.005$) difference in the hybridization profiles for DNA from the cells, treated with the two doses of cisplatin, was measured, with the lower dose exhibiting the more intense hybridization by 40%.

To assay for gene specific repair, the synchronized cells were treated with 5 uM cisplatin for 1 hr, and then allowed to grow in drug free media for 4 hr before the DNA was isolated. The results, normalized against the respective unplatinated controls, of filter hybridizing the T4 polymerase treated DNA against a labeled DHFR or δ-globin gene probe are shown in figure 2.

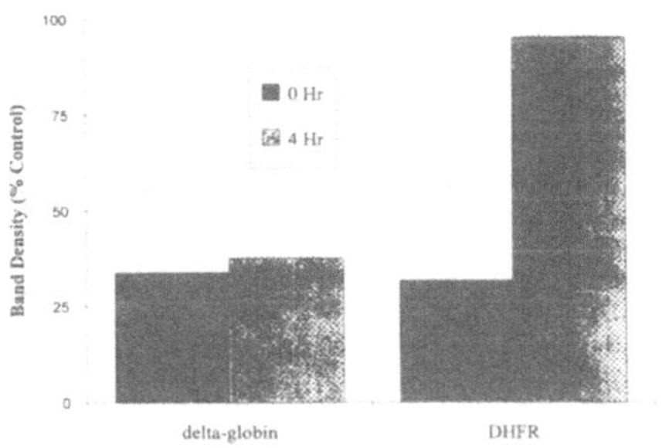

Figure 2. Graph comparing the hybridization intensities of human DNA, isolated from synchronized G_1 phase cells treated with 5 uM cisplatin, then allowed 0 or 4 hr of repair, probed for δ-Globin and DHFR.

After 4 hr of cellular repair, the degree of binding for the DHFR gene probe was significantly higher than that of the δ-globin gene probe, figure 2. At the DHFR gene locus there appears to be very rapid removal of cisplatin adducts from the transcribed strand during early G_1 phase, while at the non transcribed δ-globin gene locus cisplatin adducts appear to persist.

DISCUSSION

The enzyme T4 DNA polymerase carries an endogenous 3'-5' exonuclease activity that, in the absence of dNTPs, hydrolyzes both single and double stranded DNA, and there is a decreasing efficiency of strand digestion with increasing drug dose. The fraction of DNA fragments that are free of cisplatin adducts in the flanking region, from the restriction endonuclease cut site to the distal end of the target gene sequence, determines the proportion of molecules that can be made single stranded by incubation with T4 DNA polymerase. Once these molecules are enzymatically made single stranded in the flanking region, they are capable of binding a labeled probe molecule, and in doing so, contribute to the hybridization signal, figure 1.

This *in vitro* assay may give information as to the extent at which particular DNA lesions are detected and repaired *in vivo*. Like T4 polymerase,[13] mammalian DNA polymerase ε, required for long-patch DNA repair, possess endogenous exonuclease activity that hydrolyses both single and double stranded DNA substrates, with single-stranded DNA being degraded more rapidly. Mammalian DNA polymerase δ, implicated in DNA excision repair, is highly dependent on a cofactor, proliferating cell nuclear antigen (PCNA), for its significant exonuclease activity. The proteins coded for by the T4 phage genes 43 (DNA polymerase), 45, and 44/62 substitute *in vitro* for mammalian pol δ, PCNA and RF-C (replication factor C), respectively.[13] Lesions that block the exonuclease activity of T4 DNA polymerase may similarly act to block mammalian DNA polymerases. If so, then a component of their repair activity may involve arrest at a DNA lesion. In the case of transcription, evidence that polymerase arrest is important for DNA repair has appeared recently.[14-16]

The rapid and dramatic level of preferential repair that we measure here may be due to several factors. First, higher drug doses may kill, where by using drug doses near the IC_{50} point we may be allowing cells to manifest physiologically relevant levels of DNA repair. In addition, by focusing on repair in G_1 phase of the cell cycle, when genes are being actively transcribed, we may be detecting higher levels of transcription coupled DNA repair that is gene specific. In particular, this may be the case for a gene like DHFR that is involved in nucleotide synthesis, and whose expression may act to replenish the nucleotide pool during DNA repair processes.

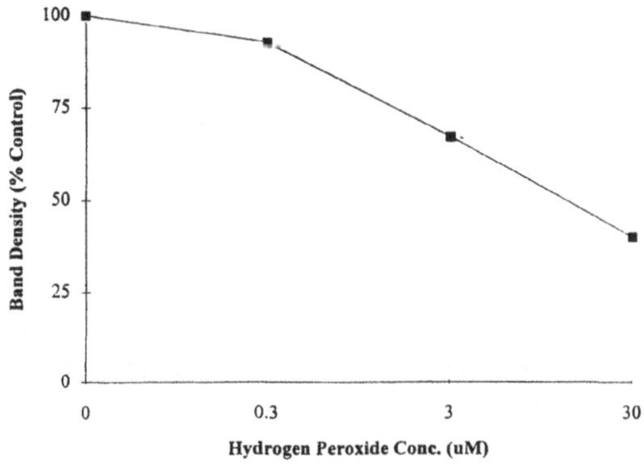

Figure 3. Intensity of hybridization for the DHFR gene probe to DNA isolated from synchronized G_0 phase WI-38 human diploid fibroblasts exposed to oxidative stress from hydrogen peroxide. The cells were synchronized in G_0 by starvation in media containing 0.2% FBS, and then these cells were treated with hydrogen peroxide concentrations of 0 mM, 0.3 mM, 3 mM, and 30 mM for 45 min. The DNA was isolated, cut with Hind III, and incubated with T4 DNA polymerase. This DNA was slot blotted, and then filter hybridized against the labeled DHFR gene probe. The intensity of hybridization varies linearly with the level of H_2O_2 treatment (r = 0.98).

PERSPECTIVE

While the repair and removal of DNA adducts induced by therapeutic agents, such as cisplatin, is of great interest, so is the repair of endogenous DNA damage. High levels of DNA oxidation occur during normal metabolism, and oxidants have long been implicated as major contributors to aging.[17] There are, to date, no studies measuring gene specific repair after DNA oxidation. The T4 DNA polymerase assay, introduced here, appears to be exquisitely sensitive to oxidative lesions, and in figure 3 dose response data is presented to demonstrate the sensitivity of the assay. The stress response after cellular exposure to H_2O_2, measured by increased levels of specific mRNA's, is seen in the 0.3-3 mM range.[18]

The expression of genes, as measured by mRNA levels, is modulated over the course of the cell cycle. Whether preferential repair targets each gene in a defined window during the cell cycle is unknown. Such a window of repair would presumably directly proceed transcription, and may be strand specific. The human diploid fibroblast WI-38 cell line has proved to be a convenient system for studying cell cycle dependent gene expression.[19, 20] Discovering the biochemical pathways that modulate DNA repair kinetics, and affect cell cycle progression is a major focus of this research.

REFERENCES

1. N.J. Rampino, An assay for determining the effects of therapeutic agents and oxidative agents on individual DNA molecules. U.S. Patent 07921182 pending.
2. N.J. Rampino and P.G. Johnston, Microfluorometric size measurements of human and phage DNA correlate with the degree of *in vitro* cisplatin treatment, Biochem. Biophys. Res. Commun. 179: 1344 (1991).
3. S.J. Loeher and L.H. Einhorn, Diagnosis and treatment. Drugs five years later. Cisplatin, Ann. Intern. Med. 100: 704 (1984).
4. V. Bohr, D. Okumoto and P.C. Hanawalt, Survival of UV-irradiated mammalian cells correlates with efficient DNA repair in essential gene, Proc. Natl. Acad. Sci. USA 83: 3830 (1986).
5. W. Zhen, C.J. Link, P.M. O'Connor, E. Reed, R. Parker, S.B. Howell and V.B. Bohr, Increased gene-specific repair of cisplatin interstrand cross-links in cisplatin-resistant human ovarian cancer cell lines, Mol. Cell Biol. 12: 3689 (1992) .
6. G. Russev and T. Boulikas, Repair of transcriptionally active and inactive genes during S and G_2 phases of the cell cycle, Eur. J. Biochem. 204: 267 (1992).
7. C. Terleth, R. Waters, J. Brouwer, and P. van de Putte, Differential repair of UV damage in Saccharomyces cerevisiae is cell cycle dependent, EMBO J. 9: 2899 (1990).
8. C. Terleth, R. Waters, J. Brouwer, and P. van de Putte, Erratum EMBO J. 3: 1228 (1992).
9. M.M. Jennerwien and A. Eastman, A polymerase chain reaction-based method to detect cisplatin adducts in specific genes, Nuc. Acids Res. 19: 6209 (1991).
10. D.P. Kalinowski, S. Iiienye and B. Van Houten, Analysis of DNA damage and repair in murine leukemia L1210 cells using a quantitative polymerase chain reaction assay, Nuc. Acids. Res. 20: 3485 (1992).
11. P.W. Doetsch, G.L. Chan and W.A. Haseltine, T4 DNA polymerase (3'-5') exonuclease, an enzyme for detecting and quantitation of stable DNA lesions: the ultraviolet light example, Nuc. Acids Res.,13: 3285 (1985).
12. R.J. Parker, A. Eastman, F. Bostick-Bruton and E. Reed, Acquired cisplatin resistance in human ovarian cancer cells is associated with enhanced repair of cisplatin-DNA lesions and reduced drug accumulation, J. Clin. Inst. 87: 772 (1991).
13. S. Linn, How many pols does it take to replicate nuclear DNA ? Cell 66: 185 (1991).
14. S. Buratowski, DNA repair and transcription: The helicase connection, Science 260: 37 (1993).
15. L. Schaeffer, R.Roy, S. Humbert, V. Moncollin, M. Vermeulen, J.H.J. Hoeijmakers, P. Chambon and J-M. Egly, DNA repair helicase: A component of BTF2 (TFIIH) basic transcription factor, Science 260: 58 (1993).
16. C.P. Selby and A. Sancar, Molecular mechanism of transcription-repair coupling, Science 260: 53 (1993).
17. B.N. Ames and M.K. Shigenagal, Oxidants are a major contributor to aging, Annal. N.Y. Acad. Sci. 663: 85 (1992)
18. P. Amstad G. Krupitza and P. Cerutti, Mechanism of c-fos induction by active oxygen, Cancer Res. 52: 3952 (1992).
19. T. Seshadri and J. Campisi, Repression of c-fos transcription and an altered genetic program in senescent human fibroblasts, Science 247: 205, 1990.
20. V.J. Cristofalo, R.J. Pignolo and M.O. Rotenberg, Molecular changes with *in vitro* cellular senescence, Annal. NY Acad. Sci. 663: 187, 1992.

PART IV

MITOSIS: INDUCTION AND MECHANICS

20

DIRECT INHIBITION OF $p107^{WEE1}$ BY THE NIM 1/CDR 1 KINASE

Laura L. Parker, Sarah A. Walter and Helen Piwnica-Worms

Beth Israel Hospital and
Department of Microbiology and Molecular Genetics
Harvard Medical School
Building D2, Room 143
200 Longwood Ave
Boston MA 02215

INTRODUCTION

The G2 to M phase transition in eukaryotes is regulated by the synergistic and opposing activities of a cascade of distinct protein kinases and phosphatases. This cascade converges on $p34^{cdc2}$, a serine/threonine protein kinase that regulates entry into mitosis (reviewed in [1]). The kinase activity of $p34^{cdc2}$ is stimulated both by its association with cyclin B and by its phosphorylation on threonine 161 [2-11]. In Schizosaccharomyces pombe, inactivation of the $p34^{cdc2}$/cyclin B complex occurs by phosphorylation of tyrosine 15 by $p107^{wee1}$ [8, 12]. In human cells, $p34^{cdc2}$/cyclin B kinase activity is negatively regulated by phosphorylation on tyrosine 15 (by p49WEE1Hu, the human homolog of $p107^{wee1}$ [13]) and on threonine 14 [14, 15]. The action of the wee 1 kinase is opposed by the action of the cdc25 phosphatase. Cdc25 dephosphorylates $p34^{cdc2}$ on tyrosine 15 and on threonine 14 resulting in the activation of the $p34^{cdc2}$/cyclin B complex [16-21].

Much less is known about the regulatory signals upstream of cdc25 and wee 1. Phosphorylation of the Xenopus cdc25 phosphatase has been shown to contribute to its activation [22-24]. Virtually nothing is known about the regulation of the wee 1 kinase. Genetic studies in S. pombe have identified a

The Cell Cycle: Regulators, Targets, and Clinical Applications
Edited by V.W. Hu, Plenum Press, New York, 1994

gene, cdr 1, that functions to transduce nutritional signals to proteins involved in mitotic size control [25]. cdr 1 is predicted to encode a protein of 67 kDa with the highest degree of homology to the serine/threonine class of protein kinases [25]. cdr 1 is the full-length gene of a previously cloned mitotic inducer denoted nim 1 [26]. Genetic studies in fission yeast indicates that nim 1/cdr 1 is a positive regulator of mitosis as its overexpression results in a smaller cell size at division whereas mutations in nim 1/cdr 1 lead to an increase in cell size at division [25, 26]. With respect to cell size, mutations in wee 1 are epistatic to nim 1/cdr 1 mutations indicating that nim 1/cdr 1 acts upstream of wee 1, most likely as a negative regulator of wee 1.

To biochemically characterize the nim1/cdr1 protein, we overproduced it in bacteria. We demonstrate that the nim 1/cdr 1 protein possesses intrinsic serine-, threonine-, as well as tyrosine- kinase activities. Furthermore, purified bacterially-produced nim 1/cdr 1 kinase directly phosphorylates and inactivates $p107^{wee1}$ *in vitro.*, whereas phosphatase treatment restores the kinase activity of $p107^{wee1}$. These results demonstrate that nim1/cdr1 functions as a positive regulator of mitosis by directly phosphorylating and inactivating the mitotic inhibitor $p107^{wee1}$.

METHODS

All procedures relating to recombinant baculovirus generation and propagation; expression and purification of proteins from insect cells and bacteria; kinase assays; two-dimensional phosphopeptide mapping; and phosphoamino acid analysis have been described previously [8, 11-13, 20, 27].

Bacterial expression of the nim 1/cdr 1 kinase

pGST-nim 1/cdr 1 was created by digesting plasmid D3 (gift of P. Young, contains cdr 1 in puc118) with Nde I, inserting a Bam HI linker, and excising a 2.8 Kb Bam HI/Eco RI fragment containing cdr 1. cdr 1 was then inserted in frame into pGEX1 (Pharmacia).

A mutant encoding a kinase-deficient form of $p67^{nim1/cdr1}$ was created by polymerase chain reaction amplification of the unique 170bp Sph I-Xho I fragment of nim 1/cdr 1 using the oligonucleotides 5' GCCAAGCATGCTAAAACTGGTGATTTGGCTGCCATC<u>AGA</u>ATTATCCC and 5' CGTACTCGAGTGCTAAGTAC. The underlined codon substitutes arginine for lysine at position 41. The mutated fragment was digested with Sph I and Xho I and was cloned into Sph I and Xho I digested GST-nim 1/cdr 1. Transformants were screened for activity in bacteria.

GST-$p67^{nim1/cdr1}$ was produced in bacteria and precipitated with glutathione agarose using methods described previously [13]. Labelling of bacterial GST-$p67^{nim1/cdr1}$ both *in vivo* [13] and *in vitro* were performed as described previously [8] except that 10 mM manganese chloride was included in the kinase reactions performed *in vitro.*

Generation of baculoviral expression vectors

A baculoviral vector encoding GST-$p37^{wee\ 1KD}$(the kinase domain of p107^{wee1}) was created by changing the second Nco I site of pGEX2T-Wee1 [13] to BamH1 using the linker 5′ CATGGCGGATCCGC. The Bam HI flanked kinase domain of wee 1 was then inserted into the Bam H1 site of GST-pVL (a baculoviral expression vector containing the GST gene).

A baculoviral vector encoding the kinase-deficient mutant of wee 1 fused to GST was created as follows: plasmid pWG-L596 [28] was digested with Kpn I and Bgl II, and the 1.2kb fragment substituting leucine for lysine 596 was excised and cloned into the Kpn I/Bgl II sites of pGEX2T-Wee1 [13]. The GST-^{wee1}(Leu 596) containing plasmid was then digested with Xba I and Nhe I and the 3.5 Kb fragment containing wee1(Leu 596) fused to GST was inserted into the Xba I site of pVL1393 [27]. Recombinant baculoviruses were generated as described previously [27].

Kinase Assays

GST-p107^{wee1} and GST-p37^{wee1}KD were produced in insect cells and affinity purified on glutathione agarose as described previously [12]. GST-p107^{wee1} coupled to glutathione agarose was then incubated with either bacterially produced GST-p67$^{nim1/cdr1}$ or a kinase-deficient mutant of GST-p67$^{nim1/cdr1}$ and kinase assays were performed *in vitro* as described previously [8, 12] except that 10 mM manganese chloride was substituted for magnesium chloride.

To assay for the ability of GST-p107^{wee1} to phosphorylate the p34^{cdc2}/cyclin B complex, glutathione agarose bound GST-p107^{wee1} was incubated with bacterially produced GST-p67$^{nim1/cdr1}$ or a kinase-deficient mutant of GST-p67$^{nim1/cdr1}$ and 1 mM ATP for 40 minutes at 30°C. The reactions were washed and then were incubated with either purified p34^{cdc2} (Arg 33)/GST-cyclin B complexes or with a lysate containing p34^{cdc2}(Arg 33)/GST-cyclin B complexes (this resulted in p34^{cdc2}(Arg 33)/GST-cyclin B complexes binding to the same agarose beads as GST-p107^{wee1}). The complexes were washed and kinase assays were performed *in vitro* [8, 12].

Phosphatase assays

Samples were washed twice with buffer consisting of 50 mM Tris pH 8.5 and 1 mM DTT, 10 units of alkaline phosphatase (Boehringer Mannheim) were added and the reactions were incubated at 30°C for 30 minutes.

RESULTS

To analyze the biochemical activities associated with the nim 1/cdr 1 gene product, we produced it in bacteria as a fusion with the glutathione-S-transferase protein (GST). The resulting fusion protein (GST-p67$^{nim1/cdr1}$) migrated on SDS-polyacrylamide gels with an apparent molecular size of 97 kDa. GST-p67$^{nim1/cdr1}$ was phosphorylated both *in vivo* and *in vitro* (Fig. 1).

Phosphoamino acid analysis indicated the presence of primarily phosphoserine, although phosphothreonine and phosphotyrosine were also present (Fig. 1). Substitution of arginine for lysine at position 41 generated a kinase-deficient mutant of the nim 1/cdr 1 gene product (Fig. 1).

To determine whether p107^{wee1} is a direct substrate of the nim 1/cdr 1 kinase, we performed phosphorylation reactions *in vitro* with purified proteins (Fig. 2). The kinase-deficient mutant of GST-p107wee1 (GST-p107^{wee1}(Leu 596))

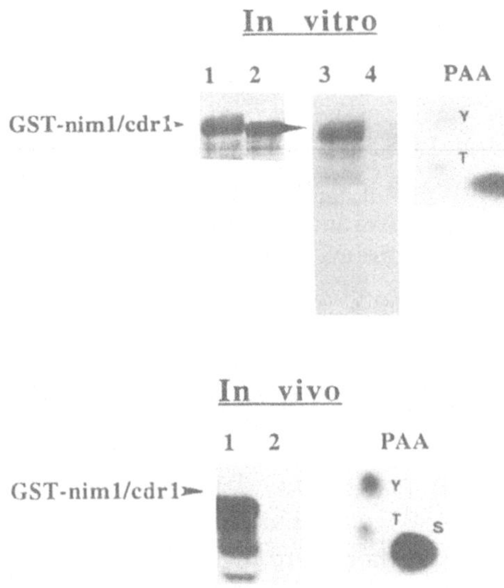

Figure 1. Expression and phosphorylation of the nim 1/cdr 1 protein produced in bacteria.

In vivo: Bacteria expressing either GST-p67$^{nim1/cdr1}$ (lane 1) or GST (lane 2) were incubated with {32P}-orthophosphate. Proteins were precipitated with glutathione agarose, resolved by SDS-PAGE on a 7.5% gel and visualized by autoradiography. PAA: Two dimensional phosphoamino acid analysis of GST-p67$^{nim1/cdr1}$ labelled *in vivo*. S, phosphoserine, T, phosphothreonine, Y, phosphotyrosine.
In vitro: Bacteria expressing either GST-p67$^{nim1/cdr1}$ (lanes 1, 3) or a kinase-deficient mutant of GST-p67$^{nim1/cdr1}$ (lanes 2 and 4) were lysed and recombinant proteins were precipitated with glutathione agarose. Kinase assays were performed *in vitro*, resolved by SDS-PAGE on a 7.5% gel, and visualized either by Coomassie blue staining (lanes 1 and 2) or by autoradiography (lanes 3 and 4).
PAA: Two dimensional phosphoamino acid analysis of GST-p67$^{nim1/cdr1}$ labelled *in vitro*. S, phosphoserine; T, phosphothreonine; Y, phosphotyrosine.
Reprinted with permission from Nature[29].

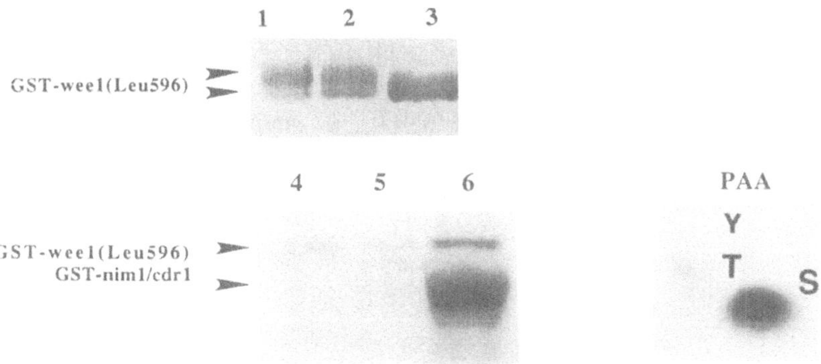

Figure 2. Phosphorylation and inactivation of GST-p107^{wee1} *in vitro* by bacterial GST-p67$^{nim1/cdr1}$.

Lanes 1-3: Insect cells expressing GST-p107^{wee1}(Leu 596) were lysed and the protein was precipitated using glutathione agarose. Kinase reactions were then performed in 1 mM ATP with bacterial GST-p67$^{nim1/cdr1}$ for either 60 min.(lane 1) or for 30 min. (lane 2). Alternatively, p107^{wee1} was precipitated with glutathione agarose and left untreated (lane 3). The proteins were resolved by SDS-PAGE on a 7% gel and were visualized by immunoblotting with anti-p107^{wee1} serum.

Lanes 4-6: Purified GST-p107^{wee1}(Leu 596) was incubated with either buffer alone (lane 4); a kinase-deficient mutant of GST-p67$^{nim1/cdr1}$ (lane 5); or kinase-active GST-p67$^{nim1/cdr1}$ (lane 6) in the presence of gamma 32P-labelled ATP. Proteins were resolved by electrophoresis through a 10% gel and were visualized by autoradiography. PAA: Two dimensional phosphoamino acid analysis of GST-p107^{wee1}(Leu 596) labelled *in vitro*. S, phosphoserine; T, phosphothreonine; Y, phosphotyrosine. Reprinted with permission from Nature[29].

was used in these experiment to eliminate signals due to the autophosphorylation of GST-$p107^{wee1}$. A shift in the electrophoretic mobility of GST-$p107^{wee1}$(Leu 596) was observed upon its incubation with bacterially produced GST-$p67^{nim1/cdr1}$ *in vitro* (Fig. 2, lanes 1-3). In addition, when GST-$p107^{wee1}$(Leu 596) was assayed either alone (Fig. 2, lane 4) or with a kinase-deficient mutant of GST-$p67^{nim1/cdr1}$ (produced in bacteria and affinity purified using glutathione agarose) (Fig. 2, lane 5) little phosphorylation of $p107^{wee1}$ was observed. In contrast, when GST-$p107^{wee1}$(Leu 596) was incubated with kinase-active GST-$p67^{nim1/cdr1}$, $p107^{wee1}$ phosphorylation was observed (Fig. 2, lane 6). Phosphoamino acid analysis again revealed phosphoserine and low levels of phosphothreonine. Similar results were obtained when the kinase domain of $p107^{wee1}$ (GST-$p37^{wee1}$KD) was substituted for $p107^{wee1}$(Leu 596). Incubation of GST-$p37^{wee1}$KD with bacterially produced GST-$p67^{nim1/cdr1}$ *in vitro* resulted both in the phosphorylation of GST-$p37^{wee1}$KD (Fig. 3, lanes 3-5) and a shift in its electrophoretic mobility (Fig. 3, lanes 1-2).

Figure 3. Phosphorylation of the kinase domain of $p107^{wee1}$ *in vitro* by $p67^{nim1/cdr1}$.

Lanes 1-2: Insect cells expressing GST-$p37^{wee1}$KD were lysed and the protein was isolated using glutathione agarose. Affinity purified GST-$p37^{wee1}$KD was then incubated with either a kinase-deficient mutant of GST-$p67^{nim1/cdr1}$ (lane 1) or kinase-active GST-$p67^{nim1/cdr1}$ (lane 2) in the presence of 1 mM ATP. Proteins were resolved by electrophoresis through a 7% gel and were visualized by immunoblotting with anti-$p107^{wee1}$ serum.

Lanes 3-5: Insect cells expressing GST-$p37^{wee1}$KD were lysed and the protein was isolated using glutathione agarose. Kinase reactions were then performed in the presence of either buffer alone (lane 3), a kinase-deficient mutant of GST-$p67^{nim1/cdr1}$ (lane 4) or kinase-active GST-$p67^{nim1/cdr1}$ (lane 5). The proteins were resolved by SDS-PAGE on a 9% gel and were visualized by autoradiography. Reprinted with permission from Nature [29].

To test the effects of nim1/cdr1-mediated phosphorylation on the activity of $p107^{wee1}$, phosphorylated $p107^{wee1}$ was monitored for its ability to phosphorylate its physiological substrate ($p34^{cdc2}$ in complex with cyclin B) *in vitro* . As evident in Fig. 4, there was a stoichiometric shift in the electrophoretic mobility of GST-$p107^{wee1}$ upon phosphorylation by kinase-active GST-$p67^{nim1/cdr1}$ (lane 7) but not by the kinase-deficient form of GST-$p67^{nim1/cdr1}$ (lane 5). Concomitant with the phosphorylation of $p107^{wee1}$ *in vitro* by GST-$p67^{nim1/cdr1}$ was a loss in its ability to phosphorylate $p34^{cdc2}$ (lanes 2 and 4). Phosphatase treatment restored both the electrophoretic mobility and the kinase activity of $p107^{wee1}$ (lanes 3 and 6).

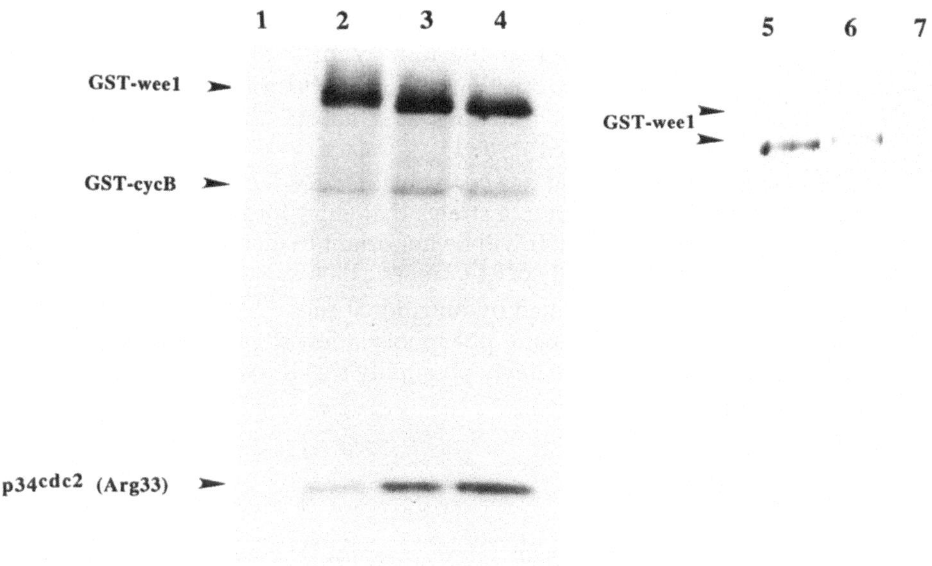

Fig. 4. Inactivation of GST-p107^{wee1} *in vitro* by GST-p67$^{nim1/cdr1}$.

Glutathione agarose-bound GST-p107^{wee1} was incubated with either kinase-active (lanes 2, 3, 6, 7) or kinase-deficient GST-p67$^{nim1/cdr1}$ (lanes 4, 5) and kinase assays were performed. Reactions were washed into phosphatase buffer in either the absence (lanes 2, 7) or in the presence of alkaline phosphatase (lanes 3-6). One half of each reaction was assayed for its ability to phosphorylate p34^{cdc2}(Arg 33)/cyclin B complexes *in vitro* (lanes 2-4) while the other half was monitored for the electrophoretic mobility of p107^{wee1} (lanes 5-7). Purified p34^{cdc2}(Arg 33)/cyclin B complexes were also assayed alone (lane 1). Proteins were resolved by SDS-PAGE on a 10% gel and visualized by autoradiography (lanes 1-4) or on a 7.5% gel and visualized by immunoblotting with anti-p107^{wee1} serum (lanes 5-7). Reprinted with permission from Nature[29].

SUMMARY

The results presented in this report demonstrate that nim 1/cdr 1 encodes a kinase that autophosphorylates on serine, threonine and tyrosine residues. In addition, direct biochemical evidence is presented to demonstrate that $p107^{wee1}$ is both phosphorylated and functionally inactivated by the nim 1/cdr 1 kinase. We have previously demonstrated that phosphorylation of $p34^{cdc2}$ on tyrosine 15 by $p107^{wee1}$ completely inhibits the activity of the $p34^{cdc2}$/cyclin B complex [8, 12]. Here, we demonstrate that the nim 1/cdr 1 kinase phosphorylates $p107^{wee1}$ and thereby ablates the activity of $p107^{wee1}$ towards $p34^{cdc2}$. Thus, the nim 1/cdr 1 gene product functions as a mitotic inducer by negatively regulating the mitotic inhibitor $p107^{wee1}$.

The biochemical data that has been obtained for the cdc2, wee 1 and nim 1/cdr 1 gene products clearly demonstrates that entry into mitosis is controlled by a cascade of protein kinases. It will be important to identify and characterize the regulators upstream of nim 1/cdr 1. Genetic data suggest that the activity of the nim 1/cdr 1 kinase is regulated by nutritional cues [25]. In addition, the nim 1/cdr 1 kinase contains consensus phosphorylation sites for several distinct protein kinases leaving open the likely possibility that it too will be regulated by phosphorylation.

ACKNOWLEDGMENTS

We thank Paul Young for his gift of plasmid D3 containing nim 1/cdr 1. This work was supported by NIH grant GM47017 to H.P-W., and NIH Public Health Service Award CA08894-01A1 to L. L. P.. H. P.-W. is a Pew Scholar in the Biomedical Sciences.

REFERENCES

1. P. Nurse, Universal control mechanism regulating onset of M-phase. *Nature.* 344:503 (1990).

2. G. Draetta, F. Luca, J. Westendorf, L. Brizuela, J. Ruderman, and D. Beach, cdc2 protein kinase is complexed with both cyclin A and B: evidence for proteolytic inactivation of MPF. *Cell.* 56:829 (1989).

3. D. Desai, Y. Gu and D.O. Morgan, Activation of human cyclin-dependent kinases in vitro. *Mol. Biol. Cell.* 3:571 (1992).

4. B. Ducommun, P. Brambilla, M.-A. Felix, B.R. Franza Jr., E. Karsenti, and G. Draetta, cdc2 phosphorylation is required for its interaction with cyclin. *EMBO J.* 10:3311 (1991).

5. K.L. Gould, S. Moreno, D.J. Owen, S. Sazer, and P. Nurse, Phosphorylation at Thr167 is required for *Schizosaccharomyces pombe* $p34^{cdc2}$ function. *EMBO J.* 10:3297 (1991).

6. J. Minshull, R. Golsteyn, C.S. Hill, and T. Hunt, The A- and B-type cyclin associated cdc2 kinases in Xenopus turn on and off at different times in the cell cycle. *EMBO J.* 9:2865 (1990).

7. A.W. Murray, M.J. Solomon, and M.W. Kirschner, The role of cyclin synthesis and degradation in the control of maturation promoting factor activity. *Nature.* 339:280 (1989).

8. L.L. Parker, S. Atherton-Fessler, M.S. Lee, S. Ogg, F.L. Falk, K.I. Swenson, and H. Piwnica-Worms, Cyclin promotes the tyrosine phosphorylation of p34cdc2 in a wee1+ dependent manner. *EMBO J.* 10:1255 (1991).

9. M.J. Solomon, M. Glotzer, T.H. Lee, M. Philippe, and M.W. Kirschner, Cyclin activation of p34cdc2. *Cell.* 63:1013 (1990).

10. M.J. Solomon, T. Lee, and M.W. Kirschner, The role of phosphorylation in p34cdc2 activation: identification of an activating kinase. *Mol. Biol. Cell.* 3:13 (1992).

11. S. Atherton-Fessler, L.L. Parker, R.L. Geahlen, and H. Piwnica-Worms, Mechanisms of p34cdc2 regulation. *MCB.* 13:1675 (1993).

12. L.L. Parker, S. Atherton-Fessler, and H. Piwnica-Worms, p107wee1 is a dual-specificity kinase that phosphorylates p34cdc2 on tyrosine 15. *Proc. Natl. Acad. Sci. USA.* 89:2917 (1992).

13. L.L. Parker, and H. Piwnica-Worms, Inactivation of the p34cdc2/cyclin B complex by the human WEE1 tyrosine kinase. *Science.* 257:1955 (1992).

14. W. Krek, and E.A. Nigg, Differential phosphorylation of vertebrate p34cdc2 kinase at the G1/S and G2/M transitions of the cell cycle: identification of major phosphorylation sites. *EMBO J.* 10:305 (1991).

15. W. Krek, and E.A. Nigg, Mutations of p34cdc2 phosphorylation sites induce premature mitotic events in HeLa cells: evidence for a double block to p34cdc2 kinase activation in vertebrates. *EMBO J.* 10:3331 (1991).

16. U. Strausfeld, J.C. Labbe, D. Fesquet, J.C. Cavadore, A. Picard, K. Sadhu, P. Russell, Dephosphorylation and activation of a p34cdc2/cyclin B complex in vitro by human CDC25 protein. *Nature.* 351:242 (1991).

17. J. Gautier, M.J. Solomon, R.N. Booher, J.F. Bazan, and M.W. Kirschner, cdc25 is a specific tyrosine phosphatase that directly activates p3cdc2. *Cell.* 67:197 (1991).

18. K. Galaktionov, and D. Beach, Specific activation of cdc25 tyrosine phosphatases by B-type cyclins: evidence for multiple roles of mitotic cyclins. *Cell.* 67:1181 (1991).

19. A. Kumagai, and W.G. Dunphy, The cdc25 protein controls tyrosine dephosphorylation of the cdc2 protein in a cell-free system. *Cell.* 64:903 (1991).

20. M.S. Lee, S. Ogg, M. Xu, L.L. Parker, D.J. Donoghue, J.L. Maller, and H. Piwnica-Worms, cdc25+ encodes a protein phosphatase that dephosphorylates p34cdc2. *Mol. Biol. Cell.* 3:73 (1992).

21. J.B.A. Millar, C.H. McGowan, G. Lenaers, R. Jones, and P. Russell, p80cdc25 mitotic inducer is the tyrosine phosphatase that p34cdc2 kinase in fission yeast. *EMBO J.* 10:4301 (1991).

22. T. Izumi, D.h. Walker, and J.L. Maller, Periodic changes in phosphorylation of the Xenopus cdc25 phosphatase regulate its activity. *Mol. Biol. Cell.* 3:929 (1992).

23. A. Kumagai, and W.G. Dunphy, Regulation of the cdc25 protein during the cell cycle in Xenopus extracts. *Cell.* 70:139 (1992).

24. I. Hoffmann, P.R. Clarke, M.J. Marcote, E. Karsenti, and G. Draetta, Phosphorylation and activation of human cdc25-C by cdc2- cyclin B and its involvement in the self-amplification of MPF at mitosis. *EMBO J.* 12:53 (1993).

25. H. Feilotter, P. Nurse, and P. G. Young, Genetic and Molecular Analysis of cdr1/nim1 in Schizosaccharomyces pombe. *Genetics.* 127:309 (1991).

26. P. Russell, and P. Nurse, The mitotic inducer nim+1 functions in a regulatory network of protein kinase homologs controlling the initiation of mitosis. *Cell.* 49:569 (1987).

27. H. Piwnica-Worms, Expression of proteins in insect cells using baculovirus vectors, *in*: "Current Protocols in Molecular Biology," F. Ausubel, R. Brent, R. Kingston, D. Moore, J. Seidman, J. Smith, and K. Struhl, ed., Greene Publishing Associates, Brooklyn, New York (1990).

28. P. Russell, S. Moreno, and S. Reed, Conservation of mitotic controls in fission and budding yeasts. *Cell.* 57:295 (1989).

29. L. L. Parker, S. A. Walter, P. G. Young, and H. Piwnica-Worms. Phosphorylation and inactivation of the mitotic inhibitor WEE1 by the nim1/cdr1 kinase. *Nature* 363:736 (1993).

21

A HUMAN PHOSPHOTYROSINE PHOSPHATASE ASSOCIATED WITH M PHASE-PROMOTING FACTOR

Robert A. Schlegel

Department of Molecular and Cell Biology
Penn State University
University Park, PA 16802

INTRODUCTION

Prior to the discovery that cdc2 kinase was the catalytic subunit of M-phase promoting factor (MPF) (Arion et al., 1988), evidence had suggested that MPF might be a protein kinase (Laskey et al., 1983; Schlegel et al., 1987). Accordingly, the protein kinases present in extracts of human mitotic cells were compared with those present in interphase extracts using a native gel assay (Halleck et al., 1987). Antibodies prepared against two of the several mitosis-specific kinases that were identified each recognized the same protein with a molecular weight of 65 kDa, called p65 (Meikrantz et al., 1990). However, the protein was neither specific to mitotic cells, nor was it a protein kinase. Rather, it had a different molecular form at mitosis (disulfide-linked homodimer) than in interphase (reduced monomer), and was a tyrosine (and threonine) phosphatase (Meikrantz et al., 1991b).

cdc2 kinase is found in several forms in extracts of human mitotic cells: a free form which elutes from a gel filtration column at a molecular weight of 30-40 kDa, and a complexed form eluting with a molecular weight of greater than 200 kDa (Draetta and Beach, 1988; Brizuela et al., 1989), the latter form representing the MPF histone H1 kinase complex containing cyclin as well as cdc2. That p65 is also a part of this complex is argued by its co-immunoprecipitation with cyclin-dependent kinase, specifically at mitosis; its co-immunoprecipitation with cyclin B, specifically at mitosis; its clearance from mitotic extracts by the cyclin-dependent kinase-specific ligand p13; and the ability of an anti-p65 immunoaffinity column to remove 60% of H1 kinase activity from human mitotic extracts (Meikrantz et al., 1990, 1991a). Based on these results, the mitosis-specific kinases identified using native gels could represent different electrophoretic forms of MPF, at least two of which contain p65.

DEPHOSPHORYLATION AND ACTIVATION OF CYCLIN-DEPENDENT H1 KINASE BY p65

MPF is activated at mitosis by removal of inhibitory phosphates on tyrosine and threonine residues in the ATP-binding site of cdc2 (Nurse, 1990), which nonetheless do not prevent ATP binding (Atherton-Fessler et al., 1993); p65 seems a likely candidate for the phosphatase responsible. This possibility was tested in an homologous system, using p65 and cyclin-dependent kinase, both purified from human mitotic extracts. The monomeric form of cdc2 is phosphorylated and inactive toward histone H1, but displays casein kinase

The Cell Cycle: Regulators, Targets, and Clinical Applications
Edited by V.W. Hu, Plenum Press, New York, 1994

activity (Draetta and Beach, 1988; Brizuela et al., 1989). On this basis it was purified approximately 1000-fold; Table 1 presents the properties of the 34 kDa protein isolated (Meikrantz et al., 1991a; Meikrantz et al., submitted). Because none of the reagents used to characterize the protein are specific for particular cyclin-dependent kinases, at present the exact identity of the kinase is not known. In fact, more than one cyclin-dependent kinase could be present in the preparation, since the reagents used are not necessarily recognizing the same 34 kDa protein.

Table 1. Properties of p34 preparation

34 kD protein the only protein detected by silver staining
34 kD protein the only ATP-binding protein detected
eluted from gel filtration column at 30-40 kD
no p65 or cyclin B detected immunologically
no cdc25 detected immunologically
casein kinase activity, but no histone kinase activity detected
34 kD protein phosphorylated on tyrosine
34 kD protein reacted with
PSTAIR antibody
J4 and JP4 monoclonal antibodies to cdc2
34 kD protein cleared by p13 beads

When purified human p65 was added to the purified cyclin-dependent kinase preparation, H1 kinase activity appeared in a dose-dependent manner, concomitant with tyrosine dephosphorylation of the kinase (Meikrantz et al., submitted). Stimulation of H1 kinase activity was about 30-fold at a mole ratio of between 1 and 2 moles of p65 per mole of cyclin-dependent kinase, and was about 1/3 that reported for mitosis-specific H1 kinase purified from HeLa (Brizuela et al., 1989) or CHO cells (Woodford and Pardee, 1986). At greater mole ratios, the apparent H1 activity declined, likely resulting from dephosphorylation of H1 by p65 (Meikrantz et al., 1991b). Both activation of H1 kinase and tyrosine dephosphorylation were preventable by addition of vanadate, arguing that dephosporylation was responsible for activation.

p65 VERSUS cdc25

The product of the cdc25 gene is a phosphatase which dephosphorylates and activates cdc2 (Dunphy and Kumagai, 1991; Gautier et al., 1991), provoking the question of the relationship between p65 and cdc25. Both p65 and human cdc25 associate with cyclin-dependent kinase/cyclin complexes (Galaktionov and Beach, 1991). Both have dual specificity for phosphotyrosine and phosphothreonine (Millar et al., 1991b). However, there are significant differences between the two molecules, summarized in Table 2.

Three cdc25 homologs have been identified in human cells, cdc25C, B and A (Sadu et al., 1990; Galaktionov and Beach, 1991; Nagata et al., 1991), which code for proteins with molecular weights of 52/55, 70 and 75 kDa. Whereas cdc25C is hyperphosphorylated at mitosis, resulting in a shift in electrophoretic mobility of 10 kDa (Hoffmann et al., 1993), p65 is poorly phosphorylated, if phosphorylated at all (Meikrantz et al., submitted). In contrast, at mitosis p65 dimerizes, whereas such behavior has not been noted for cdc25. Perhaps the sharpest contrast is that the reducing agent DTT inhibits the phosphatase activity of p65 (Meikrantz et al., 1991b), whereas it stimulates the activity of (*Xenopus*) cdc25 (Dunphy and Kumagai, 1991).

Table 2. Comparison of p65 and cdc25

p65	cdc25
65 kD	52/55, 70, 75 kD
poorly phosphorylated	highly phosphorylated at M
homodimer	----
inhibited by DTT	stimulated by DTT
nuclear, chromosomal	nuclear, non-chromosomal
O-glycosylated	----

Though the extracts used to generate p65 antisera were cytoplasmic, and the extracts examined for p65 were cytoplasmic, p65 immunoreactivity is found primarily in the nucleus (Meikrantz et al., 1991b). This is also the case for cdc25C (Millar et al., 1991a). However, whereas p65 co-localizes with chromosomes, cdc25C does not. It is interesting to note that a recently described Drosophila tyrosine phosphatase can localize either to the nucleus or the cytoplasm depending on alternate splicing (McLaughlin and Dixon, 1993). Finally, because of its chromosomal location, p65 was examined for and found to be O-glycosylated (Meikrantz et al., 1991b), perhaps explaining why it is refractory to staining with Coomassie Blue or silver, making detection, other than by immunological means, difficult.

Because of these several significant differences it is not likely that p65 and cdc25 are the same molecule. However, there is evidence to suggest that there is a cdc2 dephosphorylation pathway independent of cdc25 (Lundgren et al., 1991; Matsumoto and Beach, 1991), and recently direct evidence has been provided that fission yeast have a second tyrosine phosphatase, isolated as a high copy suppressor of a cdc25 mutation, which contributes to dephosphorylation of cdc2 (Millar et al., 1992). Based on additional genetic studies, there may be yet another alternate tyrosine phosphatase as well (Millar et al., 1992). Whether p65 is related to any of these phosphatases, including cdc25, will require isolation and sequencing of p65, or isolation and sequencing of the gene coding for it. Both of these approaches are underway.

REFERENCES

Arion, D., Meijer, L., Brizuela, L., and Beach, D., 1988, cdc2 is a component of the M phase specific histone H1 kinase: evidence for identity with MPF, *Cell* 55:371-378.

Atherton-Fessler, S., Parker, L.L., Geahlen, R.L., and Piwnica-Worms, H., 1993, Mechanisms of $p34^{cdc2}$ regulation, *Mol. Cell. Biol.* 13:1675-1685.

Brizuela, L., Draetta, G., and Beach, D., 1989, Activation of human CDC2 protein as a histone H1 kinase is associated with complex formation with the p62 subunit, *Proc. Natl. Acad. Sci. USA* 86:4362-4366.

Draetta, G., and Beach, D., 1988, Activation of cdc2 protein kinase during mitosis in human cells: cell cycle-dependent phosphorylation and subunit rearrangement, *Cell* 54:17-26.

Dunphy, W.G., and Kumagai, A., 1991, The cdc25 protein contains an intrinsic phosphatase activity, *Cell* 67:189-196.

Galaktionov, K., and Beach, D., 1991, Specific activation of cdc25 tyrosine phosphatases by B-type cyclins: evidence for multiple roles of mitotic cyclins, *Cell* 67:1181-1194.

Gautier, J., Solomon, M.J., Booher, R.N., Bazan, J.F., and Kirschner, M.W., 1991, cdc25 is a specific tyrosine phosphatase that directly activates $p34^{cdc2}$, *Cell* 67:197-211.

Hallcck, M.S., Lumley-Sapanski, K., and Schlegel, R.A., 1987, Mitosis-specific cytoplasmic protein kinases, *in*: "Molecular Regulation of Nuclear Events in Mitosis and Meiosis," R.A. Schlegel, M.S. Hallcck, P.N. Rao, eds., Academic Press, NY.

Hoffmann, I., Clarke, P.R., Marcote, M.J., Karsenti, E., and Draetta, G., 1993, Phosphorylation and activation of human cdc25-C by cdc2-cyclin B and its involvement in the self-amplification of MPF at mitosis, *EMBO J.* 12:53-63.

Laskey, R.A., 1983, Phosphorylation of nuclear proteins, *Philos. Trans. R. Soc., London*, Ser. B 302:143-150.

Lundgren, K., Walworth, N., Booher, R., Dembski, M., Kirschner, M., and Beach, D., 1991, mik1 and wee1 cooperate in the inhibitory tyrosine phosphorylation of cdc2, *Cell* 64:1111-1122.

Matsumoto, T., and Beach, D., 1991, Premature initiation of mitosis in yeast lacking RCC1 or an interacting GTPase, *Cell* 66:347-360.

McLaughlin, S., and Dixon, J.E., 1993, Alternative splicing gives rise to a nuclear protein tyrosine phosphatase in *Drosophila*, *J. Biol. Chem.* 268:6839-6842.

Meikrantz, W., Suprynowicz, F.A., Halleck, M.S., and Schlegel, R.A., 1990, Identification of mitosis-specific p65 dimer as a component of human M phase-promoting factor. *Proc. Natl. Acad. Sci. USA* 87:9600-9604.

Meikrantz, W., Feldman, R.P., Sladicka, M.M., Ho, D., Krupnick, J., Anderson, K., and Schlegel, R.A., 1991a, Isolation of mitotic $p34^{cdc2}$ apoenzyme from human cells, *FEBS Lett.* 291:192-194.

Meikrantz, W., Smith, D.M., Sladicka, M.M., and Schlegel, R.A., 1991b, Nuclear localization of an *O*-glycosylated protein phosphotyrosine phosphatase from human cells, *J. Cell Sci.* 98:303-307.

Millar, J.B.A., Blevitt, J., Gerace, L., Sadhu, K., Featherstone, C., and Russell, P., 1991a, $p55^{CDC25}$ is a nuclear protein required for the initiation of mitosis in human cells, *Proc. Natl. Acad. Sci. USA* 88:10500-10504.

Millar, J.B.A., McGowan, C.H., Lenaers, G., Jones, R., and Russell, P., 1991b, $p80^{cdc25}$ mitotic inducer is the tyrosine phosphatase that activates $p34^{cdc2}$ kinase in fission yeast. *EMBO J.* 10:4301-4309.

Millar, J.B.A., Lenaers, G., and Russell, P., 1992, *Pyp3* PTPase acts as a mitotic inducer in fission yeast, *EMBO J.* 11:4933-4941.

Nagata, A., Igarashi, M., Jinno, S., Suto, K., and Okayama, H., 1991, An additional homolog of the fission yeast $cdc25^{+}$ gene occurs in humans and is highly expressed in some cancer cells, *New Biol.* 3:959-968.

Nurse, P., 1990, Universal control of cell size at cell division in yeast, *Nature* 344:503-508.

Sadhu, K., Reed, S.I., Richardson, H., and Russell, P., 1990, Human homolog of a fission yeast mitotic inducer is predominantly expressed in G_2, *Proc. Natl. Acad. Sci. USA* 87:5139-5143.

Schlegel, R.A., Halleck, M.S., and Rao, P.N., 1987, "Molecular Regulation of Nuclear Events in Mitosis and Meiosis," Academic Press, NY.

Woodford, T.A., and Pardee, A.B., 1986, Histone H1 kinase in exponential and synchronous populations of Chinese hamster fibroblasts, *J. Biol. Chem.* 261:4669-4676.

22

THE LOCALISATION OF HUMAN CYCLINS AND CDKS IN THE CELL CYCLE

Jonathon Pines

Wellcome/CRC Institute
Tennis Court Rd.
Cambridge UK

INTRODUCTION

One of the emerging concepts in the cell cycle field is that progress through the cycle is determined at discrete points, called checkpoints, by a family of closely related protein kinases. This family has been christened the 'cyclin-dependent kinases' (CDKs), because they are only activated upon binding a member of the cyclin family. Cyclins appear to be required both to activate and to target their protein kinase partner to particular subcellular compartments. I have been attempting to define which part of the cyclin is responsible for its targetting function.

CYCLINS

The cyclin family of proteins undergo a cell cycle-dependent variation in level, and all bear a significant degree of homology to one another in an ~200 amino acid region called the cyclin box. Cyclins were first identified by Tim Hunt in developing sea urchin and clam eggs, as proteins that were strongly synthesised during interphase and then very specifically and rapidly destroyed at each mitosis. Two proteins of apparent Mr 58,000 and 62,000 were cyclically synthesised and destroyed, and these were named A and B-type cyclins (Evans et al., 1983).

These cyclins are now called "mitotic cyclins", to distinguish them from the second class of cyclins, START cyclins or CLNs, that reach their peak amount in G1 phase of the cell cycle. CLNs were first identified as mutants in budding yeast which affected the size at which the cell divided. Upon sequencing the CLN genes were found to encode proteins distantly related to the mitotic cyclins, and the homologous region was called the cyclin box. (Reviewed in Reed, 1991.)

The mitotic and START cyclins differ in their overall structure and this relates to differences in their mode of destruction. Mitotic cyclins have approximately 200 amino acids N-terminal to the cyclin box required for their degradation in mitosis; when this region is removed, the truncated cyclin cannot be degraded and this has the effect of blocking a cell in mitosis. The region required for mitotic destruction has been further delineated to a small region around a conserved arginine residue about 40 amino acids from the N-terminus of the protein (Glotzer et al., 1991). The destruction of the mitotic cyclins involves the ubiquitin-dependent proteolysis pathway. In contrast START cyclins mostly extend C-terminal to the cyclin box. This part of the protein is rich in proline, glutamic acid, serine and threonine arranged as "PEST" sequences that have been correlated with rapidly turned over proteins. START cyclins are indeed very short-lived proteins throughout the cell cycle, such that their levels are determined by changes in the rate of transcription of their cognate mRNAs. The C-terminal PEST-rich region appears to be responsible for the proteins' rapid tunover; without the C-terminus the START cyclins become much more stable proteins. However, unlike their

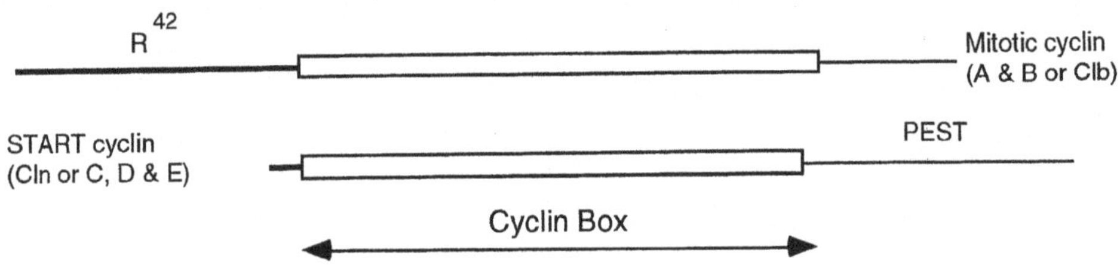

Figure 1. Schematic of mitotic and START cyclin primary structure.

mitotic counterparts, an increase in the amount of CLN protein does not block the cell in the cell cycle, rather it accelerates the cell's progress through the START checkpoint (Cross, 1988). Putative START cyclins have recently been isolated from human cells. These are called the C-, D- and E-type cyclins. (Reviewed in Lew and Reed, 1992.)

Cyclins Bind and Activate a Family of Protein Kinases

Despite their structural differences, in yeast both the mitotic and the START cyclins bind with and activate *cdc2* (called *CDC28* in budding yeast), but at different stages in the cell cycle. However, in multicellular organisms it appears that the the cell cycle is regulated by a family of structurally related protein kinases that need to bind cyclins to be activated. These are called the cyclin-dependent kinases, or CDKs. (Reviewed in Pines and Hunter, 1991a). So far 12 different human cDNAs have been identified which encode proteins that are closely related to cdc2. Of these, 3 cDNAs, cdc2 (CDK1), CDK2 and CDK3, are all able to substitute for a defective cdc2 protein when introduced into yeast (Meyerson et al., 1992). At present the role of CDK3 in the human cell cycle is unknown, but CDK2 is required for DNA replication and CDK1 is required for mitosis. Three other CDKs, CDK4, CDK5 and CDK6 are found in association with the D-type cyclins and may be involved in the regulation of G1 phase (Matsushime et al., 1992; Xiong et al., 1992).

The 2 different mitotic cyclins are localised to different parts of the cell

The A-type cyclins perform a dual role in the cell cycle; they are required both for DNA replication (Girard et al., 1991; Pagano et al., 1992) , and for entry into mitosis (Lehner and O'Farrell, 1989; Pagano et al., 1992). A-type cyclins are nuclear proteins from the time of their synthesis at the beginning of S phase (Pines and Hunter, 1991b). A-type cyclins bind to both CDK2, and later in the cell cycle to CDK1 as well (Pagano et al., 1992), although it is not clear whether the cyclin A-CDK1 complexes have a different function compared with the cyclin A-CDK2 complexes.

By contrast B-type cyclins are restricted to the regulation of mitosis. In association with CDK1 they form the primary mitosis specific protein kinase (Gautier et al., 1990; Labbé et al., 1989). Although the B-type cyclins bind with CDK1 in late S phase and G2 phase this complex is prevented from becoming an active kinase through phosphorylation of CDK1 (Gould and Nurse, 1989; Krek and Nigg, 1991; Morla et al., 1989). CDK1 is phosphorylated on the threonine and tyrosine residues that are in the ATP binding domain (T14 and Y15). These residues are phosphorylated by the wee1 and mik1 protein kinases (Featherstone and Russell, 1991; Lundgren et al., 1991; Russell and Nurse, 1987b), whose activity is modulated by upstream protein kinases such as nim1 according to the integrity of the DNA (Russell and Nurse, 1987a). Once the cell is ready to enter mitosis, the wee1 and mik1 protein kinase activities are down-regulated (Smythe and Newport, 1992) and CDK1 is dephosphorylated by the cdc25 protein (Kumagai and Dunphy, 1991; Millar et al., 1991; Strausfeld et al., 1991). cdc25 also forms part of the feedback pathway between DNA replication and mitosis. Cells will not enter mitosis until their DNA is completely replicated (Dasso and Newport, 1990), but this negative inhibition of mitosis by unreplicated DNA requires an active cdc25 protein (Enoch and Nurse, 1990).

However, there is a further regulation on the activation of the cyclin B-CDK1 complex. Cyclin B-CDK1 accumulates in the cytoplasm once cyclin B synthesis begins in late S phase. At the beginning of prophase the cyclin B-CDK1 complex moves into the nucleus, before the breakdown of the nuclear lamina (Gallant and Nigg, 1992; Ookata et al., 1992; Pines and Hunter, 1991b). This behaviour is a conserved feature of the B-type cyclins from starfish to man, and may be another means by which the cyclin B-CDK1 complex is prevented from being prematurely activated, or at the least from gaining access to some important mitotic substrates such as the nuclear lamins (Ookata et al., 1992).

The N-termini of cyclins A and B in part determine their location in the cell

In order to identify those parts of the A and B-type cyclins involved in their subcellular localisation I have generated various domain swaps between the 2 proteins. It is apparent that exchanging the N-termini of the A and B-type cyclins is sufficent to alter their location in the cell. (See Table 1.)I have subsequently used various deletion mutants to attempt to define the region of the N-terminus that is responsible, and to understand whether this the effect is due to a cytoplasmic anchor or to a nuclear transport signal.

It is unlikely that the N-terminus of cyclin A encodes a nuclear transport signal for 2 reasons. The first is that the cyclin box of cyclin A alone is a nuclear protein, therefore it does not need any signal from the N-terminus to be transported into the nucleus. The second reason is that sequential deletion of the N-terminus of cyclin A attached to the cyclin box of cyclin B has no effect on the nuclear location of the protein. Thus the default condition may be for the cyclin box-CDK complex to be nuclear.

In contrast there are data consistent with there being a cytoplasmic anchor in the N-terminus of cyclin B. Sequential deletion of the N-terminus of cyclin B attached to the cyclin box of cyclin A eventually results in the protein moving from the cytoplasm into the nucleus. However, that part of the cyclin B N-terminus whose removal results in a nuclear protein has also been implicated in binding to CDK1. Thus any model invoking a cytoplasmic anchor in cyclin B would necessarily include a contribution from CDK1. Perhaps the tertiary structure of cyclin B is altered when this part of the N-terminus is removed such that it is unable to bind to either CDK1 or to its cytoplasmic anchor.

Table 1. The localisation of cyclin A-B chimaeras

MRAIL motif	N-terminus	Cyclin Box	CDK bound	Localisation
MRAIL	Cyclin A	Cyclin A	CDK2	Nuclear
MRAIL	Cyclin B	Cyclin B	CDK1	Cytoplasm
MRAIL	Cyclin A	Cyclin B	CDK1	Nuclear
MRAIL	Cyclin B	Cyclin A	CDK2	Cytoplasm
MKAIL	Cyclin A	Cyclin A	None	Nuclear/Cytoplasmic
MKAIL	Cyclin B	Cyclin B	None	Cytoplasm
MKAIL	Cyclin A	Cyclin B	None	Nuclear
MKAIL	Cyclin B	Cyclin A	None	Cytoplasm

In an attempt to determine whether the cyclin must bind to its CDK in order to be correctly localised, I have generated point-mutantions in the cyclin box of the chimaeras that have been shown to be required to bind to the CDK subunit. In particular I have altered the critical arginine residue in the conserved MRAIL motif to a lysine. I find that this has no effect on the location of either of the 2 chimaeras nor of cyclin B, but has some effect on the nuclear localisation of cyclin A. (See Table 1.) However, I have not yet been able to determine whether this mutation prohibits the nuclear localisation of the B-type cyclins at mitosis.

CONCLUSION

In conclusion, my results thus far show that the localisation of the mitotic cyclins is determined in large part by residues in the N-terminus of the B-type cyclins. In addition, the nuclear localisation of the A-type cyclins is perturbed when the cyclin is unable to bind to a CDK. This suggests that it is the cyclin A-CDK2 complex that is transported into the nucleus rather than the individual components. Future experiments will attempt to define further the residues in cyclin B required for its cytoplasmic localisation, and whether cyclin B also needs to be bound to its CDK in order to move into the nucleus at mitosis.

REFERENCES

Cross, F., 1988, DAF1, a mutant gene affecting size control, pheremone arrest and cell cycle kinetics of S. cerevisiae., *Mol. Cell. Biol.* 8: 4675.

Dasso, M. and Newport, J. W., 1990, Completion of DNA replication is monitored by a feedback system that controls the initiation of mitosis in vitro: studies in Xenopus, *Cell* 61: 811.

Enoch, T. and Nurse, P., 1990, Mutation of fission yeast cell cycle control genes abolishes dependence of mitosis on DNA replication, *Cell* 60: 665.

Evans, T., Rosenthal, E. T., Youngblom, J., Distel, D. and Hunt, T., 1983, Cyclin: A protein specified by maternal mRNA in sea urchin eggs that is destroyed at each cleavage division, *Cell* 33: 389.

Featherstone, C. and Russell, P., 1991, Fission yeast p107wee1 mitotic inhibitor is a tyrosine/serine kinase, *Nature* 349: 808.

Gallant, P. and Nigg, E. A., 1992, Cyclin B2 undergoes cell cycle-dependent nuclear translocation and, when expressed as a non-destructible mutant, causes mitotic arrest in HeLa cells, *J Cell Biol* 117: 213.

Gautier, J., Minshull, J., Lohka, M., Glotzer, M., Hunt, T. and Maller, J. L., 1990, Cyclin is a component of MPF from Xenopus, *Cell* 60: 487.

Girard, F., Strausfeld, U., Fernandez, A. and Lamb, N. J. C., 1991, Cyclin A is required for the onset of DNA replication in mammalian fibroblasts, *Cell* 67: 1169.

Glotzer, M., Murray, A. W. and Kirschner, M. W., 1991, Cyclin is degraded by the ubiquitin pathway, *Nature* 349: 132.

Gould, K. L. and Nurse, P., 1989, Tyrosine phosphorylation of the fission yeast *cdc2*$^{+}$ protein kinase regulates entry into mitosis., *Nature* 342: 39.

Krek, W. and Nigg, E. A., 1991, Differential phosphorylation of vertebrate p34cdc2 kinase at the G1/S and G2/M transitions of the cell cycle: identification of major phosphorylation sites, *EMBO Journal* 10: 305.

Kumagai, A. and Dunphy, W. G., 1991, The cdc25 protein controls tyrosine dephosphorylation of the cdc2 protein in a cell-free system, *Cell* 64: 903.

Labbé, J. C., Picard, A., Peaucellier, G., Cavadore, J. C., Nurse, P. and Doree, M., 1989, Purification of MPF from starfish: identification as the H1 histone kinase p34cdc2 and a possible mechanism for its periodic activation, *Cell* 57: 253.

Lehner, C. F. and O'Farrell, P. H., 1989, Expression and function of Drosophila cyclin A during embryonic cell cycle progression, *Cell* 56: 957.

Lew, D. and Reed, S., 1992, A proliferation of cyclins, *Trends Cell. Biol.* 2: 77.

Lundgren, K., Walworth, N., Booher, R., Dembski, M., M., K. and Beach, D., 1991, mik1 and wee1 cooperate in the inhibitory tyrosine phosphorylation of cdc2., *Cell* 64: 1111.

Matsushime, H., Ewen, M. E., Strom, D. K., Kato, J. Y., Hanks, S. K., Roussel, M. F. and Sherr, C. J., 1992, Identification and properties of an atypical catalytic subunit (p34PSK-J3/cdk4) for mammalian D type G1 cyclins, *Cell* 71: 323.

Meyerson, M., Enders, G. H., Wu, C. L., Su, L. K., Gorka, C., Nelson, C., Harlow, E. and Tsai, L. H., 1992, A family of human cdc2-related protein kinases, *Embo J* 11: 2909.

Millar, J. B. A., McGowan, C. H., Lenaers, G., Jones, R. and Russell, P., 1991, p80^{cdc25} mitotic inducer is the tyrosine phosphatase that activates p34^{cdc2} kinase in fission yeast, *EMBO J.* 10: 4301.

Morla, A., Draetta, G., Beach, D. and Wang, J. Y. J., 1989, Reversible tyrosine phosphorylation of cdc2: Dephosphorylation accompanies activation during entrance into mitosis., *Cell* 58: 193.

Ookata, K., Hisanaga, S.-i., Okano, T., Tachnibana, K. and Kishimoto, T., 1992, Relocation and distinct subcellular localization of p34^{cdc2}-cyclin B complex at meiosis reinitiation in starfish oocytes, *EMBO J.* 5: 1763.

Pagano, M., Pepperkok, R., Verde, F., Ansorge, W. and Draetta, G., 1992, Cyclin A is required at two points in the human cell cycle., *EMBO J.* 11: 961.

Pines, J. and Hunter, T., 1991a, Cyclin-dependent kinases: a new cell cycle motif?, *Trends Cell Biol.* 1: 117.

Pines, J. and Hunter, T., 1991b, Human cyclins A and B are differentially located in the cell and undergo cell cycle dependent nuclear transport, *J. Cell Biol.* 115: 1.

Reed, S. I., 1991, G1-specific cyclins: in search of an S-phase promoting factor, *Trends. Genet.* 7: 95.

Russell, P. and Nurse, P., 1987a, The mitotic inducer nim1+ functions in a regulatory network of protein kinase homologs controlling the initiation of mitosis, *Cell* 49: 569.

Russell, P. and Nurse, P., 1987b, Negative regulation of mitosis by wee1+, a gene encoding a protein kinase homolog, *Cell* 49: 559.

Smythe, C. and Newport, J. W., 1992, Coupling of mitosis to the completion of S phase in Xenopus occurs via modulation of the tyrosine kinase that phosphorylates p34cdc2, *Cell* 68: 787.

Strausfeld, U., Labbé, J. C., Fesquet, D., Cavadore, J. C., Picard, A., Sadhu, K., Russell, P. and Dorée, M., 1991, Dephosphorylation and activation of a p34cdc2/cyclin B complex in vitro by human cdc25 protein, *Nature (London)* 351: 242.

Xiong, Y., Zhang, H. and Beach, D., 1992, D type cyclins associate with multiple protein kinases and the DNA replication and repair factor PCNA, *Cell* 71: 504.

23

REGULATION OF NUCLEAR ENVELOPE ASSEMBLY AND DISASSEMBLY BY ARF AND OTHER GTP-BINDING PROTEINS

Annette L. Boman, Kathleen M.C. Sullivan, and Katherine L. Wilson

Department of Cell Biology and Anatomy
Johns Hopkins University School of Medicine
725 N. Wolfe Street
Baltimore, MD 21205

INTRODUCTION

At the onset of mitosis in higher eukaryotes the nuclear envelope, ER, and Golgi complex are partially or completely converted to small vesicles. After mitosis, these biochemically distinct vesicles must fuse selectively to reassemble each organelle. The potential similarities between this process and the process of vesicular transport through the secretory pathway were pointed out by Warren (1985). Secretion involves the regulated formation of coated vesicles at one compartment and their fusion with a specific target membrane (Rothman and Orci, 1992). Secretion ceases during mitosis (Hesketh et al., 1984; Colman et al., 1985), indicating that vesicle formation and/or vesicle fusion are temporarily inhibited. Warren suggested that a mitotic block to vesicle fusion would be sufficient to cause organelles to vesiculate. Newport and Spann (1987) showed that mitotic disassembly of nuclear membranes required an unknown stoichiometric factor(s), perhaps coat proteins. However, direct evidence for the mechanism of mitotic disassembly of the nuclear envelope, or its relationship to membrane traffic, was lacking.

We have studied nuclear envelope assembly and disassembly *in vitro* using *Xenopus* egg extracts, which are a rich source of all nuclear components except DNA (see Wiese and Wilson, 1993). To form a functional nucleus, we reconstitute the fractionated egg cytosol and membranes with demembranated sperm chromatin. We can assay three stages of assembly by light microscopy: binding of preformed nuclear-specific vesicles to the chromatin, fusion of the bound vesicles to form an intact nuclear envelope, and continued vesicle fusion leading to nuclear envelope growth. Once nuclei have formed, the extract can be converted to mitosis by adding recombinant cyclin B. This causes complete nuclear breakdown. The *Xenopus* extracts are thus well suited for analyzing both assembly and mitotic disassembly of the nuclear envelope.

The Cell Cycle: Regulators, Targets, and Clinical Applications
Edited by V.W. Hu, Plenum Press, New York, 1994

We showed that GTPγS inhibited the fusion of nuclear vesicles. This inhibition was caused by the irreversible binding of ARF, a small GTPase, to nuclear vesicles in the presence of GTPγS. Our data and other observations discussed below lead us to propose that ARF may normally interact with nuclear membranes during mitosis to mediate nuclear envelope disassembly.

ARF—A ROLE IN THE NUCLEAR ENVELOPE

ARF (ADP-ribosylation factor) is an abundant, 20 kD GTP-binding protein that associates transiently with membranes; it is soluble when GDP-bound, and binds membranes when GTP-bound (Kahn et al., 1991). ARF proteins are covalently modified by N-terminal myristoylation. The interaction of ARF with membranes requires trypsin-sensitive, membrane-associated proteins (Serafini et al., 1991). Cells stained by indirect immunofluorescence show heavy ARF labelling of the Golgi complex and lighter staining of other organelles (Stearns et al, 1990). At least six ARF cDNAs have been isolated, but the specific localization and activity of different ARF proteins (which can be up to 96% identical at the amino acid level) are only beginning to be investigated.

We have shown that nuclear vesicle fusion is inhibited by non-hydrolyzable GTP analogs, including GTPγS (Boman et al., 1992a). When GTPγS was added at the beginning of the assembly reaction, nuclear vesicles bound to the chromatin but did not fuse. Further analysis showed that nuclear vesicles were irreversibly inhibited for fusion if they were preincubated with GTPγS and purified ARF1 protein (Boman et al., 1992b); this ARF-GTPγS binding to nuclear vesicles somehow prevented them from fusing.

ARF-GTPγS had similar inhibitory effects on ER-to-Golgi and intra-Golgi vesicular transport. Furthermore, ARF-GTPγS-arrested Golgi membranes accumulated the non-clathrin coat proteins specific for Golgi transport vesicles, αβγδ-COPs (coatomers) (Melançon et al., 1987; Waters et al., 1991; Orci et al., 1993). ARF activation (GTP binding and membrane association) is required for coatomer assembly onto Golgi membranes, and for the binding of related coat proteins to endosomes and the *trans* Golgi network (Donaldson et al., 1992a; Wong and Brodsky, 1992; Robinson and Kreis, 1992; Narula et al., 1992). These data suggested that ARF-GTPγS inhibits transport by "locking" specific coats onto Golgi (and other) membranes.

We were interested in whether nuclear vesicles were similarly coated in response to ARF activation by GTPγS. Since no nuclear vesicle coat protein had been described, we used a monoclonal antibody to βCOP for immunoblotting experiments. We found that ARF accumulated on nuclear vesicles incubated with cytosol and GTPγS as previously described (Boman et al., 1992b), but βCOP did not accumulate (Boman, unpublished data). This suggested that a unique set of coat proteins (not recognized by βCOP antibodies) may associate with ARF-GTPγS-arrested nuclear vesicles. Alternatively, ARF-GTPγS may inhibit nuclear vesicle fusion by a mechanism that does not involve coat-recruiting activity. If a nuclear vesicle coat does exist, we would not expect coat assembly to be part of the interphase fusion pathway for preformed nuclear vesicles. We speculate, however, that

ARF (and coats?) may normally interact with nuclear membranes during mitosis, and that GTPγS is forcing this interaction at the wrong time. It will be important to identify the putative coats on ARF-GTPγS-arrested, and mitotic, nuclear vesicles. Such coats could be mitosis-specific, or could be the same COP-like coats used on ER transport vesicles (Stenbeck et al., 1992; Hosobuchi et al., 1992).

ARF activity is controlled by at least two proteins (see Figure 1): a guanine nucleotide exchange factor (GNEF) and a GTPase activating protein (GAP). (R. Kahn, pers. comm.) The GNEF stimulates soluble ARF to exchange its GDP for GTP and associate with the membrane. In response to an unknown signal, the GAP activates GTP hydrolysis by ARF, and ARF-GDP is then released to the cytosol. The fungal metabolite

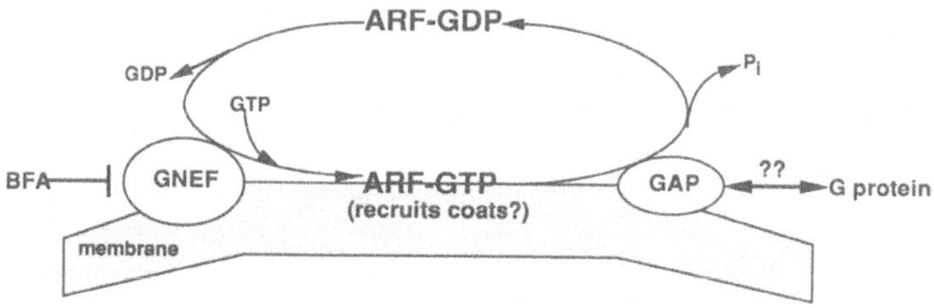

Figure 1: Proposed ARF cycle.
This diagram illustrates a proposed ARF activation cycle. GNEF, guanine nucleotide exchange factor; GAP, GTPase activating protein; BFA, brefeldin A. GNEF and GAP are proposed to be membrane proteins, and ARF-GTP associates with membranes. See text for details.

brefeldin A (BFA) prevents ARF binding to Golgi membranes. BFA is now thought to inactivate the GNEF (Helms and Rothman, 1992; Donaldson et al., 1992b; Randazzo et al., 1993) so that ARF is constitutively GDP-bound and therefore unable to bind membranes.

Nuclear assembly reactions also respond to BFA (Boman et al., 1992b). If vesicles are first exposed to BFA before adding GTPγS, ARF does not bind and the vesicles fuse normally. The effect of BFA on nuclear vesicles (preventing ARF binding) parallels its action on Golgi membranes, indicating that ARF activity on nuclear vesicles may also be regulated by a putative nuclear GNEF.

G PROTEIN REGULATION AND SIGNALLING

ARF was originally identified as a protein cofactor required for cholera toxin to enzymatically ADP-ribosylate the α-subunit of the stimulatory G protein, $G_{s\alpha}$ (Kahn and Gilman, 1984). This suggested that ARF might normally interact with heterotrimeric G proteins, but the purpose of this putative interaction was unknown. Several lines of evidence indicate that heterotrimeric G proteins regulate one or more steps in interphase vesicle transport (reviewed by Bomsel and Mostov, 1992). First, the $G_{\alpha i3}$ subunit has been immunolocalized at the Golgi complex (Stow et al., 1991). Second, G_α subunits (but not ARF, ras, rho, or rabs) are activated by aluminum fluoride ion (AlF_n) (Sternweis and Gilman, 1982; Kahn, 1991), which inhibits both Golgi transport (Melançon et al., 1987) and nuclear vesicle fusion (Boman, unpublished data) *in vitro*. AlF_n treatment causes ARF to accumulate on Golgi and nuclear vesicles in the absence of GTPγS, suggesting that a G protein may control GTP hydrolysis by ARF either directly or indirectly (see Fig. 1). (Note that AlF_n also inhibits phosphotransferase enzymes [Kahn, 1991] and may therefore inhibit vesicle fusion by a mechanism unrelated to G proteins.) Finally, addition of purified $G_{\beta\gamma}$ subunits of heterotrimeric G proteins inhibits Golgi transport and endosome fusion *in vitro* (Donaldson et al., 1991; Colombo et al., 1992). Excess $G_{\beta\gamma}$ subunits are thought to bind and inactivate G_α. These findings suggest that ARF activity is regulated by a G protein. The common involvement of ARF and putative G proteins supports the idea that shared mechanisms may regulate both interphase vesicle traffic and post-mitotic nuclear vesicle fusion.

Recent results from our lab raise the intriguing possibility that G proteins could also regulate a later step in the fusion pathway. BAPTA, a calcium buffer that does not significantly change the cytosolic free Ca^{2+} concentration, inhibits nuclear vesicle fusion *in vitro* (Sullivan et al., 1993). Order-of-addition studies showed that BAPTA arrests vesicles at a later stage than does ARF-GTPγS. BAPTA-arrested vesicles are no longer able to bind ARF in the presence of GTPγS, suggesting that the GNEF may be switched off. BAPTA inhibition is partially reversed by IP_3, a second messenger that stimulates Ca^{2+} release from lumenal stores. Furthermore, nuclear vesicle fusion is inhibited by heparin, which is a potent antagonist of Ca^{2+} release by IP_3 receptors located on nuclear and ER membranes (Ghosh et al., 1988; Meldolesi et al., 1992). These results suggest that IP_3 receptors and Ca^{2+} transients are required for nuclear vesicle fusion.

There are two major signal transduction pathways at the plasma membrane that can lead to IP_3 production and Ca^{2+} release: G protein coupled receptors and tyrosine kinase-linked receptors (Berridge, 1993). We do not yet know which, if either, pathway is used to signal Ca^{2+} mobilization during nuclear vesicle fusion. However, it is tempting to speculate that ARF activity (vesicle binding, coating, uncoating) and IP_3 production (which may mobilize Ca^{2+} to trigger fusion) might be sequentially regulated by signalling through one or more G proteins. We can now design experiments to test if G proteins are used as control points for cell cycle regulation of vesicle formation and fusion.

REFERENCES

Berridge, M.J., 1993, Inositol trisphosphate and calcium signalling, *Nature*. 361:315-325.

Boman, A.L., Delannoy, M.R. and Wilson, K.L., 1992a, GTP hydrolysis is required for vesicle fusion during nuclear envelope assembly in vitro, *J. Cell Biol.* 116:281-94.

Boman, A.L., Taylor, T.C., Melançon, P. and Wilson, K.L., 1992b, A role for ADP-ribosylation factor in nuclear vesicle dynamics, *Nature*. 358:512-4.

Bomsel, M. and Mostov, K., 1992, Role of heterotrimeric G proteins in membrane traffic, *Mol. Biol. Cell*. 3:1317-1328.

Colman, A., Jones, E.A. and Heasman, J., 1985, Meiotic maturation in Xenopus oocytes: a link between the cessation of protein secretion and the polarized disappearance of Golgi apparati, *J. Cell Biol.* 101:313-318.

Colombo, M.I., Mayorga, L.S., Casey, P.J. and Stahl, P.D., 1992, Evidence of a role for heterotrimeric GTP-binding proteins in endosome fusion, *Science*. 255:1695-1697.

Donaldson, J.G., Cassel, D., Kahn, R.A. and Klausner, R.D.,1992a, ADP-ribosylation factor, a small GTP-binding protein, is required for binding of the coatomer protein beta-COP to Golgi membranes, *Proc. Natl. Acad. Sci. (USA)*. 89:6408-6412.

Donaldson, J.G., Finazzi, D. and Klausner, R.D., 1992b, Brefeldin A inhibits Golgi membrane-catalysed exchange of guanine nucleotide onto ARF protein, *Nature*. 360:350-352.

Donaldson, J.G., Kahn, R.A., Lippincott, S.J. and Klausner, R.D., 1991, Binding of ARF and beta-COP to Golgi membranes: possible regulation by a trimeric G protein, *Science*. 254:1197-1199.

Ghosh, T.K., Eis, P.S., Mullaney, J.M., Ebert, C.L. and Gill, D.L.,1988, Competitive, reversible, and potent antagonism of inositol 1,4,5-trisphosphate-activated calcium release by heparin, *J. Biol. Chem.* 263:11075-11079.

Helms, J.B. and Rothman, J.E., 1992, Inhibition by brefeldin A of a Golgi membrane enzyme that catalyses exchange of guanine nucleotide bound to ARF, *Nature*. 360:352-4.

Hesketh, T.R., Beaven, M.A., Rogers, J., Burke, B. and Warren, G.B., 1984, Stimulated release of histamine by a rat mast cell line is inhibited during mitosis, *J. Cell Biol.* 98:2250-2254.

Hosobuchi, M., Kreis, T. and Schekman, R., 1992, SEC21 is a gene required for ER to Golgi protein transport that encodes a subunit of a yeast coatomer, *Nature*. 360:603-5.

Kahn, R. and Gilman, A.G., 1984, Purification of a membrane protein required for cholera toxin-dependent ADP-ribosylation of the stimulatory, regulatory component of adenylate cyclase, *J. Biol. Chem.* 259:6228-6234.

Kahn, R.A., 1991, Fluoride is not an activator of the smaller (20-25 kDa) GTP-binding proteins, *J. Biol. Chem.* 266:15595-15597.

Kahn, R.A., Kern, F.G., Clark, J., Gelmann, E.P. and Rulka, C., 1991, Human ADP-ribosylation factors: a functionally conserved family of GTP-binding proteins, *J. Biol. Chem.* 266:2606-14.

Melançon, P., Glick, B.S., Malhotra, V., Weidman, P.J., Serafini, T., Gleason, M.L., Orci, L. and Rothman, J.E., 1987, Involvement of GTP-binding "G" proteins in transport through the Golgi stack, *Cell*. 51:1053-1062.

Meldolesi, J., Villa, A., Volpe, P. and Pozzan, T., 1992, Cellular sites of IP_3 action. *In*: "Advances in Second Messenger and Phosphoprotein Research, Vol. 26, Inositol Phosphates and Calcuim Signalling." J. W. Putney, ed. (New York, NY, Raven Press), pp. 187-208.

Narula, N., McMorrow, I., Plopper, G., Doherty, J., Matlin, K.S., Burke, B. and Stow, J.L., 1992, Identification of a 200-kD, brefeldin-sensitive protein on Golgi membranes, *J. Cell Biol.* 11: 27-38.

Newport, J. and Spann, T., 1987, Disassembly of the nucleus in mitotic extracts: membrane vesicularization, lamin disassembly, and chromosome condensation are independent processes, *Cell*. 48:219-230.

Orci, L., Palmer, D.J., Ravazzola, M., Perrelet, A., Amherdt, M. and Rothman, J.E., 1993, Budding from Golgi membranes requires the coatomer complex of non-clathrin coat proteins, *Nature*. 362:648-652.

Randazzo, P.A., Yang, Y.C., Rulka, C. and Kahn, R.A., 1993, Activation of ADP-ribosylation factor by Golgi membranes: evidence for a brefeldin A- and protease-sensitive activating factor on Golgi membranes, *J. Biol. Chem.* in press.

Robinson, M.S. and Kreis, T.E., 1992, Recruitment of coat proteins onto Golgi membranes in intact and permeabilized cells: effects of brefeldin A and G protein activators, *Cell.* 69:129-138.

Rothman, J.E. and Orci, L., 1992, Molecular dissection of the secretory pathway, *Nature.* 355:409-415.

Serafini, T., Orci, L., Amherdt, M., Brunner, M., Kahn, R.A. and Rothman, J.E., 1991, ADP-ribosylation factor is a subunit of the coat of Golgi-derived COP-coated vesicles: a novel role for a GTP-binding protein, *Cell.* 67:239-253.

Stearns, T., Willingham, M.C., Botstein, D. and Kahn, R.A., 1990, The ADP-ribosylation factor (ARF) is physically and functionally associated with the Golgi complex, *Proc. Natl. Acad. Sci. (USA).* 87:1238-1242.

Stenbeck, G., Schreiner, R., Herrmann, D., Auerbach, S., Lottspeich, F., Rothman, J.E. and Wieland, F.T., 1992, Gamma-COP, a coat subunit of non-clathrin-coated vesicles with homology to Sec21p, *FEBS Letters.* 314:195-198.

Sternweis, P.C. and Gilman, A.G., 1982, Aluminum: a requirement for activation of the regulatory component of adenylate cyclase by fluoride, *Proc. Natl. Acad. Sci. (USA).* 79:4888-4891.

Stow, J.L., de Almeida, J., Narula, N., Holtzman, E.J., Ercolani, L. and Ausiello, D.A., 1991, A heterotrimeric G protein, G alpha i-3, on Golgi membranes regulates the secretion of a heparan sulfate proteoglycan in LLC-PK1 epithelial cells, *J. Cell Biol.* 114:1113-1124.

Sullivan, K.M.C., Busa, W.B. and Wilson, K.L., 1993, Calcium mobilization is required for nuclear vesicle fusion in vitro: implications for membrane traffic and IP_3 receptor function, *Cell.* 74:in press.

Warren, G., 1985, Membrane traffic and organelle division, *Trends Bioch. Sci.* 10:439-443.

Waters, M.G., Serafini, T. and Rothman, J.E., 1991, 'Coatomer': a cytosolic protein complex containing subunits of non-clathrin-coated Golgi transport vesicles, *Nature.* 349:248-251.

Wiese, C. and Wilson, K., 1993, Nuclear membrane dynamics, *Curr. Opin. Cell Biol.* 5:in press.

Wong, D.H. and Brodsky, F.M., 1992, 100-kD proteins of Golgi- and trans-Golgi network-associated coated vesicles have related but distinct membrane binding properties, *J. Cell Biol.* 117:1171-9.

24

POSSIBLE ROLE OF THE MULTI CATALYTIC PROTEINASE (PROTEASOME) IN REGULATING OF THE CELL CYCLE

A. Amsterdam[1], F. Pitzer[2], U. Santarius[2], A. Dantes[1] and W. Baumeister[2]

[1]Department of Hormone Research, the Weizmann Institute of Science, Rehovot, 76100, Israel and
[2]Department of Molecular Structure, the Max Planck Institute for Biochemistry, D-8033, Martinsried, F.R.G.

ABSTRACT

We have investigated the intracellular distribution of proteasomes (multi catalytic proteinase) during the somatic cell cycle and during differentiation. The granulosa cell system was chosen for investigating the spatial and temporal distribution pattern of proteasomes, since stable lines derived from primary cells have recently been established in our laboratory. Moreover, these lines can be stimulated to differentiate by substances that elevate their intracellular cAMP levels. We found a dramatic accumulation of proteasomes in metaphase and early anaphase, the stages at which cyclin A and B_1 are degraded. Moreover, only a low concentration of proteasomes could be observed in cells that were stimulated to differentiate. A preferential high concentration of proteasomes was clearly evident in close proximity to spindle microtubules in dividing cells, with significantly lower concentrations of proteasomes outside the spindle apparatus. In contrast, no spatial correlation was evident between proteasomal and microtubular localization in nondividing cells. The modulation of proteasome distribution in mitosis suggests that they play a regulatory role during the cell cycle.

PROTEOLYTIC ACTIVITY IS NEEDED FOR THE PROGRESSION OF THE CELL CYCLE

The pioneering work of Hunt and his colleagues demonstrated that cyclins are degraded during mitosis (1,2). The experiments were performed by resolving sea urchin fertilized egg proteins extracted at different stages of the cell cycle by gel electrophoresis. Pines and Hunter (3), in complementary experiments performed recently, demonstrated the disappearance of cyclin A and B_1 in the intact cell at metaphase and early anaphase by immunofluorescence microscopy. These phenomena, as well as the dissolution of the nuclear membrane in late prophase and the disappearance of the spindle apparatus in telophase suggest that degradative enzymes such as proteases are involved in the modulation of the cell cycle. Such a destructive enzyme must have access to specific substrates outside lysosomes. This makes the possibility of lysosomal enzyme participation in such processes unlikely. In contrast, the major nonlysosomal proteolytic activity in eukaryotic cells is ascribed to a 700 kDa (19S) enzyme complex; it is present in both the cytosol and the nucleus (Fig. 1) of all eukaryotic cells from yeast to man (4). Interestingly, we found by

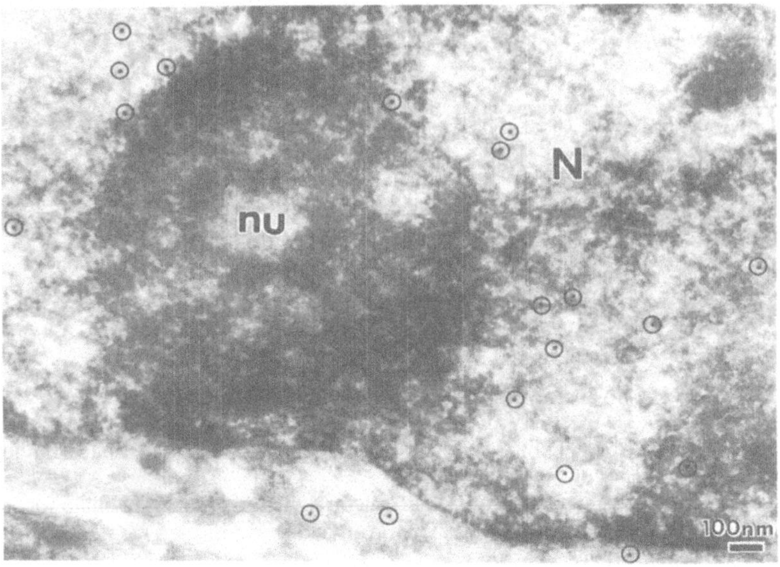

Fig. 1. Localization of proteasomes in the nucleus of a SV40 transformed granulosa cell. Cells were fixed with 2% glutaraldehyde in PBS, pH 7.4. The pellet of cells was embedded in sucrose and frozen in liquid nitrogen. Cryosections were incubated with antiserum to granulosa cell proteasomes and subsequently with goat anti-rabbit antibodies coupled to 15 mM gold particles. The section was stained with uranyl acetate before being examined under the Phillips 410 electron microscope. Gold particles are confined to the nucleus matrix but are excluded from the nucleolus. N, nucleus; nu, nucleolus.

high resolution immunoelectron microscopy that proteasomes were present in the nuclear matrix but were excluded from the nucleolus (Fig. 1). It was recently shown that proteasomes also exist in primitive eukaryotes such as *Dictyostellum discoideum* and in the archaebacterium *Thermoplasma acidophilum* (5).

The proteasome is a 19S cylinder-shaped particle composed of 8-15 different subunits, all in a molecular weight range between 20 and 35 kDa. The proteasome contains at least three distinct endopeptidase sites (6). Proteasomes from the archaebacterium *Thermoplasma* are built from only two different subunits (α and β), yet the complex is almost identical in size and shape (5,7-9) with the eukaryotic form. For *Drosophila* it was shown that the subunit composition can vary during development, i.e. that from a larger repertoire of 20-30 subunits, specific ones are recruited at different stages of development (10,11).

ROLE OF PROTEASOMES IN ANTIGEN PROCESSING MACHINERY AND IN DEGRADATION OF UBIQUITIN CONJUGATED PROTEINS

From the localization of some genes encoding proteasomal subunits in the region of the major histocompatibility complex (MHC) class II (12-14) and from the transcriptional up-regulation of the genes encoding the MHC glycoprotein, the peptide transporter genes (15-16) and the proteasomal genes, it has been concluded that the proteasome is part of the antigen processing machinery (17). It has been proposed that proteasomes generate the peptide fragments from the intact antigen which are then transported into the endoplasmic reticulum.

Increasing evidence has strongly suggested that the proteasome represents the catalytic core of an even larger (1,500 kDa) proteolytic complex which degrades ubiquitin conjugated proteins (18-19). Incorporating a proteolytically active polypeptide into a larger multisubunit complex opens up a number of possibilities for the regulation of its activity, including the binding of small ligands, interactions with inhibitor and activator proteins, and targeting. Nevertheless, the proteolytic targets *in vivo*, and hence the precise role of the proteasomes in intracellular breakdown, remain unclear. Also, it is not known, how intracellular protein turnover is regulated during different stages of the cell cycle and during differentiation and transformation.

IMMORTALIZED GRANULOSA CELLS AS A MODEL FOR INVESTIGATING THE REGULATION OF THE CELL CYCLE

We have chosen to use ovarian granulosa cells as a model to study the possible role of proteasomes in regulating the cell cycle. The ovarian granulosa cells nurse the oocyte and are the main source for production of the female hormone progesterone (20-22). They proliferate rapidly during follicular development. However, when these cells are stimulated to differentiate, rapid cell division stops concomitantly with massive production of progesterone characteristic of luteal tissue (20-22). Many studies in the last decade, including our own, demonstrated that granulosa cell differentiation can be stimulated *in vitro* in primary cultures by the coordinated effect of peptides and steroid hormones, growth factors and extracellular matrix (20-22). More recently, we have established monoclonal granulosa cell lines by co-transfection of primary cells with SV40 and the Ha-ras oncogene (23-25). Induction of differentiation, as indicated by expression of mitochondrial steroid enzymes and synthesis of progesterone, were achieved, concomitantly with repression of cell proliferation, by stimulation of the transformed cells with 8-Br-cAMP, or substances elevating intracellular cyclic AMP, such as gonadotropic hormones, forskolin, cholera toxin, and the *Bordetella pertussis* invasive adenylate cyclase (23-26). We demonstrated that highly purified proteasomal preparations obtained in significant quantities could be used for further detailed structure-function analysis (27). Moreover, analyzing the distribution of proteasomes during the cell cycle of these cells suggested an important role for proteasomes in cell cycle regulation.

SPATIAL AND TEMPORAL DISTRIBUTION OF PROTEASOMES DURING THE CELL CYCLE

In order to follow the intracellular distribution of proteasomes in immortalized granulosa cells during mitosis and during cAMP induced differentiation, as revealed by immunofluorescence microscopy, cell cultures were fixed with formaldehyde, permeabilized briefly with Triton X-100, stained with specific antibodies to rat muscle or rat liver proteasomes and subsequently with goat antirabbit IgG coupled to rhodamine.

In interphase, proteasomes were localized in small clusters throughout the cytoplasm and the nuclear matrix. In prophase, a substantial increase in proteasomal staining was observed in the perichromosomal area. A dramatic increase occurred in metaphase and in early anaphase; the chromosomes remained unstained (Fig. 2). In late anaphase, intensive staining remained associated mainly with the spindle fibers. In telophase and early interphase of the daughter cells, intensive staining of proteasomes persisted in the nuclei.

In contrast, in cells stimulated to differentiate by forskolin, which substantially elevates intracellular cyclic AMP in these cell lines, only weak staining of proteasomes was revealed in both the nucleus and cytoplasm. Double staining of nondividing cells with antibodies to proteasomes and to β-tubulin did not show colocalization of proteasomes and microtubules (Fig. 3). In contrast, dividing cells showed a preferential concentration of proteasomes around spindle microtubules, during metaphase and anaphase (27). These observations were confirmed using a laser confocal microscope where serial thin optic section could be obtained, from which 3 dimensional images could be reconstructed (Fig. 2).

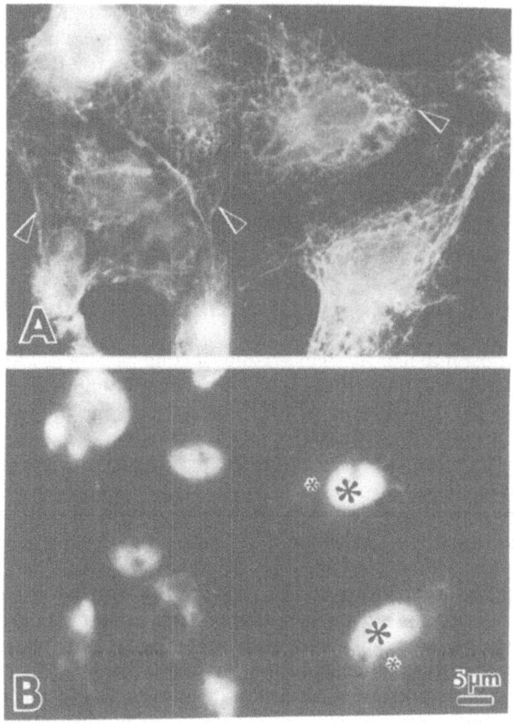

Fig. 2. Localization of microtubules (A) and proteasomes (B) in interphase cells. Primary granulosa cells were fixed with 3% formaldehyde, permeabilized with 1% Triton X-100 (27) and double labeled as follows: Cells were first incubated with mouse monoclonal antibodies to β tubulin and goat anti-mouse antibodies coupled to rhodamine (A). Subsequently, the cells were incubated with rabbit polyclonal antibodies to rat proteasomes and with goat anti rabbit antibodies coupled to fluorescein (B). Cells were visualized by fluorescence microscopy (Photomicroscope III, Zeiss). No correlation is visible between the localization of microtubules in the cytoplasm of the cultured cells (arrowheads) and the localization of proteasomes in nuclei of cells (asterices) and to a lesser extent in the cytoplasm (small asterices).

Cyclin A and B_1 are believed to act as regulatory elements during mitosis. Moreover, cyclin A was recently shown to be destroyed during metaphase and cyclin B_1 at the metaphase-anaphase transition (3). We can not exclude the possibility that cyclins are degraded by ubiquitin-dependent proteolysis. The dramatic accumulation of proteasomes we observed coincides with the degradation of cyclins A and B_1. Therefore, we suggest that a high proteasome concentration in the vicinity of the spindle apparatus is important to ensure the timely degradation of ubiquitinated cyclins, necessary for termination of mitosis.

The involvement of proteasomes in control of cell division is also supported by findings in other laboratories that a high content of proteasomes or messenger RNA coding for the proteasomal proteins is characteristic of rapidly multiplying embryonic cells and of cancerous tissues (11,28).

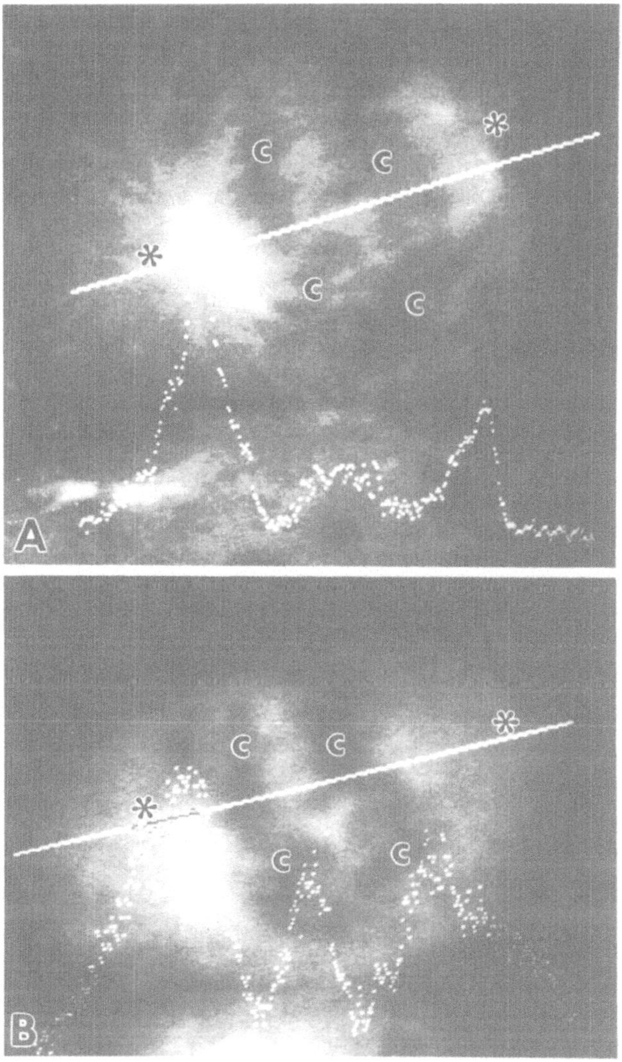

Fig. 3. Colocalization of spindle microtubules (A) and proteasomes (B) in ovarian granulosa cells in early anaphase visualized using a confocal laser scanning microscope (CLSM). SV40-transformed granulosa cells were double stained as described in the legend to Figure 2. Cells were visualized in a fluorescence CLSM (Leitz, CMUE 101); the entire cell depth was scanned in 16 serial optical sections (1.0 μm each) and the integrated picture was resolved on the television screen. Spindle microtubules are aligned toward the centrosomes in the two poles of the spindle apparatus (asterices). Chromosomes are negatively stained (c). The white bar (20 μm long) indicates the line of optical densitometer tracing which yielded the curve in the bottom part of Fig. 3A. Proteasomes show a similar but not an identical pattern of distribution as the spindle microtubules (Fig. 3B). The chromosomes are negatively stained (c) and the densitometer tracing curve shows peaks at similar locations to that of microtubule assembly.

It is generally believed that proteins conjugated to ubiquitin can be degraded in the cell via the nonlysosomal route after forming a 26 S complex which contains the 19 S proteasomal protein (18,19,29). Moreover, *Thermoplasma acidophilum* proteasomes were recently found to partially degrade *in vitro* ubiquitin-associated proteins (30). Therefore, it will be important in the future to synchronize cell division in order to isolate proteasomes during specific stages of the cell cycle in order to examine whether a 26S complex containing ubiquitinated cyclin A or B_1 can be identified. Moreover, immunofluorescence studies should be extended to colocalization of proteasomes *in situ* with ubiquitin conjugated cyclins. Such studies may lead to a better understanding of how the important metabolic activity of the proteasomes participates in controlling the cell cycle.

ACKNOWLEDGMENTS

The authors wish to thank Dr. A.M. Kaye for helpful discussion; Dr. B. Dahlmann for generous support of antibodies to rat liver and rat muscle proteasomes; Mr. S. Himmelhoch for cryosection preparation; and Mrs. Malka Kopelowitz for excellent secretarial assistance. The work was supported by grants from the Minna and James Heineman Foundation, München, and from the Joseph and Ceil Mazer Center for Structural Biology and the Helen and Milton A. Kimmelman Center for Biomolecular Structure and Assembly at the Weizmann Institute of Science. AA is the incumbent of the Joyce and Ben B. Eisenberg Professorial Chair in Molecular Endocrinology and Cancer Research.

REFERENCES

1. T. Evans, E.T. Rosenthal, J. Youngblom, D. Distel, and T. Hunt, *Cell* 33:389 (1983).

2. T. Hunt, *Curr. Opin. Cell Biol.* 1:268 (1989).

3. J. Pines, and T. Hunter, *J. Cell Biol.,* 115:1 (1991).

4. A.P. Arrigo, M. Simon, J.L. Darlix, and P.F. Spahr, *J. Molec. Evol.* 25:141 (1987).

5. B. Dahlmann, F. Kopp, L. Kuehn, B. Niedel, G. Pfeifer, R. Hegerl, and W. Baumeister, *FEBS Letts.* 251:125 (1989).

6. A.J. Rivett, *Biochem. J.* 263:625 (1989).

7. W. Baumeister, B. Dahlmann, R. Hegerl, F. Kopp, and G. Pfeifer, *FEBS Letts.* 241:239 (1988).

8. A. Grziwa, W. Baumeister, B. Dahlmann, and F. Kopp, *FEBS Lett.* 290:186 (1991).

9. G. Püller, S. Weinkauf, L. Bachmann, S. Müller, A. Engel, R. Hegeri, and W. *EMBO J.* 11:1607 (1992).

10. C. Haass, B. Pesold-Hurt, G. Multhaup, B. K., and P.M. Kloetzel, *Gene* 90:235 (1990).

11. U. Klein, M. Gerold, and P.M. Kloetzel, *J. Cell Biol.* 111:2275 (1990).

12. C.K. Martinez, and J.J. Monaco, *Nature* 353:664 (1991).

13. A. Kelly, S.H. Powis, R. Glyme, E. Radley, S. Beck, and J. Trowsdale, *Nature* 353:667 (1991).

14. R. Glynne, S.H. Powis, S. Beck, A. Kelly, L.A. Kerr, and J. Trowsdale, *Nature* 353:357 (1991).

15. J. Trowsdale, I. Hanson, I. Mockridge, S. Beck, A. Townsend, and A. Kelly, *Nature* 348:741 (1990).

16. T. Spies, M. Bresnahan, S. Bahram, D. Arnold, G. Blanck, E. Mellins, D. Pious, and R. DeMars, *Nature* 348:744 (1991).

17. M. Robertson, *Nature* 353:300 (1991).

18. A. Hershko, and A. Ciechanover, *Ann. Rev. Biochem.* 61:761 (1992).

19. J. Driscoll, and A.L. Goldberg, *J. Biol. Chem.* 265:4789 (1990).

20. A.J.W. Hsueh, E.Y. Adashi, P.B.C. Jones, and J. Welsh T.H., *Endocr. Rev.* 5:76 (1984).

21. A. Amsterdam, and S. Rotmensch, *Endocr. Rev.* 8:309 (1987).

22. A. Amsterdam, S. Rotmensch, and A. Ben-Ze'ev, *TIBS* 14:377 (1989).

23. A. Amsterdam, A. Zauberman, G. Meir, O. Pinhasi-Kimhi, B.S. Suh, and M. Oren, *Proc. Natl. Acad. Sci. U.S.A.* 85:7582 (1988).

24. I. Hanukoglu, B.S. Suh, S. Himmelhoch, and A. Amsterdam, *J. Cell Biol.* 111:1373 (1990).

25. B.S. Suh, R. Sprengel, I. Keren-Tal, S. Himmelhoch, and A. Amsterdam, *J. Cell Biol.* 119:439 (1992).

26. A. Amsterdam, I. Hanukoglu, B.S. Suh, I. Keren-Tal, D. Plehn-Dujowich, R. Sprengel, H. Hennert, and J.F. Strauss III, *J. ster. Biochem. Molec. Biol.* 43:875 (1992).

27. A. Amsterdam, F. Pitzer, and W. Baumeister, *Proc. Natl. Acad. Sci. U.S.A.* 90:99 (1993).

28. H.O. Kanayama, K. Tanaka, M. Aki, S. Kagawa, H. Miyaji, M. Satoh, F. Okada, S. Sato, N. Shimbara, and A. Ichihara, *Cancer Res.* 51:6677 (1992).

29. E. Orino, K. Tanaka, T. Tamura, S. Sone, T. Ogura, and A. Ichihara, *FEBS Lett.* 284:206 (1991).

30. T. Wenzel and W. Baumeister, *FEBS Lett.* in press.

25

CYTOSTELLIN: A NUCLEAR PROTEIN THAT REDISTRIBUTES TO PERIPHERAL CYTOSKELETAL LOCATIONS DURING MITOSIS AND G1

Stephen L. Warren,[1] David B. Bregman,[1] Yi Li[1] and Lei Du[2]

[1]Departments of Pathology and [2]Genetics
Yale University School of Medicine
New Haven, CT 06510

INTRODUCTION

Transcription and splicing of pre-mRNA transcripts are closely linked spatiotemporally and occur in discrete sites of the interphase nucleus termed "transcript domains"[1-4]. Fluorescent in situ hybridization has revealed linear "tracks" of a single gene's unspliced and spliced polyA-RNAs emerging from only one or two transcript domains per nucleus, indicating that each transcript is synthesized, processed and exported on a solid phase structure, located at a specific intranuclear position[2,3,6]. More recently, multiple transcription and splicing factors (e.g. SC35,[5] RNA polymerase II and heterogeneous ribonucleoproteins [hnRNPs]) were shown to be coordinately recruited to transcript domains, in response to new gene transcription[6].

Although some of the proteins acting in trancription domains have been defined, little is known about the mechanism by which proteins are targeted to these discrete intranuclear sites. Described below is a novel ~240 kDa protein, cytostellin[7], that can localize in the peripheral cytoskeleton, but also in nuclear transcript domains. Significantly, cytostellin redistributes between these two compartments at specific times during the cell cycle. Cytostellin and spliceosome assembly factor)[5] co-localize in transcript domains during most of interphase. At the onset of mitosis, cytostellin and SC-35 exit the nucleus on microscopically-visible particles. The cytostellin/SC-35-containing particles bind to the peripheral cytoskeleton during mitosis and G1, and they reappear in the nucleus sometime during G1. The ability of the particles to redistribute during G1 from the peripheral cytoskeleton to sites of active gene transcription and pre-mRNA processing in the nucleus suggests that they may be a shuttling mechanism that physically links activities in these two subcellular compartments. This system may be relevant to all eukaryotic cells, since cytostellin has been demonstrated in all cells tested from humans to yeast (including S. cerevisiae and S. pombe).

The Cell Cycle: Regulators, Targets, and Clinical Applications
Edited by V.W. Hu, Plenum Press, New York, 1994

MANY CYTOSTELLIN-CONTAINING PARTICLES EXIT THE NUCLEUS AT THE ONSET OF MITOSIS

Cytostellin is a ~240 kDa protein originally found in Madin-Darby Canine Kidney (MDCK) cells[7], and it is predominantly intranuclear as determined by immunoblotting nuclear and cytoplasmic fractions (Fig 1).

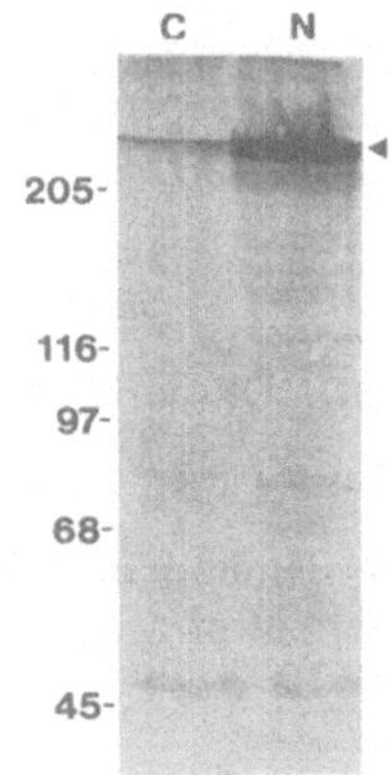

Figure 1. Cytostellin is predominantly nuclear in MDCK cells. Nuclear and cytoplasmic fractions of MDCK cells were prepared, solubilized in SDS sample buffer, electrophoresed on a 7% polyacrylamide gel, transferred to a nitrocellulose membrane and immunoblotted with monoclonal antibody H5. The ~240 kDa protein indicated by a caret is cytostellin. The immunoblot was developed using biotinylated Goat anti-mouse IgM, followed by an avidin-alkaline phosphatase conjugate.

However, cytostellin undergoes continuous redistribution during the cell cycle[7]. Figure 2 shows MDCK cells stained simultaneously with the DNA-binding fluor, 4',6- diamidino-2-phenylindole (DAPI) and anti-cytostellin Mab H5. MDCK cells in the left panels of Figure 2 reveal the chromosomal DNA, and the same cells are shown in the right panels, which reveal cytostellin immunoreactivity. The state of chromosomal condensation, and the contours of the nuclear periphery revealed by DAPI staining indicate the mitotic stage. Cytostellin immunoreactivity in interphase nuclei shows fine punctate staining as described above, but there is also a diffuse component (Figure 2A, right panel; arrows indicate punctate "dots"). In Figure 2 the cells were fixed *prior* to staining, so the diffusely staining fraction of cytostellin is not extracted; consequently, the nuclear dots are less outstanding than those shown in Figure 5. Cytostellin dots are separated from the nuclear periphery by a continuous, submembranous zone which follows the contours of the nuclear envelope (Figure 2, right panel A, bracket). The nucleolus and nuclear periphery do not stain.

Early prophase is identified by an intense, beaded appearance of the DAPI-stained nuclei, reflecting the onset of chromosomal condensation (Figure 2, left panel B, arrowheads). Coincidentally, cytostellin immunoreactivity intensifies (Figure 2, right panel B, arrows). Mid-prophase cells are identified by increasing condensation of the chromosomes, and a loss of the smooth edges at the nuclear periphery indicating early disassembly of the envelope (Figure 2, left panel C, arrows). Coincident with

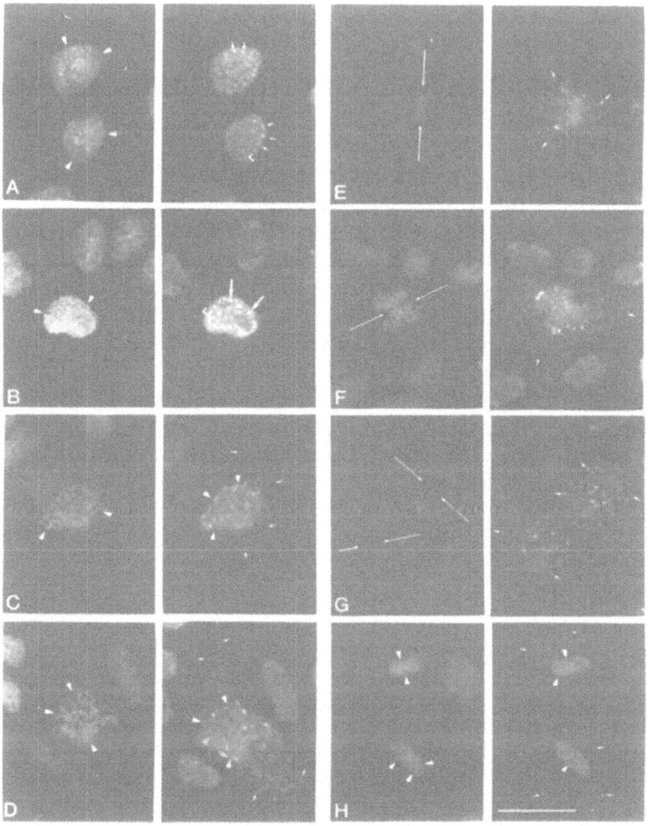

Figure 2. Cytostellin undergoes continuous redistribution during the cell cycle. Indirect immunofluorescence was performed on MDCK cells fixed with 1.7 % paraformaldehdye and permeabilized with 0.5% (vol/vol) Triton X-100 in PBS. The cells were reacted sequentially with: Mab H5 (right panels), biotinylated Goat anti-mouse IgM, avidin-rhodamine conjugate and then DAPI (left panels). A, non-dividing cells; B, early prophase; C, mid prophase; D, late prophase; E, metaphase; F, early anaphase; G, late anaphase; H, telophase. Arrowheads indicate periphery of nucleus. Small arrows indicate cytostellin bodies. Long arrows indicate the plane of metaphase plate. Brackets indicate margin under nuclear periphery which lacks discrete cytostellin immunoreactivity. All panels are presented at the same magnification. Bar in panel H, 25 uM.

these changes, a few cytostellin particles begin to move away from the chromosomes and spindle apparatus (Figure 2, right panel C, arrowheads). By late prophase, multiple cytostellin-containing particles are widely dispersed to positions throughout the cell (Figure 2, right panel D, arrows). The cytostellin particles remain widely dispersed throughout the cell during metaphase, anaphase and telophase (Figure 2, right panels E-H, arrows). The number of mitotic cytostellin-containing particles greatly excedes the number of cytostellin dots observed in interphase cell nuclei (Compare Figures 2D-G to Figure 2A).

THE PARTICLES CONTAIN CYTOSTELLIN AND SPLICEOSOME ASSEMBLY FACTOR, SC-35

Mab-35, directed against spliceosome assembly factor[5], stains the cytostellin-containing particles throughout mitosis. A pair of cells undergoing cytokinesis is shown to illustrate this point (Figure 3A, Mab SC-35; Figure 3B, Mab H5). Cytostellin and SC-35 co-localize in the mitotic particles, but there is an interesting difference in their staining patterns. In the three non-dividing cells surrounding this mitotic pair, SC35 immunoreactivity is visible (Figure 3A, arrowheads), but cytostellin immunoreactivity is not (Figure 3B, arrowheads). The level of cytostellin protein is constant throughout the cell cycle (Figure 4), and cytostellin's H5 epitope is "masked" (i.e. inaccessible to antibodies) in these cells. In fact, the masking of cytostellin's H5 epitope is a hallmark of cells in late G1 and early S phases (unpublished data).

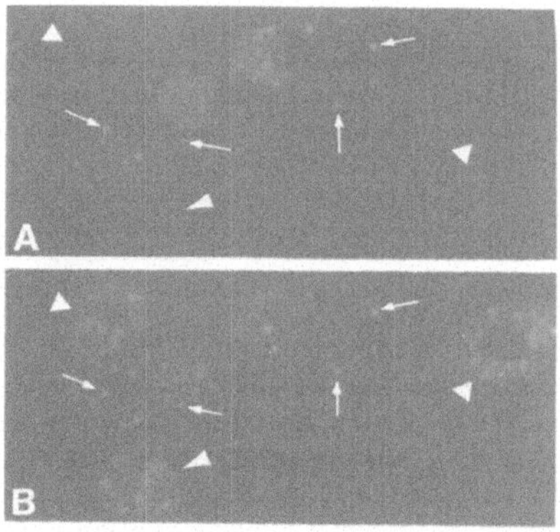

Figure 3. Spliceosome assembly factor (SC-35) and cytostellin co-localize on the mitotic particles. Indirect immunofluorescence was performed on the cells after fixation with 1.7 % paraformaldehdye and permeabilization with 0.5% (vol/vol) Triton X-100 in PBS. A pair of cells undergoing cytokinesis is shown. A, Mab H5; B, Mab SC-35. Arrows, particles that immunostain with Mab SC-35 and Mab H5. Arrowheads, nuclei of non-dividing MDCK cells. Note that after nuclear redistribution, cytostellin immunoreactivity disappears, whereas SC35 is visible. Co-localization of cytostellin and SC-35 in nuclear transcription domains of interphase nuclei can be revealed by "unmasking" cytostellin's H5 epitope with a Triton X-100 extraction prior to fixation as shown in Figure 5.

The dot-like pattern of intranuclear cytostellin immuofluroescence (Figure 2A) appears only in a subset of nuclei. Thus, G1 cell cycle arrest (e.g. serum starvation) leads to "masking" of the cytostellin's H5 epitope in all nuclei (similar to the cells in Figure 3), whereas late S- and G2 phase cells tend to have more obvious nuclear cytostellin dots (unpublished data). Finally, the diffuse component of intranuclear cytostellin immunofluorescence may partially obscure the dot-like pattern in some cells (Figure 2A). The dots are best visualized if MDCK cells are extracted with 1%

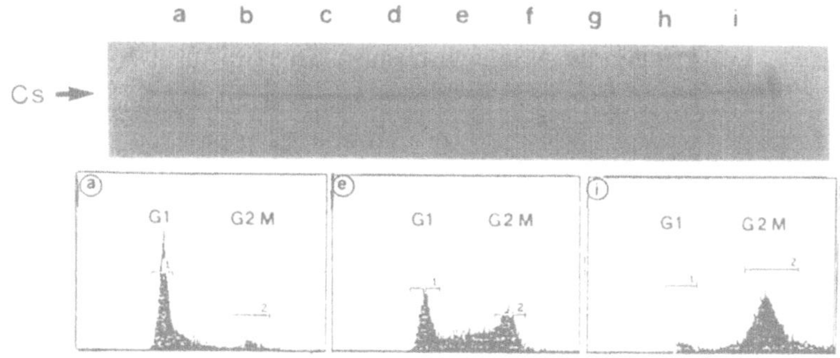

Figure 4. Cytostellin levels remain constant during the cell cycle. HeLa cells were elutriated as described[12] into nine fractions. Fractions were normalized by cell number, solubilized by boiling in SDS sample buffer, electrophoresed on a 7% polyacrylamide gel, transferred to a nitrocellulose membrane and immunoblotted with Mab H5. The immunoblot was developed using biotinylated Goat anti-mouse IgM, followed by an avidin-alkaline phosphatase conjugate. Below are representative flow cytometric histograms to show that elutriation effectively prepared fractions enriched in G1, S and G2/M cells.

Triton X-100 prior to fixation (Fig 5). This procedure extracts the diffuse cytostellin staining component and "unmasks" cytostellin's Mab H5 epitope, revealing ~30-50 transcript domains (Figure 5). These discrete cytostellin "dots" which are inside the nuclear periphery, but excluded from the nucleoli. Similar results have been obtained using cycling rodent and human cells *in vitro* (unpublished results). These dots coincide with the distribution of SC-35 and the Sm proteins[8], shown by others[1-5] to mark transcript domains (our unpublished results).

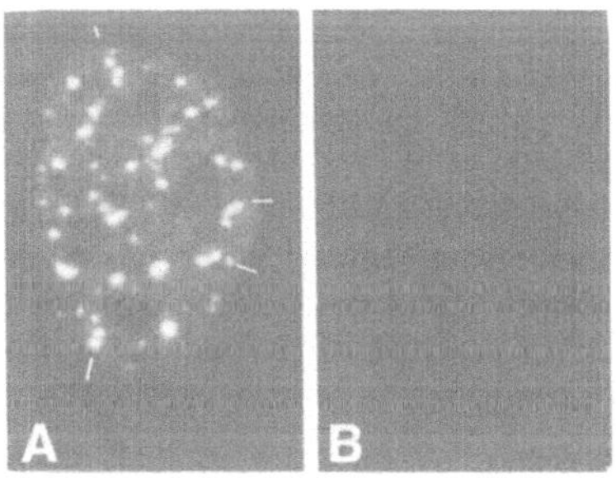

Figure 5. Cytostellin localizes to discrete "dots" in MDCK cell nuclei. MDCK cells were first extracted with 1% Triton X-100 (vol/vol) and then fixed with 1.7 % paraformaldehyde in PBS. The first antibody was either anti-cytostellin Mab H5 (A) or a control IgM (B). The second antibody was biotinylated Goat anti-mouse IgM and the tertiary reagent was an avidin- fluorescein isothiocyanate (FITC) conjugate. Bars indicate the discrete intranuclear dots that stain with Mab H5.

CYTOSTELLIN-CONTAINING PARTICLES ARE BOUND TO THE PERIPHERAL CYTOSKELETON DURING MITOSIS AND G1

Significantly, many cytostellin-containing particles remain outside the nucleus *after* the nuclear lamina has reassembled[9-11]. To illustrate this, a postmitotic cell pair was co-immunostained with Mab H5 and anti-lamin Abs (Figure 6A). In fact, cytostellin particles remain outside the nucleus well beyond the relatively early G1 timepoint shown in Figure 6, but they redistribute nearly completely to the nucleus by S-phase as demonstrated by anti-bromodeoxyuridine (BrdU) + anti-cytostellin immunofluorescence following 5' bromodeoxyuridine (BrdU) labelling (unpublished data).

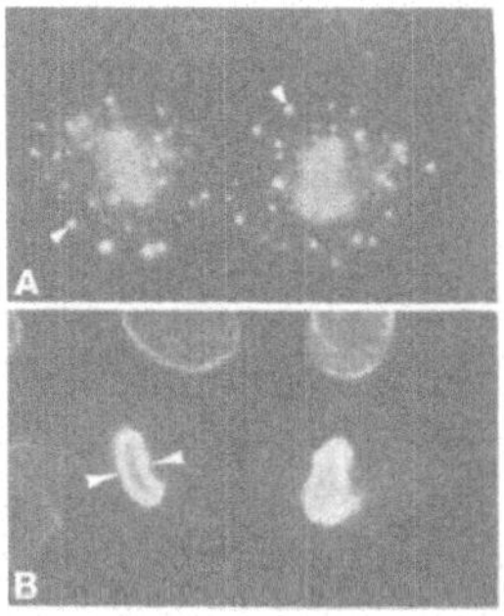

Figure 6. Cytostellin remains outside the nucleus after the nuclear lamina has reformed. Indirect immunofluorescence was performed on the cells after fixation with 1.7 % paraformaldehdye and permeabilization with 0.5% (vol/vol) Triton X-100 in PBS. Shown is a pair of MDCK cells. A, Mab H5; B, anti-Lamin[11]. Arrows mark the nuclear laminae that have formed in the nascent nuclei. Note that the same cells are replete with cytostellin-containing particles outside the nuclei.

Cytostellin may be recycled to the nucleus each cell cycle, or it may be degraded at the end of mitosis, and resynthesized during G1. No change in steady-state levels of cytostellin have been detected during the cell cycle (Figure 4), and cytochalasin B treatment leads to sequestration (without degradation) of a fraction of cytostellin in the cell periphery following mitosis (see below). These observations are consistent with the idea that cytostellin is recycled to the nucleus during G1. Experiments are underway to directly test this hypothesis. Regardless of whether cytostellin is recycled or resynthesized each cell cycle, it presumably enters the nucleus via the nuclear pores during G1.

To determine operationally whether these discrete cytostellin-containing particles interact with the cytoskeleton, cells were treated with 1 % Triton X-100, which extracts soluble proteins, leaving an insoluble "Triton X-100 resistant cytoskeleton." A substantial fraction of cytostellin resists 1 % Triton X-100 extraction indicating an association with cytoskeletal structures (unpublished experiments). Additional evidence of a cytoskeletal association was obtained using cytochalasin B, a compound that disrupts actin filaments[13]. Cytokinesis depends upon an intact actin cytoskeleton, so binucleated cells result if cell division occurs in the presence of this drug (Figure 7; Nuclei are labeled "1" and "2"). Cytochalasin B treatment led to a multifocal accummulation of cytostellin arranged in a ring at the edge of nearly every cell in the culture (Figure 7, arrows). These cytostellin-containing foci resist extraction with 1% Triton X-100 (data not shown), and

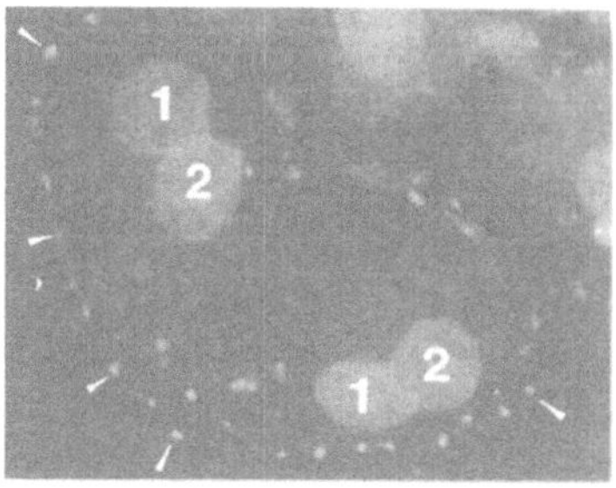

Figure 7. A fraction of cytostellin remains outside the nucleus in cytochalasin B-treated cells. Indirect immunofluorescence with anti-cytostellin Mab H5 was performed on cytochalasin B-treated cells after fixation with 1.7 % paraformaldehdye and permeabilization with 0.5% (vol/vol) Triton X-100 in PBS. Cytochalasin B treatment blocks cytokinesis and results in the accummulation of binucleated cells. The digits "1" and "2" indicate the nuclei of each binucleate cell. Arrowheads indicate the cytostellin-staining foci arranged in a ring at the cell periphery.

they co-localize with actin (Figure 8). It's presently unclear whether cytostellin binds directly to the actin cytoskeleton, but cytostellin's nuclear redistribution following mitosis depends upon an intact actin cytoskeleton. Taken toghether, these data suggest that cytostellin-containing particles are associated with the peripheral cytoskeleton during mitosis and G1.

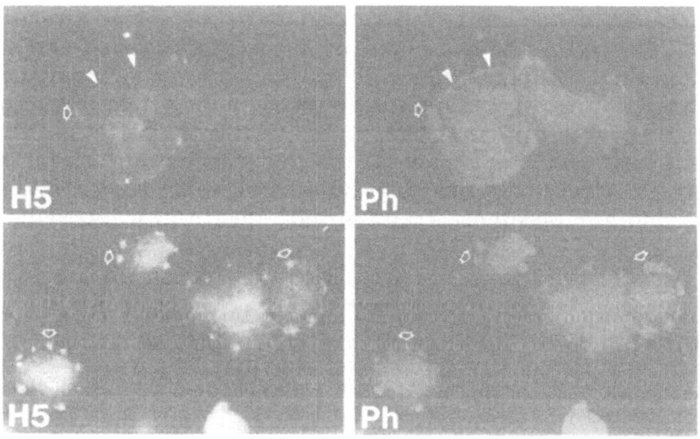

Figure 8. Cytostellin co-localizes with actin following cytochalasin B treatment. Indirect immunofluorescence was performed on cytochalasin B-treated cells after fixation with 1.7 % paraformaldehdye and permeabilization with 0.5% (vol/vol) Triton X-100 in PBS. Ph = phalloidin-rhodamine, an actin-binding alkaloid; H5 = anti-cytostellin Mab H5 Arrows = "blobs" containing actin and cytostellin.

ALL EUKARYOTIC CELLS PROBABLY HAVE CYTOSTELLIN

Total proteins from several types of eukaryotic cells were immunoblotted with Mab H5 (Figure 9). The cytostellin molecules in amphibian and fish are greater than 240 kDa, those in mammals, birds, insects and nematodes are approximately 240 kDa, and the yeast protein is approximately 210 kDa. All of these eukaryotic homologues also react with another anti-cytostellin Mab (H14) which binds to a separate epitope on cytostellin. Cytostellin has been remarkably conserved during evolution.

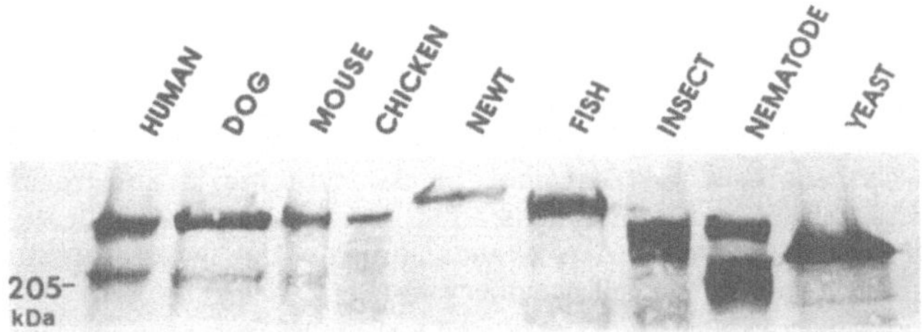

Figure 9. Evolutionary Conservation of Cytostellin. Immunoblot analysis of proteins derived from multiple eukaryotic organisms. Cells and tissues were solubilized in SDS sample buffer and the proteins were resolved on a 7% polyacrylamide gel, transferred to a nitrocellulose membrane and immunoblotted with Mab H5. The ~210 kDa band present in the mammalian species is believed to be a proteolytic breakdown product.[7] The multiple bands in the insect and nematode specimens may represent proteolytic degradation fragments, isoforms of cytostellin or cross-reactive proteins. The relative molecular mass standard is myosin, which migrates at 205 kDa.

DISCUSSION

The data presented above demonstrate that SC-35 and cytostellin co-distribute to two subcellular compartments: during mitosis and G1 they associate with the peripheral cytoskeleton, and during most of interphase they are located in nuclear transcript domains. A direct interaction between SC-35 and cytostellin has not been revealed (unpublished experiments), but their continuous co-localization in particles and in transcript domains implies that they are components of intact subnuclear structures that are maintained throughout the cell cycle. The SC-35/cytostellin-containing particles in mitotic/G1 cells are much more numerous than the transcript domains from which they arose (~20-40 per cell), so it appears that each transcript domain becomes "fragmented" into many particles (depicted in Figure 10).

One possibility is that the cytostellin/SC-35-containing particles merely serve as repositories for transcript domain proteins during mitosis: i.e., they're tethered to peripheral cytoskeletal structures while chromosome segregation occurs in the center of the dividing cell. But this simple "repository model" does not explain why the cytostellin-containing particles remain bound to the peripheral cytoskeleton well into G1---after nuclear lamina reassembly and after most other nuclear proteins have redistributed to the nucleus. SC-35-containing cytoplasmic speckles in mitotic cells, and its relatively delayed postmitotic nuclear redistribution has been observed previously in HeLa cells.[16] We find that the SC-35/cytostellin particles are much larger in MDCK cells than HeLa cells, and that their cytoskeletal association is is much more straightfoward to demonstrate in MDCK cells.

The delayed redistribution of cytostellin-containing particles to the nucleus might be related to a specific function performed while they are "docked" at the peripheral cytoskeleton during G1. Particles that are bound to the peripheral cytoskeleton are potentially accessible to signalling activities occuring at the adjacent membrane Figure 10, lower left panels), and it is during G1 that a host of extracellular ligands (e.g. growth factors) can regulate a cell's commitment to enter S phase.[14,15] One intriguing possibility is that the cytostellin containing particles comprise a shuttling mechanism for delivery of signalling molecules to sites of gene trancription during G1. An essential feature of such a shuttling system would be the *exchange* of proteins between the particles and the cytoskeleton. Sofar no proteins have been found to "jump" from the cytoskeleton to the particles, but the expression of membrane-associated src tyrosine kinases in MDCK cells can induce cytostellin to relocate to the peripheral cytoskeleton *without* the splicing protein SC-35 (unpublished results). Thus, cytostellin is a dynamic component of the particles, capable of associating with the cytoskeleton independent of the SC-35-containing particles. We are currently testing a host of extracellular ligands for the ability to: (1) induce cytostellin relocation from the particles to the cytoskeleton, and (2) stimulate the exchange of proteins between cytostellin/SC-35-containing particles and the membrane-associated cytoskeleton during G1. This approach will hopefully reveal why cytostellin-containing particles remain associated with the cytoskeleton well into the G1 cell cycle phase, and whether they participate in the exchange of proteins between these two subcellular compartments.

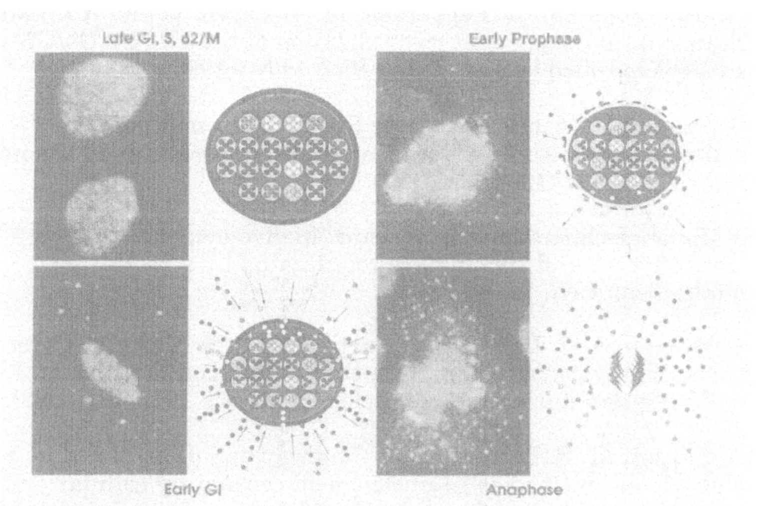

Figure 10. Model of cell cycle-regulated nucleocytoplasmic shuttling of cytostellin-containing particles. Four cell cycle stages are represented, each comprised of two panels: Left panels, photomicrographs of MDCK cells; Right panels, interpretive drawings of the cells. Follow the stages clockwise. Upper left: Cell nuclei, late G1, S, G2/M phases. Upper right: Early Prophase. Lower right: Anaphase. Lower Left: Early G1. Symbols: Large circles: nuclei surrounded by nuclear lamina and envelope; Medium-sized circles: transcript domains comprised of cytostellin/SC-35 particles; Small circles: particles containing cytostellin and splicing protein SC 35. Radially oriented lines: cytoskeletal structures to which particles are bound. Arrows indicate the proposed direction of particle movement. Interpretations of cells in late G1, S and G2/M, early prophase and anaphase are supported by data shown here and elsewhere.[7] Movement of cytostellin-containing particles back into the nucleus during G1 ---depicted by arrows---is speculative.

ACKNOWLEDGEMENTS

We thank Frank McKeown and Tom Maniatis for providing anti-lamin and anti-SC-35 antibodies, respectively. This research was funded by Awards from the National Institutes of Health (K08-CAO-1339), the March of Dimes (91-0647) and the Donaghue Medical Research Foundation to S. L. Warren, and from the The Anna Fuller Foundation and National Institutes of Health (F32 CAO9281) to D. B. Bregman.

REFERENCES

1. D. L. Spector, Higher order nuclear organization: three dimensional distribution of small nuclear ribonucleoprotein particles, *Proc Natl Acad, Sci. USA,* 87: 147(1990).

2. Y. Xing, C.V. Johnson, P. R. Dobner and J. B. Lawrence, Higher level organization of individual gene transcription and RNA splicing, *Science,* 259:1326-30 (1993).

3. K. C. Carter, D. Bowman, W. Carrington, K. Fogarty, J.A. McNeil, F. S. Fay, and J. B. Lawrence, A Three-dimensional view of Precursor messenger RNA metabolism within the Mammalian Nucleus, *Science,* 259: 1330 (1993).

4. M. Carmo-Fonseca, R. Pepperkok, M. T. Carvalho and A.I. Lamond, Transcription-dependent co-localization of the U1, U2,U4/U6 and U5 snRNPs in coiled bodies, *J. Cell Biol.* 117: 1(1992).

5. X-D. Fu, and T. Maniatis, Factor Required for mammalian spliceosome assembly is localized to discrete regions in the nucleus, *Nature,* 343:437 (1990).

6. L. F. Jimenez-Garcia, and D. Spector, In vivo evidence that transcription and splicing are coordinated by a recruiting mechanism, *Cell.* 73: 47(1993) .

7. S. L. Warren, A. S. Landolfi, C. Curtis, and J. S. Morrow, Cytostellin: A novel, highly conserved protein that undergoes continuous redistribution during the cell cycle, *J. Cell Sci.* 103: 381 (1992).

8. E. A. Lerner, M. R. Lerner, C. A. Janeway and J. A. Steitz, Monoclonal antibodies to nucleic acid-containing cellular constituents: probes for molecular biology and autoimmune disease, *Proc Natl Acad, Sci. USA.* 78: 2737 (1981).

9. L. Gerace, and B. Burke, Functional organization of the nuclear envelope, *Ann. Rev. Cell Biol..* 4, 335 (1988).

10. J. W. Newport, and D. J. Forbes, The nucleus: structure, function, and dynamics, *Ann. Rev. Biochem..* 56, 535 (1987).

11. R. Heald, and F. McKeown, Mutations of phosphorylation sites in lamin A that prevent nuclear lamina dissassembly in mitosis. *Cell,* 61: 579 (1990).

12. M. G. Kauffman, S. J. Noga, T. J. Kelly, and A. D. Donnenberg, Isolation of Cell cycle fractions by Counterflow centrifugal elutriation, *Analyt. Biochem.* 191:41(1990).

13. H. Ohmori, S. Toyama, and T. Toyama, Direct proof that the primary action of cytochalasin on cell motility processes is Actin, *J Cell Biol.* 116: 933 (1992).

14. A. B. Pardee, G1 events and regulation of cell proliferation, *Science* 246:603(1989).

15. S. I. Reed, The role of p34 kinases in the G1 to S-phase transition, *Ann Rev Cell Biol.* 8: 529 (1992).

16. D.L. Spector, X-D Fu, and T. Maniatis, Associations between distinct pre-mRNA splicing components and the cell nucleus, *EMBO J,* 10: 3467 (1991)

26

EVIDENCE FOR M-PHASE-SPECIFIC MODIFICATION OF A GAP JUNCTION PROTEIN

Han-qing Xie and Valerie W. Hu

Department of Biochemistry and Molecular Biology
The George Washington University, Medical Center
Washington DC, 20037

INTRODUCTION

Gap junctions are intercellular junctions which permit free diffusion of ions and small molecules (up to about 1000 daltons) through proteinaceous channels connecting two adjacent cells (1). These channels are formed when hexamers of connexin (cx) proteins in each of the apposing plasma membranes become aligned, thereby creating a continuous aqueous passageway connecting the cells' cytoplasm. This continuity of cell interiors lends itself to metabolic cooperation and has been postulated to be a principal means of maintaining tissue homeostasis (2). Aside from the proposed homeostatic function of gap junctions, these structures have been thought to play a major role in the control of cell growth and differentiation (1, 2). In fact, gap junction-mediated intercellular communication (GJIC) has been demonstrated to be regulated by a variety of growth signals, such as epidermal growth factor, tumor promoters, as well as viral and cellular *src* proteins (3). Moreover, the involvement of gap junctions in growth regulation is further implicated by the discovery of connexin molecules as candidate tumor suppressor proteins (4), as well as the demonstration that transfection of connexins into tumor cells reduces their tumorigenicity *in vivo* and/or *in vitro* (5, 6, 7). For normal cells, cell division represents the most common and universal process in which growth controls must be exerted. Several laboratories, including ours, have reported on cell cycle-dependent GJIC using various methods, including electronmicroscopic and immunostaining analyses, Northern hybridization, dye transfer, and GAP-FRAP (4, 8, 9, 10, 11, 12). Thus, it appears that cell cycle-dependent regulation of gap junctions is a common phenomenon. Yet the molecular basis(es) for this phenomenon has not been fully elucidated. In at least one study, differential transcription was observed for cx26 and cx43 in normal human mammary epithelial cells as a function of cell cycle progression (4).

In earlier studies, we observed that GJIC in normal human umbilical vein endothelial cells (HUVEC) was dependent on the cell cycle, reaching a minimum as the cells approached the G2/M transition (12). The following studies were an attempt to identify the molecular origins of this cell cycle dependence. In the process, we have identified a mitosis-specific form of cx43.

The Cell Cycle: Regulators, Targets, and Clinical Applications
Edited by V.W. Hu, Plenum Press, New York, 1994

METHODS

Cell Culture

HUVEC, strain HX1, was isolated from a fresh human umbilical cord according to established procedures (13). HUVEC strain H101 was a kind gift from Dr. Thomas Maciag at the American Red Cross. Both strains were cultured according to standard methods (14).

Cell Synchronization

HUVEC HX1 cells with population doubling level (PDL) less than 20 were synchronized by double thymidine block. Cells were treated with 5 mM thymidine in normal media for 12 hours, followed by normal media for 6 hours, and again by thymidine-containing media for 18 hours. Synchronized cells were collected at 0, 3, 6, and 12 hours after release from the double thymidine block. Mitotic cells were obtained by treating HX1 cells with 0.4 ug/ml nocodazole for five hours. The nocodazole was added three hours after removal from double thymidine block. Floating and loosely attached mitotic cells were collected by gentle aspiration with a pipette. Mitotic cells in Figure 3B were obtained after treating HX1 cells with 0.5 ug/ml nocodazole overnight.

Fluorescence image analysis of propidium iodide (PI)-stained cells was used to verify cell synchronization. Cells were fixed and stained according to established procedures (15). The PI-fluorescence per cell was quantitated using an ACAS 570 Interactive Laser Cytometer (Meridian Instruments), and analyzed by ACAS image analysis software.

Northern Analysis of Cx43

Total RNA from synchronized HUVEC was extracted using RNAzol™B (CINNA/BIOTECX), and quantitated spectrophotometrically at OD_{260}. Total RNA (10 ug per lane) was separated on 1% formaldehyde-agarose gels, and transferred to nitrocellulose membranes. Human cx43-specific probe was derived from a cx43 plasmid provided by Dr. Glenn I. Fishman (16) and radiolabelled using a random primed DNA labelling kit (USB). Prehybridization and hybridization were carried out in 5X SSC, 2X Denhardt's solution, 1% SDS, 50% formamide with 100 ug/ml denatured salmon sperm DNA at 42°C, and followed by three 15-minute washes in 0.2X SSC, 0.2% SDS at 55°C, as suggested by Dr. Fishman. Gamma-actin probe, was labelled and hybridized similarly with the same membrane.

Western Analysis of Cx43

Synchronized cells were lysed directly in culture dishes; the samples were applied to 10% SDS-PAGE gels and the separated proteins were transferred to nitrocellulose membranes according to standard procedures (17). Protein was quantitated using a Pierce BCA protein assay kit with acetone precipitation. The membranes were first blocked with 5% nonfat dry milk in PBS, then incubated with a 1:1,000 dilution of mouse monoclonal anti-cx43 antibody (Zymed), followed by incubation with a 1:100 dilution of rabbit anti-mouse IgG antibody and a 1:10,000 dilution of HRP-conjugated goat anti-rabbit IgG antiserum. HRP-labelled bands were detected by chemiluminescence using an Amersham ECL kit according to manufacturer's recommendations.

RESULTS AND DISCUSSION

The new strain HX1 was derived for cell cycle studies because the original strain, H101 at PDL 30, was difficult to synchronize (maximum synchrony obtained by double thymidine block was 50%). By comparison, HX1 at PDL 10-20 could be synchronized to 70% (Fig. 1). The progression of these cells through cell cycle phases was followed by fluorescence image analysis of propidium iodide-stained anchored cells in culture dishes using an ACAS 570 interactive laser cytometer. The time-dependence of PI-fluorescence changes in the cell population following release from a double thymidine block is shown in Figure 1. The greatest percentage of cells with 4n DNA content appeared at 6 hours post-release, suggesting a predominant G2 fraction at that time.

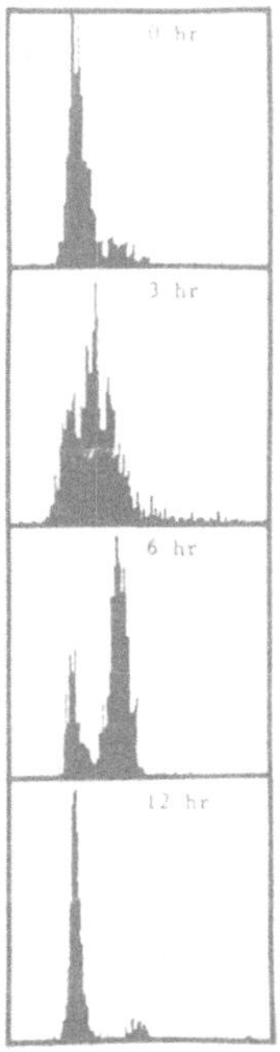

Figure 1. Propidium iodide fluorescence profile of HUVEC at 0, 3, 6, 12 hours after release from double thymidine block, corresponding to G1, S, G2, and early G1 phase, respectively.

HUVEC (HX1) cells synchronized by double thymidine block were analyzed for cx43 mRNA and protein levels at 0, 3, 6, and 12 hours after thymidine washout. These samples were taken to represent cells in G1, S, G2, and early G1 phases, respectively, as shown in Figure 1. Figure 2 shows a representative Northern blot of cx43 from samples in the different cell cycle compartments. As shown, the relative amounts of cx43 message (normalized by the amount of gamma-actin in the respective samples) vary between 1.2- and 2-fold as compared to the amount detected in the G2 sample. Similarly, Western blots show a fluctuation in cx43 protein level but the pattern differs somewhat from that observed for the mRNA levels (Fig. 3). In particular, the minimum cx43 protein levels occur in S phase while the minimum mRNA levels were detected in G2 with maximum message levels appearing in S phase. Such a discrepancy might relate to the translational activity of the cx43 message or the stability of cx43 message and cx43 protein or both, in different phases of the cell cycle. Our previous studies with strain H101 indicated cell cycle-dependent GJIC that reached a minimum level approximately 6-8 hours after thymidine washout and rose back to the G1 level by the 10 hour timepoint (12). However, the HX1 strain does not show such a cell cycle dependence (unpublished data). Since HUVEC has been recently reported to contain cx37 in addition to cx43 (18), cell cycle-dependent gap junction activity may reflect changes in mRNA or protein levels of either one (or both) of these connexins. Differences in the cell cycle dependence of strains HX1 and H101 may thus be the result of differential expression of these two connexins.

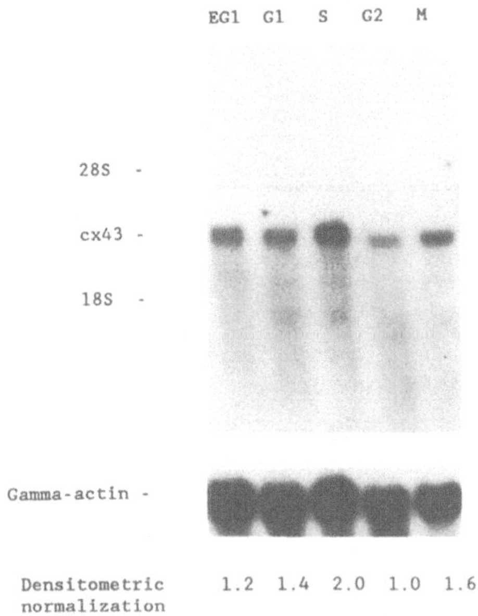

Figure 2. **Northern analysis of synchronized HUVEC cells. EG1, G1, S, G2, M refer to early G1, G1, S, G2, and M phase, respectively.**

The most significant finding of the present studies is the observation that M-phase cells contain an altered form of the cx43 protein. A slower migrating species in the M phase sample in Figure 3A was identified as being specific to cells derived from a mitotic shake-off in contrast to cells remaining anchored to the dish after shake-off (Fig. 3B). In the expanded gel of Figure 3B, several bands of cx43 are detected in the anchored cells whereas only a single high molecular weight species is observed in the mitotic cells. While the nature of the modification to cx43 in this high molecular weight species is unknown, preliminary studies in this laboratory suggest the addition of phosphate. Given that mitotic cells must round up and separate prior to cell division, it is likely that the modification of cx43 has a role in the separation or disassembly of gap junctions connecting adjacent cells. Indeed, serine phosphorylation of cx43 has been shown to play a role in gap junction assembly (19). Thus, it is not inconceivable that additional phosphorylation, possibly related to $p34^{cdc2}$ kinase activity, might be involved in junctional disassembly in mitotic cells.

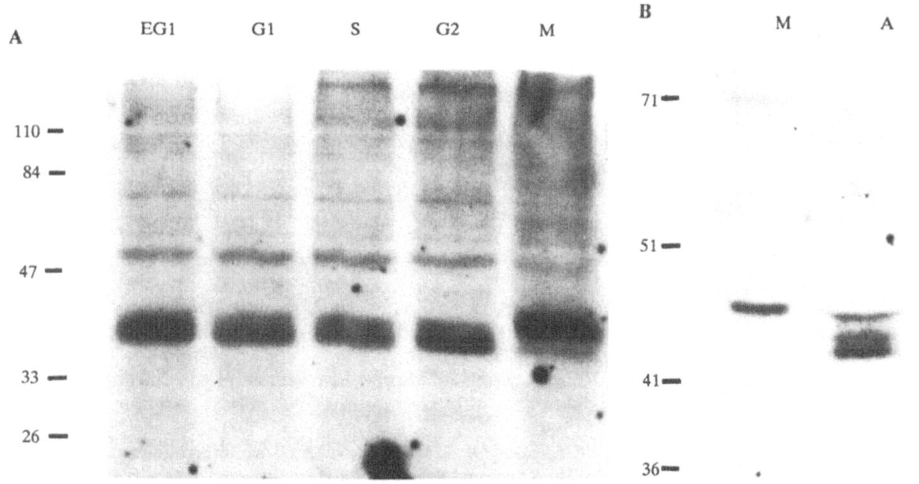

Figure 3. (A): Western analysis of cx43 in HUVEC cells after release from double thymidine block. EG1, G1, S, G2, M refer to early G1, G1, S, G2, and M phase, respectively. Positions of prestained molecular weight markers are indicated (in kD). The strong bands between 33 kD and 47 kD molecular weight markers are specifically recognized by anti-cx43 antibody. Samples were run on a Bio-Rad minigel. (B): Western analysis of cx43 in mitotic (M) and the remaining anchored (A) cells after mitotic shakeout. Prestained molecular weight markers are indicated. Samples were run on an expanded gel (16 cm).

SUMMARY

We have shown that a modified form of cx43 is expressed in HUVEC undergoing mitosis. This mitosis-specific species migrates more slowly on an SDS-PAGE gel than other non-mitotic forms of cx43 and may be the result of cell cycle-dependent phosphorylation. To our knowledge, this is the first demonstration of a cell cycle-related change in connexin structure. The exact nature of this modification and its functional relevance remains to be determined.

ACKNOWLEDGEMENT

The authors thank Dr. Glenn I. Fishman, Dept. of Medicine and Molecular Genetics, Albert Einstein School of Medicine, Yeshiva University for providing the initial cx43 clone, Dr. Thomas Maciag at the American Red Cross for providing HUVEC strain H101 cells, Dr. Ram Shukla for helping with the cx43 plasmid amplification and purification, and Matthew Rounesville in Dr. Ajit Kumar's lab in The George Washington University for providing gamma actin probe. We are also grateful to the Glenn Foundation for Medical Research for a stipend in support of Han-qing Xie.

REFERENCES

1. W. R. Loewenstein, Junctional Intercellular Communication and the Control of Growth, *Biochimica et Biophysica Acta*, 560:1-65, 1979.
2. W. R. Loewenstein, Junctional Intercellular Communication: The Cell to Cell Membrane Channel, *Physiological Reviews*, 61:829-913, 1981.
3. R. Azarnia, S. Reddy, T. E. Kmiecik, D. Shalloway, W. R. Loewenstein, The cellular *src* gene product regulates junctional cell-to-cell communication, *Science*, 239:398-401, 1988.
4. S. W. Lee, C. Tomasetto, D. Paul, K. Keyomarsi, R. Sager, Transcriptional downregulation of gap junction protein blocks junctional communication in human mammary tumor cell lines, *Journal of Cell Biology*, 118:1213-1221, 1992.
5. B. Rose, P. P. Mehta, W. R. Loewenstein, Gap-junction protein gene suppresses tumorigenicity, *Carcinogenesis*, 14:1073-1075, 1993.
6. P. P. Mehta, A. Hotz-Wagenblatt, B. Rose, D. Shalloway, W. R. Loewenstein, Incorporation of the gene for a cell-cell channel protein into transformed cells leads to normalization of growth, *Journal of Membrane Biology*, 124:207-225, 1991
7. B. Eghbali, J. A. Kessler, L. R. Reid, C. Roy, D. C. Spray, Involvement of gap junctions in tumorigenesis: transfection of tumor cells with connexin 32 cDNA retards growth *in vivo*, *Proc. Natl. Acad. Sci.*, 88:10701-10705, 1991.
8. J. W. Su, L. G. J. Tertoolen, S. W. de Laat, W. J. Hage, A. J. Durston, Intercellular communication is cell cycle modulated during early Xenopus laevis development, *Journal of Cell Biology*, 110:115-121, 1990.
9. R. E. Gordon, B. P. Lane, M. Michael, Regeneration of rat tracheal epithelium: change in gap junctions during specific phases of the cell cycle, *Experimental Lung Research*, 3:47-56, 1982
10. A. G. Yee, J. P. Revel, Loss and reappearance of gap junctions in regenerating liver, *Journal of Cell Biology*, 78:554-564, 1978.
11. L. S. Stein, J. Boonstra, R. C. Burgardt, Reduced cell-cell communication between mitotic and nonmitotic coupled cells, *Experimental Cell Research*, 198:1-7, 1992.
12. H. Xie, R. Huang, V. W. Hu, Intercellular communication through gap junctions is reduced in senescent cells, *Biophysical Journal*, 64:45-47, 1991.
13. E. A. Jaffe, R. L. Nachman, C. G. Becker, C. R. Minick, Culture of human endothelial cells derived from umbilical veins, *Journal of Clinical Investigation*, 52:2745-2756, 1973.
14. J. A. M. Maier, T. Hla, T. Maciag, Cyclooxygenase is an immediate-early gene induced by interleukin-1 in human endothelial cells, *Journal of Biological Chemistry*, 265:10805-10808, 1990.
15. A. Krishan, Rapid flow cytofluorometric analysis of mammalian cell cycle by propidium iodide staining, *Journal of Cell Biology*, 66:188-193, 1977.
16. G. I. Fishman, D. C. Spray, L. A. Leinwand, Molecular characterization and functional expression of the human cardiac gap junction channel, *Journal of Cell Biology*, 111:589-598, 1990.
17. J. Sambrook, E. F. Fritsch, T. Maniatis, "Molecular Cloning", second edition, Cold Spring Harbor Laboratory Press, New York, 1989.
18. K. E. Reed, E. M. Westphale, D. M. Larson, H. Wang, R. D. Veenstra, E. C. Beyer, Molecular cloning and functional expression of human connexin37, an endothelial cell gap junction protein, *Journal of Clinical Investigation*, 91:997-1004, 1993.
19. L. S. Musil, B. A. Cunningham, G. M. Edelman, D. A. Goodenough, Differential phosphorylation of the gap junction protein connexin43 in junctional communication-competent and -deficient cell lines, *Journal of Cell Biology*, 111:2077-2088, 1990.

27

ANALYSIS OF CENTROSOME REPLICATION EVENTS IN MAMMALIAN CELLS

Ron Balczon[1], Liming Bao[1], Warren E. Zimmer[1], Kevin Brown[2], Raymond P. Zinkowski[3], and B.R. Brinkley[4]

[1]Department of Structural and Cellular Biology
University of South Alabama, Mobile, AL 36688
[2]Department of Biological Chemistry, Johns Hopkins
University School of Medicine, Baltimore, MD 21205
[3]Department of Cell Biology, University of Alabama
at Birmingham, Birmingham, AL 35294
[4]Department of Cell Biology, Baylor College of Medicine
Houston TX 77030

INTRODUCTION

The centrosome complex plays a fundamental role in cell division events. The centrosome, which nucleates and organizes all of the cellular microtubules, must be duplicated once, and only once, during each cell cycle. At the onset of M-phase the replicated daughter centrosomes then separate and move to opposite ends of the cell where the centrosomes serve as the spindle poles. In this capacity, the mitotic centrosomes nucleate the spindle microtubules that are responsible for chromosome segregation. Clearly, the fidelity of chromosomal segregation is dependent upon the regulated replication of the centrosome complex in each cell cycle.

Relatively little is known about the mechanisms that cells use to replicate centrosomes. In higher mammalian cells, centrosome replication has been well defined at the morphological level. Ultrastructural studies, which have used centriole doubling as a landmark, have demonstrated that centrosome replication begins at the G_1/S boundary, continues through S phase, and is completed in G_2 phase of the cell cycle (Brinkley, 1985; Vandré and Borisy, 1989). Experimentation has shown that the doubling of centrioles in somatic animal cells can occur in the absence of DNA synthesis but requires protein synthesis (Phillips and Rattner, 1976; Rattner and Phillips, 1973). Moreover, studies by Maniotis and Schliwa (1991) have demonstrated that unlike embryonic cells, in which centrosome replication can occur de novo, centriole duplication in somatic mammalian cells is dependent upon the presence of a pre-existing centriole. Presumably, the centriole provides some information, or "seeding capacity", that is required for centriole replication.

Embryonic cells have been used by several investigators to study centrosome duplication events. These experiments have shown that multiple rounds of centrosome doubling occurred in the presence of either DNA synthesis or protein synthesis inhibitors (Sluder and Lewis, 1987; Raff and Glover, 1988; Gard et al., 1990; Sluder et al., 1990). Although these studies have shown that centrosome replication can be uncoupled from the cell cycle in embryonic cells, the results tell us little about the regulation of centrosome replication in somatic cells. The reason is that embryonic cells have large cytoplasmic pools of proteins that can be used for the multiple rounds of centrosome replication which have been observed following various experimental treatments. It seems unlikely that proliferating somatic cells have large pools of precursor proteins.

An approach that may prove quite useful for unraveling the events of centrosome replication is the analysis of spindle pole mutants in yeast. Two classes of yeast mutants have been identified that provide information on the regulation of centrosome duplication. One class of mutants encodes for positive regulators of spindle pole body replication and includes the gene products of the cdc 31 mutant (Baum et al., 1986) and the KAR 1 mutant (Rose and Fink, 1987). These gene products are thought to influence either the structure or function of the spindle pole body. A second type of yeast mutant is thought to encode genes whose products act as negative regulators of spindle pole doubling. In this mutant, termed ESP 1 (for Extra Spindle Poles), multiple rounds of spindle pole doubling occurred in the absence of nuclear division. These mutants should be most informative in unraveling the process of centrosome doubling.

In a previous study, a system was developed for inducing the experimental detachment of kinetochores from mammalian chromosomes (Brinkley et al., 1988; Zinkowski et al., 1989). In this system, cells that were arrested at the G_1/S boundary of the cell cycle using hydroxyurea were induced to enter M phase by the addition of caffeine as described initially by Schlegel and Pardee (1986). When these cells, which were called MUG cells (MUG = **M**itosis with **U**nreplicated **G**enomes), were observed it was determined that the kinetochores had dissociated from the remainder of the chromosomes. Moreover, the detached kinetochores were able to undergo the entire repertoire of mitotic movements. A second observation from those original studies was that many of the MUG cells contained multipolar spindles suggesting that centrosome doubling had occurred in the hydroxyurea-arrested cells. Here we report that multiple rounds of centrosome duplication occurred in CHO cells that were arrested by hydroxyurea. These results suggest that centrosome replication can be completely dissociated from other cell cycle events. This model appears to be an excellent experimental system for investigating the regulation of centrosome replication events in the somatic mammalian cells.

METHODS

Cell culture: CHO cells were used for all experiments. Cells were grown in McCoy's 5A medium supplemented with 10% fetal bovine serum, 2 mM glutamine, 1 mM sodium pyruvate, and 0.1 mM minimal essential amino acids. Cultures were maintained in a 5% CO_2 environment.

For centrosome replication studies, the cells were arrested at the G_1/S boundary of the cell cycle using previously published procedures (Zinkowski et al., 1989). Briefly, cells were trypsinized and then plated into culture medium supplemented with 2 mM hydroxyurea (HU). Cells then were collected at various times after plating and centrosomes were analyzed. For some immunofluorescence experiments, HU-arrested cells were treated with 5 mM caffeine for 4 hrs to induce mitosis. Caffeine had no affect on the number of centrosomes produced (data not shown), but centrosomes were better separated in caffeine-induced mitotic cells which made centrosome counting easier. For some experiments, CHO cells were arrested at the G_1/S boundary of the cell cycle using 5 μg/ml aphidicolin.

Immunofluorescence: Mitotic cells were collected and centrifuged onto poly-lysine coated coverslips. The coverslips were fixed in -20°C MeOH and then rinsed in PBS containing 0.1% Triton X-100. The coverslips then were treated with either monoclonal anti-α-tubulin (Sigma) diluted 1:200 in PBS or SPJ human autoimmune anticentrosome serum (Balczon and West, 1991) diluted 1:1000 in PBS for 45 min at room temperature. Following a brief PBS rinse, the coverslips were incubated with either FITC-antimouse IgG or FITC-antihuman IgG (both from Boehringer-Mannheim) diluted 1:20 in PBS for 45 mins at room temperature. Following a 5 min PBS rinse, the coverslips were mounted in PBS: glycerol (1:1) containing 25 μg/ml Hoechst 33258. The cells then were observed with a Zeiss Axiovert 35 M microscope and photographed using T-Max 400 film.

RESULTS

In previous experiments a method was developed that resulted in the detachment of kinetochores from chromosomes (Brinkley et al., 1988; Zinkowski et. al., 1989). Breifly, cells arrested at the G_1/S boundary of the cell cycle by treatment with HU were induced to enter mitosis prematurely after exposure to caffeine. Cells treated in this fashion entered mitosis without having completed either S or G_2 phases of the cell cycle. Such cells were called MUG cells (Mitosis with Unreplicated Genomes). A high percentage of the MUG cells exhibited multipolar spindles indicating that centrosome replication may have continued to occur in the HU-arrested state. To investigate this possibility, a series of experiments was performed to determine whether centrosome replication was occurring in the HU-arrested cells. Initially, CHO cells were treated for varying lengths of time prior to induction of mitosis using caffeine. Figure 1 shows cells that were arrested in HU for either 20 or 50 hrs before the addition of caffeine. When

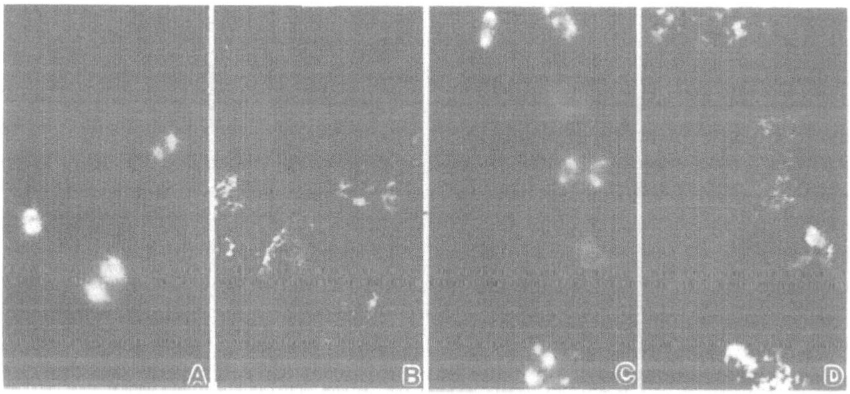

Figure 1. Antitubulin immunofluorescence staining of MUG cells. A and C are antitubulin images. B and D are the corresponding Hoechst-stained cells. A and B. MUG cells that were treated with caffeine after a 20 hr arrest with HU. Although most of the cells are still bipolar, occasionally a multipolar spindle was detected. Note that in MUG cells the chromatin is highly fragmented (B). Fragmented chromatin can be used as a landmark for MUG cells. C and D. MUG cells that were treated for 50 hrs prior to caffeine treatment. Most of the cells contained multipolar spindles (C) and fragmented chromatin (D).

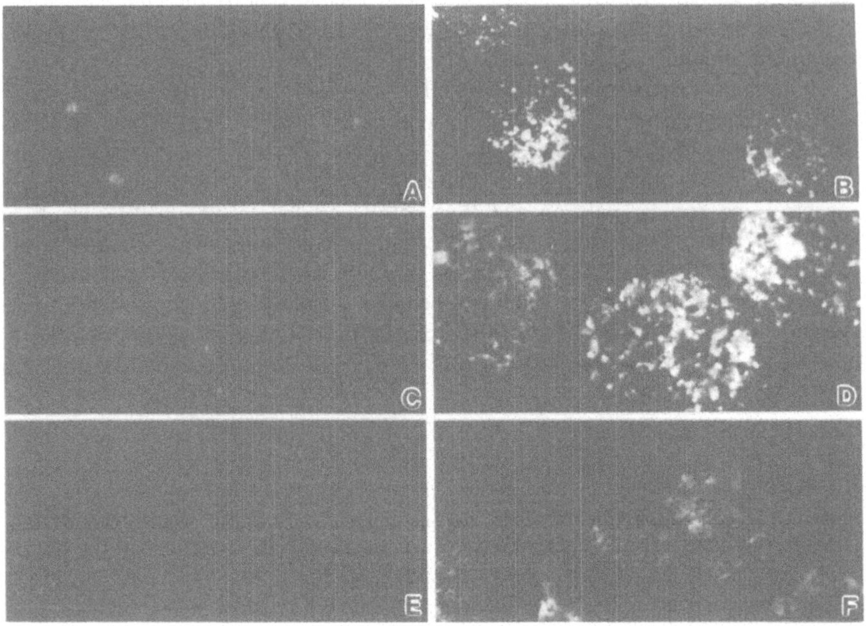

Figure 2. Staining of MUG cells with human autoimmune anticentrosome serum. A, C, and E are cells that were arrested with HU for either 20, 40, or 60 hrs, respectively, prior to the addition of caffeine. B, D, and F are the corresponding Hoechst images. Note that the number of anticentrosome-reactive foci increased in cells with increasing length of HU-arrest.

mitotic cells were fixed and processed for antitubulin immunofluorescence microscopy a low percentage of cells exhibited multipolar spindles after a 20 hour treatment with HU. However, cells treated for 50 hrs with HU prior to the onset of mitosis exhibited numerous spindle poles supporting the notion that several rounds of centrosome replication had occurred in the cells while blocked with HU. As previously reported (Brinkley et al., 1988), the chromatin of MUG cells was highly fragmented. Thus, chromatin fragmentation was found to be a convenient morphological marker for MUG cells in all subsequent experiments.

Experiments were performed to determine whether each of the spindle poles in the MUG cells contained one or more centrosomes. For these studies, the previous experiment was repeated, but the MUG cells were stained with anticentrosome autoantibodies. Figure 2 shows MUG cells that were processed for immunofluorescence after either 20, 40, or 60 hrs of arrest with HU. Cells that were arrested for only 20 hrs in HU contained predominantly one or two centrosomes, while cells that were maintained for progressively longer periods in HU contained increasing numbers of centrosomes (Fig. 2). Direct counts of the number of centrosomes per cell determined that cells that were arrested for 20, 40, and 60 hrs with HU contained 2.45 ± 0.62, 4.16 ± 1.02, and 6.44 ± 1.52 anticentrosome-reactive foci, respectively.

Electron microscopy was performed to determine whether the multiple centrosomes that were observed in the MUG cells were due to centrosome replication or

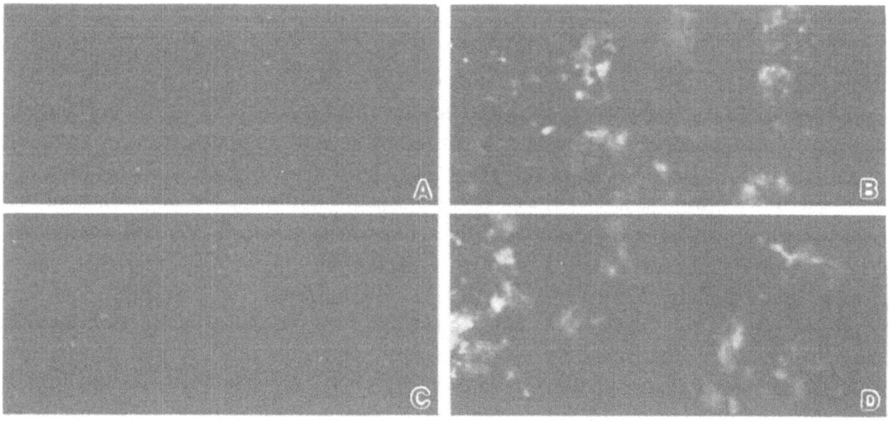

Figure 3. Staining of MUG cells with anticentrosome antiserum following treatment with either HU or aphidicolin. A and C are immunofluorescence images and B and D are the corresponding Hoechst-stained images. Note that multiple centrosomes can be detected in CHO cells following either treatment.

to fragmentation of centrosomes. For these studies, serial sections were cut and observed. Analysis of interphase cells that had been arrested with HU for 60 hrs prior to fixation and processing for EM demonstrated that the treated cells contained numerous intact, complete centrosomes as determined by centriole doubling (not shown). These results demonstrated that centrosome doubling, and not centrosome fragmentation, had occurred during the G_1 arrest that was induced by HU-treatment.

A final set of experiments was performed to investigate whether the doubling of centrosomes that was detected was caused by the HU or by blockage at the G_1/S boundary of the cell cycle. For these experiments, cells were arrested for 40 hrs at the G_1/S boundary of the cell cycle using either HU or aphidicolin. Like HU, aphidicolin also blocks cells at the G_1/S boundary of the cell cycle, but this arrest is through a different mechanism of inhibition (Nishioka et al., 1984). Following treatment, cells were driven into M-phase by caffeine addition and then processed for immunofluorescence microscopy using SPJ anticentrosome serum. As shown in figure 3, multiple centrosomes were detected in both HU and aphidicolin-arrested CHO cells. These results demonstrated that the uncoupling of centrosome replication was due to arrest at the G_1/S phase boundary of the cell cycle and not to an effect on centrosomes by HU.

DISCUSSION

Centrosomes undergo replication once, and only once, during each cell cycle. Little is known about either the mechanisms or the regulation of centrosome replication in somatic cells. Here we describe an experimental system where centrosome replication has been uncoupled from other cell cycle events. In CHO cells that were arrested with HU, multiple rounds of centrosome replication occurred in the complete absence of cycles of either DNA synthesis or mitosis. The conclusion that the numerous spindle poles that were detected in MUG cells was due to centrosome replication and not to simple fragmentation of centrosomes was supported by serial section EM analysis which demonstrated that HU-arrested cells contained numerous complete and mature centriole pairs (not shown). Therefore, CHO cells arrested at the G_1/S boundary of the cell cycle should provide an excellent experimental system for investigating the molecular events of centrosome replication and mechanisms used by cells to regulate centrosome doubling.

The results reported here represent an extension of earlier work by Rattner and Phillips (1973) who first reported that procentriole formation and elongation could occur in the absence of DNA synthesis. In this study, we have arrested cells with HU for periods of time that corresponded to over to 4 cell cycles in the CHO line that was used. In these treated cells, multiple centrosomes were detected demonstrating that several rounds of centrosome replication had occurred during blockage at the G_1/S boundary. These results show that not only will cells form a procentriole without a detectable S phase as demonstrated by Rattner and Phillips (1973), but that several cycles of centrosome replication will occur during a G_1/S blockage. Moreover, these results differ from those obtained using invertebrate eggs because eggs contain large pools of proteins that can be recruited for the multiple rounds of centrosome replication that have been detected in invertebrate and vertebrate eggs following various experimental treatments (Sluder and Lewis, 1987; Raff and Glover, 1988; Gard et al., 1990). It seems most likely that the cultured cells that were used for these studies would need to synthesize the centrosome proteins that would be necessary for the multiple rounds of centrosome duplication that occurred. The fact that the synthesis of centrosomal components is turned on during the G_1/S blockage and apparently is not turned off during the HU-arrest period raises interesting questions regarding the cell cycle-regulated transcription and translation of centrosome components.

Direct centrosome counts showed that many of the cells exhibited odd numbers of centrosomes. One might have expected a geometric progression of centrosomes from 1-2-4-8-16-etc. However, this did not appear to be the case in HU-arrested cells. One possible explanation is that centrosomes were too closely packed in HU-treated cells and could not be counted accurately. If this was the case, then two closely-spaced centrosomes might have only been scored as a single centrosome by immunofluorescence. However, this possibility was not supported by serial section electron microscopic analysis of MUG cells in which cells with odd numbers of centrosomes were observed routinely. A second possibility is that the replication of centrosomes occurred asynchronously in the experimental system that was used. According to this scenario, some of the centrosomes in a cell would be replicated while other centrosomes within the same cytoplasm would be unduplicated. The reasons for the proposed asynchronicity of centrosome doubling are unknown.

In summary, an experimental system has been developed in which centrosome doubling was uncoupled from other cell cycle progression events. This system should be useful for investigating the positive and negative governing mechanisms used by somatic cells to direct the tightly regulated replication of the centrosomes that occurs each cell cycle.

ACKNOWLEDGEMENTS

We would like to thank Ms. Sheila White for typing this manuscript. This work was funded by NIH grants GM46453 to R.B. and CA41424 to B.R.B.

REFERENCES

Balczon, R., and West, K., 1991, The identification of mammalian centrosomal antigens using human autoimmune anticentrosome antisera, *Cell Motil. Cytoskel.* 20:121.

Baum, P., Furlong, C., and Byers, B., 1986, Yeast gene required for spindle pole body duplication: homology of its product with Ca^{2+}-binding proteins, *Proc. Natl. Acad. Sci. USA* 83:5512.

Baum, P., Yip, C., Goetsch, L., and Byers, B., 1988, A yeast gene essential for regulation of spindle pole duplication, *Mol. Cell Biol.* 8:5386.

Brinkley, B.R., 1985, Microtubule organizing centers, *Ann. Rev. Cell Biol.* 1:145.

Brinkley, B.R., Zinkowski, R.P., Mollon, W.L., Davis, F.M., Pisegna, M.A., Pershouse, M., and Rao, P.N., 1988, Movement and segregation of kinetochores experimentally detached from mammalian chromosomes, *Nature* 336:251.

Gard, D.L., Hafezi, S., Zhang, T., and Doxsey, S.J., 1990, Centrosome duplication continues in cycloheximide-treated Xenopus blastulae in the absence of a detectable cell cycle, *J. Cell Biol.* 110:2033.

Maniotis, A., and Schliwa, M., 1991, Microsurgical removal of centrosomes blocks cell reproduction and centriole generation in BSC-1 cells, *Cell* 67:495.

Nishioka, D., Balczon, R., and Schatten, G. 1984, Relationships between DNA synthesis and mitotic events in fertilized sea urchin eggs: aphidicolin inhibits DNA synthesis, nuclear breakdown, and proliferation of microtubule organizing centers, but not cycles of microtubule assembly, *Cell Biol. Int. Reps.* 8:337.

Phillips, S.G., and Rattner, J.B., 1976, Dependence of centriole formation on protein synthesis, *J. Cell Biol.* 70:9.

Raff, J.W., and Glover, D.M., 1988, Nuclear and mitotic cycles continue in Drosphila embryos in which DNA synthesis is inhibited with aphidicolin, *J. Cell Biol.* 107:2009.

Rattner, J.B., and Phillips, S.G., 1973, Independence of centriole formation and DNA synthesis, *J. Cell Biol.* 57:359.

Rose, M.D., and Fink, G.R., 1987, KAR 1, a gene required for function of both intranuclear and extranuclear microtubules in yeast, *Cell* 48:1047.

Schlegel, R., and Pardee, A.B., 1986, Caffeine-induced uncoupling of mitosis from the completion of DNA replication in mammalian cells, *Science* 232:1264.

Sluder, G., and Lewis, K., 1987, Relationship between nuclear DNA synthesis and centrosome reproduction in sea urchin eggs, *J. Exp. Zool.* 244:89.

Sluder, G., Miller, F.J., Cole, R., and Rieder, C.L., 1990, Protein synthesis and the cell cycle: centrosome reproduction in sea urchin eggs is not under translational control, *J. Cell Biol.* 110:2025.

Vandré, D.D., and Borisy, G.G., 1989, The centrosome cycle in animal cells, In: "Mitosis: Molecules and Mechanisms", J.S. Hyams and B.R. Brinkley, eds, Academic Press, San Diego, CA.

Zinkowski, R.P., McCune, S.L., Balczon, R.D., Rao, P.N., and Brinkley, B.R., 1989, The centromere and aneuploidy: caffeine-induced detachment and fragmentation of kinetochores of mammalian chromosomes, In: "Mechanisms of Chromosome Distribution and Aneuploidy", M. Resnick and B. Vig, eds., Alan R. Liss, Inc., New York, NY.

28

REQUIREMENTS FOR MICROTUBULE POLYMERIZATION AND A CALCIUM SURGE FOR THE METAPHASE-TO-INTERPHASE TRANSITION IN MATURE MOUSE OOCYTES

Ruth M. Moses and Yoshio Masui

Department of Zoology
University of Toronto
25 Harbord St.
Toronto, Ont. Canada M5S 1A1

ABSTRACT

In order to determine the requirements for the transition from metaphase to interphase, mature mouse oocytes, in which meiosis is arrested at metaphase II, were treated with the parthenogenetic agents, the calcium ionophore, A23187, or the protein synthesis inhibitor, cycloheximide (CHX), in conjunction with the microtubule inhibitor, colcemid, or the calcium chelator, 1,2-bis(o-aminophenoxy)ethane-N,N,N'N'-tetraacetic acid (BAPTA). We found that, whereas oocytes treated with either A23187 or CHX formed nuclei, colcemid-treated oocytes remained at metaphase when exposed to either of the activating agents. However, when colcemid-treated oocytes were exposed to A23187 and CHX in combination, nuclei formed. When oocytes pre-treated with BAPTA were exposed to either of the parthenogenetic drugs, alone or in combination, they remained arrested at metaphase. This suggests that a calcium increase is required for the exit from metaphase in all these cases. However, the presence of microtubules is required for the exit from metaphase only when induced by either A23187 or CHX, but not when these drugs are used in combination.

INTRODUCTION

Although much progress has been made recently in identifying the biochemical events which occur when a cell proceeds from interphase to metaphase, the signal transduction pathway that causes the transition from metaphase to interphase is not well understood. The mouse oocyte is a suitable system in which to study this pathway, since in the mature oocyte meiosis is arrested at metaphase of the second meiotic division. Following activation, which can be induced under controlled experimental conditions, by fertilization or a parthenogenetic stimulus, meiosis is completed, the chromosomes decondense, and the oocyte enters interphase. The nuclear changes which occur can be readily observed in cytologic preparations.

The mouse oocyte will undergo parthenogenetic activation in response to a number of physical and chemical stimuli[1]. By using chemical agents with known mechanisms of action, some insight has been gained into possible requirements for the release from metaphase arrest. An increase in intracellular calcium at the time of fertilization is thought to be the universal signal for the resumption of meiosis[2]. Similarly, inducing a calcium surge by treatment with a calcium ionophore causes mouse oocyte activation[3]. Because mouse oocytes are activated by protein synthesis inhibitors, metaphase arrest may be maintained by the activity of a short-lived protein(s)[4].

In contrast, BAPTA has been shown to prevent the calcium surge that occurs at the time of fertilization or in response to a calcium ionophore, and to prevent the resumption of meiosis[5]. Microtubule inhibitors have been found to prevent decondensation of oocyte and sperm chromatin following *in vitro* fertilization[6-8]. Whether microtubules are required for the generation of signals such as calcium release or the termination of synthesis of short-lived proteins, or whether they are required for the transduction of these signals to cause chromosome decondensation, is not known. The present study was undertaken to investigate the requirements for calcium release and role of microtubules during activation. Thus, by using activating agents with specific mechanisms of action, along with inhibitors which prevent activation, we have identified some of the requirements for the transition from metaphase to interphase in the mouse oocyte.

MATERIALS AND METHODS

Oocytes

Cumulus-enclosed ovulated oocytes were recovered from the oviducts of 9-week old female CD-1 mice, which had received intraperitoneal injections of 5 I.U. pregnant mare's serum (PMS) and human chorionic gonadotrophin (HCG), 65-69 and 17 hours, respectively, prior to sacrifice by cervical dislocation. The cumulus masses were dispersed by incubation for 8-10 minutes in 0.1% hyaluronidase in culture medium (Minimal Essential Medium with Earle's salts (MEM), supplemented with 0.23 mM pyruvate, 75 mg/l penicillin G, 50 mg/l streptocmycin sulfate and 5% fetal bovine serum or 4 mg/ml bovine serum albumin) at 37°C in an atmosphere of 5% CO_2 in air. The cumulus-free oocytes were collected and rinsed three times in fresh culture medium prior to further incubation in welled-culture dishes (Falcon 3037).

Drugs

Stock solutions of: calcium ionophore A23187 5 mg/ml; the cell permeant acetoxymethyl ester (AM) form BAPTA 25 mM; in dimethyl sulfoxide (DMSO); CHX 20 mM in water; colcemid 5 mM in MEM; were stored at -20°C. Prior to each experiment, the drugs were diluted into culture medium in the following concentrations: A23187 5 uM; BAPTA-AM 20 uM; cycloheximide 400 uM; colcemid 50 uM.

Activation and Inhibitor Treatments

Oocyte activation was carried out as follows: i) 5 minutes exposure to A23187, followed by transfer to drug-free medium for further culture or ii) continuous exposure to CHX. Where the two activating agents were used in combination, oocytes were treated with A23187 and then transferred to CHX-containing culture medium.

When colcemid was used in conjunction with activation treatments, oocytes were cultured in medium containing colcemid for 30 minutes prior to, and throughout, the activation treatment and subsequent culture period. Some oocytes were exposed to BAPTA-AM for 30 minutes, then returned to drug-free culture medium prior to the activation treatments.

Control oocytes were exposed to medium containing the same concentration of DMSO as that found in the drug-containing solutions, for the same length of time as experimental oocytes, and then returned to drug-free culture medium.

Cytology

Oocytes were cultured for 8 hours from the time that they were first exposed to the activating agents in order to allow sufficient time for pronuclear formation to occur. Oocytes were then fixed in the culture dishes, by the addition to the culture medium of an equal volume of fixative (75ml ethanol, 10ml formalin, 5ml glycerol, 10ml acetic acid, and 1g cetylpyridium chloride) (Moses and Masui manuscript submitted). Fixed oocytes were processed for microscopic observation using a procedure similar to that described by Tarkowski[9]. Briefly, a drop of filtered egg white was placed on a microscope slide coated with Albumin Fixative (20% egg white, 80% glycerol). The fixed oocytes were placed in the drop of egg white and immediately all excess fluid surrounding the oocytes was removed using a mouth pipette. The slide was exposed to the vapour of a 3:1 ethanol:acetic acid mixture for 3-5 minutes, and then transferred to 70% ethanol. The specimens were stained with Erlich's Haematoxylin, dehydrated through an alcohol series, transferred to xylene and mounted in Protexx (Scientific Products) or Permount (Fisher Scientific).

RESULTS

Over 90% of control oocytes remained at metaphase throughout the culture period, while nuclei formed in over 85% of oocytes treated with either A23187 or CHX (Figures 1A & B and 2A). When oocytes were exposed to colcemid, the metaphase spindle disappeared, and the chromosomes became dispersed into one or more clumps at the periphery of the oocyte (Figure 1C). When colcemid-treated oocytes were exposed to either one of the activating agents alone, nuclei formed in less than 10% (Figure 1D). However, when colcemid-treated oocytes were exposed to the combination of A23187 and CHX, nuclear formation resulted in over 90% of oocytes (Figure 2B). In these cases, often multiple nuclei were present in a single oocyte, as each group of dispersed chromosomes decondensed to form a small nucleus (Figure 1E). In contrast, when oocytes were treated with BAPTA-AM prior to exposure to either of the activating agents alone or in combination, they remained at metaphase (Figure 2C).

DISCUSSION

The present study has demonstrated that the presence of microtubules is required for the exit from metaphase following exposure to either A23187 or CHX alone. However, if the activating agents are used in combination, nuclear formation can occur in the absence of microtubules. When an increase in intracellular calcium is prevented, oocytes fail to exit from metaphase whether the activating agents are used alone or in combination.

Although spindle microtubules are required for chromosome movement during anaphase, the persistence of metaphase in the absence of microtubule assembly cannot be explained by the lack of anaphase movements. In fact, chromosomes do decondense to form nuclei in the oocytes lacking microtubules if they are treated with two activating agents in combination. As has been observed previously in other types of dividing cells, disruption of the mitotic spindle may delay, but does not necessarily prevent chromosome decondensation[10,11]. Thus, as well as

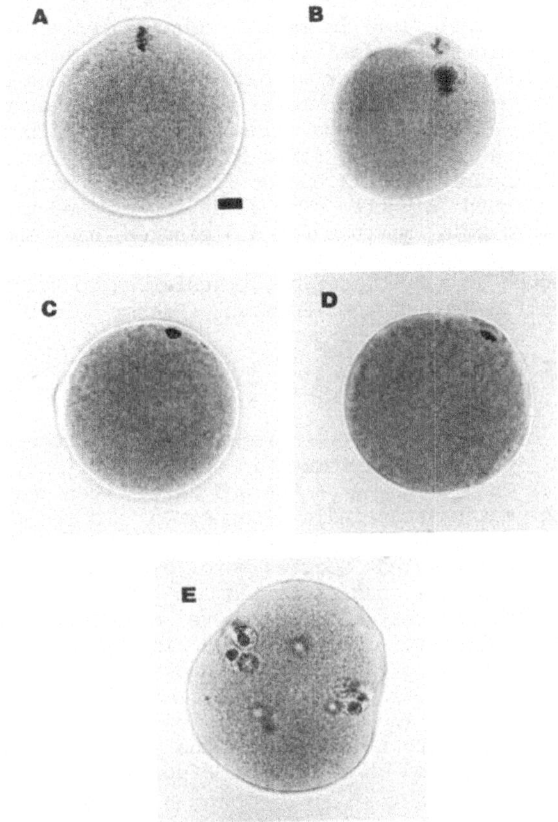

Figure 1. Bar = approximately 10 microns. A. Oocyte at metaphase II. Note chromosomes arranged on the metaphase plate. B. Oocyte at interphase following exposure to CHX. Note large single nucleus and extruded second polar body. C. Oocyte following exposure to colcemid. Note clumped chromosomes at periphery of oocyte. D. Oocyte following exposure to colcemid and A23187. E. Oocyte following exposure to colcemid, A23187 and CHX. Note multiple small nuclei.

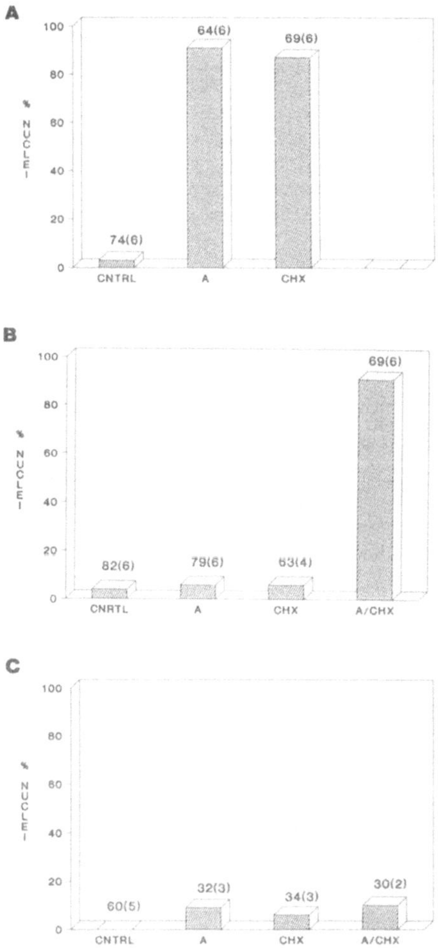

Figure 2. Nuclear formation in oocytes following various treatments. The numbers above each bar are the total number of oocytes used, and in parentheses, the number of experiments performed. A. A23187(A) or CHX. Control (CNTRL) oocytes were transferred to medium containing 0.05% DMSO for 5 minutes and then returned to drug-free culture medium. B. Colcemid + A23187, CHX, or A23187 + CHX (A/CHX). Control oocytes were kept in colcemid for the entire culture period. C. BAPTA + A23187, CHX, or A23187 + CHX. Control oocytes were exposed to 50 uM BAPTA-AM for 30 minutes and then returned to drug-free culture medium.

guiding the correct segregation of chromosomes at anaphase, microtubules must serve some other function.

Since the BAPTA-treated oocytes remained at metaphase, this suggests that an increase in intracellular calcium is a necessary component for the exit from metaphase, and that the inhibition of synthesis of the putative short-lived protein that maintains metaphase arrest is not sufficient. However, a calcium surge alone may also not be sufficient, as in the case of the colcemid-treated oocytes. Although it is not known for certain whether or not microtubule depolymerization prevents a calcium increase following exposure to the ionophore, this is unlikely, since the ionophore is known to abolish the selective permeability of biological membranes to ions, and calcium was present in the external medium at all times. Since the colcemid-treated oocytes formed nuclei only when treated with both activating agents, it is possible that the calcium was increased only under these conditions. However, it is more likely that in addition to an increase in intracellular calcium, the inhibition of synthesis or inactivation of some protein(s) is required for nuclear formation. If so, microtubules may function to relay the signals, so that a calcium surge results in the inhibition of synthesis of some protein(s), and *vice versa*, possibly by bringing interacting molecules in close proximity.

It must be kept in mind that, since oocytes are arrested at metaphase of meiosis, the requirements for the transition from metaphase to interphase as determined in this study, may represent a special case. Although the mechanism for the maintenance of metaphase arrest in mouse oocytes is not known, the calcium surge at the time of fertilization may act by triggering the destruction of the cytoplasmic activity that maintains metaphase[3]. Metaphase is also prolonged in most mitotic cells in the absence of correct spindle assembly, but the mechanism for the maintenance of, and the release from, mitotic metaphase arrest may be different. We plan to investigate this question in the near future.

ACKNOWLEDGMENTS

This research was supported by NSERC Grant A5855 to Y.M. and by a MRC Fellowship to R.M.M.

REFERENCES

1. M.H. Kaufman. "Early Mammalian Development: Parthenogenetic Studies." Cambridge University Press, Cambridge. (1983).
2. R.A. Steinhardt, D. Epel, E.J. Carroll Jr. and R. Yanagimachi, Is calcium ionophore a universal activator for unfertilised eggs? *Nature* 252:41-43 (1974).
3. Y. Masui, P.G. Meyerhof, M.A. Miller, and W.J. Wasserman, Roles of divalent cations in maturation and activation of vertebrate oocytes, *Differentiation*. 9:49-57 (1977).
4. G. Siracusa, D.G. Whittingham, M. Molinaro, and E.Vivarelli, Parthenogenetic activation of mouse oocytes induced by inhibitors of protein synthesis, *J. Embryol. exp. Morphol.* 43:157-166 (1978).
5. D.G. Kline and J.T. Kline, Repetitive calcium transients and the role of calcium in exocytosis and cell cycle activation in the mouse egg, *Dev. Biol.* 149:80-89 (1992).
6. G. Schatten, C. Simmerly, and H. Schatten, Microtubule configurations during fertilization, mitosis, and early development in the mouse and the requirement for egg microtubule-mediated motility during mammalian fertilization, *Proc. Nat. Acad. Sci.* 82:4152-4156 (1985).

7. H. Schatten, C. Simmerly, G. Maul, and G. Schatten, Microtubule assembly is required for the formation of the pronuclei, nuclear lamin acquisition, and DNA synthesis during mouse, but not sea urchin, fertilization, Gamete Res. 23:309-322 (1989).
8. B. Maro, M.H. Johnson, M. Webb, and G. Flach, Mechanism of polar body formation in the mouse oocyte: an interaction between the chromosomes, the cytoskeleton and the plasma membrane, J. Embryol. exp. Morphol. 92:11-32 (1986).
9. A.K. Tarkowski, Development of single blastomeres, in "Methods in Mammalian Embryology," J.C. Daniel, ed., Freeman, San Francisco (1971).
10. A.L. Kung, S.W. Sherwood, and R.T. Schimke, Cell line-specific differences in the control of cell cycle progression in the absence of mitosis, Proc. Nat. Acad. Sci. 87:9553-9557 (1990).
11. C.L. Reider and R.E. Palazzo, Colcemid and the mitotic cycle, J. Cell Sci. 102:387-392 (1992).

PART V

CELL CYCLE REGULATION IN DEVELOPMENT

29

Mos PROTO-ONCOGENE AND CELL CYCLE REGULATION

George F. Vande Woude, Taesaeng Choi, Renping Zhou,
Monica Murakami, Wayne Matten, and Kenji Fukasawa

ABL-Basic Research Program
NCI-Frederick Cancer Research and Development Center
P.O. Box B
Frederick, MD 21702

INTRODUCTION

We have investigated the mechanisms by which the *mos* oncogene affects the phenotypes and genetic instability of transformed cells. Our laboratory discovered that the *mos* proto-oncogene, as a regulator of chromosome partitioning during meiosis, is required for the formation of an unfertilized egg (Sagata *et al.*, Nature 355:519-525, 1988; Yew *et al.*, Current Opin. in Gen. & Dev. 3:19-25, 1993). These observations led to our hypothesis that inappropriate expression of the Mos meiotic program in somatic cells results in cellular transformation by the *mos* oncogene.

Overexpression of the *mos* oncogene in mouse fibroblasts results in cellular growth arrest and the generation of two population of cells. While 70% of the arrested cells have a 2C DNA content, 30% are 4C. The arrested cells exhibit inappropriate chromosome condensation and the 4C cells are binucleated, indicating these cells have undergone DNA replication and karyokinesis, but not cell cytokinesis. In collaboration with Xiao-Min Wang and Gary Borisy (McArdle Laboratories, Madison, WI), we found that introduction of Mos protein in prometaphase epithelial cells results in a lengthened metaphase characterized by severely disorganized chromosome alignment and partitioning. The severity of metaphase disruption is directly proportional to the level of injected Mos protein. In these experiments Mos also exhibit karyokinesis in the absence of cytokinesis.

We have gained insights on the biochemical basis of Mos-induced growth arrest and chromosomal disorganization. Initially, we examined the activity of two factors - MPF and MAP kinase - both of which are activated by Mos during meiosis (Sagata *et al.*, Nature 355:519-525, 1988; Posada *et al.*, Mol. Cell. Biol. 7:2489-2498, 1992). Maturation of M-phase promoting factor (MPF), a complex of the $p34^{cdc2}$ protein kinase and cyclin B, has been long recognized as a regulator of cell cycle progression events, including chromosome condensation. In cells overexpressing or transformed by Mos, we have found that $p34^{cdc2}$ is in its active , dephosphorylated form; however, the level of $p34^{cdc2}$ is reduced (relative to control cells) and its kinase

The Cell Cycle: Regulators, Targets, and Clinical Applications
Edited by V.W. Hu, Plenum Press, New York, 1994

activity is undetectable. In contrast, Mos-overexpressing and transformed cells express high levels of mitogen activated protein (MAP) kinase; and, in *mos*-transformed cells, the level of MAP kinase is directly proportional to both the level of Mos expression and the extent of morphological transformation. MAP kinase has been implicated in the regulation of the interphase-metaphase transition of microtubule arrays; therefore, constitutively-expressed MAP kinase could mediate the microtubule disorganization found in cells overexpressing or transformed by Mos. In addition, we have used monoclonal antibodies which recognize phosphorylated proteins specific to mitotic or meiotic, but not interphase cells, to examine the expression of these epitopes in *mos*-transformed cells. While the actively growing, *mos*-transformed cells exhibited a normal cell cycle distribution, we have detected in extracts of these cells levels of mitosis-specific antigens comparable to that observed in extracts derived from a pure population of mitotic cells.

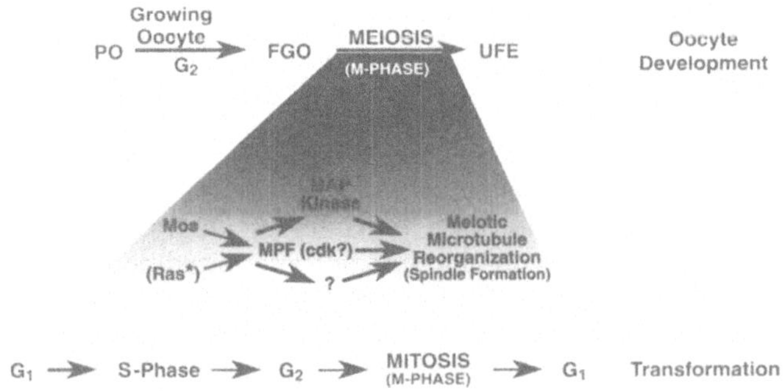

Fig. 1. Mos in oocyte development and transformation (CD). The Mos product is a regulator of meiotic maturation. The association of Mos with tubulin, MPF p34^{cdc2} (and other cdk family members?), as well as with MAP-kinase activation suggests that Mos may contribute to the reorganization of microtubules that leads to spindle formation and the partitioning of chromosomes that occurs during meiosis I and II. The participation of Mos in other meiotic activities, such as nuclear envelope breakdown and chromosome condensation, is not excluded (as indicated by the question mark). The Ras oncoprotein, Ras*, can mimic the activity of Mos during oocyte maturation and in embryonic cleavage-arrest assays. We suggested that the M-phase activity of Mos (or Ras*), inappropriately expressed during interphase of the somatic cell cycle, is responsible for the phenotype of transformed cells. Abbreviations include PO, primordial oocyte; FGO, fully grown oocyte; UFE, unfertilized egg.

Karyokinesis in the absence of cytokinesis, dephosphorylated but kinase-inactive p34^{cdc2}, elevated MAP kinase activity, expression of M-phase-specific phosphorylated antigens, and high levels of Mos protein are landmarks of the interval between Meiosis I and II during oocyte maturation. The presence of these histological and biochemical markers in *mos*-transformed cell supports our proposal that superimposition of the *mos* meiotic program on somatic cells provides a molecular explanation for the phenotypes of transformed cells (Fig. 1).

ACKNOWLEDGEMENTS

Research sponsored by the National Cancer Institute, DHHS, under contract No. NO1-CO-74101 with ABL. The contents of the publication do not necessarily reflect the views or policies of the Department of Health and Human Services, nor does mention of trade names, commercial products, or organizations imply endorsement by the U.S. Government.

REFERENCES

1. N. Sagata, M. Oskarsson, T. Copeland, J. Brumbaugh, and G.F. Vande Woude, The c-*mos* proto-oncogene product functions during meiotic maturation in *Xenopus* oocytes, Nature 355:519 (1988).

2. N. Yew, M. Strobel, and G.F. Vande Woude, Mos and the cell cycle: The molecular basis of the transformed phenotype, Current Opin. in Gen. & Dev. 3:19 (1993).

3. J. Posada, N. Yew, N.G. Ahn, G.F. Vande Woude, and J.A. Cooper, Mos stimulates MAP kinase in *Xenopus* oocytes and activates a MAP kinase *in vitro*, Mol. Cell. Biol. 7:2489 (1992).

30

ALTERED REGULATION OF CELL CYCLE GENES AND PROTEINS IN SENESCENT HUMAN DIPLOID FIBROBLASTS

Gretchen H. Stein,[1] Linda F. Drullinger,[1] Emma Lees,[2]
Steven I. Reed[3] and Vjekoslav Dulić[3]

[1]Department of Molecular, Cellular and Developmental Biology
University of Colorado
Boulder, CO 80309-0347
[2]Massachusetts General Hospital Cancer Center
Charlestown, MA 02129
[3]Scripps Research Institute
La Jolla, CA 92037

INTRODUCTION

Human diploid fibroblasts have a finite proliferative lifespan at the end of which they are unable to enter S phase in response to mitogenic stimulation even though they remain alive for many months.[1] The G1 arrest state of senescent cells has much in common with the G1 arrest state of early passage quiescent cells, because serum stimulation induces the expression of many early to mid-G1 genes in both senescent and quiescent cells. For example, mitogen-stimulated senescent cells are similar to mitogen-stimulated quiescent cells in their expression of c-myc, c-jun and c-H-ras.[2,3,4] Nevertheless, mitogen-stimulated senescent cells are unable to enter S phase, whereas mitogen-stimulated early passage quiescent cells enter S phase approximately 12-18 hours after stimulation. As a means of investigating the molecular basis for the failure to enter S phase in senescent human fibroblasts, we have sought to identify molecules and/or functions that are deficient in the mitogen response pathways in senescent cells. Since the control of cell proliferation in eukaryotes from yeast to man involves the regulated synthesis, activation and degradation of a family of cyclins, which interact with the Cdc2/CDC28 family of cyclin-dependent kinases (Cdk's),[5,6] we have been analysing the amount and the activity of several cyclins and Cdk's in senescent and quiescent human fibroblasts.[7,8]

Lack of Cyclin A, Cyclin B1 and Cdc2 in Senescent Cells

In our earlier studies, we have shown that senescent human fibroblasts lack cyclin A, cyclin B1 and Cdc2 mRNA and/or protein both before and after mitogen stimulation[7], whereas these genes are expressed beginning in late G1 or S phase in mitogen-stimulated quiescent cells. The lack of cyclin A in mitogen-stimulated senescent cells may be part of

The Cell Cycle: Regulators, Targets, and Clinical Applications
Edited by V.W. Hu, Plenum Press, New York, 1994

the molecular basis for the failure to synthesize DNA in these cells because several studies have indicated that disruption of cyclin A function can inhibit DNA replication in mammalian cells.[9,10] In contrast, neither cyclin B1 (a mitotic cyclin) nor Cdc2 appears to play an essential role in the initiation or execution of S phase.[5,6,11] In this report, we have focused our attention on the G1 cyclins that act earlier in the cell cycle, i.e., during the G_0/G1 to S phase transition, because senescent cells are characterized by their inability to enter S phase.

Role of G1 Cyclins in Passage of START or the R Point

Genetic and molecular studies of the cell cycle in the yeast *S. cerevisiae* have identified the cell cycle control point in G1, called START, at which cells become committed to entry into S phase and execution of the cell division cycle.[6,12] Cells must grow to a certain size to execute START, and both nutrient limitation and mating pheromone arrest cells at START. Passage of START requires the combined action of the CDC28 protein kinase and at least one of the three G1 cyclins (CLN1, CLN2 and CLN3) that have been identified in *S. cerevisiae*.[12] Accumulation of G1 cyclins appears to be rate limiting for execution of START because cells that overexpress a G1 cyclin are smaller and have shorter G1 phases than wild type cells.[13] Conversely, cells with partial CLN deficiencies (e.g., a $CLN1^-$, $CLN2^-$ double mutant), proliferate slowly, become very large, and have the morphology of cells blocked at START. As human fibroblasts approach the end of their proliferative lifespan, they too become enlarged and their G1 periods lengthen.[14] These observations suggested that senescent human fibroblasts might be deficient in their accumulation or activation of G1 cyclins.

Several lines of evidence make it plausible that human cells have an activity comparable to START. First, mammalian cells have a restriction point (R point) several hours before entry into S phase, at which time the cells are committed to entry into S phase regardless of mitogen deprivation, cell crowding or nutrient limitation.[15] Second, progress through G1 up to the R point requires a high level of protein synthesis and the accumulation of a labile protein(s), whereas after the R point, the cells are relatively insensitive to protein synthesis inhibition.[16] Third, candidate G1 cyclins have been identified in human cells by a) their ability to rescue *S. cerevisiae* that were deficient in all three CLN proteins, b) their modest homology to other cyclins, c) their periodic accumulation in the G1 phase of the cell cycle and d) their ability to form complexes with members of the Cdc2/CDC28 family of kinases.[17,18,19,20,21,22] Fourth, constitutive overexpression of the human G1 cyclin, cyclin E, in rat and human fibroblasts shortened the duration of G1 in those cells and decreased their size, suggesting that cyclin E levels are rate-limiting for G1 progression in mammalian cells.[23] Taken together, these data suggest that the R point could be analogous to START and the unknown labile proteins that must accumulate before the R point could be G1 cyclins.[6,12]

Lack of Phosphorylation of the Retinoblastoma Protein in Senescent Cells

The retinoblastoma protein (pRb) is a tumor suppressor that acts as an inhibitor of entry into S phase in its underphosphorylated state.[24] pRb is underphosphorylated in quiescent cells, senescent cells and cycling cells in early G1 phase, and becomes phosphorylated several hours before S phase in serum-stimulated quiescent cells and cycling cells, but not in serum-stimulated senescent cells.[25,26,27,28] Therefore, the failure to phosphorylate pRb in late G1 is an important reason that senescent HDF fail to enter S phase in response to mitogenic stimulation. Although the kinase or kinases responsible for pRb phosphorylation are not known, cyclin E and cyclin A associated kinases are candidates for this function because ectopically expressed cyclin E or cyclin A stimulates the phosphorylation and inactivation of pRb *in vivo*.[29] However, the cyclin E-associated kinase is a more likely candidate for the endogenous G1 phase pRb kinase because it normally becomes activated at approximately the same time that pRb becomes phosphorylated. These data provide a further rationale for determining whether the accumulation of cyclin E and/or cyclin E-associated kinase activity is deficient in mitogen-stimulated senescent cells.

RESULTS AND DISCUSSION

Increased Accumulation of Cyclin D1 and Cyclin E mRNA in Senescent Cells

By Northern blot analysis, we have found that unstimulated senescent cells contain an overabundance of cyclin D1 and cyclin E transcripts in comparison to unstimulated quiescent cells.[8] Cyclin D1 mRNA was elevated more dramatically (3-6 fold) than was cyclin E mRNA (2-4 fold), and both types of mRNA showed a progressive increase as old fibroblasts approached the end of their lifespan (Fig. 1). Thus, the increases in cyclin D1 and cyclin E transcripts occur concurrently with the declining proliferative capacity of late passage human fibroblasts. The overabundance of both cyclin D1 and cyclin E mRNA in unstimulated senescent cells suggests that downregulation of these G1 cyclins may be deficient in senescent HDF, perhaps owing to their inability to progress through the cell cycle.

Serum stimulation caused a 3-4 fold increase of cyclin D1 and cyclin E transcripts during G1 phase in early passage quiescent cells, but only a 1.4 fold increase in cyclin D1 and cyclin E mRNA's in senescent cells (data not shown).[8] Nevertheless, the amount of cyclin D1 mRNA in stimulated senescent cells was always greater than or equal to the amount in stimulated quiescent cells owing to the high constitutive level of cyclin D1 mRNA in unstimulated senescent cells. In contrast, cyclin E mRNA levels in stimulated senescent cells were less than or equal to the amount in stimulated quiescent cells because cyclin E was elevated less dramatically in unstimulated senescent cells.

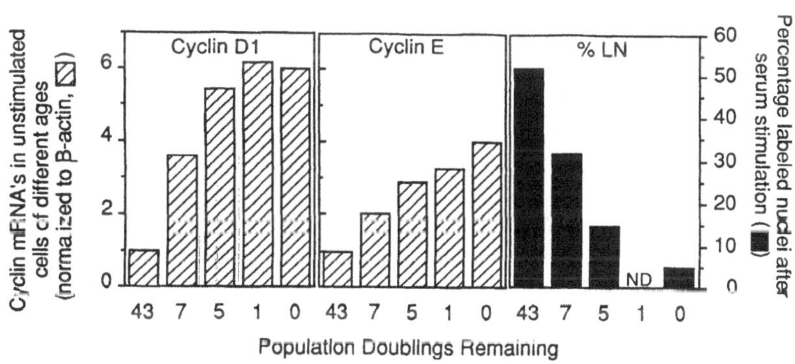

Figure 1. Steady state amounts of cyclin D1 and cyclin E transcripts in unstimulated quiescent and senescent IMR-90 human diploid fibroblasts. Cells were considered senescent when they had 0 population doublings remaining, as well as <5% ^{3}H-thymidine labeled nuclei during a 24 hour period following serum stimulation. Data were obtained by Northern blot analysis of poly A$^+$ mRNA prepared from IMR-90 cells of various ages, as previously described[7,8].

Increased Accumulation of G1 Cyclin Proteins in Senescent Cells

The levels of cyclins D1 and E, as well as cyclin A, were analyzed in quiescent and senescent IMR-90 cells before and after serum stimulation. In quiescent cells, amounts of all three cyclins were low relative to asynchronously replicating cells (Fig. 2A).[8] As expected upon serum stimulation, cyclins D1 and E accumulated during the G1 phase of the cell cycle while accumulation of cyclin A began during S phase and reached maximal levels in G2. Cyclins A and D1 increased approximately 15 fold, while cyclin E underwent only a 3-4 fold increase (Fig. 2A,C). However, in addition to this increase in cyclin E protein accumulation in late G1 and S phase (12h and 18h after stimulation, respectively) there was also a characteristic shift of a portion of the cyclin E molecules toward reduced electrophoretic mobility at these times (Fig. 2A,C). In studies to be reported elsewhere, we (Dulic *et al.*, submitted) have shown that the more slowly migrating forms of cyclin E are phosphorylated. Consequently, cyclin E may also be regulated by this post-translational modification during late G1 and early S phase.

Unstimulated senescent fibroblasts contained 10-15 times as much cyclin D1 and cyclin E, respectively, as did unstimulated quiescent cells (Fig. 2A,C). Following serum stimulation, cyclins D1 and E increased by approximately 20-50% in senescent cells. Thus, both before and after serum stimulation, senescent cells contained at least as much cyclin D1 and cyclin E as did stimulated quiescent cells.[8] The 10-fold increase in cyclin D1 protein was not too surprising given the 3-6 fold elevated levels of cyclin D1 mRNA observed. However, the 15-fold elevation of cyclin E protein was even greater relative to the modest 2-4 fold elevation of cyclin E mRNA seen in unstimulated senescent cells, suggesting that cyclin E protein might be more stable in senescent cells. In contrast to their overabundance of the G1 cyclins, both stimulated and unstimulated senescent cells contained extremely low levels of cyclin A, as do unstimulated quiescent cells.

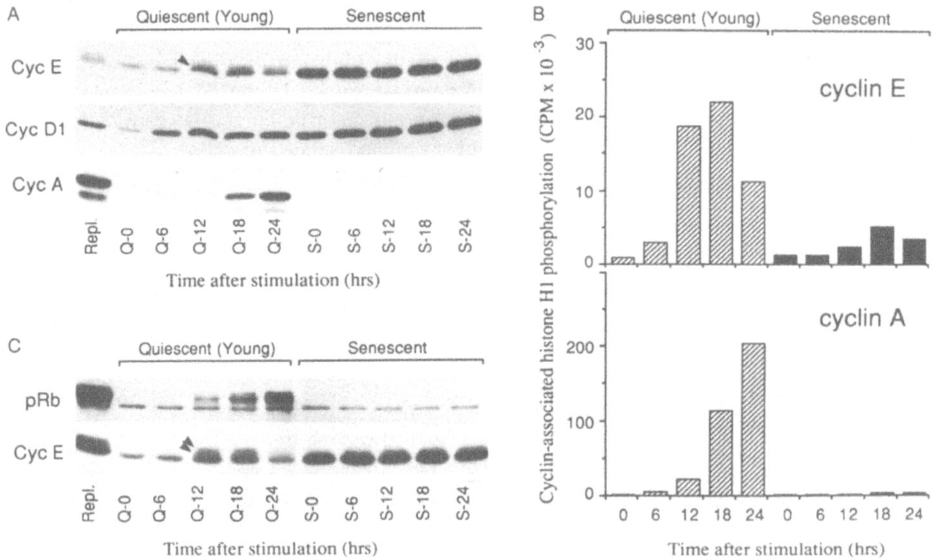

Figure 2. G1 cyclin proteins in quiescent and senescent cells at various times after serum stimulation. (A) Western blots of cell lysates from IMR-90 cells in different growth states were fractionated on 11% SDS-PAGE and immunoblotted sequentially with antibodies to cyclins E, D1 and A. (B) Anti-cyclin A and anti-cyclin E immune complexes were immunoprecipitated from aliquots of the same cell extracts used for part A. Kinase activity of the complexes was measured using histone H1 as the substrate. (C) Western blots, prepared using 8% SDS-PAGE to enhance resolution of the phosphorylated forms of cyclin E, were immunoblotted sequentially for cyclin E and pRb. Phosphorylated forms of cyclin E are indicated by arrowheads.

Low Cyclin E-associated Kinase Activity in Senescent Cells

As discussed earlier, senescent cells are restrained from progression through the cell cycle, possibly due to an inability to phosphorylate the retinoblastoma protein pRb (Fig. 2C).[28] Moreover, several lines of evidence suggest that the cyclin E-associated protein kinase may be responsible for the phosphorylation of pRb in mid- to late G1 phase.[29] Thus, it was surprising to find high levels of cyclin E protein in senescent cells. Consequently, we compared the cyclin E-associated kinase activity in serum-stimulated early passage quiescent cells versus serum-stimulated senescent cells (Fig. 2B).[8] Kinase levels were low in quiescent and early G1 (6h after stimulation) cells but increased dramatically when the cells reached late G1 and early S phase (12-18h after stimulation). The kinetics of activation of the cyclin E-associated kinase activity in early passage human fibroblasts is in agreement with results obtained previously using synchronized HeLa and Manca cells.[20,21] These data also show directly that cyclin E-associated kinase activity increased concomitantly with phosphorylation of pRb in stimulated quiescent human fibroblasts (compare Fig. 2B and Fig. 2C). In contrast, senescent cells had low levels of cyclin E-associated kinase activity both before and after serum stimulation (Fig. 2B). Thus, the high levels of cyclin E protein in senescent cells were not paralleled by elevated cyclin E-associated kinase activity. These data suggest that the lack of phosphorylation of pRb in serum-stimulated senescent cells could be a consequence of the lack of cyclin E-associated kinase activity in these cells. Finally, no cyclin A-associated kinase activity was detectable in the senescent cells, as expected given the lack of cyclin A in senescent cells.

Abundant Cyclin E/Cdk2 Complexes in Senescent Cells

We investigated the basis for the low levels of cyclin E-associated kinase activity in senescent cells by analyzing the degree to which cyclin E was complexed to its functional partner Cdk2. Cyclin E binds to two major SDS-PAGE mobility forms of Cdk2 (Fig. 3A,B), as was shown previously in HeLa cells.[20] The slower mobility form corresponds to unphosphorylated Cdk2, which is the inactive form of this kinase.[20,30] The faster mobility form is phosphorylated on Thr160 and is the potentially active form when complexed with an appropriate cyclin.[30] Cyclin E associates predominantly with the slower migrating form in quiescent (Q-0h) and early G1 cells (Q-6h) that have low kinase activity. At the G1/S boundary (Q-12h), the complexes have increased kinase activity and an increased proportion of the Thr160-phosphorylated form of Cdk2 (Fig. 2B and Fig. 3A).[8] Cyclin E also forms abundant complexes with Cdk2 in senescent cells (Fig. 3A,B). Although unphosphorylated Cdk2 predominated in these complexes, the absolute amount of Thr160-phosphorylated Cdk2 associated with cyclin E was still high owing to the overall abundance of cyclin E/Cdk2 complexes in senescent cells. For example, the amount of Thr160-phosphorylated Cdk2 complexed to cyclin E was at least as high in immunocomplexes from 12h-stimulated senescent cells (S-12h) as in 12h-stimulated quiescent cells (Q-12h), even though the kinase activity of these same immunocomplexes was only 11% as high in S-12h as in Q-12h (compare Fig. 2B and Fig. 3A, which represent results derived from aliquots of the same cyclin E immune complex preparations). These data indicate that senescent cells do not lack cyclin E-associated kinase activity because their cyclin E fails to bind significant amounts of the potentially active form of Cdk2.

Cdc25 Phosphatase Treatment of Cyclin E/Cdk2 Complexes from Senescent Cells

Cdk2 that is phosphorylated on Thr160 can be inhibited by additional phosphorylations on Tyr15 and possibly Thr14[30]. Therefore, this mechanism might account for the lack of kinase activity of the cyclin E/Cdk2 complexes from senescent human fibroblasts.

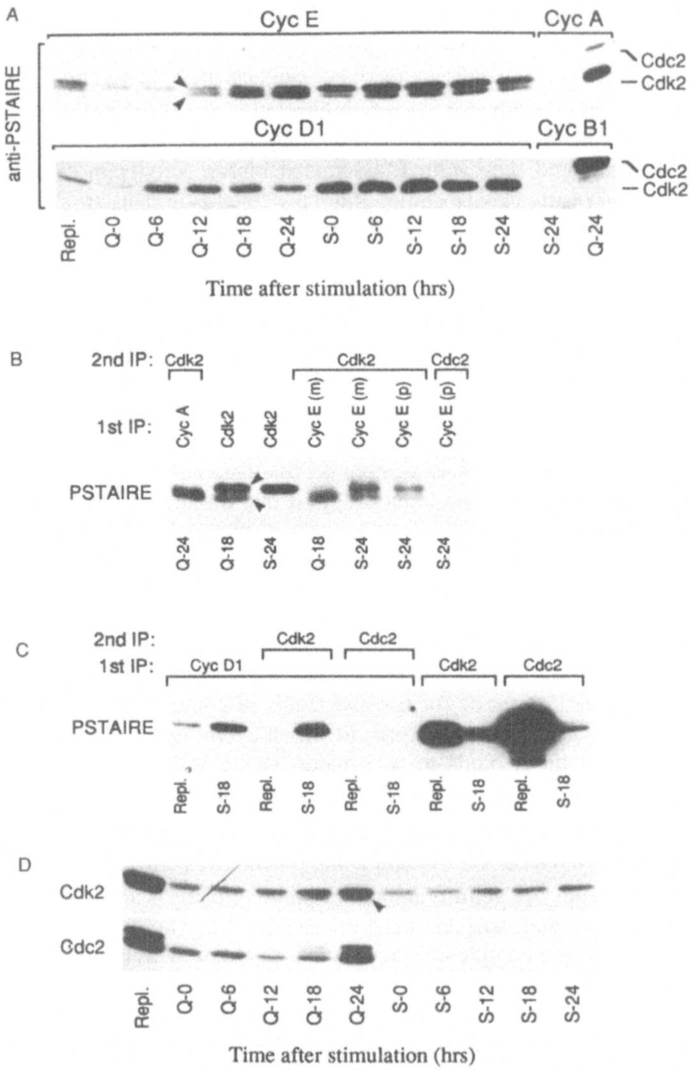

Figure 3. Association of cyclin E and cyclin D1 with Cdk2 in quiescent and senescent human fibroblasts. (A) Aliquots of the cyclin D1 and E immunoprecipitates prepared for Fig. 2B, were fractionated on 11% SDS-PAGE and immunoblotted with anti-PSTAIRE monoclonal antibody. The slower and higher mobility forms of cyclin E-associated Cdk2 represent unphosphorylated and phosphorylated forms of Cdk2, respectively. (B)Single and double immunoprecipitation experiments with antibodies to cyclin E, cyclin A and/or Cdk2 were followed by immunoblot analysis with anti-PSTAIRE antibody to demonstrate that cyclin E is complexed with both forms of Cdk2 in senescent cells. Arrowheads mark the active (faster migrating) and inactive (slower migrating) forms of the Cdk2 doublet. (C) Anti-cyclin D1 immunoprecipitates (1st IP) prepared from lysates of early passage replicating cells and stimulated senescent cells (S-18 h) were divided into 3 equal parts and left untreated or reimmunoprecipitated with antibodies to Cdk2 or Cdc2 (2nd IP), prior to immunoblotting with anti-PSTAIRE antibody. Anti-Cdk2 and anti-Cdc2 immune complexes prepared from equal amounts of the original cell lysates are also included on the immunoblots such that each lane on the immunoblot derives from the same initial amount of cell lysate protein. (D) Western blots prepared as in Fig. 2A were probed sequentially for Cdk2 and Cdc2. An arrowhead marks the Thr160-phosphorylated form of Cdk2.

To test this hypothesis, we attempted to activate senescent cyclin E/Cdk2 complexes by treating them with bacterially produced Cdc25 phosphatase.[31] Cdc25 has been shown to promote kinase activity of Cdc2[32,33] and Cdk2[30,34] by dephosphorylating Tyr15 and Thr14. Cdc25 treatment caused no significant increase in the H1 kinase activity of cyclin E/Cdk2 complexes from stimulated senescent cells (Fig. 4, S-12h).[8] In contrast, Cdc25 treatment caused a 50% increase in the kinase activity of cyclin E/Cdk2 complexes from comparably treated quiescent cells (Q-12h). Since most of the cyclin E/Cdk2 complexes from the Q-12h stimulated cells may already have been in the active state (see Fig. 2B), a relatively modest increase in Cdk2 activation was not unexpected in this situation. On the other hand, the same Cdc25 phosphatase caused a 400% increase in the kinase activity of cyclin E immune complexes from thymidine block-arrested HeLa cells, which demonstrates that Cdc25 can activate Cdk2 that is bound to cyclin E (data not shown).[8] As a final postive control for the activity of our Cdc25 phosphatase, we showed that it caused a 200% increase in the activity of cyclin B1 immunoprecipitates from asynchronously replicating fibroblasts, which contain both active and inactive forms of Cdc2 (Fig. 4). Taken together, these data imply that the low kinase activity of cyclin E/Cdk2 complexes in senescent cells is not due to inhibitory phosphorylations on Tyr15 and Thr14.

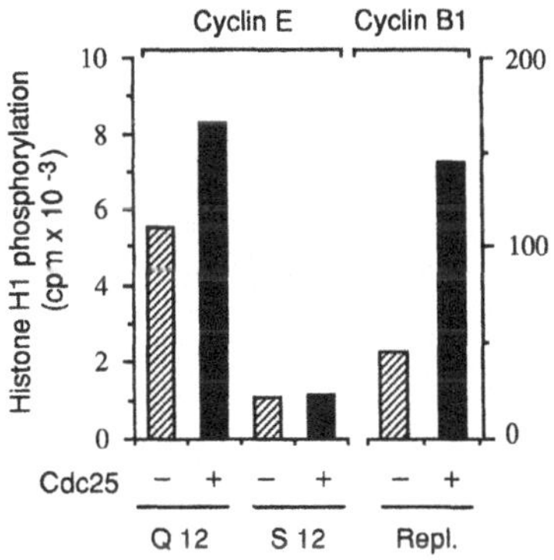

Figure 4. Cdc25 phosphatase treatment of cyclin E/Cdk2 complexes from senescent versus quiescent cells. Cyclin E/Cdk2 complexes from stimulated quiescent (Q-12h) and senescent (S-12h) cells were incubated with and without GST-Cdc25 phosphatase prior to assay for histone H1 kinase activity.[8] Cyclin B1 immunoprecipitates from replicating cells were subjected to the same treatment as a control for the efficiency of the GST-Cdc25 phosphatase.

Lack of Phosphorylated Cyclin E in Senescent Cells

Serum stimulation of quiescent cells induces the appearance of electrophoretically retarded phosphorylated forms of cyclin E concomitant with the activation of cyclin E-associated kinase activity (see Q-12h and Q-18h in Fig. 2). In contrast, serum stimulated senescent cells are deficient in both the appearance of phosphorylated cyclin E and the activation of cyclin E-associated kinase activity (Fig. 2). Since the deficiency in the phosphorylation of cyclin E is the primary difference that we have observed in the composition of cyclin E/Cdk2 complexes in stimulated senescent cells versus stimulated quiescent cells, we suggest the following hypothesis: 1) cyclin E phosphorylation is necessary for the activity of cyclin E/Cdk2 complexes and 2) stimulated senescent cells are deficient in the kinase activity of cyclin E/Cdk2 complexes because they fail to phosphorylate cyclin E. Phosphorylation of cyclin E and phosphorylation of pRb occur coordinately in stimulated quiescent cells (Fig. 2C), as expected if phosphorylation of cyclin E is necessary for the cyclin E/Cdk2 complex to act as a G1 phase pRb kinase.

Complexes of Cyclin D1 with Unphosphorylated Cdk2 in Senescent Cells

Cyclin D1 is known to form complexes with three different Cdk's in mammalian cells, i.e., Cdk2, Cdk4 and Cdk5,[22,35] but paradoxically, no one has been able to detect histone H1 kinase activity of cyclin D1 immune complexes from mammalian cell lysates.[35,36,37] Therefore, we were not able to determine whether cyclin D1 present in senescent cells was associated with an active kinase. Nevertheless, we investigated whether cyclin D1 could form a complex with a PSTAIRE-containing Cdk in senescent and quiescent cells. Our results show that cyclin D1 forms a complex with unphosphorylated Cdk2 in both senescent and quiescent cells.[8] Specifically, the PSTAIRE-containing Cdk complexed with cyclin D1 in both types of cells has the same electrophoretic mobility as unphosphorylated Cdk2 (Fig. 3A,C), whereas Cdk5 (p31) migrates more rapidly than Cdk2[22] and Cdk4 is not recognized by anti-PSTAIRE antibody.[38] Furthermore, double immunoprecipitation experiments with anti-cyclin D1 antibody followed by anti-Cdk2 antibody indicate that unphosphorylated Cdk2 accounts for most, if not all, of the PSTAIRE-containing Cdk associated with cyclin D1 in both stimulated senescent cells (Fig. 3C) and stimulated quiescent cells (data not shown).

There are two obvious differences in the behavior of the cyclin D1/Cdk2 complexes in senescent versus quiescent cells. First, in quiescent cells, the cyclin D1/Cdk2 complexes were serum-inducible and periodic, i.e., complex formation was maximal by 6 hours after stimulation and had decreased by 24 hours (Fig. 3A), whereas there was little effect of serum stimulation on the already abundant cyclin D1/Cdk2 complexes in senescent cells. Second, the amount of Cdk2 complexed to cyclin D1 was 3-4 fold as great in stimulated senescent cells as in stimulated quiescent cells even though the amount of cyclin D1 was similar in both cases. Indeed, at least 75% of the total Cdk2 in stimulated senescent cells (S-18) was complexed with cyclin D1 (Fig. 3C).[8] At this point, we cannot say whether the increased association of unphosphorylated Cdk2 with cyclin D1 in senescent cells adversely affects G1 progression in these cells because we do not know the physiological role of cyclin D1/Cdk2 complexes in mammalian cells. However, it is clear that excessive Cdk2 binding to cyclin D1 in senescent cells does not lead to a deficiency in the amount of Cdk2 associated with cyclin E in those cells, even though they have only about half as much total Cdk2 as do early passage quiescent cells (Fig. 3D).[8] This is consistent with the observation (data not shown) that only a minor part of total Cdk2 is associated with cyclin E, the less abundant of the two G1 cyclins in senescent human fibroblasts.

The finding that cyclin D1 forms complexes only with unphosphorylated, inactive Cdk2 may explain why cyclin D1 immunoprecipitates from both replicating and senescent cells fail to phosphorylate histone H1 (V.D. and S.I.R., unpublished results), which is normally a good substrate for Cdk2. In addition, it has been shown recently that insect cells infected with Cdk2 and cyclin D1 failed to phosphorylate a GST-Rb fusion protein even though Cdk2 was fully active with all other cyclins tested in this system.[37,39] This result could also be a consequence of the lack of the Thr160-phosphorylated, potentially active form of Cdk2 in cyclin D1/Cdk2 complexes. It is possible that the cyclin D1/Cdk2 complex is not recognized by CAK, a Cdc2/Cdk2 activating kinase that phosphorylates Cdc2 on Thr161 and Cdk2 on Thr160.[40].

Hypothesis: Senescent Cells Fail to Pass the R point

Serum stimulation of senescent human fibroblasts induces them to undergo a number of the same prereplicative events as do serum-stimulated early passage quiescent cells.[2,3,4] However, a small subset of key events are deficient in serum-stimulated senescent cells,[3,7,8,28,41,42,43] and these probably contribute to the inability of the senescent cells to enter S phase. Analysis of the events that have been compared in stimulated senescent cells and stimulated quiescent cells (Fig. 5) shows that senescent cells are able to carry out almost all of the early to mid-G1 events analysed, but are preferentially deficient in the late G1 events. Passage of the R point, which implies commitment to the cell division cycle, may be the trigger for the expression of late G1 genes whose products are necessary for further traverse of the cell cycle. The late G1 genes whose transcripts fail to accumulate in stimulated senescent cells fall in this category because they encode proteins that function in DNA synthesis and/or mitosis. These data suggest the possibility that stimulated senescent cells are unable to pass the R point.

Although it has been hard to define the R point in molecular terms, regulation of the similarly defined START point in *S. cerevisiae* involves activation of the cyclin-dependent kinase CDC28 by G1 cyclins.[12] The kinetics of accumulation of cyclin E-associated kinase activity in the mammalian cell cycle are consistent with an analogous role for this mammalian G1 cyclin in R point regulation.[20,21] Thus, the lack of cyclin E-associated kinase activity in senescent cells could contribute to an inability to pass the R point in these cells.

Another clue to the possible basis for R point control comes from analysis of pRb. A number of genetic and biochemical lines of evidence suggest the possibility that pRb is a negative regulator of cell cycle progression at the R point.[24] The hypothesized role of pRb in the R point mechanism may be mediated by the ability of underphosphorylated pRb to bind and sequester transcription factors such as E2F, thought to be required for late G1 transcription of a number of S phase genes that include dihydrofolate reductase, thymidine kinase and DNA polymerase α.[44] The critical phosphorylations of pRb occur late in G1 at the approximate time that the cells traverse the R point[25,26,27], and these phosphorylations fail to take place in serum-stimulated senescent cells.[28] Thus, we propose a model in which senescent cells are deficient in a small subset of critical early to mid-G1 events, which are sufficient to impair the activation of the cyclin E/Cdk2 kinase. Failure to activate the cyclin E/Cdk2 kinase would leave pRb in its underphosphorylated, pre-R point inhibitory state, thereby blocking late G1 events necessary for entry into S phase.[8]

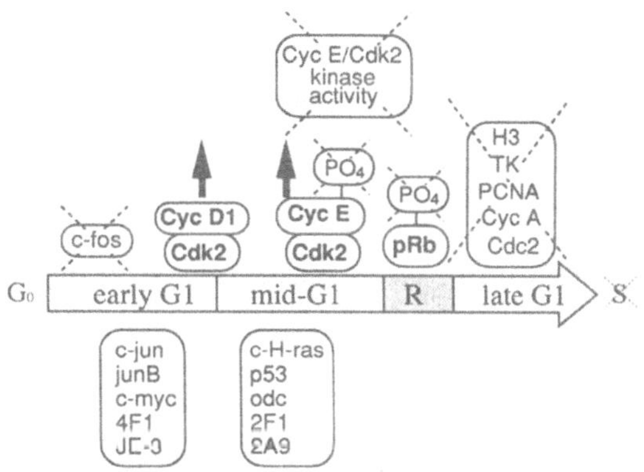

Figure 5. Summary of serum-inducible events that have been compared in senescent versus early passage quiescent human fibroblasts. Serum-stimulated quiescent human fibroblasts carry out all of the diagrammed events, which represent induction of mRNA or protein accumulation unless otherwise indicated, e.g., Cyc E/Cdk2 kinase activity and phosphorylation of Cyc E and pRb. Although serum stimulation induces many of the same early to mid G1 events in senescent cells (2,3,4), stimulated senescent cells are deficient in the events crossed out with dashed X's (3,7,8,28,41,42,43). Senescent cells also have an overabundance of cyclin D1/Cdk2 and cyclin E/Cdk2 complexes both before and after serum stimulation, as indicated by the bold arrows. Placement of the phosphorylation of pRb at the R point is part of our model, but is not proven. H3, histone H3; TK, thymidine kinase; PCNA, proliferating cell nuclear antigen; odc, ornithine decarboxylase. (This figure is reproduced from Dulic et al.[8], with permission).

ACKNOWLEDGMENTS

We express our sincere appreciation to J. Pines, C. McGowan, and M. Yamashita for antibodies to cyclin A, cyclin B1 and Cdc2, and PSTAIRE, respectively; to J. Millar and C. McGowan for purified GST-Cdc25; to M Beeson for excellent technical assistance; and to D.J. Lew for helpful discussions. This work was supported by grant 3165 from the Council for Tobacco Research to S.I.R. and V.D.; by a Fellowship from the Science and Engineering Council to E.L.; and by NIH grants AG00947 and GM46006 to G.H.S. and S.I.R., respectively.

REFERENCES

1. Goldstein, S., (1990). Replicative senescence: The human fibroblast comes of age. *Science* 24, 1129-1133.

2. Rittling, S.R., Brooks, K.M., Cristofalo, V.J., & Baserga, R., (1986). Expression of cell cycle-dependent genes in young and senescent WI-38 fibroblasts. *Proc. Natl. Acad. Sci. USA* 83, 3316-3320.

3. Seshadri, T., & Campisi, J. (1990). Repression of c-fos transcription and an altered genetic program in senescent human fibroblasts. *Science* 247, 205-209.

4. Pignolo, R.J., Cristofalo, V.J., & Phillips, P.D. (1990). Altered expression of AP-1 binding activity but not c-jun or junB mRNA expression in senescent WI-38 cells. *J. Cell Biol.* 111, 498a.

5. Nurse, P. (1990). Universal control mechanism regulating onset of M-phase. *Nature* 344, 503-508.

6. Reed, S.I. (1992). The role of p34 kinases in the G1 to S-phase transition. *Annu. Rev. Cell Biol.* 8, 529-561.

7. Stein, G.H., Drullinger, L.F., Robetorye, R.S., Pereira-Smith, O.M., & Smith, J.R. (1991). Senescent cells fail to express the CDC2 gene in response to mitogen stimulation. *Proc. Natl. Acad. Sci. USA* 88, 11012-11016.

8. Dulič, V., Drullinger, L.F., Lees, E., Reed, S.I., & Stein, G.H. (1993). Altered regulation of G1 cyclins in senescent human diploid fibroblasts: Accumulation of inactive cyclin E/Cdk2 and cyclin D1/Cdk2 complexes. *Proc. Natl. Acad. Sci. USA*, in press.

9. Pagano, M., Pepperkok, R., Verde, F., Ansorge, W., & Draetta, G. (1992). Cyclin A is required at two points in the human cell cycle. *EMBO J.* 11, 961-971.

10. Girard, F., Strausfeld, U., Fernandez, A., & Lamb, N. (1991). Cyclin A is required for the onset of DNA replication in mammalian fibroblasts. *Cell* 67, 1169-1179.

11. Fang, F. & Newport, J.W. (1991). Evidence that the G1-S and G2-M transitions are controlled by different cdc2 proteins in higher eukaryotes. *Cell* 66, 731-742.

12. Reed, S., Wittenberg, C., Lew, D., Dulič, V., & Henze, M. (1992). G1 control in yeast and animal cells. *Cold Spring Harbor Symp. Quant. Biol.* LVI, 61-67.

13. Hadwiger, J.A., Wittenberg, C., Richardson, H.E., De Barros Lopes, M., & Reed, S.I. (1989). A family of cyclin homologs that control the G1 phase in yeast. *Proc. Natl. Acad. Sci. USA* 86, 6255-6259.

14. Rabinovitch, P. (1983). Regulation of human fibroblast growth rate by both noncycling cell fraction and transition probability is shown in 5-bromodeoxyuridine followed by Hoechst 33258 flow cytometry. *Proc. Natl. Acad. Sci. USA* 80, 2951-2955.

15. Pardee, A.B. (1989). G1 events and regulation of cell proliferation. *Science* 246, 603-608.

16. Rossow, P.W., Riddle, V.G.H., & Pardee, A.B. (1979). Synthesis of labile, serum-dependent protein in early G1 controls animal cell growth. *Proc. Natl. Acad. Sci. USA* 76: 4446-4450.

17. Lew, D.J., Dulić, V., & Reed, S.I. (1991). Isolation of three novel human cyclins by rescue of G1 cyclin (Cln) function in yeast. *Cell* 66, 1197-1206.

18. Xiong, Y., Connolly, T., Futcher, B., & Beach, D. (1991). Human D-type cyclin. *Cell* 65: 691-699.

19. Koff, A., Cross, F., Fisher, A., Schumacher, J., Leguellec, K., Philippe, M. & Roberts, J.M. (1991). Human cyclin E, a new cyclin that interacts with two members of the *CDC2* gene family. *Cell* 66: 1217-1228.

20. Dulić, V., Lees, E., & Reed, S.I. (1992). Association of human cyclin E with a periodic G1-S phase protein kinase. *Science* 257, 1958-1961.

21. Koff, A., Giordano, A., Desai, D., Yamashita, K., Harper, J.W., Elledge, S., Nishimoto, T., Morgan, D.O., Franza, B.R., & Roberts, J.M. (1992). Formation and activation of a cyclin E-cdk2 complex during the G1 phase of the human cell cycle. *Science* 257, 1689-1694.

22. Xiong, Y., Zhang, H. & Beach, D. (1992). D type cyclins associate with multiple protein kinases and the DNA replication and repair factor PCNA. *Cell* 71, 505-514.

23. Ohtsubo, M., & Roberts, J.M. (1993). Cyclin-dependent regulation of G1 in mammalian fibroblasts. *Science* 259, 1908-1912.

24. Cobrinik, D., Dowdy, S.F., Hinds, P.W., Mittnach, S., & Weinberg, R.A. (1992). The retinoblastoma protein and the regulation of cell cycling. *Trends Biochem. Sci.* 17, 312-315.

25. Buchovich, K., Duffy, L.A., & Harlow, E. (1989). The retinoblastoma protein is phosphorylated during specific phases of the cell cycle. *Cell* 58: 1097-1105.

26. Chen, P.-L., Scully, P., Shew, J.-Y., Wang, J.Y.J., & Lee, W.-H., (1989). Phosphorylation of the retinoblastoma gene product is modulated during the cell cycle and cellular differentiation. *Cell* 58, 1193-1198.

27. DeCaprio, J.A., Furukawa, Y., Ajchenbaum, F., Griffin, J.D., & Livingston, D.M. (1992). The retinoblastoma-susceptibility gene product becomes phosphorylated in multiple stages during cell cycle entry and progression. *Proc. Natl. Acad. Sci. USA* 89: 1795-1798.

28. Stein, G.H., Beeson, M., & Gordon, L. (1990). Failure to phosphorylate the retinoblastoma gene product in senescent human fibroblasts. *Science* 249: 666-669.

29. Hinds, P.W., Mittnach, S., Dulić, V., Arnold, A., Reed, S.I., & Weinberg, R.A. (1992). Regulation of retinoblastoma protein functions by ectopic expression of human cyclins. *Cell* 70, 993-1006.

30. Gu, Y., Rosenblatt, J., & Morgan, D.O. (1992). Cell cycle regulation of Cdk2 activity by phosphorylation of Thr160 and Tyr15. *EMBO J.* 11, 3995-4005.

31. Millar, J.B.A., McGowan, C.H., Lenaers, G., Jones, R., & Russell, P., (1991). $p80^{cdc25}$ mitotic inducer is the tyrosine phosphatase that activates $p34^{cdc2}$ kinase in fission yeast. *EMBO J.* 10, 4301-4309.

32. Dunphy, W.G., & Kumagai, A., (1991). The cdc25 protein contains an intrinsic phosphatase activity. *Cell* 67, 189-196.

33. Millar, J.B.A., & Russell, P., (1992). The cdc25 M-phase inducer: An unconventional protein phosphatase. *Cell* 68, 407 410.

34. Sebastian, B., Kakizuka, A., & Hunter, T., (1993). Cdc24M2 activation of cyclin-dependent kinases by dephosphorylation of threonine-14 and tyrosine-15. *Proc. Natl. Acad. Sci. USA* 90, 3521 3524.

35. Matsushime, H., Ewen, M.E., Strom, D.K., Kato, J., Hanks, S.K., Roussel, M.F., & Sherr, C.J. (1992). Identification and properties of an atypical catalytic subunit ($p34^{PSK-J3}$/cdk4) for mammalian D type G1 cyclins. *Cell* 71, 323-334.

36. Lucibello, F.C., Sewing, A., Brüsselbach, S., Bürger, C., & Müller, R., (1993). Deregulation of cyclins D1 and E and suppression of cdk2 and cdk4 in senescent human fibroblasts. *J. Cell Sci.* 105, 123-133.

37. Ewen, M.E., Sluss, H.K., Sherr, C.J., Matsushime, H., Kato, J.-y., & Livingston, D.M., (1993). Functional interactions of the retinoblastoma protein with mammalian D type cyclins. *Cell* 73, 487-497.

38. Meyerson, M., Enders, G.H., Wu, C.-L., Su, L.-K., Gorka, C., Nelson, C., Harlow, E. & Tsai, L.-H. (1992). A family of human cdc2-related protein kinases. *EMBO J.* 11, 2909-2917.

39. Dowdy. S.F., Hinds, P.W., Louie, K., Reed, S.I., Arnold, A., & Weinberg, R.A., (1993). Physical interaction of the retinoblastoma protein with human D cyclins. *Cell* 73, 499-511.

40. Poon, R.Y.C., Yamashita, K., Adamczewski, J.P., Hunt, T., & Shuttleworth, J. (1993). Cdc2-related protein $p40^{MO15}$ is the catalytic subunit of a protein kinase that can activate $p33^{cdk2}$ and $p34^{cdc2}$. *EMBO J.*, in press.

41. Chang, C. Phillips, P., Lipson, K.E., Cristofalo, V.J., & Baserga, R. (1991). Senescent human fibroblasts have a post-transcriptional block in the expression of the proliferating cell nuclear antigen gene. *J. Biol. Chem.* 266, 8663-8666.

42. Zambetti, G., Dell'Orco, R., Stein, G. & Stein, J. (1987). Histone gene expression remains coupled to DNA synthesis during in vitro cellular senescence. *Exp. Cell Res.* 172, 397-403.

43. Chang, Z.-F., & Chen, K.Y. (1988). Regulation of ornithine decarboxylase and other cell cycle-dependent genes during senescence of IMR-90 human diploid fibroblasts. *J. Biol. Chem.* 263, 11431-11435.

44. Nevins, J.R. (1992). E2F: A link between the Rb tumor suppressor protein and viral oncoproteins. *Science* 258, 424-429.

31

CELL PROLIFERATION AS A BIOMARKER OF AGE AND DEVELOPMENT

M.H. Lu, S.F. Ali, D.S. He, A. Turturro and R.W. Hart

National Center for Toxicological Research
Food and Drug Administration
3900 NCTR Road
Jefferson, AR 72079

ABSTRACT

Changes in the proliferative capacity of cells can serve as a biomarker of development, age, and carcinogenesis. Perturbation of the cell cycle is considered a toxic endpoint. A paucity of data is available on the cell cycle as a function of age. In order to understand the etiology of ontogenesis and carcinogenesis, it is essential that background information be established on cell proliferation. The brain is known to have a very high level of proliferation during the early stages of development which decreases with age. The present study was designed to establish background information on cell proliferation, by cell cycle analysis, in the brain of normal male rats during development. Brain tissues of Sprague-Dawley rats (newborn, 21 days, 6 months, and one year of age) were dissected to obtain the cerebrum, cerebellum, and brain stem. Cellular proliferation activity was determined by flow cytometry. The percentage of S-phase cells was used to establish the proliferative index (PI). Our results indicate that there are differences in proliferation among the three brain regions in all age groups except for the 6-month old animals. In all cases, cell proliferation was highest in newborn rats. Proliferation activity in the cerebrum of newborn and 21-day old rats was equivalent (17%); after 21 days of age, proliferative activity decreases and maintains the same low rate (about 6.5%) until at least one year of age. The proliferation activity of cerebellum (21% at birth) decreases in animals at day 21 (7%) and the decrease continues until 6 months of age at which time it remains constant (5%) for at least one year. The proliferative activity of brain stem decreases from 16% to 5% immediately after birth and maintains the same low rate from day 21 to one year of age.

The Cell Cycle: Regulators, Targets, and Clinical Applications
Edited by V.W. Hu, Plenum Press, New York, 1994

INTRODUCTION

Changes in the cell cycle[1] can serve as biomarkers of both aging[2,3,4] and carcinogenesis[1]. Perturbation of the cell cycle in a tissue can be considered as a toxicological endpoint[5,6,7]. While a numerous studies have reported on various macromolecular (DNA, RNA, and protein) changes[8,9,10,11] as a function of the cell cycle, few have evaluated the inherent rate of cellular proliferation of tissues *in vivo*. Basic information on cell cycle of many tissues is lacking. Due to the critical role of cellular proliferation in ontogenesis and carcinogenesis, it is important that such information be established. In teratogenesis, for example, toxic chemicals may inhibit cell proliferation and result in reduced cell numbers through intrauterine growth disturbances (aminopterin, chromosomal defect, radiation, and rubella) thereby leading to malformations[12,13]. In carcinogenesis, other chemicals may promote initiated cells to proliferate and form tumors[14].

On the other hand, from a therapeutic standpoint, many agents including the antitumor drugs exert their biological effects by either impairing cell division or permanently blocking DNA replication and cellular proliferation. An understanding of cellular kinetics is important since many drugs act specifically on proliferating cells and are harmless to those in the resting phase. Some drugs act in a phase-dependent manner, affecting those cells in one particular phase of the cell cycle and not other phases[12]. For instance, methotrexate, hydroxyurea, 5-fluorouracil and 6-mercaptopurine act specifically on S-phase-cells, while vinblastine and colchicine act on cells in mitosis phase[12]. Thus, understanding how cellular proliferation is altered as a function of age and development in normal tissue is critical to a better understanding of not only the aging process, but the process of carcinogenesis and chemotherapy as well. Data on cell cycle of normal tissues is also important to risk assessment[15].

The present study was undertaken to establish background information on cell proliferation in the brain of normal rats during development. We investigated cell proliferative activity in the brain of normal male Sprague-Dawley rats of different ages by cell cycle analysis and determined the proliferative activity of different brain regions.

MATERIALS AND METHODS

A. Animals and Dissection Procedures: Male Sprague-Dawley rats at different developmental stages were used. They were obtained from the NCTR (National Center for Toxicological Research) breeding colony, raised at 23°C and 50% relative humidity in a specific pathogen free (SPF) facility and were maintained in a 12-hour light-dark cycle daily. The rats were housed singly in plastic cages with hardwood chip bedding and metal tops. All animals were provided a normal diet (NIH-31) and water *ad libitum*. The different age groups of rats were newborn [postnatal day-1 (PND-1)], 21-days (PND-21), 6 months, and one year old. The animals were killed by decapitation, and the brains were removed, blotted, and chilled. Dissections were performed on an ice-cooled plastic plate. Brain tissues were dissected, according to the methods of Glowinski and Iversen[16] and Robertson *et al.*[17], at sacrifice to obtain cerebrum, cerebellum, and brain stem from rats of all four age groups. Tissues harvested were preserved in a small plastic vial (14 mm x 50 mm, Walter Sarstedt, Inc.,

Princeton, NJ) with citrate-sucrose-DMSO (dimethylsulfoxide) buffer solution[18], frozen in liquid nitrogen, and stored in a -80°C freezer until assayed.

B. Cell Cycle Analysis by Flow Cytometry: Nuclear suspensions of the brain were made from frozen tissues as described previously[3] by trypsinization of the suspension. Nuclear DNA was stained with propidium iodide (Calbiochem-Behring Diagnostics, La Jolla, CA) for DNA/cell cycle analysis according to the modified method of Vindelov *et al.*[18]. The prepared samples were kept at 4°C and analyzed on the flow cytometer within one hour of addition of the staining solution. Cellular DNA content was measured by a FACScan Flow Cytometer (Becton Dickinson, San Jose, CA) equipped with an Argon-laser emitting at 488 nm to excite the DNA-associated propidium iodide to fluoresce at a wavelength of 630 nm. Histograms of 10,000 cells were recorded for each sample at a flow rate of 100-200 cells/second. Data collection was performed using the Consort 30 Data Acquisition Program (Becton Dickinson) and an HP310 Computer (Hewlett Packard, Palo Alto, CA). Analysis of the DNA histogram was performed using Becton Dickinson's DNA/Cell cycle Analysis Software, Revision B, Based on the SFIT polynomial model[19]. Populations of nuclei in the various cell cycle phases (based on DNA content) were assumed to represent the cell populations and were expressed as percentage of the entire population of the histogram obtained. Cell proliferation is expressed in terms of the total percentage of S-phase cells[20], i.e., the percentage of S-phase cells in the cell cycle was used as the proliferative index (PI).

C. Statistical Analysis: Unpaired Student's t-test was used for the comparison of the percentage of cells in each phase of the cell cycle between groups. The PI's of cerebrum, cerebellum, and brain stem were compared statistically within each age group to establish any regional differences. A p-value of ≤ 0.05 was considered a statistically significant difference.

RESULTS

Table I summarizes the cell proliferative data from three different regions of the rat brain as a function of age as determined by flow cytometric cell cycle analysis. Except for the 6-month old group, differences in proliferative activity among the three brain regions were observed in each age group. While the PI's in all three regions of brain differed significantly from one another in newborn rats, the PI's in the cerebellum and brain stem were essentially equivalent. In 21-day old rats, the PI's differed significantly among brain regions. In 1-year old rats, the PI of the cerebrum was significantly greater than that of the cerebellum and brain stem. The PI's in the brain stem and cerebrum of 1-year rats were similar. Proliferative activity among the age groups showed that in the cerebrum the PI's of newborn (PND-1) and 21-day old (PND-21) rats were similar. The same result was obtained when comparing 6-month and 1-year old rats. However, all other comparisons of any two age groups in the cerebrum show significant differences. In cerebellum, there were significant differences between the PI of all age groups except when comparing the results from the 6-month and 1-year old groups. In brain stem, the PI's of 21-day, 6-month, and 1-year old groups were comparable, but the values of all three groups were significantly different from the PI of the newborn group.

Table 1. Cell proliferation by cell cycle analysis in different regions of rat brain during development.

Brain Regions	% S-phase cells in the cell cycle (mean ± S.E.M.)			
	PND-1 (Newborn)	PND-21 (21-day old)	6-month old	1-year old
Cerebrum	17.29±1.41[a,h,i] (n=7)	16.83±2.96[c,d,j,k] (n=12)	6.00±0.58[h,j] (n=4)	6.88±1.17[f,g,i,k] (n=16)
Cerebellum	20.71±1.95[a,b,l,m,n] (n=7)	7.44±0.46[c,e,l,o,p] (n=16)	4.50±0.29[m,o] (n=4)	4.75±0.56[f,n,p] (n=16)
Brain Stem	16.14±1.85[b,q,r,s] (n=7)	5.63±0.42[d,e,q] (n=16)	5.25±0.63[r] (n=4)	4.69±0.53[g,s] (n=16)

Values having at least one similar superscript are significantly different ($p < 0.05$). Significant of cell proliferation activity are noted both by column and by row. Comparisons of the values among the same column revealed the regional effect on cell proliferation for a specific age group except for the 6-month old group. Comparisons of the values among the same row revealed the age effect on cell proliferation for a specific region of brain. S.E.M.= standard error of mean. PND denotes postnatal day. n=number of animals used in each group.

DISCUSSION

Our data indicates that in the rat, cellular proliferation in the brain is age and region dependent. Similar to the findings of Fish and Winick[21], we found that proliferation activity differed among the three brain regions studied, thereby supporting the belief that different regions of brain undergo different patterns of growth and development.

In cerebrum, the proliferative activity is high during the first three weeks of life and declines after this time, maintaining the same low level of activity throughout all other time periods examined. Mandel *et al.*[22] reported that in cerebrum, DNA synthesis continues for at least 21 days after birth. Winick and Noble[23] reported that in the brain of rats cell division is over by the twenty-first day. Our data is in agreement with these findings. It is reasonable to assume that since the neuronal cell proliferation ceases prior to birth[24], the S-phase cells (6-7%) seen in the cerebrum of the 6-month and 1-year old rats in the present study are the glial cell population.

In cerebellum, proliferative activity is highest at birth, decreases at 21 days, further decreases by 6 months, and then remains constant from that point forward. Both neuronal and glial cells of the cerebellum are known to continue to divide in the postnatal period[11,24]. The S-phase population observed (5% of total cell population) in cerebellum and brain stem of the 6-month and 1-year old rats is a mixture of neuronal and glial cells. The synthesis of DNA in rat cerebellum was found to stop at about 17 days postnatal[21]. These findings are in agreement with Altman's finding[25] in which the external granular layer of the cerebellum of rats exhibited considerable proliferative activity from 6 to 14 days and showed no DNA synthesis by the 20th day. Cell proliferative activity in cerebellum was found to completely disappear with no mitotic activity by PND-21 in controls[26] but not in the underfed animals[27]. This indicates that dietary restriction might extend the duration of the proliferative activity of cerebellar cells during the early development.

In brain stem, proliferative activity is highest in newborn and decreases rapidly from that point forward, maintaining the same low rate throughout the rest of the study. Winick *et al.*[28] found that in the brain stem of rats, the total cell number increases up to 14 days of age. Our results did not contain data on the first and second weeks of life. Judging from the findings of Winick *et al.*[28] and our present results, one can infer that the decrease of cell proliferation in brain stem of rats begins after 14 days of age.

The newborn rats were found to have the highest percentage of S-phase cells in all three brain regions compared to the other age groups. This clearly indicates that the brain of newborn rats had the highest level of cell proliferation compared to other age groups. This finding also agrees with the fact that most species undergo the most brain growth around birth[10].

In all three regions of rat brain, the PI decreases from birth to one year of age. The decreasing rate of cell proliferation seen in brain clearly reflects the existence of a postnatal developmental process. Having found that cellular proliferative activity of all three regions of brain decreases between the first and the 20th day of age, we have initiated further studies to determine the profile of brain cell proliferation at 7 and 14 days of age. In order to better understand the role of cell proliferation in brain development, a study of S-phase cell distributions of neuronal and glial cells[29] in different brain regions should also be undertaken.

REFERENCES

1. J.W. Gray, F. Dolbeare, M.G. Pallavicini, W. Beisker, and F. Waldman, Cell cycle analysis using flow cytometry. Int. J. Radiat. Biol. 49:237 (1986).
2. P.D. Bowman. Aging and the cell cycle *in vivo* and *in vitro*, in: Handbook of cell biology and aging, V.J. Cristofolo, R.C. Adelman, and G.S. Roth, eds., CRC Press, Boca Raton, FL pp.117 (1985).
3. M.H. Lu, W.G. Hinson, A. Turturro, J. Anson, and R.W. Hart, Cell cycle analysis in bone marrow and kidney tissues of dietary restricted rats. Mech. Ageing Devel. 59:111 (1991).
4. M.H. Lu, W.G. Hinson, A. Turturro, W.G. Sheldon, and R.W. Hart, Cell proliferation by cell cycle analysis in young and old dietary restricted mice. Mech. Ageing Devel. 68:151 (1993).
5. B. Barlogie, B. Drewinko, J. Schumann, and E.J. Freireich, Pulse cytophotometric analysis of cell cycle perturbation with bleomycin *in vitro*. Cancer Res. 36:1182 (1976).
6. O. Sletvold and D. Laerum, Alterations of cell cycle distribution in the bone marrow of aging mice measured by flow cytometry. Exp. Gerontol. 23:43 (1988).

7. K. Yamada, M. Ohtsu, M. Sugano, and G. Kimura, Effects of butyrate on cell cycle progression and polyploidization of various types of mammalian cells. Biosci. Biotech. Biochem. 56:126 (1992).
8. S. Reinis and J.M. Goldman, The development of the brain: biological and functional perspectives, Charles C. Thomas Publisher, Springfield, IL pp.123 (1980).
9. S.J. Rozovski and M. Winick, Nutrition and cellular growth, in: Nutrition: pre- and postnatal development, M. Winick, ed., Plenum Press, New York, pp.61 (1979).
10. G.A. Dhopeshwarkar, Nutrition and brain development, Plenum Press, New York, pp.13 (1983).
11. R. Balazs, T. Jordan, P.D. Lewis, and A.J. Patel, Undernutrition and brain development, in: Human growth, vol. 3, Neurobiology and nutrition, F. Falkner and J.M. Tanner, eds., Plenum Press, New York, pp.415 (1979).
12. H. Tuchmann-Duplessi, Drug effect on the fetus, AIDS Press, Sydney, pp.63 (1975).
13. D.A. Karnofsky, Mechanisms of action of certain growth-inhibiting drugs, in: Teratology: principles and techniques, J.G. Wilson and J. Warkanay, eds., The University of Chicago Press, Chicago, IL pp.185 (1965).
14. S.M. Cohen and L.B. Ellwein, Cell proliferation in carcinogenesis. Science 249:1007 (1990).
15. U.S. Interagency Staff Group on Carcinogens, Chemical carcinogenesis: a review of the science and its associated principles. Environ. Health Perspect. 67:201 (1986).
16. J. Glowinski and L.L. Iversen, Regional studies of catecholamines in the rat brain, I. The disposition of [^{3}H]norepinephrine, [^{3}H]dopamine and [^{3}H]dopa in various regions of the brain. J. Neurochem. 13:655 (1966).
17. R.T. Robertson, J. Zimmer, and B.H. Gahwiler, Dissection procedures for preparation of slide cultures, in: A dissection and tissue culture manual of the nervous system, A. Schahar, J. de Vellis, A. Vernadakis, and B. Haber, eds., Alan R. Liss, Inc., New York, pp.1 (1989).
18. L.L. Vindelov, I.J. Christensen, and N.I. Nissen, A detergent-trypsin method for the preparation of nuclei for flow cytometric DNA analysis. Cytometry 3:323 (1983).
19. P.N. Dean, A simplified method of DNA distribution analysis. Cell Tissue Kinetics 13:299 (1980).
20. S.L. Cohen, A.W. Rademaker, H.R. Salwen, W.A. Franklin, F. Gonzales-Crussi, S.T. Rosen, and K.D. Bauer, Analysis of DNA ploidy and proliferative activity in relation to histology and N-myc-amplification in neuroblastoma. Am. J. Pathol. 136:1043 (1990).
21. I. Fish and M. Winick, Cellular growth in various regions of the developing rat brain. Pediatr. Res. 3:407 (1969).
22. P. Mandel, H. Rein, S. Harth-Edel, and R. Mardell, Distribution and metabolism of ribonucleic acid in the vertebrate central nervous system, in: Comparative neurochemistry, D. Richter, ed., Pergamon Press, London, pp.153 (1964).
23. M. Winick and A. Noble, Quantitative changes in DNA, RNA, and protein during prenatal and postnatal growth. Devel. Biol. 12:451 (1965).
24. M. Winick, Normal cellular growth of the brain, in: Malnutrition and brain development, Chapter 2, Oxford University Press, New York, pp.35 (1976).

25. J. Altman, Autoradiographic and histological studies of postnatal neurogenesis, III. Dating the time of production and onset of differentiation of cerebellar microsomes in rats. J. Comp. Neurol. 136:26 (1969).
26. M.G. Deo, V. Bijlani, and V. Ramalingaswami, Nutrition and cellular growth and differentiation, in: Growth and development of the brain: Nutrition, genetics, and environmental factors, M.A.B. Brazier, ed., Raven Press, New York, pp.1 (1975).
27. G. Gopinath, V. Bijlani, and M.G. Deo, Undernutrition and developing cerebellar cortex in the rat. J. Neuropathol. Exp. Neuro. 35:125 (1976).
28. M. Winick, J.A. Brasel, and P. Rosso, Nutrition and cell growth, in: Nutrition and development, M. Winick, ed., John Wiley and Sons, New York, pp.49 (1972).
29. A.M. Giuffrida, A. Hamberger, I. Serra, and E. Geremia, Effects of undernutrition on nucleic acid synthesis in neuronal and glial cells from different regions of developing rat brain. Nutr. Metab. 24:189 (1980).

32

HEAT SHOCK GENES AND CELL CYCLE REGULATION DURING EARLY MAMMALIAN DEVELOPMENT

David Walsh, Li Zhe, Frank Zeng, Wu Yan and Karen Li

Mammalian Development Unit B19
Department Veterinary Science
University of Sydney
NSW 2006 Australia

INTRODUCTION

The formation of the mammalian forebrain is one of the least known processes in embryology and is extremely sensitive to heat shock[1]. The structure of the early mammalian head and forebrain is predetermined in the neural plate during gastrulation. The development of the vertebrate nervous system begins with the induction of the neural plate on the dorsal surface of the embryo near the completion of gastrulation. Neural tube closure is one of the most critical stages of early embryogenesis. Heat shock interrupts neuroectoderm differentiation and can result in cell death and major developmental defects of the face and brain.A process closely linked to neural induction is the regionalization of the neural ectoderm along the anteriorposterior (AP) axis into prospective forebrain, midbrain hindbrain and spinal cord.

One of our major interests has been to study how and why the heat shock genes are activated and how they relate to normal embryonic development and neuronal induction and development[2,3]. Heat shock can result in severe developmental craniofacial defects in all species[4]. The major developmental deformities are a result of cell death following heat shock, but details of why the cells die remain unknown.

WHOLE EMBRYO CULTURE

In embryogenesis, stability of critical developmental proteins during cellular proliferation, differentiation and migration is critical. The chaperone functions underscore the biological importance of the hsp during development. By using a whole embryo culture system, normal gastrulation in mammalian development can be observed[2,5] (Fig. 1). The embryo culture system allows a controlled and precise heat shock to be applied at specific stages of neural development. Maternal influences are removed and the embryonic response to heat shock and neural plate development can be studied directly. Induction of the neuroectoderm and closure of the neural

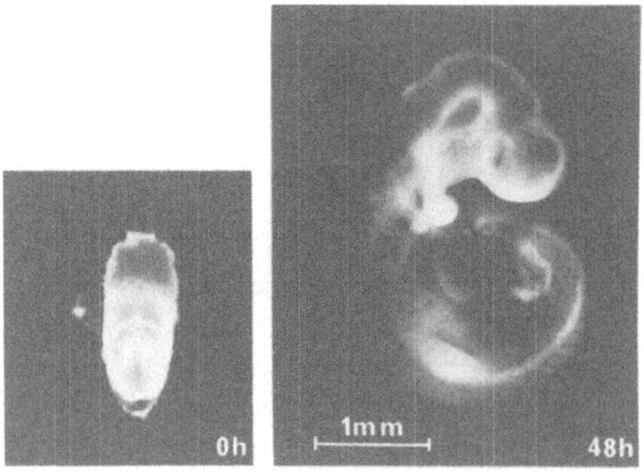

Fig. 1. Developmental stages of in vitro 9.5-11.5 gd whole rate embryo in culture. The period of embryo culture embraces most of the major events of organogenesis, with the embryo developing from the relatively undifferentiated headfold stage (0 hr) to have a beating heart, prominent optic vesicles, and a closed neural tube at the end of culture (48 hr).

tube occurs between days 9.5 to 10.5 in rat embryos. We have examined the heat shock response using known lethal and nonlethal heat shock regimes in whole cultured rat embryos at the pre-somite neural plate stage.

Heat shock causes cell death and irreversible damage to the neural plate and our aims were to study hsp gene expression in association with cell cycle changes. The influence of hsp gene expression on cell fate at specific phases of the cell cycle in relation to either cell death or acquired thermotolerance were investigated.

When the early neural plate of mammalian embryos is exposed to heat shock during gastrulation, severe developmental defects of the forebrain and eye follow[6] (Fig. 2). Interruption or inhibition of gene expression and protein synthesis in the developing neural plate could result in neuroectoderm cell death. Heat shock induction and the heat shock response however need not cause developmental defects[1] (Fig. 2b). A mild heat shock while inducing hsp can result in a phenomenon termed thermotolerance (Fig. 2c) whereby cells appear protected against a further heat stress that would be normally lethal in unprotected embryos (Fig. 2d).

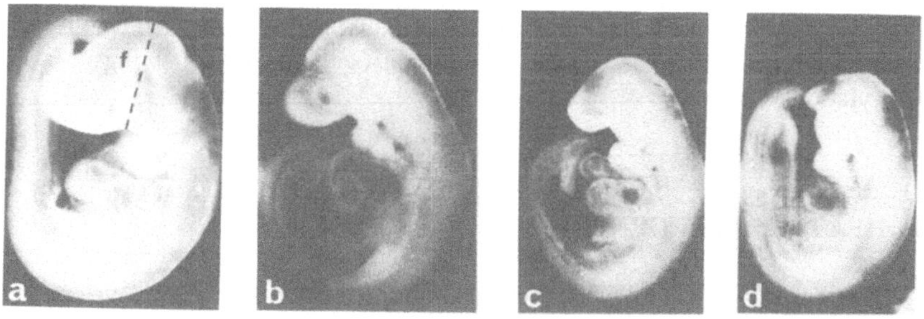

Fig. 2. Rat embryos at 11.5 days were cultured for 48 h after various heat shock regimens at the neural plate stage. **a:** Control embryo with 24 somites, well-developed forebrain (f) and eye, midbrain, and hindbrain. **b:** Embryos exposed to a mild heat shock of 42°C for 10 min with 22 somites representing a general growth delay of 4 h. No obvious developmental defects were observed. **c:** Embryos subjected to 42°C for 10 min, followed by a recovery period of 60 min at 38.5°C, then a further heat shock of 43°C for 7.5 min. Although growth delay was observed, there were no morphological defects. **d:** Embryos exposed to a 43°C for 7.5 min, displaying developmental defects and absence of both the eye and forebrain. The midbrain, branchial arch, and hindbrain were unaffected.

HEAT SHOCK GENES AND THE CELL CYCLE

During early formation of the CNS, control of the cell cycle length is critical for cell proliferation and closure of the neural tube. The normal cell cycle time in neuroectoderm cells is 7-8 hr[8,25.] Throughout organogenesis a general lengthening of the cell cycle is observed particularly in G1 and G2 phases. Our interest has been to study the role of hsp in cell cycle regulation since heat shock causes rapid changes to the cell cycle phases at both G1 and G2[7]. Acquired thermotolerance and hsp induction is associated with a cell cycle delay with cells accumulating at the G1/S and the S/G2 boundaries[8]. The heat shock response appears to lengthen the cell cycle time and thermotolerance is associated with this. During neural tube closure many developmental genes such as the homeobox-containing genes are expressed to establish regional patterning of the peripheral nervous system and craniofacial structures[9]. The ramifications of heat shock on cell cycle changes and the expression, control and regulation of these genes is not known.

The questions we would like to ask here is 1. are heat shock genes cell cycle regulators? 2. what is the association between cell cycle changes and hsp expression 3. how does this alter cell differentiation and cell fate?

THE HEAT SHOCK RESPONSE

The heat shock response is a well-studied biological phenomenon that is ubiquitous to all living cells and organisms[10]. The heat shock genes and their encoded proteins are highly conserved through evolution. Studies on the expression of hsp genes have provided evidence of a complex pattern of regulation. The expression of the human hsp70 gene is regulated in a cell cycle dependent manner[11,12] and is responsive to many stresses such as heat shock, serum and heavy metal stimulation[13,14]. The hsp genes are induced by a trans-acting heat shock transcription factor (HSF) following phosporylation[15].

THE HEAT SHOCK GENES

Some major hsp families are activated during early mammalian development. The hsp90 family is associated with membrane stability and glucocorticoid receptor functions[16,17]. Analysis of hsp90 in early neural plate development revealed that hsp90 expression may be required to maintain cells in G0 and could affect cell progression at the start of the cell cycle[7]. The hsp70 family comprises the heat shock cognate, hsc73 that is constitutive during neural tube closure, an inducible hsp71[2,18] and a hsp72 that is developmentally expressed during differentiation of the mammalian germ line[19]. Recently hsp47 was identified in early developing embryos[20] and is reported to be a specific chaperone for collagen IV[21]. A small hsp27 also has a role in development and is involved in the organisation of several cytoskeletal proteins[22]. Cyclin degradation is also the key step controlling exit from mitosis and progression into the next cell cycle. During the course of degradation, cyclin forms conjugates with ubiquitin in anaphase and is susceptible to proteolysis[23].All the hsps appear to have important 'chaperone' and protective roles[29] that would be required for many embryological events including neurogenesis. We propose that the function for hsp chaperones during mammalian development may dramatically affect the control of the cell cycle checkpoints and ultimately cell fate of the neuroectoderm.

The heat shock response also involves transcriptional activation mediated by the HSF. In unstressed and early embryonic cells HSF is present in both the cytoplasm and nucleus in a monomeric form that as no DNA binding activity. At critical stages of development and in response to heat shock and other stresses the HSF forms a trimer and accumulates in the nucleus. When the HSF trimer becomes activated it binds to a specific 5' consensus sequence of DNA [GAAnnTTCnnGAA] known as the heat shock element (HSE). The hsp70 members may also be important autoregulators of their own negatively regulated hsp70 gene[24]. The hsp71 may respond under stress to the increased damaged, misfolded and abnormal proteins in the cell that affect HSF binding and activation by phosphorylation of the hsp70 complex. The recent cloning and identification of a family of three HSF genes also offers an explanation for heat shock gene expression and induction during specific stages of development and differentiation. The activation of the HSF1 gene by different chemical and physical stresses and the developmental expression of the HSF2 gene responding to specific developmental stresses could explain the modulation of the heat shock response in embryos[24].

METHODS AND RESULTS

Using a whole in vitro embryo culture 9.5 day neural plates from the uterus of rats were removed and cultured in serum for 48 hr (Fig. 1). The morphological induction of neuroectoderm in the developing neural plate and the formation of the mid-, hind- and forebrain following heat shock were examined. Our previous studies with rat embryos in culture have shown that a temperature of 43°C for 7.5 min causes severe brain abnormalities (Fig. 2d) resulting from death of specific progenitor cells in the anterior neuropore of the neural plate[1]. If embryos are exposed to 42°C for 10 min alone (Fig. 2b) or a thermotolerant heat shock of 42°C before an exposure to 43°C for 7.5 min (Fig. 2c) embryos do not develop craniofacial defects compared to controls (Fig. 2a). Acquired thermotolerance is associated with hsp gene expression and lasts for 6-7 hr corresponding to neuroectoderm cell cycle time.

RESULTS

Morphological Study

The single layer of the neuroectoderm cells differentiates into anterior forebrain and posterior mid and hindbrain during gastrulation. During this period the nuclei from the neuroectoderm migrate from the base membrane at Go/G1 to the ventricular surface where they undergo G2+M. The nuclei of the neuroectoderm accumulate on the basement membrane following heat shock and progress through S phase only after recovery from heat shock. Mild heat shock can also result in a phenomenon termed acquired thermotolerance or protection from known lethal agents. The molecular controls and regulating mechanisms causing thermotolerance are unknown. We have examined the association of the hsp gene expression with rapid cell cycle changes and the ability of the embryo to recover and return to the normal body plan following heat shock. Following heat shock the neuroectoderm of the neural plate undergoes rapid cell cycle change with regulation at the Go/G1 and G2 boundary. The function of hsp in these changes is not clear. All hsps are expressed and appear to be tightly regulated through the cell cycle.

CELL FLOW CYTOMETRY

Hsp Expression and Cell Death

Why do neuroectoderm cells in the anterior neuropore region die following severe heat shock? Neuroectoderm cells respond to mild heat shock with acquired thermotolerance which protects the cells from further heat shock. The fate of cells after heat shock depends on how they respond and the length and state of inhibition.

The DNA distribution patterns of neuroectoderm cells were analysed by cell flow cytometry after the three heat shock regimes and mRNA identified at Go/G1, early S phase, late S phase and G2+M phases. Progress of the cells in association with hsp expression through the cell cycle was studied. Thermotolerance (42°C) results in cells accumulating in G1 and a delay of cells entering and progressing through S phase to G2+M. This results in lengthening of S phase by 2-3 hr.

Neuroectoderm cells heat shocked during G2+M die within 3-4 hr. Cells exposed to heat shock in Go/G1 and mid to late S phase die within 12-15 hr. The types of cell death observed were apoptosis and mitotic damage. Apoptosis is a complex set of events including activation of cell surface movement and membrane disruption, alterations in cytoarchitecture, chromatin hypercondensation and DNA degradation. Heat shock appears to prevent the cells from completing mitosis irrespective of where the cells are in the cell cycle[7]. Cell death seen in embryos appears to depend on the time and length of inhibition. Apoptosis occurs after a prolonged state of cell arrest after a 43°C heat shock resulting in a specific 2-3 hr G1 block. After a mild heat shock of 42°C (thermotolerant) a short cell cycle delay of 2-3 hr in G1 is observed and cell death occurs after 12-15 hr; neuroectoderm cell replacement may occur explaining observed thermotolerance. It appears that once cells are committed to a cell cycle (start) an apoptotic program is initiated and cancelled at the time of successful exit from mitosis[26]. This suggests that one determinant of cell death could be the time from cell cycle arrest to the induction of the apoptotic program. This may be influenced by the heat shock response and result in cell cycle delay.

Another interesting question is the ability of the cells to undergo apoptosis. Although suicide genes such as the ced and Bcl2 genes have been reported similar events at neurulation in mammals has not. Apoptosis may be a safety mechanism during development to insure cell death if progression through a cell cycle is inhibited. The regulated expression of hsps during the cell cycle at various phases could play an important role in the fate and recovery of neuroectoderm cells during early mammalian embryo development. Using flow cytometry and a cell cycle inhibitor (ICRF 159) to block the cycle at mitosis progression and changes to the cell cycle phases were determined (Fig. 3). The thermoprotective 42°C regime delayed cells at the G1 boundary and delayed entry to S phase. Cell progression was slowed through S. A synchronized wave of cells from Go/G1 entered S phase and progressed slowly through the cell cycle. A teratogenic heat shock of 43°C blocked the cell cycle at G1 for 2-3hr but allowed cells to complete mitosis. This resulted in S phase depletion[7]. Thermotolerance resulted in elevated expression of all hsps in all phases of the cell cycle.

Neuroectoderm cells are most susceptible to a lethal heat shock during late S phase and G2+M[7]. If the neural plate is exposed to heat shock during other stages of the cell cycle a dissociation of orderly cycle progression and aberrant mitotic events (mal motisis) occur. This results in aberrant segregation of chromosomes into daughter cells. These cells are unable to progress through a subsequent cell cycle because of incomplete chromosome complements and die. A sensing mechanism termed a mitotic checkpoint in yeast prevents or delays mitosis[27] if chromosomes are broken, if DNA synthesis is not completed, or if spindles are incorrectly formed. In mammalian neuroectodermal cells similar mechanisms of mitotic checkpoints have not been identified. Differentiating neuroectoderm cells delayed at the G1 and/or G2 boundary appear to survive. Differences in the susceptibility to cell death in the anterior neuroectoderm compared with that of the posterior region could be due to differences in mitotic checkpoint mechanisms. These mitotic checkpoints would have important survival benefits to the developing embryo and allow the cells to withstand short term stress.

Hsp Expresion and Cell Death

Neuroectoderm cells display a longer thermoprotection when heat shocked at 42°C in G1 and early S phase compared with the other stages of the cell cycle. The hsp71 normally expressed only during G2+M is expressed throughout all cell cycle phases. In thermoprotected cells the hsp kinetics delays the cell cycle by slowing cell progression at both G1/S and the S/G2+M interface thus prolonging S phase. Under stress delay of the cell cycle delayed may prevent the initiation of DNA synthesis at the sensitive stages of the cycle.

The Regulation of Hsp Expression in the Cell Cycle

The heat shock response influences the survival of differentiating neuroectoderm

TABLE 1. Heat shock proteins in early mammalian embryos.

hsp	location 37°	location 42°C	function
hsp110	nucleosomes	nucleosomes	
hsp100	lysosome	lysosome	
hsp90	cytoplasm	cytoplasm	bind to actin filaments bind to steroid receptors
grp78	lysosome	lysosome	attach to nascent proteins glucose deprivation response
grp75	mitochondria	mitochondria	
hsc73	cytoplasm ATP/activation	cytoplasm nuclear	uncoating of clathrin cages protein degredation/transport
hsp71	cell cycle(G2+M)	lysosome nuclear	bind to new protein from ribosomes bind to HSF/DNA binding
hsp60	mitochondria	mitochondria	intermembrane protein transport
hsp47	Endoplasmic reticulum	Endoplasmic reticulum	collogen transformation, mRNA splicing, cell division
hsp27	cytoplasm golgi bodies	nuclear cytoplasm	DNA regulation
ubiquitin	cytoplasm	cytoplasm	protein degradation

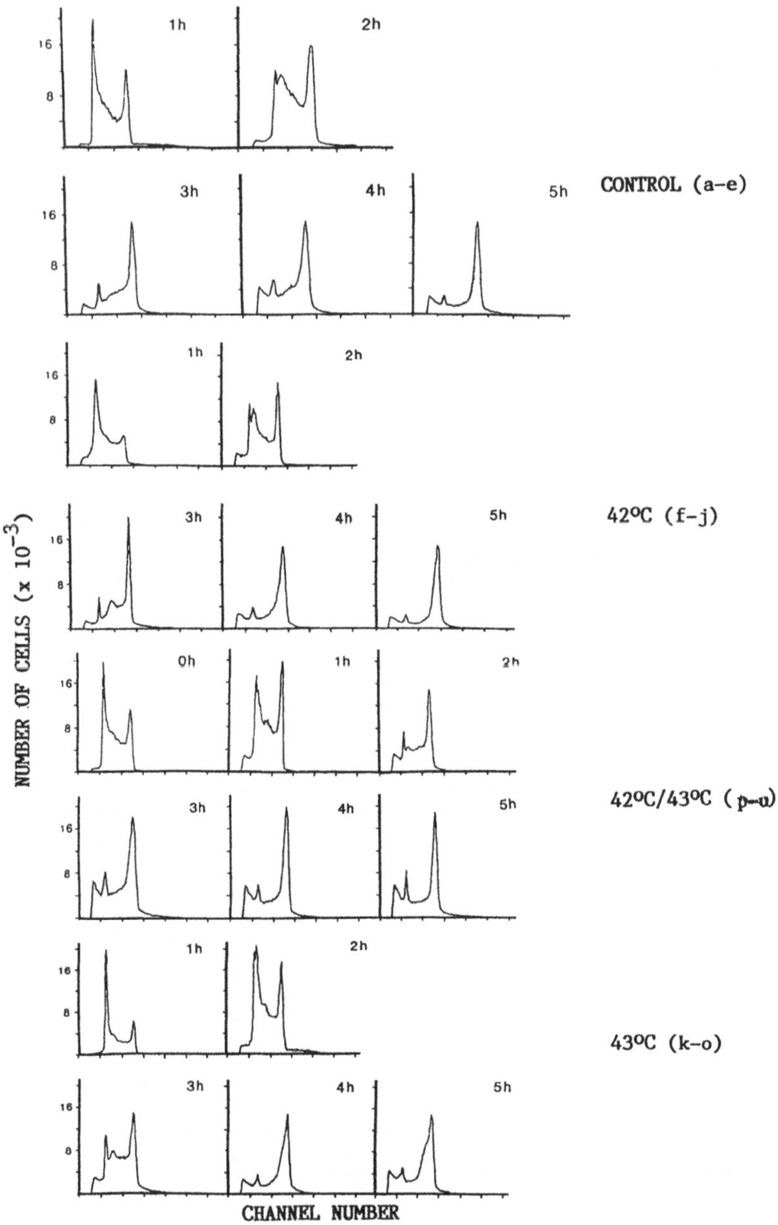

Fig. 3. DNA histrograms at times after heat shock and ICRF 159. Controls, a-e; mild heat shock, f-i; teratogenic heat shock, j-m. Percentages of cells in the 2C peak in b, f, and j estimated at 11.6%, 23.4% , and 25.8%, respectively, and illustrate relative rates of exit from the 2C peak into S phase at 1 h after the treatments. Induced-thermotolerance heat shock with 1-h interval and ICRF 159 given immediately after the mild component: n-q, showing cycling after the mild component; induced-thermotolerance heat shock with 1-h interval and ICRF 159 given immediately after the teratogenic component; The numbers in each column are hours after the first heat shock and ICRF 159. Arrows in g and h point to the synchronized peak due to the mild component; black arrows in k and l point to the synchronized peak due to the teratogenic component. Peak heights are plotted approximately relative to the proportions in the 2C peak. CV ranged from 2.2% in k to 5.6% in r.

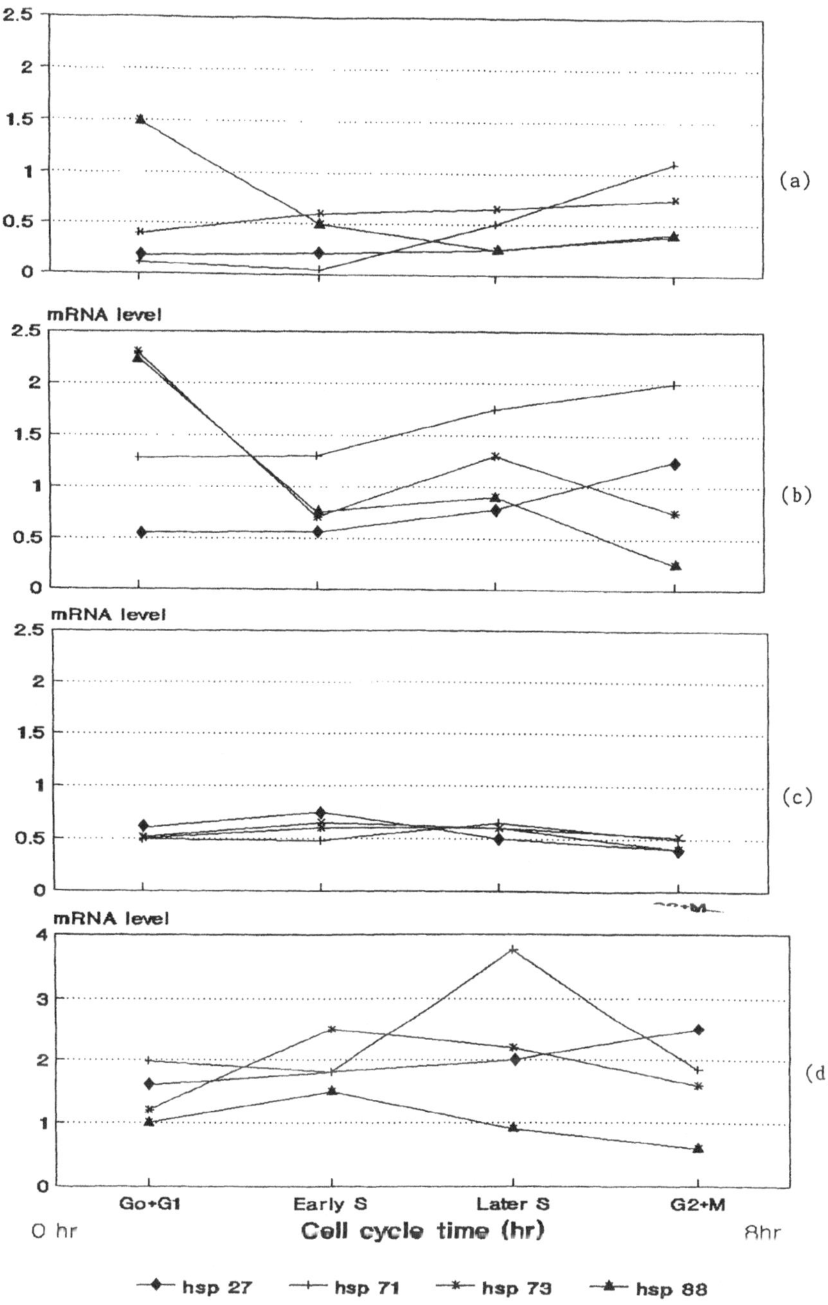

12

Fig. 4. Regulation of hsp in the cell cycle. hsp27, 71, 73, and 88 expression was determined at Go + G1, early S, late S, and G2 + M phases of the cell cycle after exposure to heat shock; after 1 h, cells were sorted into the four phases. **a:** control, **b:** 42°C, **c:** 42/43°C, **d:** 43°C for 1 h. Note: the high expression levels of hsp88 in Go and hsp71 expression in G2 + M of controls. Following a heat shock of 42°C, a fourfold increase of all hsp was observed. A heat shock of 42°C/43°C resulted in down-regulation of all mRNA hsps when compared with a 43°C heat shock. Exposure to a lethal 43°C caused a high overexpression of all hsp, especially hsp71 in late S phase.

cells. The hsp might maintain through their chaperone function a critical rate-limiting step or checkpoint in the cell cycle affecting acquired thermotolerance. The stability of many of the cyclins (ABD and E) and their degradation at specific phases is critical[28]. Cyclin B synthesis, its activation and breakdown by phosporylation are required to allow cells to progress through late S phase and complete G2+M.

The kinetics and expression of the hsp genes were examined in neuroectodermal cells by flow cytometry and Northern analysis. The levels of hsp mRNA 27, 71, 73 and 88 were identified (Fig. 4) following exposure at 42°C (non-lethal), 43°C (lethal) and 42°/43°C (thermotolerant) heat shock in the cell cycle phases. Hsp 88 expression was observed at Go in controls and hsp71 at G2+M of the normal cell cycle. Cells exposed to a thermotolerant heat shock of 42°C induced hsp71 mRNA in all phases of the cell cycle with the mRNA levels of hsp27, 73 and 88 also increased but remained relatively constant. Following a lethal heat shock dramatic changes were seen especially with enhanced hsp71 induction in late S phase, our results indicated that the transcript was not translated.

CONCLUSION

Elevated temperatures at critical stages of development in the neural plate interrupts early brain and head formation causing major craniofacial defects. Exposure of the pre-somite neural plate to heat results in death in the precursor cells of the developing forebrain and eye situated around the anterior neuropore. In response to heat shock a highly conserved set of heat shock genes are activated and heat shock proteins (hsp) are induced. Hsp gene expression is tightly regulated during the cell cycle and following heat shock rapid changes occur in the cell cycle. The heat shock response affects cell survival. Hsp induction can result in acquired thermotolerance with protection of the cells from a subsequent lethal heat shock. A delay at both G1 and G2 stages of the cell cycle accompanies thermotolerance. A lethal heat shock blocks cell progression at the G1 boundary and cells die within 3-5 hrs. The role of the hsp in neural plate formation and the mechanisms of thermotolerance may be critical for normal cellular differentiation in the developing nervous system. We conclude that:

1. The heat shock response and hsp gene expression affects the ability of neuroectoderm cells to survive.
2. Following heat shock neuroectoderm cells recover from both protein inhibition and DNA damage. These two factors regulate the recovery of the cell via the cell cycle.
3. Hsp gene expression is associated with regulation at either the G1 and/or G2 boundary of the cell cycle.
4. The hsp 'chaperone' function may affect cell cycle proteins and influence the feedback control mechanisms of the cell cycle.

ACKNOWLEDGEMENTS: Supported by an Australian NH&MRC 1992 grant.

REFERENCES

1. D.A. Walsh, K. Li, C. Crowther, D. Marsh and M.J. Edwards, Thermotolerance and heat shock response during early development of the mammalian embryo, in: "Heat Shock Development", L. Hightower and L. Nover, eds., Springer-Verlag 17:58-69 (1991).

2. D.A. Walsh, N.W. Klein, L.E. Hightower and M.J. Edwards, Heat shock and thermotolerance during early rat embryo development, *Teratology* 36:181-191 (1987).
3. D.A. Walsh, K. Li, J. Speirs, C.E. Crowther and M.J. Edwards, Regulation of the inducible heat shock 71 genes in early neural development of cultured rat embryos, *Teratology* 40:321-334 (1989).
4. N.S. Peterson and H.K. Mitchell, Environmentally induced development defects in Drosophila, in: L. Hightower and L. Nover, eds., "Heat Shock and Development" Springer-Verlag 17:29-43 (1991).
5. D.A.T. New, P.T. Coppola and S. Terry, Culture of explanted embryos in rotating tubes, *J Reprod Fertil* 35:135-138 (1973).
6. D.A. Walsh, L.E. Hightower, N.W. Klein and M.J. Edwards, The induction of the heat shock proteins during early mammalian development, Heat shock, Cold Spring Harbor Symposium 2.92 (1985).
7. D.A. Walsh, K. Li, J. Wass, A. Dolnikov, Z. Zeng Li and M.J. Edwards, Heat shock gene expression and cell cycle changes during mammalian development, *Dev Genetics*, 14 2 127-136 (1993).
8. D.A. Walsh and V.B. Morris, Heat shock affects cell cycling, *Teratology* 40:583-592 (1989).
9. A. Simeone, D. Acampora, M. Gulisano, J. Stornaiuolo and E. Boncinelli, Nested expression domains of four homeobox genes in developing rostral brain, *Nature* 358:687-690 (1992).
10. L.E. Hightower and L. Nover, eds., Heat Shock: Results and problems Cell Differentiation. Springer-Verlag, Heidelberg (1991).
11. K.L. Milarski and R.I. Morimoto, Expression of human hsp70 during the synthetic phase of the cell cycle PNAS. 83: 9517-521 (1986).
12. K.L. Milarski, W.J. Welch and R.I. Morimoto, Cell cycle-dependent association of hsp70 with specific cellular proteins, *J Cell Biol* 108:413-423 (1989).
13. R.I. Morimoto, D. Mosser, T.K. McClanahan, N.G. Theodorakis and R. Williams in Stress-induced Proteins, Morimoto, Tissieres and Georgopoulos, eds., Alan R Liss Inc pp 83-94 (1989).
14. B.J. Wu, R.E. Kingston and R.I. Morimoto, Human hsp 70 promoter contains at least two distinct regulatory domains, *PNAS* 83, 629-633 (1986).
15. C. Hueng-Sik, B. Li, Z. Lin, E. Huang and A. Liu, cAMP dependent Protein Kinase Regulate the human heat shock protein 70 gene promoter activity, *J Biol Chem* 266 18 11858-11865, (1991).
16. S. Lindquist and E.A. Craig, The heat-shock proteins, *Annu Rev Genet* 22:631-677 (1988).
17. E. Hickey, S.E. Brandon, G. Smale, D. Lloyd and L.A. Weber, Sequence and regulation of a gene encoding a human 89 kD hsp, *Mol Cell Biol* 9, 2615-26 (1989).
18. P.E. Mirkes and B. Doggett, Accumulation of heat shock 72 (hsp72) in postimplantation rat embryos after exposure to vaariousperiods of hyperthermia in vitro, *Teratology* 46. 3: 301-309 (1992).
19. Z.F. Zakeri, W.J. Welch, D.J. Wolgemuth, Characterization and inducibility of the hsp70 proteins in the male mouse germ line, *J Cell Biol* 111:1785-1792 (1990).
20. Z. Li, F. Zeng, K. Nagata and D. Walsh , Hsp47 expression during early mammalian development: Shock Proteins UOEH Japan 12. 93 (1992).
21. K. Nagata, K. Hirayoshi, M. Obara, S. Saga and K. Yamada, Biosynthesis of a novel transformation-sensitive heat shock protein that binds to collagen, *J Biol Chem* 263 17 8344-8349 (1988).
22. M. Gernold, U. Knauf, M. Gaestel, J. Stahl and P.M. Kloetzel, Tissue-Specific Distribution of Mouse Small Heat Shock Protein hsp25, *Dev Gen* 2.14:103-111 (1993).
23. M. Glotzer, A.W. Murry and M.W. Kirschner, Cyclin is degraded by the ubiquitin pathway, *Nature* 349 132 138 (1991).
24. R. Morimoto, Cells in Stress: Transcriptional Activation of Heat Shock Genes, *Science* 259 1409-1410 (1993).
25. A. MacAuley, Z. Werb and P. Mirkes, Characterization of the unusually rapid cell cycles during rat gastrulation, *Devel* 117:873-883 (1993).
26. A.L. Kung, S.W. Sherwood and R.T. Schimke, Cell line specific differences in the control of cell cycle progression in the absence of mitosis, *PNAS* 87, 9553-9557 (1990).
27. L.H. Hartwell and T. Weinert, Checkpoints: controls that ensure the order of cell cycle events, *Science* 246 629-634 (1989).
28. T. Hunt, Under arrest in the cell cycle, *Nature* 342, 483-484 (1989).
29. H. Pelham, Functions of the hsp70 protein family: An overview. In Stress Proteins in Biology and Medicine, R. Morimoto, A. Tissieres and Georgopoulos eds., Cold Spring Harbor 287-299 (1989).

33

EXPRESSION OF G1 CYCLINS DURING EARLY DEVELOPMENT OF ZEBRAFISH EMBRYOS

Anat Yarden, Zvi Kam and Benjamin Geiger

Department of Chemical Immunology, The Weizmann Institute of Science, Rehovot 76100, Israel

INTRODUCTION

The generation of multicellular organisms requires precise spatio-temporal regulation of cell divisions. The proliferation of cells within the embryo is well coordinated with morphogenesis and differentiation, and the respective regulatory mechanisms vary in different developmental stages. Progress through all phases of the cell cycle, in invertebrates as well as vertebrates, is extensively modified during early embryonic development. For example, in many species such as *Xenopus laevis* and *Drosophila melanogaster*, the early cleavages immediately following fertilization are very rapid and consist of alternating S and M phases of the cell cycle, with no detectable G1 or G2 [1,2,3]. At the midblastula transition, the cell cycle slows dramatically [2,4]. In *Drosophila*, the three cell cycles following the midblastula transition are composed of S, G2 and M phases in which the length of G2 is regulated by the product of the string gene (*cdc25*) [5]. During subsequent development in *Drosophila*, in addition to S, G2 and M, cells undergo G1, a phase in which much of the succeeding growth regulation appears to occur [6,7].

The study of cell cycle regulation during vertebrate development has been limited. We are studying this process in the zebrafish (*Brachydanio rerio*) embryo which has recently become an attractive subject for developmental studies, due to its extracorporal development, amenability to experimental manipulation and transparency [8]. During early cleavages, zebrafish blastomeres divide rapidly and synchronously [9]. At the tenth cleavage, the cell cycle lengthens and three spatially separate mitotic domains arise with distinctive cell cycle lengths and rhythm [10]. Two of these domains are extra embryonic; The yolk syncytial layer (YSL) nuclei were shown to divide most rapidly, and the enveloping epithelial layer (EVL) most slowly. The third domain which forms the entire embryo, consist of an inner mass of deep spherical cells, which divide with similar rhythms through cycle 16 [10].

We are studying the molecular mechanisms participating in the regulation of the transitions in cell cycle lengths during early zebrafish embryogenesis. Our approach has been to isolate and analyze the zebrafish homologues of cyclin molecules which have been proven to participate in regulation of cell cycle checkpoints in other organisms [11]. Here we report on studies of the pattern of expression of G1-cyclins during early zebrafish development.

The Cell Cycle: Regulators, Targets, and Clinical Applications
Edited by V.W. Hu, Plenum Press, New York, 1994

RESULTS

Cloning and Characterization of a Zebrafish Cyclin E Genomic Fragment

As a first step in the isolation of zebrafish genes which participate in cell cycle regulation during early development, we chose to isolate the cyclin E gene. Degenerate cyclin E specific oligonucleotides, which were designed on the basis of conserved regions of the human and the *Drosophila* gene products (S.I. Reed, personal communication, see Fig. 1) were used in a polymerase chain reaction with zebrafish genomic DNA as a template. A 320 bp zebrafish genomic fragment which was isolated contained a region with a deduced amino acids sequence similar to a previously described cyclin E gene product [12, 13], and a 146 bp putative intron (Fig. 1). The 5' and 3' border sequences of the putative intron agree with the eukaryote splice sequence consensus [14]. The presumed coding sequences correspond to a predicted polypeptide showing 85% homology to the human cyclin E (Fig. 2). Hence, this fragment appears to be part of the zebrafish cyclin E gene.

```
  1   CATCGGGAGACGTTTTATTTGGGTCAGGACTACTTTGATCGCTTCATGGCC
      H  R  E  T  F  Y  L  G  Q  D  Y  F  D  R  F  M  A

 52   ACACAAGAAAATGTCCTGAAAACAACGCTACAGCTTATAGGCATCTCCTGT
      T  Q  E  N  V  L  K  T  T  L  Q  L  I  G  I  S  C

103   CTCTTTATAGCTGCCAAAATGGAAGTAAGTTTTGCTTTCACATGCCCTTGCA
      L  F  I  A  A  K  M  E

154   CTTCAACCTGAACTTTTGATAAGGGCTGTAATAGTGAGTGAATGAGTGTTGT

204   TCTTGTTGTAGGGAAAATGCAGGATAATTTAAAGCTGTTTGTTTTTTTTTTT

257   TTTTTTTTTTAAGGAAATTTACCCTCCTAAGGTGCATCAGTTTGTTTATGTC
                     E  I  Y  P  P  K  V  H  Q  F  V  Y  V

309   ACCGACGGGGCCTGC
      T  D  G  A  C
```

Fig. 1. Sequence of the zebrafish cyclin E PCR product. The nucleotide sequence (top) and the predicted amino acid sequence (below) of the zebrafish cyclin E genomic fragment. The consensus splicing sequences [14] which border the putative intron, are underlined. The conserved amino-acids used for designing the oilgonucleotides (kindly provided by S.I. Reed) for the PCR reaction appear in bold type. The nucleotide sequence, in the conserved region which was used for the PCR reaction, was subsequently verified by partial sequence analysis of a zebrafish cyclin E cDNA (not shown).

```
BR        H R E T F Y L G Q D Y F D R F M A T Q E N V L K T T L Q L I
HS (147)  * * * * * * * A * * F * * * Y * * * * * * * V * * L * * * *

BR        G I S C L F I A A K M E E I Y P P K V H Q F V Y V T D G A C
HS (177)  * * * S * * * * * * L * * * * * * * L * * * A * * * * * * *
```

Fig. 2. Sequence alignment of the zebrafish (BR) and the human (HS) cyclin E genes at the cyclin box region. Amino acids identical to the zebrafish cyclin E are marked with an asterisk (*).

When used as a probe, the zebrafish cyclin E clone hybridized to a single band in Southern blot analysis of total zebrafish genomic DNA (Fig. 3). This suggests that the cyclin E is a single copy gene in zebrafish. A single 2.2 Kb RNA was detected in zebrafish total RNA analyzed by Northern analysis (Fig. 4). This mRNA is similar in size to the 2-2.5 Kb cyclin E mRNA which has been observed in HeLa cells [11]. The developmental expression of cyclin E in zebrafish was determined by Northern blot analysis. Cyclin E mRNA appeared in early embryos before the onset of transcription (2 hours post fertilization). The steady state levels of cyclin E mRNA were constant up to the end of epiboly (9 hours post fertilization) and decreased below the levels of detection thereafter (Fig. 4). The same size cyclin E mRNA was detected in total RNA from adult zebrafish. However, in two tissues which are expected to be fully differentiated and therefor not dividing (namely muscle and brain), the cyclin E mRNA was below the levels of detection (Fig. 4).

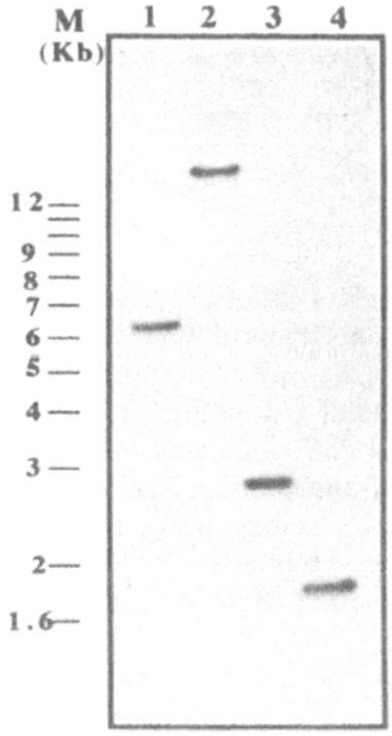

Fig. 3. Southern blot analysis of zebrafish genomic DNA: Ten micrograms of total zebrafish genomic DNA were digested with BamHI (lane 1), EcoRI (lane 2), HindIII (lane 3) or PstI (lane 4), blotted and probed with the zebrafish cyclin E genomic fragment. Molecular weight markers were run alongside the gel and are marked on the left.

Cyclin D Expression in Zebrafish Embryos

Encouraged by the high conservation of the zebrafish and human cyclin E genes, we tested whether antibodies, raised against other G1 cyclins, will react with the zebrafish homologue. We have tested anti human cyclin D1 polyclonal antibodies from two different sources, kindly provided to us by S.I. Reed (Ab 49) and G. Draetta (affinity purified anti-human cyclin D1 [15]). Protein extracts from normal human fibroblasts and from zebrafish embryos were analyzed on duplicated western blots and reacted with the cyclin D1 specific antibodies. Both antibodies reacted with the expected 36 Kd band in the human cell extract, and with a slower migrating protein (of approximately 40 Kd) in the zebrafish extract (Fig. 5). The affinity purified anti-cyclin D1 recognized additional proteins, one of the same size as the human cyclin D1 and a much slower migrating protein (not well resolved) which

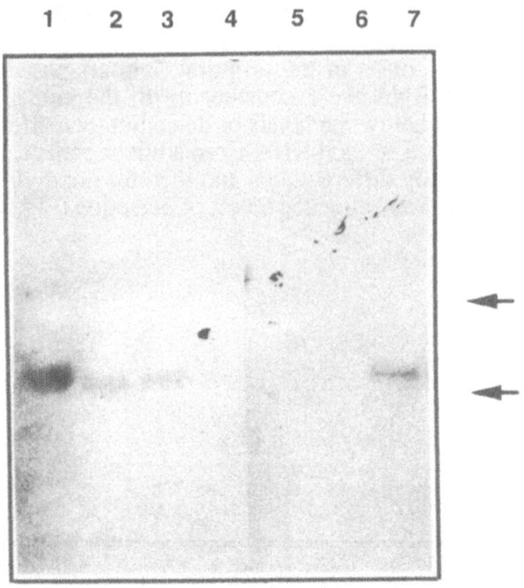

Fig. 4. Northern blot analysis of total RNA from zebrafish embryos and adults. Ten micrograms of total RNA prepared at various developmental stages of zebrafish embryos were loaded in each lane: 2 hours post-fertilization (1), 5 hours post-fertilization (2), 9 hours post fertilization (3) and 25 hours post-fertilization (4). Ten micrograms of total RNA from adult fish brain (5), muscle (6) and a whole fish (7) were analyzed in parallel. Arrows indicate the 18S and 28S ribosomal RNA migration points. The blot was probed with a hexamer labeled 240 bp DraI-BamHI fragment of the zebrafish cyclin E, which corresponds to the 5' portion of the clone (eliminating the stretch of T's in the intron). The amounts of RNA in each lane were evaluated by staining the blot with 0.04% Methylene-blue in 0.5 M NaAcetate. Lane 1 was found to contain more RNA then all other lanes which had equal amounts of RNA (not shown).

corresponds to size of the yolk proteins visualized by the Ponceau staining of the blot. The fact that the presumptive zebrafish cyclin D appears in 32-64 cell stage embryo before the onset of transcription (Fig. 5), suggests that it is translated from maternal mRNA. The presumptive zebrafish cyclin D was present at high levels at least up to the end of epiboly (data not shown).

To study the pattern of expression of the presumptive zebrafish cyclin D, we have used the affinity purified anti-human cyclin D1 antibodies for indirect immunofluorescent staining of fixed zebrafish embryos at various developmental stages. The embryos were also stained with the DNA specific dye, DAPI (4', 6'-Diamidino-2-Phenylindole), which enables visualization of all the nuclei in the embryo, and double labeled fluorescent images were taken. The cyclin D reactive proteins were found in nuclei of embryos from 30-50% epiboly, in both the deep cell layer and the YSL. Fig. 6-B shows the YSL layer of an embryo at 50% epiboly in which the immunofluorescent signal of the presumptive cyclin D was localized predominantly in nuclei. However, not all the nuclei in the YSL stained positively with anti cyclin D1 antibodies, as can be visualized by comparing the immunolabeling to the DAPI staining in the same field (Fig. 6-A). We have recently developed a computerized microscopic method to quantitate, in three dimensions, the cellular DNA content in DAPI stained samples (Yarden, Geiger & Kam, unpublished). These analysis provides the integrated DAPI fluorescence which was shown to be proportional to the total DNA content of the nucleus [16]. Such analysis indicated that the YSL nuclei, which stained positive with anti cyclin D1 antibodies, contain half the amount of DAPI fluorescence compared to nuclei which were not labeled. Therefore, we conclude that the nuclei which contain the presumptive cyclin D, in the YSL layer, are at the G1 phase of the cell cycle.

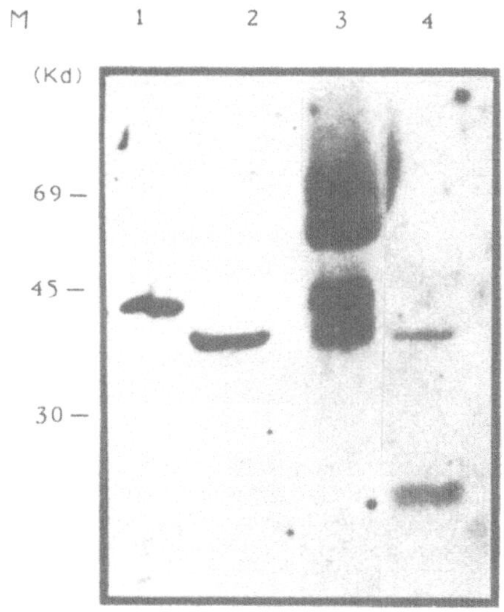

Fig. 5. Western blot analysis of zebrafish embryos and human fibroblasts. Two hundred micrograms of total 32-64 cell stage zebrafish embryo protein extracts (lanes 1, 3), or 40 micrograms of total primary human fibroblast protein extracts (lanes 2, 4), were loaded and run on 11% polyacryl-amide gel, blotted and reacted with anti-human cyclin D1 49 (from S.I. Reed; lanes 1, 2) or with affinity purified anti-human cyclin D1 (from G. Draetta; lanes 3, 4).

DISCUSSION

Two members of the G1 cyclin family, cyclin E and a presumptive cyclin D, were found to be expressed in early zebrafish embryos before the onset of transcription and the appearance of G1 phase. The level of expression of both of them was high throughout the cleavage stage and at least up to the end of epiboly. The presumptive cyclin D was found to be expressed exclusively in nuclei at G1 phase of the cell cycle, in the YSL layer of zebrafish embryos at 50 % epiboly.

Zebrafish cyclin E genomic fragment was cloned and the pattern of expression of its mRNA was analyzed in various developmental stages. Cyclin E mRNA was found to be expressed during the cleavage period of zebrafish embryos (before transcription occurs), suggesting that cyclin E mRNA levels do not fluctuate throughout the blastula stage, in contrast to the human cyclin E mRNA which fluctuates periodically through the cell cycle [12]. The presence of cyclin E mRNA at this early developmental stage is surprising, since there is no G1 phase during the early cleavages period [1]. It is possible, that cyclin E is present early in zebrafish development, but will become functional, (for example through possible association with additional proteins), in a subsequent stage. Cyclin E may be functionally active in the initial G1/S transition which occurs in the embryo, when the first G1 appears. Alternatively, it is possible that in addition to the role of cyclin E in G1 to S transition (as has been suggested in mammalian cells) cyclin E is functioning in another phase of the cell cycle during early embryogenesis. Experiments which will eliminate the cyclin E protein from zebrafish embryos during the blastula and subsequent stages, may assist in clarifying this point.

Two distinct polyclonal antibodies against the human cyclin D1 reacted with a 40 Kd protein in zebrafish protein extracts. The affinity purified anti-human cyclin D1 were reported to also recognize human cyclin D2 [15]. It is possible that the additional 36 Kd protein which reacted with those antibodies in zebrafish extracts is another type of cyclin D of zebrafish. Our initial results with the human cyclin D1 specific antibodies in zebrafish

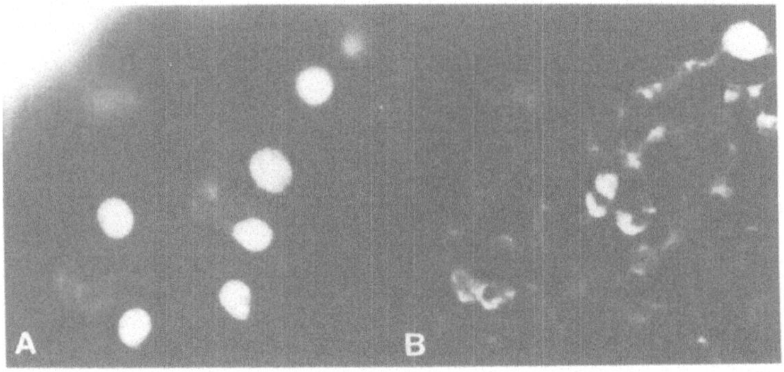

Fig. 6. Immunolocalization of the presumptive zebrafish cyclin D in the YSL layer. Zebrafish embryos at 50% epibloy were fixed and stained with DAPI (A) and with affinity purified anti human cyclin D1 antibodies followed by goat-anti-rabbit IgG conjugated to Rhodamine (B). Water immersion objective (X40, 0.75=N.A.) was used to visualize the whole embryo on a Zeiss Axiomat microscope with a CCD camera.

will be further established once the zebrafish homologue will be cloned and zebrafish specific antibodies prepared.

The specific localization of the presumptive zebrafish cyclin D in nuclei which were weakly stained with DAPI, and may correspond to the G1 phase of the cell cycle, is of interest especially in the YSL of the zebrafish embryo. The YSL is a multinucleated syncytium [17] in which all nuclei reside in the same cytoplasm. Our observation that the presumptive cyclin D is localized only in G1 nuclei, while other nuclei in the same syncytium remained negative, implies that the regulatory mechanism which governs cyclin D nuclear localization at G1 phase are present in the nuclei themselves.

ACKNOWLEDGMENTS

We thank Steven I. Reed for providing cyclin E oligonucleotides, human cyclin D1 antibodies and for communicating results prior to publication. We would also like to thank Giolio Draetta for human cyclin D1 antibodies. This study was supported in part by grants from the Minerva Foundation and the Council For Tobacco Research (B.G.).

REFERENCES

1. Graham, C.F. & Morgan, R.W. *Dev. Biol.* **14**, 439-460 (1966).
2. Newport, J. & Kirschner, M. *Cell* **30**, 675-686 (1982).
3. Edgar, B.A., Kiehle, C.P. & Schubiger, G. *Cell* **44**, 365-372 (1986).
4. Foe, V.E. *Development* **107**, 1-25 (1989).
5. Edgar, B.A. & O'Farrell, P.H. *Cell* **57**, 177-187 (1989).
6. O'Farrell, P.H., Edgar, B.A., Lakich, D. & Lehner, C.F. *Science* **246**, 635-640 (1989).
7. Glover, D.M. *Trends Genetics.* **7**, 125-132 (1991).
8. Kimmel, C.B. *Trends. Genet.* **5**, 283-288 (1989).

9. Kimmel, C.B. & Law, R.D. *Develop. Biol.* **108**, 78-85 (1985).
10. Kane, D.A., Warga, R.M. & Kimmel, C.B. *Nature* **360**, 735-737 (1992).
11. Lew, D.J. & Reed, S.I. *Trends Cell Biol.* **2**,77-81 (1992).
12. Lew, D.J., Dulic, V. & Reed, S.I. *Cell* **66**, 1197-1206 (1991).
13. Koff, A., Cross, F., Fisher, A., Schumacher, J., Leguellec, K. Phillippe, M. & Roberts, J.M. *Cell* **66**, 1217-1228 (1991).
14. Breathnach, R. & Chambon, P. *Ann. Rev. Biochem.* **50**, 349-383 (1981).
15. Baldin, V., Lukas, J., Marcote, M.J., Pagano, M. & Draetta, G. *Genes & Development* **7**, 812-821 (1993).
16. Arndt-Jovin, D.j. & Jovin, T.M. in *Fluorescence microscopy of living cells in culture, Part B, Methods in Cell Biology* (eds. Taylor, D.L.) 417-448 (Academic Press, Inc. Harcourt Brace Jovanovich, Publishers, 1989).
17. Kimmel, C.B. & Law, R.D. *Devlop. Biol.* **108**, 86-93 (1985).

34

APOPTOSIS: DEFINITION, ROLES AND REGULATION

L. E. Gerschenson, R. J. Rotello, R. Lieberman and C.-I Sze

Department of Pathology
University of Colorado School of Medicine
4200 E. Ninth Avenue, Box B216
Denver, CO 80262

CELL DEATH

The emergence of molecular biology, with the possibility of cloning genes and introducing them in cells where they are expressed, has challenged the classic concept of ***cell death***. But that concept is still valid when defined as "a state in which cells are totally incapable of any function ... ".[1] That concept has been based mostly on morphological observation.

NECROSIS

Diseases are always the result of cell injury. The consequent damage depends on the injury type, duration and intensity, and it can be repaired resulting in partial or total function restoration. However, the injury could result in death at the cellular or organ level, which could also be conducive to death of the whole organism. It is not surprising then, that cell death resulting from injury has been extensively characterized.[2] That type of pathological cell death is called ***necrosis***.

APOPTOSIS

For many years it has been known that there is a form of cell death which occurs not as the result of injury, but under physiological conditions such as deletion of cell populations in embryonic development and in growth modulation of adult tissues.[3] It is called ***apoptosis***.

The term ***programmed cell death*** has also been used, originally because of cell death presence in embryonic development[4] and later because of cell death inhibition by RNA and protein synthesis inhibitors, implying the need for gene expression and "de novo" protein synthesis in the mechanism of action of cell death.[5,6] The latter concept led some researchers to propose the existence of a "suicide program" in cells.[7] However, there is not yet definitive proof of a specific "apoptotic death program" in all cells and the above RNA and protein synthesis inhibitors have also been shown to induce apoptosis in many experimental systems.[3] It is possible that there may be genes that are "turned on or off" in response to effectors regulating cell death, as well as there are genes that are regulated by metabolic effectors such as glucose, insulin, etc. The term ***programmed cell death*** should be restricted to embryonic models, with the understanding that it implies only the existence of a developmental program, not necessarily a cell death mechanistic concept.[3,7]

The most important differences between necrosis and apoptosis are shown in Table 1.

Table 1. *General differences between apoptosis and necrosis*[a]

Characteristics	Apoptosis	Necrosis
Stimuli	Physiological	Pathological (injury)
Occurrence	Single cells	Groups of cells
Reversibility	No (after morphological changes)	Yes (up to the point of no return)
Adhesions between cells and to BM	Lost (early)	Lost (late)
Cytoplasmic organelles	Late stage swelling	Very early swelling
Lysosomal enzyme release	Absent	Present
Nucleus	Convolution of nuclear outline and breakdown (karyorrhexis)	Disappearance (karyolysis)
Nuclear chromatin	Compaction in uniformly dense masses	Clumping not sharply defined
DNA breakdown	Internucleosomal	Randomized
Cell	Formation of apoptotic bodies	Swelling and later disintegration
Phagocytosis by other cells	Present	Absent
Exudative inflammation	Absent	Present
Scar formation	Absent	Present

[a]This table summarizes in broad terms the differences between apoptosis and necrosis. There are only a few exceptions to the general concepts presented in it. BM = basement membrane.

Reprinted with permission from the FASEB Journal.

The known sequence of events in cells undergoing apoptosis is illustrated in Fig. 1.

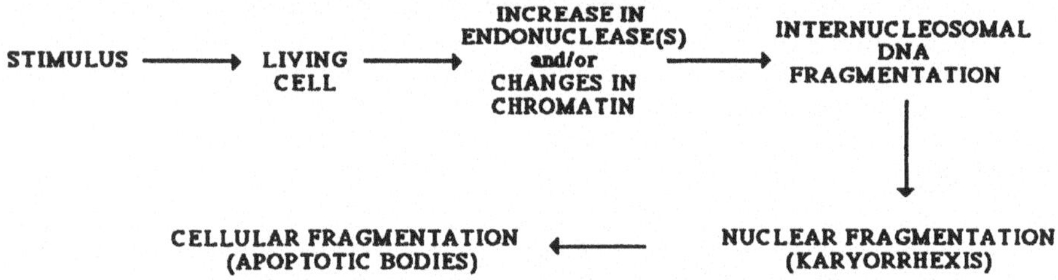

Figure 1. Sequence of events in cells undergoing apoptosis.
Reprinted with permission from the FASEB Journal.

ENDONUCLEASE ACTIVITY

Well defined events in the development of apoptosis are nuclear DNA double-strand cleavage at linker regions between nucleosomes, nuclear fragmentation and cell breakdown into membrane-bound apoptotic bodies. Our research on apoptosis induction by progesterone withdrawal in rabbit uterine epithelium has shown a parallel increase in endonuclease activity (Fig. 2), apoptotic morphologic detection and internucleosomal DNA fragmentation[3,8-10](Fig 3). The latter phenomenon has been attributed to an endonuclease activity that has been shown in lymphocytes to be activated by both Ca^{++} and Mg^{++}.

However, in other experimental systems, apoptosis appears not to be mediated by a Ca^{++}-requiring system[11, 12] (Fig. 3), and an acidic pH-dependent endonuclease activity also has been described, also related to induced apoptosis.[1,2] The existence of apoptosis in which the typical DNA ladder fragmentation may not be present has been suggested.[13,14]

The endonuclease-related DNA fragmentation is clearly a terminal event, assuring the cell demise. The events preceding it are most important, but still undefined, and may throw light on the mechanism(s) of apoptotic cell death.

GENE EXPRESSION

Research on gene expression related to apoptosis has been very active. The earlier studies were based on the hypothesis that a "genetic program" was involved in apoptosis.[5,6] The description of cell death-related genes in *C. elegans* supported the rationale behind those studies.[15-17]

Prostate gland cells undergo apoptosis after suppression of testosterone action.[18,19] Two genes appear to have increased expression in the prostate after castration; one of them encodes for a 29kd protein[20] while the other SGP-2 or TRPM-2 encodes for a sulfated glycoprotein.[21] Neither has yet been mechanistically related to apoptosis. Several proto-oncogenes have been found to be transiently expressed very early in prostatic induced apoptosis: c-fos, c-myc and hsp-70.[22] However, the same sequential gene expression paradoxically was found in prostate stimulated to proliferate.[23] One possible pitfall in these types of studies is the early increase in RNAse activity in prostate after castration,[24] which would make detection of novel mRNAs difficult.

Recent studies have shown a need for c-myc expression in induced apoptosis.[25-29] However, in myeloid leukemia cells only high deregulated c-myc levels were associated with apoptosis, while regulated c-myc levels were not. [30]

Bcl-2 gene expression appears in general to inhibit apoptosis[26-28] by unknown mechanism of action.

The tumor suppressor gene p53 has been found to play a role in the induction of apoptosis. Cells transfected with the wild type p53, as well as with a temperature-sensitive mutant p53, were found to undergo increased apoptosis.[31,32] DNA damaging agents which induce apoptosis result in accumulation of the p53 phosphoprotein in cell nuclei.[33] However, apoptosis induced by other agents appears not be regulated by p53. Adenoviruses appear to immortalize cells by mechanisms involving apoptosis inhibition. Those mechanisms involve regulation of p53 protein turnover by the two viral genes E1A and E1B.[34,35]

Changes in gene expression have not yet helped in the understanding of apoptosis. The original working hypothesis was that the genetic expression changes observed were causal. It is possible that they could really be effects of apoptosis and related to repair mechanisms.[36,37]

REGULATION

The concept that cell death can be regulated is not new. Numerous studies on necrosis have already shown sequential morphologic and biochemical changes, the existence of "a point of no return" and the involvement of Ca^{++} in its regulation.[2] Many ways of inducing apoptosis have been described[3] including radiation, heat, RNA and protein synthesis inhibitors, cancer chemotherapeutic compounds, removal of trophic hormones, etc. Biological response modifiers, such as Tumor Necrosis Factor (TNF) and Transforming Growth Factor-β1 (TGF-β1), have also been found to induce apoptosis in different experimental systems. TNF was found to induce either apoptosis or necrosis according to the target cell used.[38] TGF-β1 has interesting effects, since it is a growth promoter in

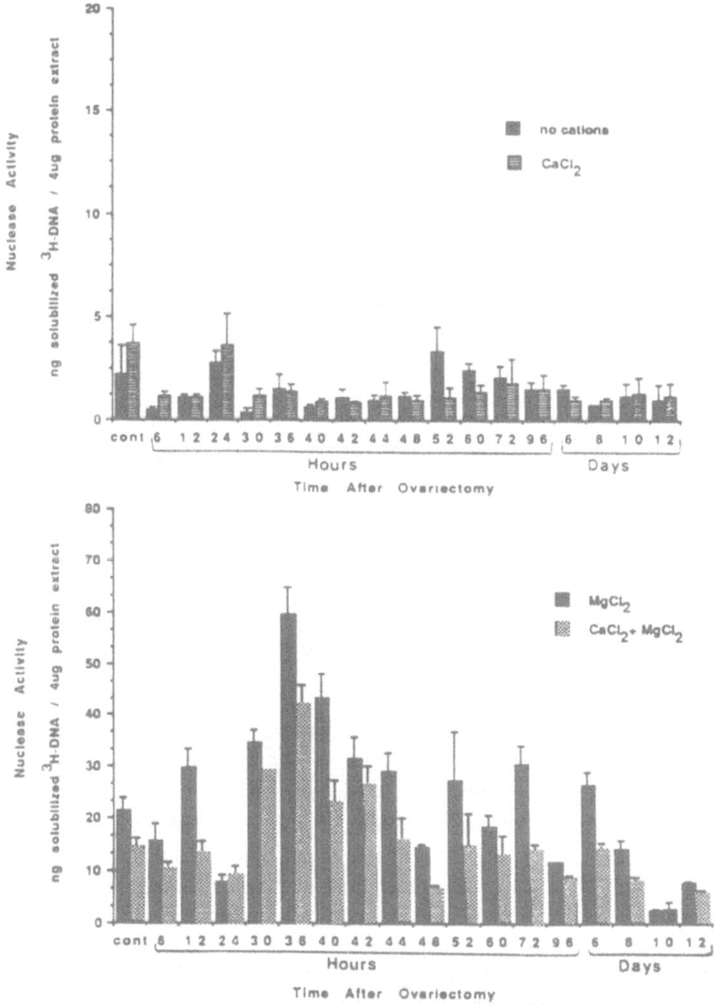

Figure 2. *Determination of endonuclease activity in protein nuclear extracts from rabbit endometrium in the absence or presence of Ca^{++} at selected times after ovariectomy employing an acid-solubilization assay. Each 40μl reaction contained: 25 mM HEPES, pH 7.5, 5 mM $CaCl_2$ or no cations, 100 μg/ml BSA, 1 mM dithiothreitol, 725 ng of Bov1 [^{3}H]DNA (4,100 cpm/μg) and 4 μg of nuclear protein extract. Reactions were incubated 40 min at 37oC and terminated by the addition of 100% (w/v) trichloroacetic acid to a final concentration of 10%. Samples were placed at 0oC for 15 min, precipitates were collected at 12,000 rpm for 10 min, and a 20 μl aliquot of each supernatant was combined with 7.5 ml of aquasol, and counted using a liquid scintillation counter.*

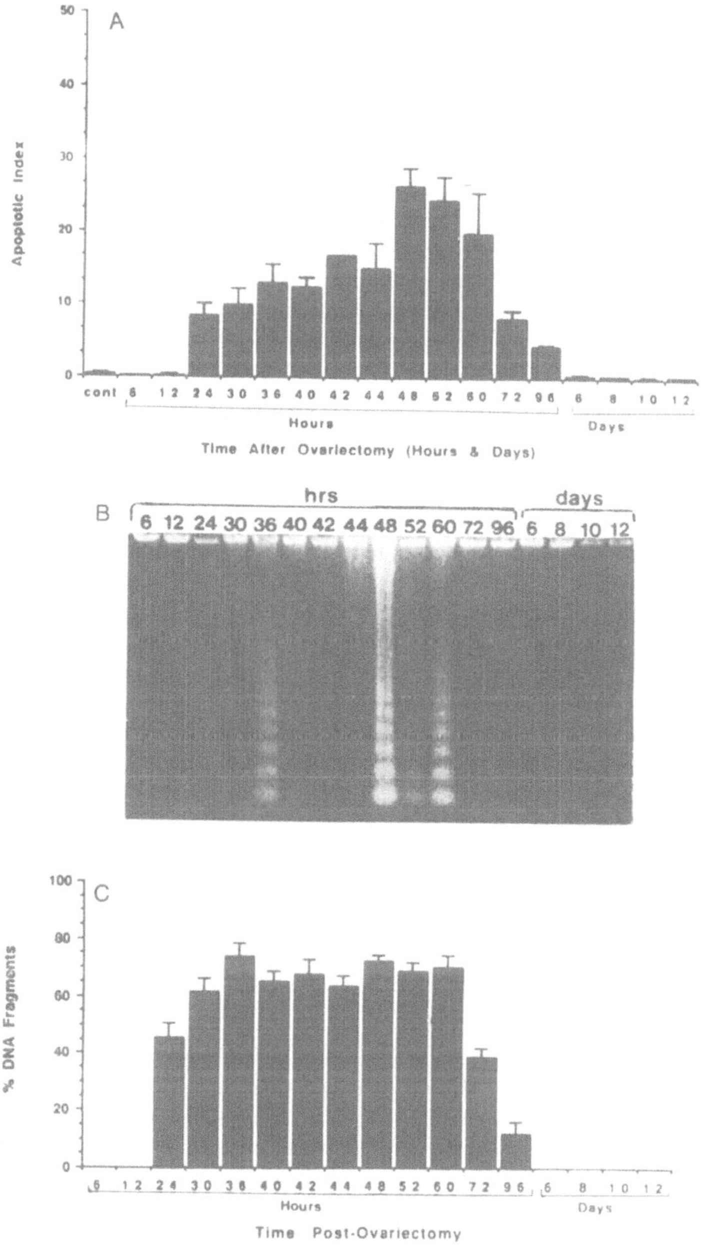

Figure 3. *Morphologic and biochemical quantitation of apoptosis in uterine epithelium at selected times post-ovariectomy.* **A:** *Uterine epithelial apoptosis was quantitated in hematoxylin and eosin stained uterine sections at selected times post-ovariectomy, mean ± SEM, n = 3.* **B:** *Twenty micrograms is isolated and purified DNA from endometrial scrapings at indicated times was separated in a 1.5% agarose gel and stained with 0.5 μm/ml ethidium bromide for 1 hour. Photographs were taken with Polaroid-type 55 film.* **C:** *The percentage of DNA fragments is shown as determined by densitometric analysis of high molecular weight DNA (>23kb) versus fragmented lower molecular weight DNA visualized on the photographic negative image of the gel.*

Reprinted with permission from the American Journal of Pathology.

fibroblasts while it inhibits growth in epithelial cells.[39] Its effect on apoptosis induction has been described in several epithelial cell types.[14,40-42] Our work has centered on its effects on uterine epithelium. We described that after addition of TGF-β1 to primary cultures of rabbit uterine epithelial cells, there was a decreased growth rate, due to decreased cell proliferation and increased apoptosis.[40] It was then suggested that the apoptotic effect of progesterone withdrawal could be elicited through TGF-β1. Later on it was shown that TGF-β1 immunostain in rabbit endometrium correlated inversely with progesterone serum levels,[43] providing support for the hypothesis. It is interesting to point out that we have described several examples of coordinated but inversed relationship between cell proliferation and apoptosis in the rabbit uterine epithelium. Increase of progesterone serum levels resulted in several fold increases in cell proliferation and decreased apoptosis, while decreases of that hormone level had opposite effects,[8,10] serum addition or withdrawal to the cultured cells resulted also in coordinated but inverse regulation of both growth parameters[44] and TGF-β1 has similar effects.

The coordinated regulation we observed in whole animals and in cultured cells suggests the existence of feedback mechanisms, which could work through paracrine or autocrine mechanism of action, involving growth factors or receptors. Several years ago we proposed such a regulation.[3,45,46] On the other hand the feedback mechanism could involve only intracellular second messengers, having opposite effects on cell proliferation vs. apoptosis pathways.

The relationship of apoptosis to the cell cycle is of great interest because of the above described growth coordination. Recent advances in the molecular understanding of the cell cycle could help provide useful information on biochemical mechanisms involved in apoptosis. On the basis of our own work[3,8,10,45-48] it would appear that differentiated, non-proliferating cells are the natural candidates for apoptosis. However, recent work using cultured cell lines showed that apoptosis induction is not dependent on growth arrest;[31] it can happen at various points of the cell cycle,[25] cell cycle inhibitors do not change apoptosis inducibility,[33] and there is no association between susceptibility to apoptosis induction and differentiation induction.[30] Therefore, it is not clear yet if there is a correlation between apoptosis and the cell cycle. However, it could be postulated that there should be coordinated regulatory mechanisms for both cell functions, if the growth of normal tissues is to be under control.

SUMMARY

It is now quite evident that there is more than one type of cell death. Necrosis is caused by pathological stimuli, while apoptosis is due to physiological stimuli. However, there is some overlapping between the two, since some types of cell injury may result in apoptosis. The morphological differences between necrosis and apoptosis are clear cut. The known biochemical differences relate to DNA degradation, which appears to be internucleosomal in most apoptosis examples and randomized in necrosis. However, there are a few studies demonstrating that apoptotic cells exhibiting clear apoptosis morphology appear not to show the typical apoptotic fragmented DNA ladder. The endonuclease(s) involved in the apoptotic DNA fragmentation has not been yet fully characterized. It is becoming evident that the DNA fragmentation is a late event in the apoptotic pathway. Gene expression is involved in apoptosis modulation, but it is not clear if it is the cause or the effect, such as repair mechanisms. It is also not clear if the gene expression changes are apoptosis-specific or if it they are also present in necrosis.

The coordination of apoptosis and cell proliferation is a most promising area of research. It should help to unravel the mechanism by which growth is regulated and it should also provide further information on regulation of both phenomena. The latter will be of great help to devise ways of improving healing, modulating unregulated neoplastic growth and preventing diseases where cell death is a primary consideration. The use of apoptosis observation for diagnosis is still underdeveloped, but it could become an important diagnostic pathology tool.

REFERENCES

1. D.G. Scarpelli and M. Chiga. Cell injury and errors of metabolism, *in*: "Pathology," W.A.D. Anderson and J.M. Kissane, eds., The C.V. Mosby Co., St. Louis, Missouri (1974).
2. B. F. Trump, I.K. Berezesky and A.R. Osornio-Vargas. The role of calcium, *in*: "Cell Death in Biology and Pathology," I.D. Bowen and R.A. Lockshin, eds., Chapman and Hall Ltd., London, England (1981).
3. L.E. Gerschenson and R.J. Rotello. Apoptosis: a different type of cell death, *FASEB J.* 6:2450 (1992).
4. R.A. Lockshin. Cell death in metamorphosis, *in*: "Cell Death in Biology and Pathology," I.D. Bowen and R.A. Lockshin, eds., Chapman and Hall Ltd., London, England (1981).
5. A.H. Wyllie, R.G. Morris, A.L. Smith, and D. Dunlop. Chromatin cleavage in apoptosis: association with condensed chromatin morphology and dependence on macromolecular synthesis, *J. Pathol.* 142:67 (1984).
6. J.J. Cohen and R.C. Duke. Glucocorticoid activation of a calcium-dependent endonuclease in thymocyte nuclei leads to cell death, *J. Immunol.* 132:38 (1984).
7. A. Alles, K. Alley, J.C. Barrett, R. Buttyan, J.A. Columbano, F.O. Cope, E.A. Copelan, R.C. Duke, P.B. Farel, L.E. Gerschenson, D. Goldgaber, D.R. Green, K.V. Honn, J. Hully, J.T. Isaacs, J.F.R. Kerr, P.H. Krammer, R.A. Lockshin, D.P. Martin, D.J. McConkey, J. Michaelson, R. Schulte-Hermann, A.C. Server, B. Szende, L.D. Tomei, T.R. Tritton, S.R. Umansky, K. Valerie, H.R. Warner. Apoptosis: a general comment. *FASEB J.* 5, 2127 (1991).
8. S. Nawaz, M.P. Lynch, P. Galand and L.E. Gerschenson. Hormonal regulation of cell death in rabbit uterine epithelium. *Am. J. Pathol.* 127:51 (1987).
9. R.J. Rotello, M.B. Hocker and L.E Gerschenson. Biochemical evidence for programmed cell death in rabbit uterine epithelium. *Am. J. Pathol.* 134:491 (1989).
10. R.J. Rotello, R.C. Lieberman, R.B. Lepoff and L.E. Gerschenson. Characterization of uterine epithelium apoptotic cell death kinetics and regulation by progesterone and RU486. *Am. J. Pathol.* 140:449 (1991).
11. M.A. Barry and A. Eastman, Endonuclease activation during apoptosis: the role of cytosolic Ca^{++} and pH. *Biochem. Biophys. Res. Commun.* 186:782 (1992).
12. M.A. Barry, J.E. Reynolds and A. Eastman. Etoposide-induced apoptosis in human HL-60 cells is associated with intracellular acidification. *Cancer Res.* 53:2349 (1993).
13. L.D. Tomei, J.P. Shapiro and F.O. Cope. Apoptosis in C3H/10T1/2 mouse embryonic cells: evidence for internucleosomal DNA modification in the absence of double-strand cleavage. *Proc. Natl. Acad. Sci. USA* 90. 853 (1993).
14. F. Oberhammer, G. Fritsch, M. Schmied, M. Pavelka, D. Printz, T. Purchio, H. Lassmann and R. Schulte-Hermann. Condensation of the chromatin at the membrane of an apoptotic nucleus is not associated with activation of an endonuclease. *J. Cell Sci*, 104:317 (1993).
15. H.M. Ellis and H.R. Horvitz. Genetic control of programmed cell death in the nematode *C. elegans*. *Cell* 44.817 (1986).
16. J. Yuan and H.R. Horvitz. The *Caernohabditis elegans* genes ced-3 and ced-4 act autonomously to cause programmed cell death. *Dev. Biol.* 138:33 (1990).

17. D.L. Vaux, I.L. Weissman and S.K. Kim. Prevention of programmed cell death in *C. elegans* by human bcl-2. *Science* 258:1955 (1992).

18. J.F.R. Kerr and J. Searle. Deletion of cells by apoptosis during castration-induced involution of the rat prostate. *Virchows Arch. Zellpathol.* 13:87 (1973).

19. B. Lesser and N. Bruchovsky. The effects of testosterone, 5-α-dihydrotestosterone and adenosine 3', 5'-monophosphate on cell proliferation and differentiation in rat prostate. *Biochem. Biophys. Acta* 308:426 (1973).

20. C. Chang, A.G. Saltzman, N.S. Sorensen, R.A. Hüpakka and S. Liao. Identification of glutathione S-transferase Ybi mRNA as the androgen-repressed mRNA by cDNA cloning and sequence analysis. *J. Biol. Chem.* 262:11901 (1987).

21. M.L. Montpetit, K.R. Lawless and M.P. Tenniswood. Androgen-repressed messages in the rat ventral prostate. *Prostate* 8:25 (1987).

22. R. Buttyan, Z. Zakeri, R. Lockshin and D. Wolgemuth. Cascade induction of c-fos, c-myc and heat shock 70K transcripts during regression of the rat ventral prostate gland. *Mol. Endocrinol.* 2:650 (1988).

23. A.E. Katz, M.C. Benson, G.J. Wise, C.A. Olsson, M.G. Bandyk, I.S. Sawczuk, P. Tomashefsky and R. Buttyan. Gene activity during the early phase of androgen-stimulated rat prostate regrowth. *Cancer Res.* 49:5889 (1989).

24. G. Engel, C. Lee and J.T. Grayhock. Acid ribonuclease in rat prostate during castration-induced involution. *Biol. Reprod.* 22:827 (1980).

25. C.I. Evan, A.H. Wyllie, C.S. Gilbert, T.D. Littlewood, H. Land, M. Brooks, C.M. Waters, L.Z. Penn, D.C. Hancock. Induction of apoptosis in fibroblasts by c-myc protein. *Cell* 69:119 (1992).

26. A.J. Wagner, M.B. Small and N. Hay. Myc-mediated apoptosis is blocked by ectopic expression of bcl-2. *Mol. Cell. Biol.* 13:2432 (1993).

27. A. Fanidy, E.A. Harrington and G.I. Evan. Cooperative interaction between c-myc and bcl-2 proto-oncogenes. *Nature* 359:554 (1992).

28. R.P. Bissonette, F. Echeverri, A. Mahboubi and D. R. Green. Apoptotic cell death induced by c-myc is inhibited by bcl-2. *Nature* 359:552 (1992).

29. D.S. Askew, R.A. Ashmun, B.C. Simmons and J.L. Cleveland. Constitutive c-myc expression in an IL-3-dependent myeloid cell line suppresses cell cycle arrest and accelerates apoptosis. *Oncogene* 6:1915 (1991).

30. J. Lotem and L. Sachs. Regulation by bcl-2, c-myc and p53 of susceptibility to induction of apoptosis by heat shock and cancer chemotherapy compounds in differentiation-competent and defective myeloid leukemic cells. *Cell Growth Differ.* 4:41 (1993).

31. E. Yonish-Rouach, D. Grumwald, S. Wilder, A. Kimchi, E. May, J.J. Lawrence, P. May and M. Oren. p53-mediated cell death: relationship to cell cycle control. *Mol. Cell. Biol.* 13:1415 (1993).

32. P. Shaw, R. Bovey, S. Tardy, R. Sahli, B. Sordat and J. Costa. Induction of apoptosis by wild-type p53 in a human colon tumor-derived cell line. *Proc. Natl. Acad. Sci. USA* 89:4495 (1992).

33. M. Fritsche, C. Haessler and G. Brandner. Induction of nuclear accumulation of the tumor-suppressor protein p53 by DNA-damaging agents. *Oncogene* 8:307 (1993).

34. M. Debbas and E. White. Wild-type p53 mediates apoptosis by EIA, which is inhibited by EIB. *Genes Dev.* 7:546 (1993).

35. S.W. Lowe and H.E. Ruley. Stabilization of the p53 tumor suppressor is induced by adenovirus 5 E1A and accompanies apoptosis. *Genes Dev.* 7:535 (1993).

36. S.W. Lowe, E.M. Schmitt, S.W. Smith, B.A. Osborne and T. Jacks. p53 is required for radiation-induced apoptosis in mouse thymocytes. *Nature* 362:847 (1993).

37. A.R. Clarke, C.A. Purdie, D.J. Harrison, R.G. Morris, C.C. Bird, M.L. Hooper and A.H. Wyllie. Thymocyte apoptosis induced by p53-dependent and independent pathways. *Nature* 362:849 (1993).

38. S.M. Lester, J.G. Wood, and L.R. Gooding. Tumor necrosis factor can induce both apoptotic and necrotic forms of cell lysis. *J. Immunol.* 141:2629 (1993).

39. H.L. Moses, E.Y. Yang and J.S. Pietenpol. TGF-β stimulation and inhibition of cell proliferation: New mechanistic insights. *Cell* 63:245 (1990).

40. R.J. Rotello, R.C. Lieberman, A.F. Purchio, and L.E. Gerschenson. Coordinated regulation of apoptosis and cell proliferation by TGF-β1 in cultured cells. *Proc. Natl. Acad. Sci. USA* 88:3412 (1991).

41. K. Yanagihara and M. Tsumuraya. Transforming Growth Factor β1 induces apoptotic cell death in cultured human gastric carcinoma cells. *Cancer Res.* 52:4042 (1992).

42. J. Lotem and L. Sachs. Hematopoietic cytokines inhibit apoptosis induced by Transforming Growth Factor β1 and cancer chemotherapy compounds in myeloid leukemic cells. *Blood* 80:1750 (1992).

43. C.-I Sze, W. Sun, R. Lieberman, T. Purchio and L.E. Gerschenson. Progesterone regulates apoptosis, Transforming Growth Factor-β1 (TGF-β1) immunostain patterns and TGF-β1 mRNA content in rabbit uterine epithelium. *FASEB J.* 7:A137 (1993).

44. L.E. Gerschenson, R.J. Rotello, R. Low, R. Lieberman, J. Rochon and A.F. Purchio. Apoptosis (programmed cell death) regulation in rabbit uterine epithelium. *FASEB J.* 5:A518 (1991).

45. M. Lynch, S. Nawaz, and L.E. Gerschenson. Evidence for soluble factors regulating cell death and cell proliferation in primary cultures of rabbit endometrial cells grown on collagen. *Proc. Natl. Acad. Sci. USA* 83:4784 (1986).

46. L.E. Gerschenson and R.J. Rotello. Apoptosis and cell proliferation are terms of the growth equation , *in* "Apoptosis: The Molecular Basis of Cell Death," L.D. Tomei and F.O. Cope, eds., Cold Spring Harbor Laboratory Press, Plainview, New York (1991).

47. J.T. Murai, C.J. Conti, I. Gimenez-Conti, D. Orlicky and L.E. Gerschenson. Temporal relationship between rabbit uterine epithelium proliferation and uteroglobin production. *Biol. Reprod.* 24:649 (1981).

48. C.J. Conti, I. Gimenez-Conti, E.A. Conner, J.M. Lehman and L.E. Gerschenson. Estrogen and progesterone regulation of proliferation, migration and loss of different target cells of rabbit uterine epithelium. *Endocrinology* 114:345 (1984).

PART VI

ROLE OF THE CELL CYCLE IN DISEASE

35

MOLECULES OF DEREGULATED CELL CYCLE CONTROL IN CANCER

Arthur B. Pardee, Qing-Ping Dou, and Khandan Keyomarsi

Cell Growth and Regulation
Dana-Farber Cancer Institute
Boston, MA 02115

INTRODUCTION

Our overall aim is to identify and characterize basic mechanisms responsible for defective growth control of cancer cells. Much evidence now supports our earlier hypothesis[1] that normal cell proliferation is controlled just prior to the sudden onset of DNA synthesis and related biochemical processes. Deregulation is dependent upon a mechanism normally controlling passage of cells through a restriction point located in late G1 phase of the cell cycle. Cyclins, cyclin dependent kinases (cdks), retinoblastoma protein (pRB) and others have been implicated in passage through G1 into S phase. We conclude that deranged expression of cyclins, particularly of cyclin E, contributes to defective growth control of cancer cells.

TRANSCRIPTION FROM THE PROMOTER REGION OF THE MOUSE THYMIDINE KINASE GENE

The question we ask is which molecules are altered to deregulate the growth of cancer cells. We start with the appearance of mouse thymidine kinase (TK) mRNA as a terminal marker of the G1/S transition. This gene is regulated before DNA synthesis starts[3]. We have discovered G1/S Yi protein complexes which bind to consensus promoter sequences in murine TK and regulate this enzyme's production. They include the retinoblastoma-like p107 protein, an E2F family DNA binding protein, cyclins A and E, and cdk kinase[4,5]. We are

The Cell Cycle: Regulators, Targets, and Clinical Applications
Edited by V.W. Hu, Plenum Press, New York, 1994

relating these complexes and their molecules to cell proliferation control.

Mutation of TK promoter sequence that binds E2F causes constitutively expressed transcription, indicating a repressing function of E2F[6]. Furthermore expression of a reporter gene (ß-globin) under control of the TK promoter is constitutive in transformed BPA31 cells. This cell cycle related transcriptional control is deranged in these tumor forming BPA31 cells[7].

We long ago proposed that control of cell proliferation is due to an unstable protein that is produced during the pre-DNA synthetic G1 phase of the cell cycle, and that this protein is either overproduced or stabilized in tumor cells resulting in defective growth control[8]. We now propose that this labile protein is a cyclin, very possibly cyclin E[9]. Cyclins are a family of proteins that regulate important (cdk) kinases. They are prime cell cycle regulators and are central to the control of major checkpoints in eukaryotic cells.

CYCLIN E IS PROBABLY THE RESTRICTION POINT PROTEIN

To connect these findings to restriction point control defects, determined by cell biology, kinetic studies were made with mouse cells on expression of cyclins and cdk kinases. These molecules were found to associate into multi-protein DNA binding complexes "Yi" as the cells passed through G1 phase into S, with different kinetics in the tumor vs. normal cells[4,5]. In particular, cyclin E and its activation of cdk2 were found to pass all three tests that we had proposed for the restriction point protein[8].

Test 1 is activation in late G1 in nontransformed cells[9]. We first performed an experiment to measure the increases of H1 kinases in late G1 associated with cyclin or cdk proteins, the first criterion for the R protein. We used antibodies against human cyclins (A, B, E) and cdks (cdc2, cdk2), as well as $p13^{suc1}$ beads which bind to several cdk/H1 kinases including cdk2, to precipitate the corresponding proteins and then measured their H1 kinase activities. All these H1K activities were low in G0 cells, but they increased at different phases of the cell cycle. CycE-H1K increased around the R point (before 12 hr) and peaked at 17 hr, at which time about 40% of nuclei were labeled by $[^3H]$-thymidine. CycA-H1K gradually increased after 12 hr, and reached its highest value at 36 hr, parallel to the kinetics of DNA synthesis as measured by nuclear labeling. In contrast, CycB-H1K did not increase until 36 hr. The cdk2-H1K increased at 12 hr and dropped after S phase (36h), supporting its interaction with both cyclin E and cyclin A. The $p34^{cdc2}$-H1K showed a similar pattern to that of CycA-H1K, suggesting that it mainly interacts with cyclin A before 36 hr. The p13-H1K had two major increases, one before 17 hr which may be derived from cyclin E, while another increase was before 36 hr which is probably from both cyclin A and cyclin B. Our data, together with other groups' results, support cyclin E/cdk2 and cyclin A/cdk2 as late G1 and S phase-specific kinases that are involved in entry of cells in S.

We then performed western blot analyses to determine whether cyclin E and cyclin A proteins accumulate in G1/S. An antiserum to human cyclin E recognized mainly two bands with approximate molecular masses of 52 and 65 kDa. p52 probably represents the mouse cyclin E protein since accumulation of this protein correlates well with CycE-H1K during the cell cycle: very low in G0 cell nuclear extract, increased at G1/S (around 12 hr), peaked at 18 hr (about 8-fold higher than in 0 hr) and decreased at 22 hr. In contrast, p65 was relatively constitutively expressed during the cell cycle, and its nature remains unknown. A mouse protein of p55, identified by an antiserum to human cyclin A, was not present in G0 cells, but increased after cells entered S phase (about 38-fold higher at 22 hr than at 0 hr), coincident with CycA-H1K. When anti-cdk2 was used, a major band of p38 with two minor bands of p30 and p41 were detected; all of them were present in G0 nuclear extracts and increased only 2-fold after cells entered S phase. These data strongly suggest that accumulation of cyclin E and cyclin A proteins at G1/S importantly regulate induction of their cdk kinases activities.

Test 2 for cyclins as R protein candidates is an excess delay of their appearance after inhibition of protein synthesis in non-transformed cells. We expected that CycE-H1K and CycA-H1K are labile in nontransformed A31 cells, the second criterion for the R protein. Indeed, after treatment with cycloheximide (CHX), we observed 2-4 fold decreases in these H1K kinases. Several hours after CHX was removed, H1K activities started to recover; further chasing revealed large increases in these activities.

The lags in both CycE-H1K and CycA-H1K activities, caused by CHX-pulse treatment, may reflect the pulse-inhibition of synthesis of unstable cyclin proteins. Indeed, even though the level of p52/cyclin E protein was very low before and during the pulse inhibition, it increased several fold after chasing; p65 was relatively unchanged during the process. Cyclin A p55 protein was almost undetectable at 12 hr and during the protein inhibition, and gradually increased several hours after removing CHX. In contrast, cdk2 was relatively constitutive during this process. Therefore, cyclin proteins are probably the labile components responsible for regulation of cdk kinases.

Test 3 is faster recovery from the CHX inhibition in transformed cells. We expected that transformed BPA31 cells would show no or less delay in H1K appearance at high levels after the pulse of CHX (the third criterion for the R protein). CHX equally inhibits protein synthesis in A31 and the transformed BPA31 cells . Comparison of cyclin E- and cyclin A-associated H1Ks between BPA31 and A31 cells gave results similar to, but more complicated than, those observed from earlier cell biology experiments. The levels of CycE-H1K and CycA-H1K at 12 hr were 2-3 fold higher in BPA31 than in A31 cells, as found in multiple experiments. Secondly, even though CHX reduced these H1Ks in both

A31 and BPA31 cells, the remaining activities after the treatment were higher in BPA31 than in A31 cells. Thirdly, both CycE-H1K and CycA-H1K recovered to high levels much earlier in BPA31 cells than in A31 cells.

To find whether there are such differences in cyclin proteins between BPA31 and A31 cells, we performed western blot analysis using nuclear extracts from the preparations. Basically, we observed similar differences in cyclin proteins, and in cdk kinases. At 12 hr, cyclin E protein was much higher in BPA31 cells than in A31 cells while cyclin A was slightly higher in BPA31 cells. Secondly, CHX treatment reduced accumulation of both cyclin E and cyclin A proteins in BPA31 cells (it was difficult to measure decreases of these proteins in A31 cells because of their low basal levels), but the remaining cyclin proteins were higher in BPA31 than in A31 cells. Thirdly, cyclin E protein recovered from the CHX inhibition several hours earlier in BPA31 than in A31 cells, probably due to its higher basal level. Though cyclin A in both BPA31 and A31 cells was at a similar level at 12 hr (G1/S) and during the CHX inhibition, it recovered to high levels several hours earlier in BPA31 than in A31 cells. In contrast to these cyclins, cdk2 protein in both BPA31 and A31 cells was at a similar level and remained relatively unchanged during the pulse-chase process. These data suggest that the higher level and/or faster recovery of cyclin E/cyclin A proteins are probably responsible for the higher level and/or faster recovery of these cdk kinases activities in BPA31 cells.

From these cell cycle and pulse-chase experiments, we conclude that accumulation of cyclin E and cyclin A proteins and induction of their dependent kinases satisfies all the three properties of the R protein, and that therefore these kinases may be both proliferation-controlling and deranged in these tumor cells. The molecular basis of the R point protein could depend upon cyclin production and inactivation.

CYCLIN E AND BREAST CANCER

Our approach to determine whether and which of these molecules are altered in cancers was to investigate their properties, production and appearance during the cell cycle in normal vs. tumor human breast cells[10]. Recently we have established new evidence correlating the deranged expression of cyclins to loss of growth control. To emphasize the role that cyclins can play in the process of tumor formation, we have provided evidence for their general derangement, not just the derangement of one or two cyclins in rare isolated cells lines and/or tumors. Using proliferating normal versus human tumor breast cells as a model system, we have described several changes that are seen in all or most breast tumor cell lines. These alterations include i) eight-fold amplification of cyclin E gene in one tumor line, 64 fold overexpression of its mRNA, and altered expression of its protein, ii) deranged expression of cyclin E protein in all (10/10) tumor cell lines studied, iii) increased cyclin mRNA stability, resulting in iv) general overexpression of mitotic cyclins and of CDC2 RNAs and proteins in 9/10 tumor lines, and v) deranged order of appearance of cyclins in synchronized tumor versus normal cells, with mitotic cyclins appearing prior to G1 cyclins. General cyclin overexpression was also detected in 3/3 breast tumor tissue samples. These multiple changes in cyclin expression in human breast cancer cells suggest that cyclins may be a new class of oncogenes. They may perform redundantly in cancer, with overexpression and/or amplification of one replacing functions of others.

Cyclin E may be a diagnostic/prognostic marker for breast cancer. The most striking abnormality was that of cyclin E expression. Cyclin E protein not only was overexpressed in all (10/10) breast tumor cells lines examine but also is present in different sizes than in normal cells. The normal cell strains we used in this study were rapidly proliferating. Thus the altered expression of cyclin E protein in tumor cell lines is not dependent on cell proliferation and represents a true difference between normal and tumor cells. These normal mammary epithelial cells were obtained from discarded tissue of reduction mammoplasty operations from three different individuals and established in culture with population doubling times of 24-27 hours. In addition, irrespective of their tumorigenicity potentials or estrogen-receptor status, we observed overexpression of one, two, or all three of the cyclin E-like proteins ranging in size from 35 to 50KD, whereas in the normal cells strains we observed only one major protein of about 50KD. Since a changed "picture" of cyclin E protein expression is observed in tumor cells, this cyclin might be used diagnostically or prognostically in breast cancer. Our preliminary results using breast cancer tissues and normal adjacent non-tumorous biopsy material is very encouraging to indicate that altered expression of cyclin E can also occur *in vivo*.

DIFFERENTIAL DISPLAY - FINDING NOVEL GENES

We have developed a general method to detect mRNAs that are expressed in a given cell type. Comparisons of displays of mRNA markers permit isolation of genes that are differently expressed, for example in normal vs. tumor cells[11]. Very many researchers are now using this Differential Display (DD) method with success, and have isolated numerous genes of interest. The basic principle is to reverse transcribe and with PCR systematically amplify subsets of mRNAs, and then distribute their 3' termini on a denaturing polyacrylamide gel. We provided methodological details and examined in depth the specificity, sensitivity, and reproducibility of the method[12]. Four anchored oligo-dT primers degenerate at the penultimate base from the 3' end in combination with 20 arbitrary decamers were able to distribute many of the mRNA species within a cell into a comprehensive 2-D array.

These results enable us to further streamline the DD tehnique and make it readily applicable to a broad spectrum of biological systems.

SUMMARY

The significance of our research is three-fold. First, these studies are very relevant to cancer because they deal with defective growth control at the molecular level. We focused upon the most critical event, namely preparation of cells for the onset of DNA synthesis (S-phase). Normal and transformed cells differ in their ability to regulate this decision. The underlying genetic and biochemical mechanism involves cyclins, cdk kinases and pRB. It is just beginning to be understood.

Secondly, this research leads to new insights into mechanism for regulating gene expression in normal cells, providing results of broad applicability.

Thirdly, use of human epithelial cells, enzymes, genes and drugs involved in therapy or diagnosis make these studies clinically relevant. For example, thymidine kinase is required for activation of several important drugs including ara-C, ddC, AZT, ganciclovir, and F-UdR. A TK specific mutation assay is used for screening chemical mutagens. TK in the blood is a diagnostic marker. TS is a target for FU and for a new drug D1694.

REFERENCES

1. A.B. Pardee. A restriction point for control of normal animal cell proliferation, Proc. Natl. Acad. Sci., USA. 71:1286 (1974).
2. A.B. Pardee. G1 events and regulation of cell proliferation, Science 246:603 (1989).
3. J.G. Gudas, J.L. Fridovich-Keil, M.W. Datta, J.Bryan, A.B. Pardee. Characterization of the murine thymidine kinase-encoding gene and analysis of transcription start point heterogeneity, Gene 118:205 (1992).
4. Q.-P. Dou, P. Markell, A.B. Pardee. Thymidine kinase transcription is regulated at G1/S phase by a complex that contains retinoblastoma-like protein and a cdc2 kinase, Proc. Natl. Acad. Sci, USA. 89:3256 (1992).
5. Q.-P. Dou, A.H. Levin, J. Wang, K. Helin, A.B. Pardee. G1/S regulated complexes containing E2F, P107, and cyclin-dependent kinases bind to the mouse thymidine kinase gene promoter, submitted to J. Biol. Chem.
6. J.L. Fridovich-Keil, P.J. Markell, J.G. Gudas, A.B. Pardee. DNA sequences required for serum-responsive regulation of expression from the mouse thymidine kinase promoter, Cell Growth and Differentiation 4:679 (1993).
7. D.W. Bradley, J.L. Fridovich-Keil, A.B. Pardee. Serum-responsive expression from murine TK gene sequences is disrupted in transformed cells, in preparation.

8. P.W. Rossow, V.F.G. Riddle, A.B. Pardee. Synthesis of labile, serum-dependent protein in early G1 controls animal cell growth, Proc. Natl. Acad. Sci. USA. 76:446 (1979).
9. Q.-P. Dou, A. Levin, S. Zhao, A.B. Pardee. Cyclin E and cyclin A as candidates for the restriction point protein, Cancer Res. 53:1493 (1993).
10. K. Keyomarsi, A.B. Pardee. Redundant cyclin overexpression and gene amplification in breast cancer cells, Proc. Natl. Acad. Sci. USA 90:1112 (1993).
11. P. Liang, L. Averboukh, K. Keyomarsi, R. Sager, A.B. Pardee. Differential display and cloning of messenger RNAs from human breast cancer vs. mammary epithelial cells, Cancer Res. 52:6966 (1992).
12. P. Liang, L. Averboukh, A.B. Pardee. Distribution of eukaryotic mRNAs by means of differential display: refinements and optimization, Nucleic Acids Res. 21:3269 (1993).

36

THE TWO AMINO TERMINAL TRANSFORMING FUNCTIONS OF THE SV40 LARGE T-ANTIGEN ARE REQUIRED TO OVERCOME P53 MEDIATED GROWTH ARREST

Robin S. Quartin and Arnold J. Levine

Department of Molecular Biology
Princeton University
Princeton, NJ 08544-1014

INTRODUCTION

About 60% of cancers from humans contain mutations at the p53 locus[1,2]. Most commonly, there is a missense mutation in one of the p53 alleles and a loss of the second allele resulting in a reduction to homozygosity of the mutant gene[3]. Thus p53 behaves like a tumor suppressor gene[4] where a loss of function (recessive to wild-type) enhances the probability of cancerous growth. Returning the wild-type p53 gene into cells that are being transformed by an oncogene[5] or are already transformed[6] blocks the transformation process and, when p53 is expressed at high levels, inhibits cell division of transformed cells. Cells transformed with a temperature sensitive p53 mutant replicate at 37-39°C, where the p53 protein is in a mutant conformation, but fail to grow at 32°C, where the p53 protein is acting as a wild-type tumor suppressor[7,8]. The wild-type p53 protein blocks progression through the cell cycle in G_1[7,8]. Thus, the wild-type protein can, under certain circumstances[9], regulate progression of cells through the cell cycle while mutant p53 proteins fail to do this. The wild-type p53 protein, but not the mutant protein, can act as a transcription factor[10-12] and so it is tempting to speculate that p53-mediated transcription of a set of genes can block progression through the cell cycle in the G_1 phase of the cycle.

The Simian Virus 40 (SV40) large T-antigen can bind to the p53 protein[13,14] and other tumor suppressor gene products[15,16], and eliminate this negative regulation of the cell cycle, resulting in the transformation of cells in culture and the initiation of tumors in animals[17]. A genetic analysis of the SV40 large T-antigen gene[15-26] has localized three regions of the T-antigen that are essential for transformation of a variety of cell types (primary and permanent cell lines) in culture. The SV40 large T-antigen is composed of 708 amino acid residues and residues 1 - 72, 105 - 114 and

The Cell Cycle: Regulators, Targets, and Clinical Applications
Edited by V.W. Hu, Plenum Press, New York, 1994

350-627 are each critical for the transformation process. Mutations in residues 1-72 are complemented by the adenovirus oncogene product E1A which binds to a cellular protein of 300 kD[27] and mutations in E1A that eliminate this protein-protein interaction fail to complement SV40 T-antigen mutants in residues 1-72[26,28]. Although it has not been demonstrated directly, it is thought that SV40 T-antigen residues bind the 300 kD protein and alter its functions resulting in cellular transformation[26]. Residues 105-114 of the SV40 large T-antigen bind to the retinoblastoma susceptibility protein (Rb)[15] which is a tumor suppressor protein regulating the transcription factor E2F which in turn regulates entry into the S-phase of the cell cycle[28]. The third region of the SV40 large T-antigen required for transformation lies between residues 350-627 (in two non-continuous regions, residues 350-450 and 533-627)[19,20,25] which is required to bind to the p53 protein. When T-antigen binds to the p53 protein, it inactivates its ability to function as a transcription factor[29] just as mutations in the p53 gene block this function. Thus T-antigen mediates its ability to transform cells by binding to three cellular proteins, 300 kD, Rb and p53, which are thought to play critical roles in regulating cell cycle events[15-29].

RESULTS

Primary rat embryo fibroblasts were transfected with cDNA or partial genomic clones containing a p53 temperature sensitive mutant (an alanine to valine change at codon 135) plus an activated *ras* oncogene[30]. At 37-39°C these cells produce transformed foci and express both the mutant p53 protein and the ras protein[7,8,30]. At 32°C the p53 protein behaves as a wild-type protein and no transformed foci are produced[7,8]. If one isolates cloned cell lines from the foci that arise at 37°C and then shifts these cells (the cell line used in these studies is T64-7B[30]) to 32°C, cell division stops at 32°C but proceeds normally at 37°C (see Figure 1). The cells blocked at 32°C are in the G_1 phase of the cell cycle[7,8] and remain viable for long periods of time. Because these cells were blocked by wild-type p53 at a specific stage in the cell cycle, experiments were designed to determine which portions of the SV40 large T-antigen were able to reverse this cell cycle block. In this fashion we could gain insights into the relationships among 300 kD, Rb and p53 during their interactions with T-antigen.

When T64-7B cells, transfected with SV40 large T-antigen, are shifted to 32°C (18 hours post transfection), clones of cells which form colonies are detected (Table 1), while in the absence of T-antigen, no colonies form. On the average about 30 clones of T64-7B cells will be produced after a SV40 T-antigen gene transfection into 2 X 10^5 cells which are then shifted to 32°C to form colonies. When the colonies grown at 32°C are picked and cloned, the cells all express the SV40 T-antigen which is bound to p53. No clones without T-antigen expression were detected in this study. Thus, the SV40 large T-antigen was able to eliminate the temperature sensitive p53-mediated G_1 block of the cell cycle and produce a cell line growing at 32°C (Table 1, lines 1,2).

Next, a variety of mutant genes coding for T-antigens defective in 300 kD binding, Rb binding, p53 binding, T-antigen nuclear localization signal, or with an altered T-antigen zinc finger normally required for T-antigen DNA binding were tested in this assay (Table 1, lines 3-7). Mutations resulting in the loss of the T-antigen nuclear localization signal or the loss of the zinc finger had little or no effect (76-87% as efficient as wild-type) upon the ability of T-antigen to reverse the p53 block to cell

Table 1. 2X105 T64-7B cells[30] were transfected with the wild-type or mutant SV40 DNA clones at 37°C and 18 hours later shifted to 32°C to test for the formation of colonies. In the absence of SV40 DNA, no colonies formed. The average number of colonies is from three separate experiments. The 300 kD and Rb mutant T-antigens failed to overcome the p53-mediated temperature sensitive block to progression through the cell cycle. These data may be explained by one of the models presented in Figure 2

T-antigen	Average number of colonies	Activity relative to wild-type
1. No DNA clone	0	0
2. Wild-type T-antigen	30.7	1.00
3. 300 kD binding mutants		
(a) 2803; linker insertion residue 35	0.3	0.01
(b) dl 1135; deletion of residues 17-27	0	0
(c) 2831; deletion of residues 5-35	1.7	0.05
4. Rb binding mutants		
(a) pVU-1; deletion of residues 106-113	0.7	0.02
(b) K1; E107K	1.0	0.03
5. p53 binding mutants		
(a) dl 1137; deletion of residues 122-708	15.7	0.51
(b) 2809; linker insertion at 409	12.0	0.39
(c) 2811; linker insertion at 424	19.3	0.63
6. Nuclear localization mutant pSVcT; K128N	23.3	0.76
7. DNA binding - zinc finger mutant 2837/p6-1; deletion of residues 168-346	26.7	0.87

division (Table 1, lines 6,7). Removal of the nuclear localization signal still leaves some T-antigen in the nucleus and does not block cellular transformation, so this result is not surprising. Three different mutations that eliminate p53 binding to T-antigen (dl 1137, 2309, 2811), including a deletion of the entire p53 binding site, resulted in T-antigen proteins that were 40-63% as efficient as the wild-type T-antigen in reversing the p53-mediated block to cell division (Table 1, line 5). Examination of these cells growing at 32°C detected the SV40 T-antigen and free p53 protein (not bound to T-antigen). This was a surprising result which indicated that T-antigen did not have to bind to p53 to bypass this p53-mediated block in the cell cycle. Rather, the two other functional regions of T-antigen (300 kD, Rb binding), which remain intact in these mutants, could play a role in the reversal of this p53-mediated block to

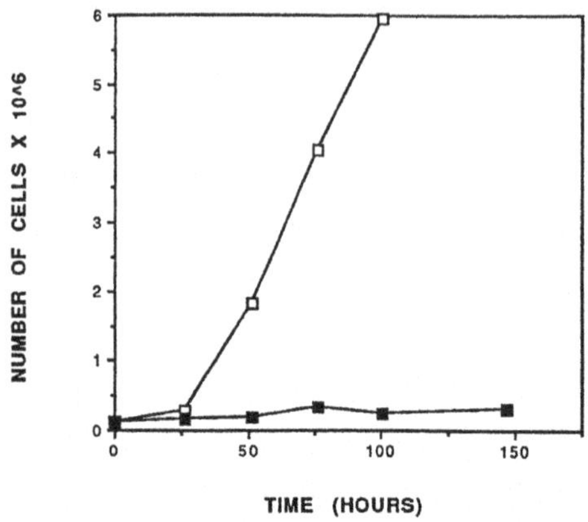

Figure 1. T64-7B cells, growth curve at 37°C and 32°C. The T64-7B cells were plated at 37°C (□-□) or 32°C (■-■) and cell counts were carried out over 150 hours. The cells fail to grow at 32°C and are blocked in the G_1 phase of the cell cycle.

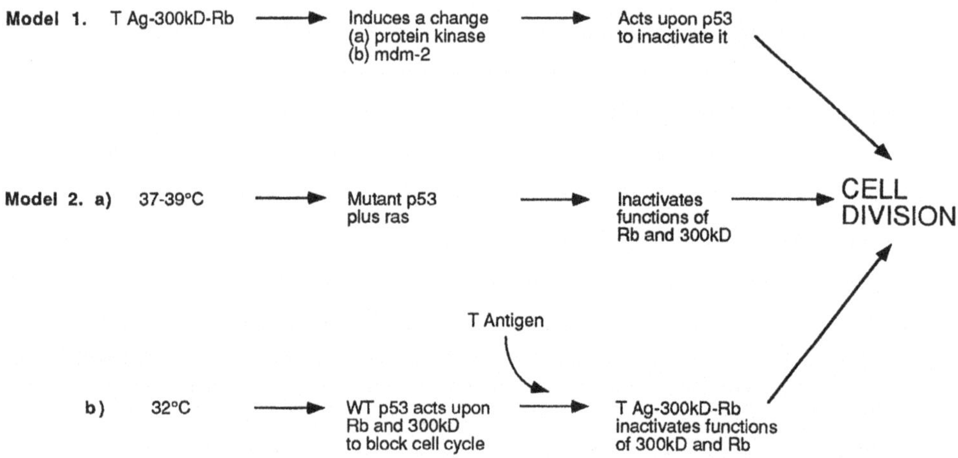

Figure 2. Two alternative models for interpreting the results and relating 300 kD, Rb and p53 gene functions. Either model suggests that these three proteins can communicate either directly or indirectly.

cell cycle progression. This was tested by employing two SV40 T-antigen mutants in the Rb binding site (pVU-1 and K1) (Table 1, line 4) and three mutants in the 300 kD binding site (2803, dl 1135 and 2831) (see Table 1, line 3). Mutations in the Rb binding site or mutations in the 300 kD binding site each destroyed T-antigen's ability to reverse the p53 block to cell division by 95-100% (Table 1, lines 3a, b, c, lines 4a, b). Complementation between the 300 kD binding mutant 2831 and the Rb binding mutant K1 gave about 25% of the wild-type T-antigen yield of colonies which was a 5- to 8-fold higher number of colonies than that observed with either mutant alone.

DISCUSSION

This genetic analysis of the ability of T-antigen to overcome a p53-mediated block in the cell cycle demonstrates that the SV40 T-antigen-300 kD-Rb complex can interact with and modulate the activity of the p53 protein. Thus, there is communication or a pathway between these proteins. There are at least two ways in which this pathway could act. The T-antigen-300 kD-Rb complex could change the enzyme activity or physiology of the cell (i.e. an altered protein kinase activity) so as to alter the p53 protein which would no longer act as a negative regulator of cell growth in G_1. The inactivation of p53 as a transcription factor could come about by phosphorylation modification or even by the induction of mdm-2[31] or some other protein which would bind to p53 and prevent it from functioning. In this case, T-antigen-300 kD-Rb complex would act via other gene functions to inactivate p53 regulatory functions.

Alternatively, the replication of these T64-7B cells could be dependent upon the mutant p53 protein at 37-39°C which is not present at 32°C so that the cells stop dividing. It has recently been shown that p53 mutant proteins have a "gain of a new function" in that they can enhance the tumorigenic potential of cells when they are expressed in these cells[32]. It has also been shown that cell lines transformed with mutant p53 plus ras require this mutant p53 protein for an optimal rate of cell division[33]. Therefore, at 32°C, the p53 "gain of function" would be lost and cells would then fail to divide because the 300 kD and Rb gene products in these cells block progression through the cell cycle in G_1. The ability of T-antigen to reverse 300 kD and Rb blocks in these cells would then be an expected outcome of these experiments.

These two models are contrasted in Figure 2. In both cases, there is a clear prediction that 300 kD and Rb on the one hand and p53 on the other must have a way to communicate with each other as the cell cycle progresses. Either T-antigen-300 kD-Rb can overcome the function of wild-type p53 (directly or more likely indirectly) or mutant p53 (plus ras) can alter the functions of Rb and 300 kD in regulating the cell cycle in a manner similar to SV40 large T-antigen-300 kD-Rb complex formation. The virtue of these models is that they make predictions which can be tested in future experiments.

REFERENCES

1. M. Hollstein, D. Sidransky, B. Vogelstein, and C.C. Harris, p53 mutations in human cancers, *Science* 253:49 (1991).

2. A.J. Levine, J. Momand, and C.A. Finlay, The p53 tumor suppressor gene, *Nature* 351:453 (1991).
3. S.J. Baker, A.C. Preisinger, J.M. Jessup, C. Paraskeva, S. Markowitz, J.K. Willson, S. Hamilton, and B. Vogelstein, p53 gene mutations occur in combination with 17p allelic deletions as late events in colorectal tumorigenesis, *Cancer Res*. 50:7717 (1992).
4. L.A. Donehower, M. Harvey, B.L. Slagle, M.J. McArthur, C.A. Montgomery Jr., J.S. Butel, and A. Bradley, Mice deficient for p53 are developmentally normal but susceptible to spontaneous tumours, *Nature* 356:215 (1992).
5. C.A. Finlay, P.W. Hinds, and A.J. Levine, The p53 proto-oncogene can act as a suppressor of transformation, *Cell* 57:1083 (1989).
6. W.E. Mercer, M. Amin, G.J. Sauve, E. Appella, S.J. Ullrich, and J.W. Romano, Wild-type human p53 is antiproliferative in SV40-transformed hamster cells. *Oncogene* 5:973 (1990).
7. D. Michalovitz, O. Halevy, and M. Oren, Conditional inhibition of transformation and of cell proliferation by a temperature-sensitive mutant of p53, *Cell* 62:671 (1990).
8. J. Martinez, I. Georgoff, J. Martinez, and A.J. Levine, Cellular localization and cell cycle regulation by a temperature sensitive p53 protein, *Genes & Develop*. 5:151 (1991).
9. M.B. Kastan, O. Onyekwere, D. Sidransky, B. Vogelstein, and R.W. Craig, Participation of p53 protein in the cellular response to DNA damage, *Cancer Res*. 51:6304 (1991).
10. G.E. Farmer, J. Bargonetti, H. Zhu, P. Friedman, R. Prywes, and C. Prives, Wild-type p53 activates transcription in vitro, *Nature* 358:83 (1992).
11. S. Kern, J.A. Pietenpol, S. Thiagalingam, A. Seymour, K. Kinsler, and B. Vogelstein, Oncogenic forms of p53 inhibit p53-regulated gene expression, *Science* 256:827 (1992).
12. G.P. Zambetti, J. Bargonetti, K. Walker, C. Prives, and A.J. Levine, Wild-type p53 mediates positive regulation of gene expression through a specific DNA sequence element, *Genes & Develop*. 6:1143 (1992).
13. D.I.H. Linzer, and A.J. Levine, Characterization of a 54K dalton cellular SV40 tumor antigen in SV40 transformed cells, *Cell* 17:43 (1979).
14. D.P. Lane, and L.V. Crawford, T antigen is bound to a host protein in SV40-transformed cells, *Nature* 278:261 (1979).
15. M.B. Ewen, J.W. Ludlow, E. Marsilio, J.A. DeCaprio, R.C. Millikan, S.H. Cheng, E. Paucha, and D.M. Livingston, An N-terminal transformation-governing sequence of SV40 large T antigen contributes to the binding of both $p110^{Rb}$ and a second cellular protein, p120, *Cell* 58:257 (1989).
16. A. Srinivasan, K.W. Peden, and J.M. Pipas, The large tumor antigen of Simian Virus 40 encodes at least two distinct transforming functions, *J. Virol.* 63(12):5459 (1989).
17. R.L. Brinster, H.Y. Chen, A. Messing, T. van Dyke, A.J. Levine, and R. Palmiter, Transgenic mice harboring SV40 T antigen genes develop characteristic brain tumors, *Cell* 37:367 (1984).
18. S. Chen, and E. Paucha, Identification of a region of simian virus 40 large tumor antigen required for cell transformation, *J. Virol.* 64:3350 (1990).
19. T.D. Kierstead, and M.J. Tevethia, Association of p53 binding and immortalization of primary C57Bl/6 mouse embryo fibroblasts by using simian

virus 40 T-antigen mutants bearing internal overlapping deletion mutations. *J. Virol.* 67:1817 (1993).
20. J.-Y. Lin, and D.T. Simmons, The ability of large T-antigen to complex with p53 is necessary for the increased life span and partial transformation of human cells by SV40, *J. Virol.* 65:6447 (1991).
21. L. Sompayrac, and K.J. Danna, A new SV40 mutant that encodes a small fragment of T antigen transforms established rat and mouse cells, *Virology* 163:391 (1988).
22. L. Sompayrac, and K.J. Danna, The amino-terminal 147 amino acids of SV40 large T antigen transform secondary rat embryo fibroblasts, *Virology* 181:412 (1991).
23. M.J. Tevethia, J.M. Pipas, T.D. Kierstead, and C. Cole, Requirements for immortalization of primary mouse embryo fibroblasts probed with mutants bearing deletions in the 3' end of SV40 gene A, *Virology* 162:76 (1988).
24. J. Zhu, P.W. Rice, L. Gorsch, M. Abate, and C.N. Cole, Transformation of a continuous rat embryo fibroblast cell line requires three separate domains of simian virus 40 large T antigen, *J. Virol.* 66:2780 (1992).
25. J.Y. Zhu, M. Abate, P.W. Rice, and C.N. Cole, The ability of simian virus 40 large T antigen to immortalize primary mouse embryo fibroblasts cosegregates with its ability to bind to p53, *J. Virol.* 65:6871 (1991).
26. P. Yaciuk, M.C. Carter, J.M. Pipas, and E. Moran, SV40 large T-antigen expresses a biological activity complementary to p300-associated transforming function of the adenovirus E1A gene products, *Mol. Cell. Biol.* 11:2116 (1991).
27. P. Yaciuk, and E. Moran, Analysis with specific polyclonal antiserum indicates that the E1A-associated 300 KDa product is a stable nuclear phosphoprotein that undergoes cell cycle phase-specific modification, *Mol. Cell. Biol.* 11:5389 (1991).
28. N. Dyson, and E. Harlow, Adenovirus E1A targets key regulators of cell proliferation, *in:* "Cancer Surveys," A.J. Levine, ed. Cold Spring Harbor Laboratory Press, Cold Spring Harbor (1992).
29. J.A. Mietz, T. Unger, J.M. Huibregtse, and P.M. Howley, The transcriptional transactivation function of wild-type p53 is inhibited by SV40 large T-antigen and by HPV-16 E6 oncoprotein, *EMBO J.* 11:5013 (1992).
30. P.W. Hinds, C.A. Finlay, and A.J. Levine, Mutation is required to activate the p53 gene for cooperation with the *ras* oncogene and transformation, *J. Virol.* 63:739 (1989).
31. J. Momand, G.P. Zambetti, D.C. Olson, D. George, and A.J. Levine, The mdm-2 oncogene product forms a complex with the p53 protein and inhibits p53 mediated transactivation, *Cell* 69:1237 (1992).
32. D. Dittmer, S. Pati, G. Zambetti, S. Chu, A.K. Teresky, M. Moore, C. Finlay, and A.J. Levine, p53 gain of function mutations, *Nature Gen.* 4:42 (1993).
33. G.P. Zambetti, D. Olson, M. Labow, and A.J. Levine, A mutant p53 protein is required for the maintenance of the transformed cell phenotype in p53 plus *ras* transformed cells, *Proc. Natl. Acad. Sci. USA* 89:3952 (1992).

37

DOWN REGULATION OF CANDIDATE TUMOR SUPPRESSOR GENES IN BREAST CANCER

Zhiqiang Zou, Anthony Anisowicz, Kristina Rafidi, and Ruth Sager

Division of Cancer Genetics
Dana-Farber Cancer Institute
44 Binney Street
Boston, MA 02115

INTRODUCTION

Changes in gene expression alter the tumorigenic potentialities of cancer cells either by increasing their oncogenic potential (oncogenes) or by decreasing it (tumor suppressor genes, here called TSGs) (1). The identification of oncogenes has been based on positive selection for loss of growth control, and has led to the recovery of numerous candidates, principally retroviral oncogenes, but the identification of TSGs based on negative selection has been meager. The need for positive selection schemes to identify TSGs has been a problem of outstanding importance in the molecular genetic analysis of cancer.

Methods of expression gene cloning approach this problem at the molecular level. Cells contain 100,000 genes in their genome, but only about 10,000 (i.e. 10%) are expressed in an average cell. Traditional methods of screening for genetic alterations in cancer or other diseases have focussed on the total genome, but expression gene cloning focusses on the expressed genes. Subtractive hybridization, for example, provides a means to recover genes that are over-expressed in one cell type compared with another (2). Recently, the development of the powerful Differential Display method of Liang and Pardee (3) has provided a highly productive method for expression gene cloning. We have applied both methods successfully and have identified, cloned and sequenced either partially or fully some 30 genes that are over-expressed in normal mammary epithelial cells grown in culture, compared with mammary carcinoma cells from cultured cell lines (4-7, and unpubl.).

The majority of these genes are either known or related to known genes, as determined by sequence comparisons in computer databanks. They have been divided into postulated functional categories: 9 genes involved in growth control including three that encode calcium-binding proteins; at least 7 genes encoding proteins involved in cytostructure or the extra-cellular matrix; one protease and three protease inhibitors; and a few less readily classifiable. In addition, there are a number of unknown genes, whose full or partial cDNA sequences offer no clues to their identity.

The Cell Cycle: Regulators, Targets, and Clinical Applications
Edited by V.W. Hu, Plenum Press, New York, 1994

Nine of these genes have been examined for expression over the cell cycle in normal mammary epithelial cells synchronized by growth factor deprivation. Five of the nine show cell cycle regulation. They are connexin 26 (5), CaNl9 (6), HBp17, elafin, and a novel dehydrogenase.

SELECTION PROCEDURES

In previous publications, candidate tumor suppressor genes were described that had been isolated by the hydroxy-apatite procedure (4-6). Additional genes to be discussed here have been isolated by the biotin-streptavidin procedure (8). Total cell RNA was isolated from growing normal (76N) and tumor (21MT-2) cells by the guanidinium isothiocyanate CsCl method, mRNA was purified by oligo dT affinity chromatography, and total RNA was fractionated on 1% agarose formaldehyde gels and transferred to nylon membranes.

A subtractive cDNA library was constructed from 76N cells and screened with total RNA from 21MT-2 cells as described (8). After two rounds of hybridization, the remaining single stranded DNA was made double-stranded and transformed into competent cells. The library was differentially screened on replica filters using 32P cDNA probes synthesized from 1 mg of mRNA from normal and tumor cells. Potential clones were confirmed by Northern hybridization.

GENE EXPRESSION IN CANDIDATE TUMOR SUPPRESSOR GENES

Using both procedures, close to 30 genes were identified as differentially expressed candidate tumor suppressor genes, following cloning and full or partial sequencing of the cDNAs. Although only a few of the genes are under investigation at this time, certain generalizations can be made.

It now seems likely that the majority of candidate TSGs detected by differential expression are down-regulated but not themselves mutated. Genes whose mutation is directly reflected in the phenotype of the cell are called Class I, whereas those whose regulation is changed by mutations occurring elsewhere are called Class II. This distinction is very important. In applications to cancer therapy, for example, genes of Class I can only be re-expressed by gene transfer, whereas Class II genes may be re-expressed by treatment with drugs, metabolites, or other factors, leading to normalization of the tumor phenotype. Furthermore, sets of Class II genes may be coordinately regulated by the same signal transduction pathway, so that the same treatments may lead to re-expression of clusters of functionally related tumor suppressor genes.

In fact, two of the genes that we have been investigating, have shown phenotypic reversion following treatment with the phorbol ester PMA, leading to their re-expression at the mRNA level in tumor cells. The genes are connexin 26, a gap junction protein, (5) and maspin, a novel member of the serpin (serine protease inhibitor) family (Z.Z. and R.S., unpubl.) In addition, we found that CaNl9, a gene encoding a small calcium-binding protein, can be re-induced in tumor cells by treatment with 5-azadeoxycytidine, presumably by inhibiting DNA methylation (6).

Most of the genes so far identified in this laboratory are involved in either one of the two major processes that go wrong in tumor cells: loss of growth control and gain of ability to invade through the basement lamina and become metastatic. The genes that are likely to be involved in loss of growth control include three that encode calcium-binding proteins (6,9), a novel cytokine called *gro*, (10), and two connexin genes that encode gap junction proteins (5). Genes that may be involved in metastasis include genes encoding components of the extra-cellular matrix and membrane receptors such as α6 integrin (7), as well as proteases, whose expression is elevated in invasion, and protease inhibitors, whose expression is decreased or lost (11).

In examining cell cycle regulation of these genes, it came as a surprise that five out of the first nine genes examined each showed cell cycle regulation, and the patterns of regulated expression were different for each of the five genes. One of them, a novel dehydrogenase is expressed during G1 and is down-regulated at or before the G1/S boundary. A second gene, elafin, exhibiting what is perhaps the most interesting pattern, is down-regulated specifically in S-phase, and the inhibition is released in G2. The third gene, encoding a heparin binding protein, is expressed during S-phase, and its expression levels off in G2. As previously published (5), Cx26, encoding a gap junction protein, shows a small induction during G1, and further elevation in late S or G2, whereas the other gap junction protein expressed in mammary epithelial cells, encoded by Cx43, is constant in mRNA expression across the cell cycle. Lastly, CaNl9, as previously reported (6), shows a major elevation of expression at or near the G1/S boundary.

In summary, some 30 candidate tumor suppressor genes in breast cancer have been selected by expression gene cloning. Most of these genes seem, from preliminary evidence, to be involved either in growth control or in processes leading to invasion and metastasis. Five of the first nine genes examined show cell cycle regulation. Of particular importance with respect to modes of therapy, most of these genes are probably Class II genes, down-regulated rather than mutated in tumor cells. As such they may be up-regulated or re-expressed in tumor cells by treatment with drugs or metabolites, leading to a relatively non-toxic form of therapy in which tumor cells are normalized rather than killed.

REFERENCES

1. R. Sager, Tumor suppressor genes: The puzzle and the promise, *Science* 246:1406 (1989).
2. M.D. Scott, H.H. Westphal, and R.W.J. Rigby, *Cell* 34:557 (1983).
3. P. Liang, L. Averboukh, K. Keyomarsi, and R. Sager, Differential display and cloning of mRNAs from human breast cancer versus mammary epithelial cells, *Cancer Res.* 52:6966-6968 (1992).
4. S. Lee, C. Tomasetto, and R. Sager, Positive selection of candidate tumor suppressor genes by subtractive hybridization, *Proc. Natl. Acad. Sci. USA* 88:2825 (1991).
5. S.W. Lee, C. Tomasetto, D. Paul, K. Keyomarsi, and R. Sager, R, Transcriptional down-regulation of gap junction proteins blocks junctional communication in human mammary tumor cell lines, *J. Cell Biol.* 118:1213 (1992).
6. S. Lee, C. Tomasetto, K. Swisshelm, K. Keyomarsi, and R. Sager, Down regulation of a new member of the S100 gene family in mammary carcinoma cells and reexpression by azadeoxycytidine treatment, *Proc. Natl. Acad. Sci. USA* 89:2504 (1992).

7. R. Sager, A. Anisowicz, P. Liang, G. Sotiropoulou, Identification by differential display of alpha 6 integrin as a candidate tumor suppressor gene, *FASEB J.* (1993) in press.
8. A. Swaroop, J. Xu, A. Neeraj, and S.M. Weissman, A simple and efficient cDNA library subtraction procedure: isolation of human retina-specific cDNA clones, *Nucl. Acid Res.* 19:1954 (1991).
9. P. Yaswen, A. Smoll, D.M. Peehl, D.K. Trask, R. Sager, and M.R. Stampfer, Down-regulation of a calmodulin-related gene during transformation of human mammary epithelial cells, *Proc. Natl. Acad. Sci. USA* 87:7360 (1990).
10. R. Sager, A. Anisowicz, M.C. Pike, P. Beckmann, and T. Smith, Structural, regulatory, and functional studies of the GRO gene and protein. In: Interleukin 8 and Related Chemotactic Cytokines, (M. Baggiolini and C. Sorg, eds.) Karger, Basel (1991).
11. L.A. Liotta, P.S. Steeg, and W.G. Stetler-Stevenson, Cancer metastasis and angiogenesis: an imbalance of positive and negative regulation, *Cell* 64:327 (1991).

38

EXPRESSION AND REGULATION OF CYCLIN GENES IN BREAST CANCER CELLS

Elizabeth A. Musgrove, Michael F. Buckley, Anna deFazio,
Colin K.W. Watts, and Robert L. Sutherland

Cancer Biology Division
Garvan Institute of Medical Research
St Vincent's Hospital
Sydney, NSW 2010, Australia

INTRODUCTION

The regulatory subunits of cell cycle-regulated kinases, cyclins, are key regulators of cell cycle progression in eukaryotic cells. Mammalian cells contain at least five cyclin classes (cyclins A to E), which reach maximum abundance at different points in the cell cycle[1]. The transcriptional activation of cyclin genes and consequent transient accumulation of different cyclin proteins which then bind to one of the cyclin-dependent kinases (cdks) to initiate phosphorylation cascades is thought to be the central mechanism for a series of control points in the mammalian cell cycle[1-3]. In synchronised or growth factor-stimulated mammalian cells, cyclins C, D1, D2, D3 and E are most abundant during G_1 phase[4-8], suggesting that these cyclins may function at G_1 control points. This is supported by recent evidence that cyclin E is rate-limiting for progression through G_1[9] and that complex formation between D type cyclins and the *cdk4* kinase is maximum in late G_1 phase[10].

In responsive tissues, including normal and neoplastic breast epithelium, the regulation of proliferation by steroids and steroid antagonists occurs by cell cycle-specific actions on cells in G_1 phase[11-13]. The central mechanism of steroid hormone action is the regulation of transcription by ligand-activated steroid hormone receptors binding to specific response elements in the regulatory regions of target genes[14]. Steroid antagonists bind to the ligand-binding domain of steroid hormone receptors and interfere with their transactivation of gene expression[15,16]. Steroid and steroid antagonist effects on proliferation are thus likely to be mediated by modulation of the expression of specific genes intimately involved in the control of cell cycle progression. The target genes responsible are unknown, but the transcriptional regulation of cyclins suggested that they might be part of the mechanism by which steroids and steroid antagonists regulate cell cycle progression.

Their central role in cell cycle control has led to the suggestion that cyclins are proto-oncogenes[1]. The cyclin D1 gene (*PRAD1, BCL1*) is located at 11q13, one of the most frequently amplified regions in human carcinomas[17] and is the favoured candidate oncogene associated with translocations involving this locus in a subset of B-cell lymphomas[18,19]. Up to 23% of breast tumour specimens have 11q13 amplification[17]

Since data on cyclin gene expression and regulation in breast cancer are limited, cyclin gene expression in 19 human breast cancer cell lines was characterised by Southern and Northern blot analysis, and the relationship between regulation of proliferation and regulation of G_1 cyclin gene expression was examined using T-47D breast cancer cells.

CYCLIN GENE EXPRESSION AND AMPLIFICATION IN HUMAN BREAST CANCER

Expression of Cyclin mRNA in Breast Cancer Cell Lines

A series of human breast cancer cell lines was screened for the expression of cyclin A, B1, C, D1 and E mRNAs[20]. These cell lines encompass a spectrum of phenotypes, ranging from estrogen receptor-positive, hormone-responsive, more differentiated cell lines to estrogen receptor-negative, hormone-independent cell lines, representative of more aggressive, poorly differentiated tumors. Although the cyclins were uniformly expressed in the majority of the cell lines, increased expression of one or more of the cyclin A, B1, D1 or E genes was found in 7/19 cell lines (Table 1). Cyclin C was not highly expressed in any of the cell lines examined.

Cyclin A expression was clearly increased in two cell lines, MDA-MB-157 and BT-549. One of these cell lines, MDA-MB-157, was the only cell line in which increased expression of cyclin E was observed. Increased expression of cyclin B1 mRNA was observed in BT-549 cells. This was an unexpected result, since the overexpression of either wild-type or truncated avian cyclin B arrests cells in mitosis, with evidence for aberrant spindle formation and eventual cell loss[21]. Thus, rather than being oncogenic, constitutive expression of cyclin B would be predicted to lead to cell cycle arrest.

Table 1. Expression and amplification of cyclin genes in 19 human breast cancer cell lines*

Estrogen receptor positive		
MCF-7M†	Cyclin D1	Overexpressed but not amplified
MDA-MB-134	Cyclin D1	Overexpressed and amplified
MDA-MB-361	Cyclin D1	Amplified but not overexpressed
ZR-75-1	Cyclin D1	Amplified but not overexpressed
BT-474, BT-483, MCF-7†, T-47D		No alteration observed
Estrogen receptor negative		
BT-549	Cyclin A	Overexpressed but not amplified
	Cyclin B1	Overexpressed but not amplified
	Cyclin D1	Decreased expression
DU-4475	Cyclin D1	Decreased expression
MDA-MB-157	Cyclin A	Overexpressed but not amplified
	Cyclin E	Overexpressed and amplified
	Cyclin D1	Decreased expression
MDA-MB-175	Cyclin D1	Overexpressed and amplified
MDA-MB-330	Cyclin D1	Overexpressed and amplified
MDA-MB-453	Cyclin D1	Overexpressed and amplified
MDA-MB-231, MDA-MB-436, MDA-MB-468, BT-20, Hs-578T, SK-BR-3		No alteration observed

* Based on data in reference 20.

† Two variants of MCF-7 cells from different sources were examined.

Increased expression of cyclin D1 was the most common alteration in cyclin gene expression noted. This gene was highly expressed in MDA-MB-134, -175, -330 and -453 cells and one of two MCF-7 variants, compared with the level of mRNA observed in the majority of the breast cancer cell lines and in two strains of normal, non-transformed breast epithelial cells[20]. Very low levels of cyclin D1 expression were noted in two breast cancer cell lines (DU-4475 and BT-549) and in HBL-100, a cell line derived from normal breast epithelial cells but known to contain SV40 sequences. None of the changes in cyclin gene expression could be accounted for by differences in the proportion of cells in S phase as assessed by reprobing filters for histone H4 expression.

Cyclin Gene Amplification in Breast Cancer Cell Lines

Southern blot analysis was used to determine if the increased expression of cyclin genes observed in the breast cancer cell lines was due to cyclin gene amplification. No gross rearrangements were noted for any of the genes examined. Amplification of the cyclin E gene was present in the MDA-MB-157 cell line, accompanying increased expression of cyclin E mRNA (Table 1). Cyclin D1 gene amplification was detected in six cell lines but amplification was not a prerequisite for, and did not always lead to, increased cyclin D1 expression (Table 1).

Cyclin Expression in Breast Cancer

In view of the dissociation between cyclin D1 amplification and mRNA expression in breast cancer cell lines, cyclin D1 expression was examined in a series of 124 well-characterised breast tumours and 16 samples of histologically normal breast tissue[20]. All tumour samples had detectable levels of cyclin D1 mRNA although large variations in the level of expression were observed. Fifty-six of the 124 tumours analysed (45%) expressed levels of cyclin D1 mRNA that exceeded the maximum level observed in sixteen specimens of histologically normal breast tissue[20]. Thus the increased expression of cyclin D1 displayed by some breast cancer cell lines is also observed *in vivo*.

CYCLIN GENE REGULATION IN T-47D BREAST CANCER CELLS

Cyclin Gene Expression after Growth Factor Stimulation

The relationship between cell cycle position and cyclin gene expression in breast cancer cells was examined using the T-47D cell line. None of the cyclin genes examined was either amplified or overexpressed in this cell line (Table 1). In serum-free medium T-47D cells become growth-arrested but re-initiate proliferation upon the addition of a single growth factor (e.g., insulin, IGF-I, bFGF, EGF), progressing synchronously through the cell cycle from early G_1[23]. Sequential induction of cyclin gene expression was observed following stimulation of growth-arrested T-47D cells with insulin (Figure 1). Cyclin D1 mRNA levels were increased 2- to 3-fold within 2 h of insulin addition while increased cyclin D3 mRNA levels were observed as cells progressed through G_1. Increased expression of cyclin E and A coincided with entry into S phase; cyclin E expression declined as cells passed through S phase but cyclin A expression remained elevated. Induction of cyclins D1, D3 and E in T-47D cells was not restricted to insulin stimulation, but was also observed after treatment with other potent breast cancer mitogens, i.e. IGF-I, fetal calf serum[22] and bFGF (unpublished data). Furthermore, the proportion of cells which later entered S phase appeared to be related to the degree of induction of cyclins D1, D3 and E[22].

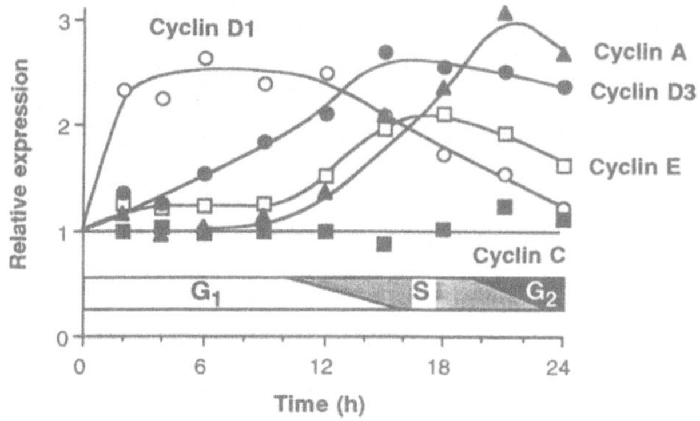

Figure 1. Effect of insulin on the expression of cyclins A, C, D1, D3 and E. T-47D cells growth arrested in serum free medium were treated with 1.7 μM insulin to re-initiate cell cycle progression. Densitometric analysis of Northern blots of total cellular RNA is shown, redrawn from reference 22. Measurement of histone H4 and cyclin B1 expression was used to monitor cell cycle progression of the synchronised cells.

Regulation of Cyclin Gene Expression by Steroid Antagonists

The induction of cyclin D1 gene expression within 2 h of mitogenic stimulation is compatible with a role for this gene in early G_1 phase, a time when breast cancer cells are sensitive to the growth-inhibitory effects of antiestrogens and antiprogestins[24-26]. Examination of cyclin expression after treatment with the antiestrogen ICI 164384, a potent inhibitor of breast cancer cell cycle progression[25-27] showed time-dependent decreases in the level of cyclin D1 and cyclin E mRNA but not cyclin D3 mRNA (Figure 2). The decreases were similar in magnitude (50-60%) to the decreases in histone H4 expression and %S phase[22]. Cyclin D1 expression began to decrease within 4 h of antiestrogen treatment, substantially preceding any change in histone H4 expression (Figure 2). Changes in cyclin E expression occurred later than changes in cyclin D1 expression, but still preceded changes in histone H4 expression (Figure 2). Thus, the regulation of cyclin D1 and E expression by ICI 164384 is not merely a consequence of growth arrest.

Changes in cell cycle phase distribution occur over a similar time-frame after antiestrogen or antiprogestin treatment[26]. However, clear differences in the regulation of G_1 cyclin genes were observed after treatment with these compounds. While cyclin D1 expression decreased after antiestrogen treatment, the antiprogestin failed to affect the expression of this gene (Figure 2). In contrast to the changes in cyclin D1 expression, cyclin D3 expression, which was unaffected by antiestrogen treatment, was markedly decreased by antiprogestin treatment, to < 50% of control (Figure 2)[22]. This decrease was apparently coincident with the decrease in histone H4 expression and was preceded by a decrease in the level of cyclin E mRNA (Figure 2).

CONCLUSIONS

Despite evidence for abnormal cyclin gene regulation in some breast cancer cell lines[28], the similarity between the patterns of cyclin gene expression in mitogen-stimulated normal mammary epithelial cells[6] and T-47D breast cancer cells[22] suggests that these cyclins play a similar role in the regulation of proliferation in both normal and neoplastic mammary epithelium. The early induction of cyclin D1 indicates that it may be intimately

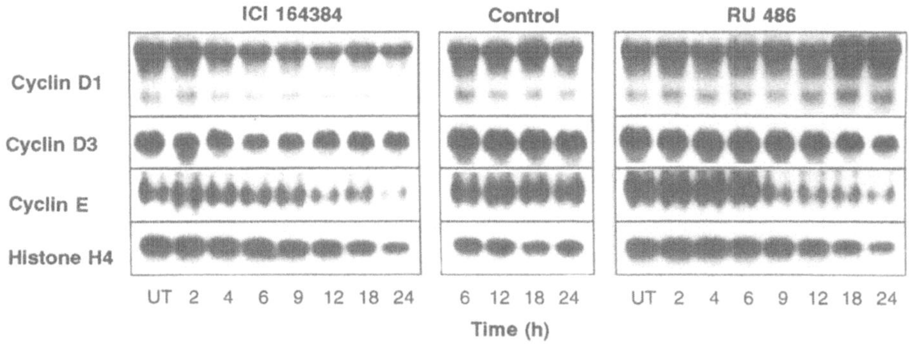

Figure 2. Effect of the estrogen antagonist ICI 164384 and the progestin antagonist RU 486 on cyclin D1, D3 and E mRNA. Triplicate flasks of T-47D cells proliferating in insulin-supplemented serum-free medium were treated with ICI 164384 (500 nM), RU 486 (100 nM) or vehicle and harvested at intervals for Northern blot analysis. UT: untreated. One of four replicate filters was probed for each mRNA species. Redrawn from reference 22.

associated with the initiation of progress through G_1 phase, a hypothesis confirmed by recent experiments in this laboratory. Regulation of cyclin D1 gene expression is a rapid response to regulation of breast cancer cell proliferation, not only after growth factor and steroid stimulation but also after antiestrogen-mediated growth inhibition[22]. Evidence that inhibition of cyclin D1 expression or function (for example, by the use of antibodies or antisense techniques)[29,30] inhibits entry into S phase suggests that the decrease in cyclin D1 expression after antiestrogen treatment may contribute to growth arrest. Furthermore, the difference in cyclin gene regulation by antiestrogens and antiprogestins suggests a mechanism by which tumors which are relatively insensitive to one compound may retain sensitivity to other hormonal therapies.

The data presented above implicate aberrant expression of several cyclin genes as a potential factor in the pathogenesis of breast cancer, but suggest that cyclin D1 in particular may be important in breast cancer cell cycle control. Overall, amplification of a cyclin gene was observed in only about half of the cell lines where the cyclin was overexpressed and conversely, there were examples of cyclin D1 gene amplification without increased cyclin expression. The latter observation is consistent with the idea that there may be more than one potential oncogene in the 11q13 amplicon. In addition to the cyclin D1 gene, this region of chromosome 11 contains the *FGF3* and *FGF4* genes, which are only rarely expressed in breast cancer[17], and *EMS1*, which, like cyclin D1, is commonly expressed[31]. Amplification of 11q13 has been associated with poor prognosis in breast cancer[32-34] but these analyses may have pooled several distinct molecular pathologies and may need to be reassessed by determining the level of expression of a number of genes within the 11q13 amplicon. However, one possibility requiring investigation is that deregulated expression of cyclin D1 may confer antiestrogen resistance, perhaps contributing to poor prognosis for a subset of patients displaying 11q13 amplification.

ACKNOWLEDGMENTS

These studies were supported by the National Health and Medical Research Council of Australia, the NSW State Cancer Council, MLC-Life Ltd and the St Vincent's Hospital Research Fund. Elizabeth Musgrove is an MLC-Life Research Fellow.

REFERENCES

1. T. Hunter, and J. Pines, Cyclins and cancer, *Cell* 66:1071 (1991).

2. S.I. Reed, The role of p34 kinases in the G1 to S-phase transition, *Annu. Rev. Cell Biol.* 8:529 (1992).

3. P. Nurse, Universal control mechanism regulating onset of M-phase, *Nature* 344:503 (1990).

4. H. Matsushime, M.F. Roussel, R.A. Ashmun, and C.J. Sherr, Colony-stimulating factor 1 regulates novel cyclins during the G1 phase of the cell cycle, *Cell* 65:701 (1991).

5. T. Motokura, T. Bloom, H.G. Kim, H. Jüppner, J. Ruderman, H. Kronenberg, and A. Arnold, A novel cyclin encoded by a *bcl1*-linked candidate oncogene, *Nature* 350:512 (1991).

6. T. Motokura, K. Keyomarsi, H.M. Kronenberg, and A. Arnold, Cloning and characterization of human cyclin D3, a cDNA closely related in sequence to the PRAD1/cyclin D1 proto-oncogene, *J. Biol. Chem.* 267:20412 (1992).

7. D.J. Lew, V. Dulic, and S.I. Reed, Isolation of three novel human cyclins by rescue of G1 cyclin (Cln) function in yeast, *Cell* 66:1197 (1991).

8. K.-A. Won, Y. Xiong, D. Beach, and M.Z. Gilman, Growth regulated expression of D-type cyclin genes in human diploid fibroblasts, *Proc. Natl. Acad. Sci. USA* 89:9910 (1992).

9. M. Ohtsubo, and J.M. Roberts, Cyclin-dependent regulation of G_1 in mammalian fibroblasts, *Science* 259:1908 (1993).

10. H. Matsushime, M.E. Ewen, D.K. Strom, J.-Y. Kato, S.K. Hanks, M.F. Roussel, and C.J. Sherr, Identification and properties of an atypical catalytic subunit ($p34^{PSK\text{-}J3}$/cdk4) for mammalian D type G1 cyclins, *Cell* 71:323 (1992).

11. R.L. Sutherland, R.R. Reddel, and M.D. Green, Effects of oestrogens on cell proliferation and cell cycle kinetics. A hypothesis on the cell cycle effects of antioestrogens, *Eur. J. Cancer Clin. Oncol.* 19:307 (1983).

12. E.A. Musgrove, and R.L. Sutherland, Steroids, growth factors and cell cycle controls in breast cancer, *in:* "Regulatory Mechanisms in Breast Cancer". M.E. Lippman, and R.B. Dickson, eds, Kluwer Academic Publishers, Boston (1991).

13. E.A. Musgrove, C.S.L. Lee, and R.L. Sutherland, Progestins both stimulate and inhibit breast cancer cell cycle progression while increasing expression of transforming growth factor α, epidermal growth factor receptor, c-*fos* and c-*myc* genes, *Mol. Cell. Biol.* 11:5032 (1991).

14. M. Beato, Gene regulation by steroid hormones, *Cell* 56:335 (1989).

15. S.E. Fawell, J.A. Lees, R. White, and M.G. Parker, Characterization and colocalization of steroid binding and dimerization activities in the mouse estrogen receptor, *Cell* 60:953 (1990).

16. N.J.G. Webster, S. Green, J.R. Jin, and P. Chambon, The hormone-binding domains of the estrogen and glucocorticoid receptors contain an inducible transcription activation function, *Cell* 54:199 (1988).

17. P. Gaudray, P. Szepetowski, C. Escot, D. Birnbaum, and C. Theillet, DNA amplification at 11q13 in human cancer: from complexity to perplexity, *Mutation Res.* 276:317 (1992).

18. C.L. Rosenberg, E. Wong, E.M. Petty, A.E. Bale, Y. Tsujimoto, N.L. Harris, and A. Arnold, *PRAD1*, a candidate *BCL1* oncogene: mapping and expression in centrocytic lymphoma, *Proc. Natl. Acad. Sci. USA.* 88:9638 (1991).

19. D.A. Withers, R.C. Harvey, J.B. Faust, O. Melnyk, K. Carey, and T.C. Meeker, Characterization of a candidate *bcl-1* gene, *Mol. Cell. Biol.* 11:4846 (1991).

20. M.F. Buckley, K.J.E. Sweeney, J.A. Hamilton, R.L. Sini, D.L. Manning, R.I. Nicholson, A. deFazio, C.K.W. Watts, E.A. Musgrove, and R.L. Sutherland, Expression and amplification of cyclin genes in human breast cancer, *Oncogene*, in press. (1993).

21. P. Gallant, and E.A. Nigg, Cyclin B2 undergoes cell cycle-dependent nuclear translocation and, when expressed as a non-destructible mutant, causes mitotic arrest in HeLa cells, *J. Cell Biol.* 117:213 (1992).

22. E.A. Musgrove, J.A. Hamilton, C.S.L. Lee, K.J.E. Sweeney, C.K.W. Watts, and R.L. Sutherland, Growth factor, steroid and steroid antagonist regulation of cyclin gene expression associated with changes in T-47D human breast cancer cell cycle progression, *Mol. Cell. Biol.* 13:3577 (1993).

23. E.A. Musgrove, and R.L. Sutherland, Acute effects of growth factors on T-47D breast cancer cell cycle progression, *Eur. J. Cancer*, in press (1993).

24. I.W. Taylor, P.J. Hodson, M.D. Green, and R.L. Sutherland, Effects of tamoxifen on cell cycle progression of synchronous MCF-7 human mammary carcinoma cells, *Cancer Res.* 43:4007 (1983).

25. E.A. Musgrove, A.E. Wakeling, and R.L. Sutherland, Points of action of estrogen antagonists and a calmodulin antagonist within the MCF-7 human breast cancer cell cycle, *Cancer Res.* 49:2398 (1989).

26. E.A. Musgrove, and R.L. Sutherland, Effects of the progestin antagonist RU 486 on T-47D cell proliferation, submitted.

27. A.E. Wakeling, and J. Bowler, Steroidal pure antioestrogens, *J. Endocrinol.* 112:R7 (1987).

28. K. Keyomarsi, and A.B. Pardee, Redundant cyclin overexpression and gene amplification in breast cancer cells, *Proc. Natl. Acad. Sci. USA.* 90:1112 (1993).

29. A. Sala, and B. Calabretta, Regulation of BALB/c 3T3 fibroblast proliferation by B-myb is accompanied by selective activation of *cdc2* and cyclin D1 expresssion, *Proc. Natl. Acad. Sci. USA.* 89:10415 (1992).

30. V. Baldin, J. Lukas, M.J. Marcote, M. Pagano, and G. Draetta, Cyclin D1 is a nuclear protein required for cell cycle progression in G_1, *Genes Dev.* 7:812 (1993).

31. E. Schuuring, E. Verhoeven, W.J. Mooi, and R.J.A.M. Michalides, Identification and cloning of two overexpressed genes, U21B31/*PRAD*1 and *EMS*1, within the amplified chromosome 11q13 region in human carcinomas, *Oncogene* 7:355 (1992).

32. H. Tsuda, S. Hirohashi, Y. Shimosato, T. Hirota, S. Tsugane, H. Yamamoto, N. Miyajima, K. Toyoshima, T. Yamamoto, J. Yokota, T. Yoshida, H. Sakamoto, M. Terada, and T. Sugimura, Correlation between long-term survival in breast cancer patients and amplification of two putative oncogene-coamplification units: *hst*-1/*int*-2 and c-*erb*B2/*ear*-1, *Cancer Res.* 49:3104 (1989).

33. A. Borg, H. Sigurdsson, G.M. Clark, M. Ferno, S.A. Fuqua, H. Olson, D. Kilander, and W.L. McGuire, Association of INT2/HST coamplification in primary breast cancer with hormone-dependent phenotype and poor prognosis, *Br. J. Cancer.* 63:136 (1991).

34. E. Schuuring, E. Verhoeven, H. van Tinteren, J.L. Peterse, B. Nunnink, F.B.J.M. Thunnissen, P. Devilee, C.J. Cornelisse, M.J. van de Vijver, W.J. Mooi, and R.J.A.M. Michalides, Amplification of genes within the chromosome 11q13 region is indicative of poor prognosis in patients with operable breast cancer, *Cancer Res.* 52:5229 (1992).

39

GENOTOXIN-INDUCED APOPTOSIS: IMPLICATIONS FOR CARCINOGENESIS

Steven R. Patierno, Lori J. Blankenship, John P. Wise, Jian Xu, Laura C. Bridgewater and Francis C.R. Manning

Department of Pharmacology, The George Washington University Medical Center 2300 Eye Street N.W., Washington DC 20037

INTRODUCTION

Apoptosis is thought to be a programmed form of cell death which, when balanced by cell proliferation, contributes to the regulation of tissue homeostasis (1). An imbalance between apoptosis and proliferation may be a factor in tumorigenesis. Apoptosis differs from necrosis both morphologically and biochemically. Necrotic cells exhibit an early loss of the ionic gradient across the cell membrane and non-cell cycle dependent random DNA degradation subsequent to release of lysosomal enzymes (1,2). In contrast, cells dying by apoptosis maintain ionic gradients and characteristically exhibit internucleosomal DNA fragmentation (IDF) before the loss of cell membrane integrity occurs (3). IDF results from the activation of a nuclear, Ca^{2+}/Mg^{2+} dependent or pH sensitive endonuclease, possibly DNase I or II (4). Apoptosis appears to be related to arrest of the cell cycle (3,5,6). Induction of apoptosis often requires a proliferative stimulus and can be thought of as process of abortive mitosis.

Recently the observations that stress-inducing conditions, including exposure to genotoxic agents (3,5-9), can induce apoptosis have raised questions about the its potential role in carcinogenesis. One obvious possibility is that apoptosis may protect an organism from carcinogenesis by destroying cells with pre-mutagenic genetic damage. It is also possible that the induction of apoptosis may enhance carcinogenesis. This may occur passively as a result of stimulating regeneration of surrounding cells which may contain induced or spontaneous non-lethal but pre-mutagenic lesions (similar to necrosis caused by non-genotoxic agents). It is also possible that the induction of apoptosis may result in the selection of cells with either an intrinsic resistance to apoptosis (i.e., cells with an altered senescence/differentiation program), or cells with an aggressive proliferative capacity which can replicate damaged DNA (i.e., survive low fidelity DNA replication).

We have been studying hexavalent chromium compounds which are carcinogenic to humans and experimental animals (10), genotoxic (11-15), and which transform cells in culture (16-18). Treatment of cultured cells with moderately toxic doses of sodium chromate results in an immediate inhibition of cell growth and macromolecular synthesis, and accumulation of cells in the S-phase of the cell cycle. Many of the cells that did progress to metaphase exhibited chromosome damage. Cell death was delayed for 48 h and was accompanied by detection of internucleosomal fragmentation of DNA. The

The Cell Cycle: Regulators, Targets, and Clinical Applications
Edited by V.W. Hu, Plenum Press, New York, 1994

nuclear matrix was highly enriched in chromium-induced DNA damage, and matrix DNA damage and repair correlated strongly with inhibition and recovery of both replication and transcription. Biologically relevant levels of chromium-DNA adducts formed on a synthetic DNA template *in vitro* inhibited progression of a DNA polymerase in a dose-dependent manner. These results suggest that matrix-associated chromium-DNA adducts may be responsible for the cell cycle delay and subsequently the induction of apoptosis, caused by chromate. Carcinogen-treated cells which overcome the cell cycle block and escape apoptosis may begin the process of multistage tumorigenesis and it will, therefore, be important to determine whether this is a stochastic or selection-driven process.

TOXICITY AND EFFECTS ON CELL GROWTH, DNA SYNTHESIS AND CELL CYCLE

Treatment of CHO cells with 150 or 300 μM sodium chromate for 2 h reduced colony forming efficiency by 46 and 92%, respectively, compared with untreated controls (13). Figure 1 shows that immediately after treatment with either dose of sodium chromate, the growth of CHO cells was suppressed. By day 3 after treatment, the number of adherent cells in the 150 μM and 300 μM treatment groups decreased by 35% and 71% respectively, indicating that considerable cell death occurred during this period. The number of adherent cells did not begin to increase until day 6, after treatment with 150 μM and 300 μM chromate. To measure DNA synthesis, cells were pulse labeled with 1 μCi/ml [methyl-^{3}H]thymidine (60-90 Ci/mmol, ICN Biomedicals, Inc., Costa Mesa, CA) for 1 h. Cells were harvested and counted using a model Zf Coulter counter. Incorporation of radioactivity into cellular macromolecules (cpm/cell) was determined by scintillation counting of acid insoluble material by the method of Lehmann and Stevens (19). Table 1 shows the residual DNA synthesis occurring in sodium chromate treated CHO cells expressed as a percentage of that occurring in logarithmic phase control cells. Treatment with 150 μM and 300 μM chromate for 2 h immediately suppressed DNA synthesis to 23% and 14% of control values, respectively. This inhibition continued for at least 3 days in both treatment groups. By day 4, there was some recovery of DNA synthesis to 49% of control values in the 150 μM cells. DNA synthesis in cells treated with 300 uM chromate remained suppressed throughout day 5.

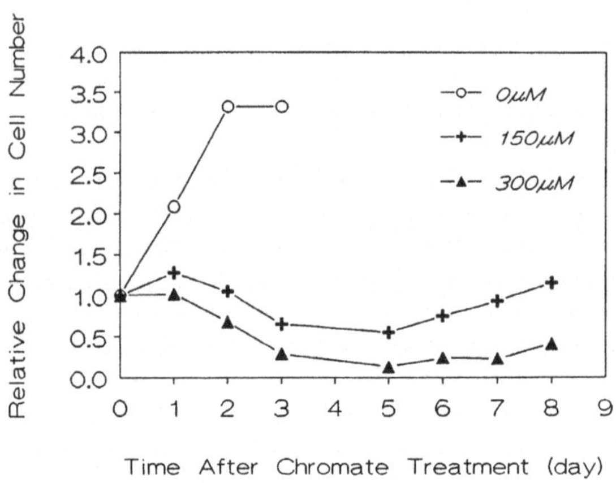

Figure 1. The change in cell number in cultures of CHO cells after treatment with chromate. Cells (1 x 10^6) were seeded into 10 cm dishes. After 48 h in culture, the cells were treated with 0 μM, 150 μM or 300 μM sodium chromate for 2 h. Harvested cells were counted with a model Zf Coulter Counter.

Table 1 Effect of Sodium Chromate on DNA Synthesis (percent of control)[a]

Days after Treatment	1	2	3	4	5
150 μM chromate	14	10	21	49	59
300 μM chromate	13	5	8	3	4

[a]Cells were treated for 2 hr and exposed to [^{3}H]thymidine for 1 hr before harvest. Incorporation (cpm/cell) in chromate treated cells is presented as a percent of that occurring in logarithmic phase control cells.

Table 2 Clastogenicity of sodium chromate[a]

Concentration of Sodium Chromate (μM)	0	150
Percent Metaphases with Damage	2 ± 1	33 ± 0

[a] Results are the average of 2 experiments (100 metaphases analyzed per experiment).

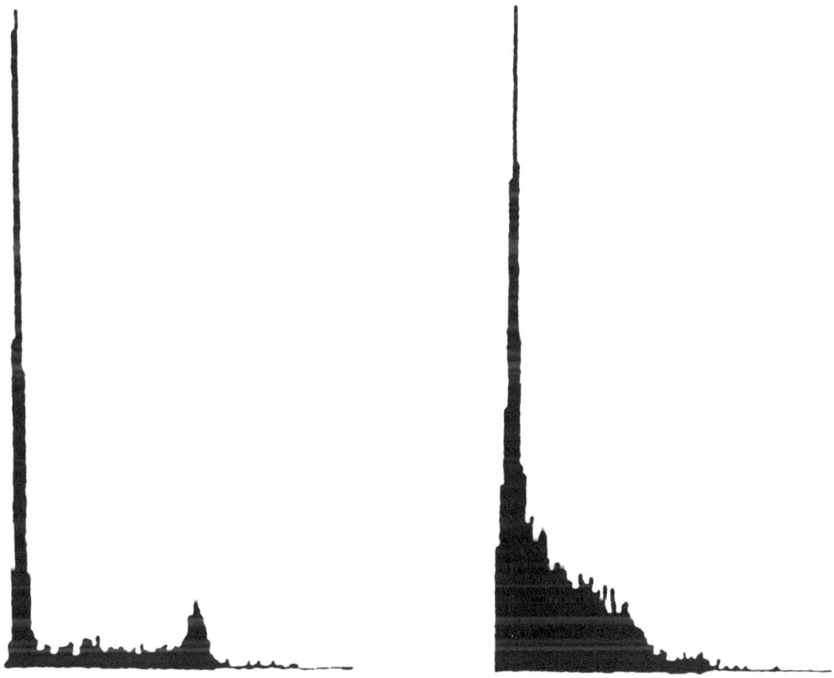

Figure 2. Effect of sodium chromate on the cell cycle. At 48 hr after treatment cellular DNA was analyzed by flow cytometry for DNA content and cell cycle distribution.

Cellular DNA was analyzed by flow cytometry at various times after treatment to determine the effect of chromate on the cell cycle. Figure 2 shows a representative experiment in which cells accumlated in S phase by 48 h after treatment with 150 uM chromate.

CHROMIUM-INDUCED CHROMOSOME DAMAGE

Chromosome preparations and scoring of individual types of damage were as described by Wise *et al.* (14). Table 2 shows that chromosomes prepared from those cells that reached metaphase 24 h after 2 h of treatment with 150 μM sodium chromate, contained high levels of damage (33% of metaphases with damage) compared with chromosomes obtained from untreated controls (2% metaphases with damage). Metaphase chromosomes could not be obtained 24 h after treatment of cells with 300 μM chromate.

ANALYSIS OF GENOMIC DNA FROM CHROMIUM-TREATED CELLS

Genomic DNA prepared from pooled unattached and adherent cells at various times after treatment with sodium chromate was analyzed by agarose gel electrophoresis (20). DNA isolated from untreated cells was high molecular weight (greater than 24 kb) and unfragmented at all times examined (Figure 3, CON lane). Dose dependent internucleosomal DNA fragmentation, indicative of apoptosis, was detectable 48 h (not shown) and 72 hr after treatment of cells with 150 μM or 300 μM chromate (Figure 3). IDF was complete by day 6.

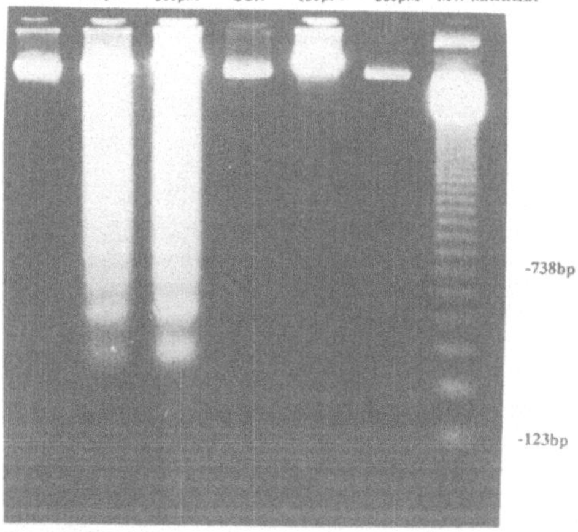

Figure 3. Analysis of genomic DNA isolated from CHO cells 3 days (72 h) and 7 days (168 h) after treatment with 0 μM, 150 μM or 300 μM chromate for 2 h. DNA was isolated from pooled adherent and non-adherent cells and analyzed on a 1.6% agarose gel.

DISTRIBUTION OF CHROMIUM-DNA ADDUCTS

In previous reports we described DNA damage induced in CHO cells by 2 h of treatment with 150 or 300 μM sodium chromate as measured by alkaline elution (13). DNA single-strand breaks and DNA-protein cross-links were detected immediately after treatment of cells with chromate. The single-strand breaks were essentially repaired within 8 h. DNA-protein cross-links were more persistent but were repaired in cells treated with 150 μM chromate by 24 h. Cells treated with the 300 μM dose contained a much reduced level of cross-links at 24 h compared with 0 h indicating that DNA repair had occurred. Chromium-DNA adducts were formed immediately after treatment of cells with [chromium-51]-labeled sodium chromate. After a slight reduction in adduct levels between 0 and 8 h post-treatment little further decrease occurred up to at least 32 h (13).

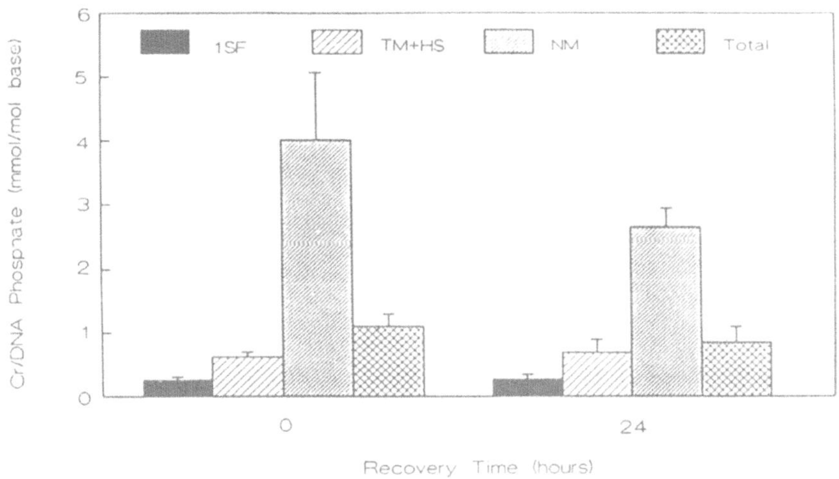

Figure 4. Cells were treated with 150 μM sodium chromate for 2 h and chromatin isolated 0 or 24 hrs after treatment. 1SF=First soluble fraction; TM+HS=bulk chromatin; NM=nuclear matrix

In order to study the distribution of adducts in different chromatin fractions, nuclei were isolated and chromatin fractionated by micrococcal nuclease digestion as previously described (21). In Figure 4 we show chromium-DNA adduct distribution in CHO cells treated with 150 μM chromate. Immediately after treatment with chromate, chromium-DNA adducts were markedly enriched in the nuclear matrix (approximately 4-fold compared with total nuclear DNA). Twenty-four h after treatment, chromium-DNA adducts were still enriched in nuclear matrix DNA but at a slightly decreased level compared with 0 h. Consistent with our previous data (13), however, there was little decrease in adduct levels in total nuclear DNA.

EFFECT OF CHROMIUM ON TRANSCRIPTION

Figure 5 shows the effect of sodium chromate on the synthesis of cytoplasmic RNA and mRNA in CHO cells expressed as a percentage of that occurring in control cells. The extent of [^{3}H]uridine incorporation into total cytoplasmic RNA and mRNA in a 1 h pulse was measured up to 32 h after treatment with sodium chromate. Treatment with 150 and 300 μM chromate suppressed total RNA synthesis by approximately 50-60% and 75-90%, respectively compared with controls (Figure 5). These concentrations also suppressed mRNA synthesis to approximately 30 and 20% of control levels, respectively. Suppression of transcription of both total and mRNA synthesis occurred immediately after treatment (13) and this suppression persisted for at least 32 h. Chromate was found to inhibit the induction of expression of the inducible GRP78 gene (13) by tunicamycin in a time and concentration dependent manner. CHO cells were exposed to 150 or 300 μM chromate for 2 h. At various times after this treatment, cultures were incubated with tunicamycin for 8 h to induce GRP78 gene expression. Increases in GRP78 mRNA levels were measured by northern blot hybridization analysis. Immediately after treatment of cells with 150 and 300 μM chromate, induction of GRP78 expression was suppressed to approximately 10% and 2% of that occurring in non-chromate treated cells, respectively (Figure 5, 8 h column). In contrast, 24 h after treatment with 150 μM chromate there was a partial recovery of GRP78 inducibility to approximately 56% of that occurring in untreated cells (Figure 5, 32 h column). There was, however, no induction of GRP78 expression in cultures treated with 300 μM chromate at this time.

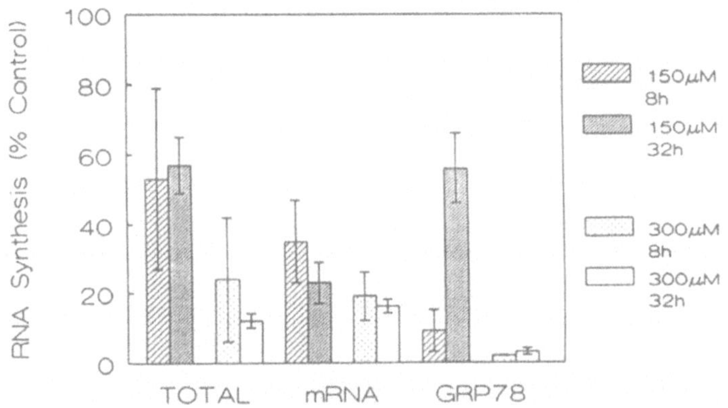

Figure 5. Cells were treated with 150 μM or 300 μM sodium chromate for 2 h and then transferred to fresh medium (0 h). For 1 h prior to the indicated times cells were incubated in the presence of [^{3}H]uridine to assess total cytoplasmic RNA synthesis and messenger RNA synthesis. Transcription occurring in chromate-treated cells was expressed as a percentage of that occurring in control cultures at the same time $\pm$ the standard error of the mean of 2-4 independent experiments. Either immediately after treatment with chromate (0 h) or 24 h later, cells were incubated with 10 μg/ml tunicamycin for 8 h to induce GRP78 expression. At the end of these incubation periods (8 and 32 h, respectively), RNA was isolated and GRP78 transcript levels were assayed by northern blot hybridization analysis. Results are expressed as a percentage of the increase in GRP78 mRNA levels occurring in non-chromate treated control cells exposed to tunicamycin $\pm$ standard error of the mean of 3 determinations.

EFFECTS OF CHROMIUM ON IN VITRO REPLICATION

To investigate the effect of chromium-DNA adducts on *in vitro* replication, a synthetic 88 bp DNA template was synthesized and cloned into the Eco R1-BamH1 site of the plasmid pSV2neo to form pSV2neoTS. The sequence of this template includes a 24 bp sequence 67% enriched in AT residues, a 17 bp sequence 100% enriched in GC residues, and a 17 bp self-complementary inverted repeat. The plasmid pSV2neoTS was linearized by digestion with Not I to generate a discrete 105 bp full length elongation product in the polymerase arrest assay. Linearized plasmid was treated at a concentration of 840 μM with 0, 35, 105, 210, 420 or 840 μM $CrCl_3$ for 16 h at room temperature. These treatments correspond to DNA-nucleotide: Cr^{3+} ratios of 1:0, 1:0.04, 1:0.13, 1:0.25, 1:0.5 and 1:1, respectively. The lowest two ratios correspond to the chromium-DNA adduct levels measured in the nuclear matrix DNA of treated cells. After removal of unbound chromium by microdialysis, the treated DNA was analyzed by polymerase arrest assay using Sequenase Version 2.0 T7 DNA polymerase. Few arrests occurred when untreated DNA was used as a template in this assay. The majority of synthesized DNA was of a size which corresponded to the expected 105 bp full-length fragment (Figure 7, control lane). Treatment of pSV2neoTS with $CrCl_3$ resulted in the dose-dependent premature replication termination (Figure 6, 35 μM to 840 μM lanes) in a sequence-specific manner, preferentially in the GC-rich (100% GC) and palindromic (65% GC) sequences.

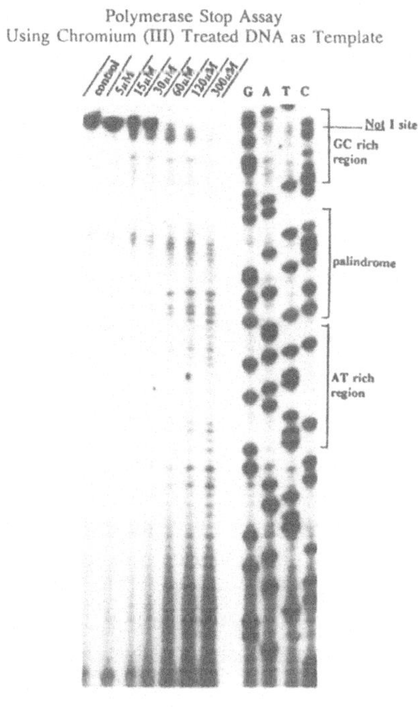

Figure 6. Analysis of chromium (III)-treated DNA by polymerase arrest assay. The plasmid pSV2neoTS was treated with the indicated concentrations $CrCl_3$ and replicated *in vitro* with Sequenase Version 2.0 T7 DNA polymerase. The products of these reactions were analyzed on a sequencing gel. Also shown are the corresponding sequencing reactions (G, A, T, and C lanes).

DISCUSSION

The data presented here have facilitated formulation of a working hypothesis concerning the mechanism of chromium-induced apoptosis and carcinogenesis. We postulate that trivalent chromium ions, having crossed the plasma membrane in the form of hexavalent chromate oxyanions, preferentially form adducts in nuclear matrix-associated DNA. These persistent, repair-resistant lesions block the progression of both DNA and

POTENTIAL ROLE OF APOPTOSIS IN CARCINOGENESIS

	INHIBITION
PROTECTION:	MECHANISM FOR CLEARING INITIATED CELLS; PREVENTING PROLIFERATION IN THE PRESENCE OF UNREPAIRED GENETIC DAMAGE.
	ENHANCEMENT
SELECTION:	OF CELLS WITH INTRINSIC RESISTANCE TO APOPTOSIS; OF CELLS WITH AN ALTERED SENESCENCE/DIFFERENTIATION PROGRAM.
SELECTION:	OF CELLS WITH A MORE AGGRESSIVE PROLIFERATIVE CAPACITY; OF CELLS THAT CAN SURVIVE LOW FIDELITY REPLICATION.
REGENERATION:	PROLIFERATION OF SURROUNDING CELLS WITH SUBLETHAL (PRE-MUTAGENIC) DAMAGE

Figure 7. Summary of the potential roles for apoptosis in carcinogenesis.

RNA polymerases thereby inhibiting both replication and transcription. This protracted inhibition of macromolecular synthesis causes cells to accumulate in S phase and ultimately activates the molecular program for apoptosis. The role of apoptosis in carcinogenesis is currently unknown but several pathways, including both inhibition and enhancement of the process of multistage carcinogenesis, are possible (Figure 7).

ACKOWLEDGEMENTS: This work was supported by NIH Grant RO1-ES-05304 and an Elaine H. Snyder Cancer Research Award. John P. Wise and Laura C. Bridgewater were supported by pre-doctoral Presidential Merit Scholarships from The George Washington University. We would like to thank Dr. David Wilkinson and Jean Rhame for their kind assistance with experiments involving flow cytometry. We would also like to acknowledge Drs. Elizabeth Woo and Louis Barrows for designing and synthesizing pSV2neoTS.

REFERENCES

1. Wyllie, A.H., Kerr, J.F.R. and Currie, A.R. Cell death: The significance of apoptosis. International Review of Cytology., 68: 251-306 (1980).
2. Kerr, J.F.R. and Harmon, B.V. Definition and incidence of apoptosis: An historical perspective. L.D. Tomei, and F.O. Cope, Ed., Apoptosis: The Molecular Basis of Cell death, p. 5-29. Plainview: Cold Spring Harbor Press, 1991.
3. Barry, M.A., Behnke, C.A. and Eastman, A. Activation of programmed cell death (apoptosis) by cisplatin, other anticancer drugs, toxins, and hyperthermia. Biochem. Pharm., 40: 2353-2362 (1990).
4. Eastman, A. and Barry, M. Platinum and other metal coordination compounds in cancer chemotherapy. Howell, S.B., Ed. Plenum Press, New York, 1991.
5. Muschel, R.J., Zhang, H.B., Iliakis, G. and McKenna, W.G. Cyclin B expression in hela cells during the G2 block induced by ionizing radiation. Cancer Res., 51: 5113-5117 (1991).
6. O'Connor, P.M., Wassermann, K., Sarang, M., Magrath, I., Bohr, V.A. and Kohn, K.W. Relationship between DNA cross-links, cell cycle, and apoptosis in Burkett's lymphoma cell lines differing in sensitivity to nitrogen mustard. Cancer Res., 51: 6550-6557 (1991).
7. Corcoran, G.B. and Ray, S.D. Contemporary issues in Toxicology. The role of the nucleus and other compartments in toxic cell death produced by alkylating hepatotoxicants. Tox. and Appl. Pharm., 113: 167-183 (1992).
8. Eastman, A. Activation of programmed cell death by anticancer agents: cisplatin as a model system. Cancer Cells., 2:275-280 (1990).
9. Kasten, M.B., Onyekewere, O., Sidransky, D., Vogelstein, B. and Craig, R.W. Participation of p53 protein in the cellular response to DNA damage. Cancer Res., 51: 6304-6311 (1991).
10. Chromium, Nickel and Welding, IARC monographs on the evaluation of carcinogenic risks to humans 49, IARC, Lyon, France (1990).
11. Sugiyama, M., Patierno, S.R., Cantoni, O., and Costa, M. Characterization of DNA lesions induced by $CaCrO_4$ in synchronous and asynchronous cultured mammalian cells. Mol. Pharm., 29: 606-613 (1986).
12. Xu, J., Wise, J.P., and Patierno, S.R. DNA damage induced by carcinogenic lead chromate particles in cultured mammalian cells. Mutat. Res., 280:129-136 (1992).
13. Manning, F.C.R., Xu, J., and Patierno, S.R. Transcriptional inhibition by carcinogenic chromate: Relationship to DNA damage. Mol. Carc., in press.
14. Wise, J.P., Leonard, J.C. and Patierno, S.R. Clastogenicity of lead chromate particles in hamster and human cells. Mutat Res 278: 69-79 (1992).
15. De Flora, S., Bagnasco, M., Serra, D. and Zanacchi, P. Genotoxicity of chromium compounds. A review. Mutat. Res., 238: 99-172 (1990).

16. Elias, Z., Poirot, O., Pezerat, H., Suquet, H., Schneider, O., Daniére, M.C., Terzetti, F., Baruthio, F., Fournier, M., and Cavelier, C. Cytotoxic and neoplastic transforming effects of industrial hexavalent chromium pigments in Syrian hamster embryo cells. Carcinogenesis, 10: 2043-2052 (1989).
17. Patierno, S.R., Banh, D. and Landolph, J.R. Transformation of C3H/10T1/2 mouse embryo cells to focus formation and anchorage independence by insoluble lead chromate but not soluble calcium chromate: relationship to mutagenesis and internalization of lead chromate particles. Cancer Res., 48: 5280-5288 (1988).
18. Landolph, J.R. Neoplastic transformation of mammalian cells by carcinogenic metal compound: cellular and molecular mechanisms. Foulkes, E.C., Ed. Biological Effects of Heavy Metals, Vol 2. CRC Press Inc, Boca Raton, Florida, 1990, p 1-18.
19. Lehmann, A.R. and Stevens, S. A rapid procedure for measurement of DNA repair in human fibroblasts and for complementation analysis of xeroderma pigmentosum cells. Mutat. Res., 69: 177-190 (1980).
20. Martikainen, P., Kyprianou, N., Tucker, R.W., Isaacs, J.T. Programmed cell death of non-proliferating androgen-independent prostatic cancer cells. Cancer Res., 51;4693-4700 (1991).
21. Obi, F.O., Ryan, A.J. and Billett, M.A. Preferential binding of the carcinogen benzol[a]pyrone to DNA in active chromatin and the nuclear matrix. Carcinogenesis, 7: 907-913 (1986).

40

MONITORING AND REPAIR OF DNA DAMAGE DURING G_2 IN RELATION TO CARCINOGENESIS

Katherine K. Sanford[1] and Ram Parshad[2]

[1]National Cancer Institute
Bethesda, MD 20892
[2]Howard University College of Medicine
Washington, D.C. 20059

The DNA of mammalian cells is continually subject to environmental and endogenous damaging agents. Radiations from the sun or other sources, chemical mutagens in the atmosphere, foods or drugs, as well as normal metabolites such as hydrogen peroxide and its derivative, the clastogenic free hydroxyl radical, can produce DNA lesions.[1,2] These include among others DNA single-strand breaks, double-strand breaks and base damage, all of which can be processed into chromatid aberrations seen at metaphase. The DNA damage, if not repaired, can have serious consequences, resulting in infidelity of replication, mutations, neoplastic transformation and even cell death. Cellular repair mechanisms involving multienzymatic steps have evolved to remove these lesions. The repair mechanisms appear to be integrated with events of the cell cycle. Monitoring and repair of DNA damage sustained during G_2 phase just before mitosis and distribution of chromosomes to daughter cells appears to play an important role in carcinogenesis. In fact, deficient repair of x-ray-induced DNA damage during G_2 seems to be a prerequisite for malignant neoplastic transformation of mouse and human cells in culture and *in vivo*. This conclusion stems from a series of observations initially on the spontaneous transformation of mouse cells and subsequently on the induced transformation of human cells in culture.

The spontaneous transformation from normal to tumorigenic state after long-term culture, first reported in rodent cells in the early 1940's by George Gey at Johns Hopkins and Wilton Earle at the National Cancer Institute, stimulated intensive efforts during the following two decades to understand mechanisms. Such transformations occurred at a relatively high frequency in cells from inbred mouse strains and also to some extent in cells from inbred strains of rat, the Chinese hamster and Syrian hamster but were exceedingly rare in cultures of human cells. Their occurrence in mouse cells initiated and maintained continuously in serum-free, chemically defined culture medium tended to eliminate chemical carcinogens and exogenous viruses as causative agents. However, three environmental factors were identified that increased their frequency or accelerated their time of occurrence. These were the use of horse serum rather than fetal bovine serum to supplement the culture medium, atmospheric oxygen tension (18%) rather than 1% in the gas phase of the cultures, and repeated exposure of cells and medium to fluorescent room lights. All three factors also produced DNA damage and a high frequency of chromosomal aberrations. The results suggested a causal relationship between chromosomal aberrations and spontaneous neoplastic transformation in culture.[3]

Experimental studies further evaluated the effect of fluorescent light exposure on production of chromatid breaks and exchanges in mouse cells. Through the use of interference filters, we found that the effective wavelength for producing chromatid breaks and exchanges was 405 nm in the visible, near-UV range. The combined impact of visible light and oxygen generated hydrogen peroxide (H_2O_2) in the culture medium. Catalase, which decomposes H_2O_2, when added to the culture medium during light exposure, almost completely prevented the chromatid breaks and exchanges. This observation and later use of mannitol, a scavenger of the free hydroxyl radicals derived from H_2O_2, implicated these photoproducts as causative agents in producing the chromatid aberrations.

Susceptibility to the light-induced chromatid breaks was found to increase with period of culture and with spontaneous malignant transformation of mouse cells as determined by sarcoma-production in syngeneic hosts. The spontaneously transformed malignant cells differed from their non-tumorigenic precursors in showing a high frequency of chromatid breaks when exposed during the late S-G_2 period of the cell cycle, *i.e.*, for 5 h prior to arrest of metaphase cells by colcemid. The precursor non-tumorigenic cells, in contrast, showed few or no chromatid aberrations during the comparable period. Similar results were obtained with human neoplastic cells compared with their nonneoplastic precursors. In addition, exposure to light for 5 h followed by light-shielding and caffeine to inhibit S-phase postreplicative DNA repair during the subsequent 15 h period resulted in a high frequency of chromatid breaks in neoplastic human cells but not in their non-transformed precursors. These experimental results suggested an increased susceptibility during both G_1 and late S-G_2 periods in human cells after their neoplastic transformation.[4,5]

To localize the susceptible period more precisely in G_2 phase, we then used x-rays rather than light as the damaging agent; x-rays also generate free hydroxyl radicals within the cell through radiolysis of water. Exposure of cells in culture to x-rays during G_2 phase produces chromatid breaks and gaps seen at the first metaphase. Chromatid breaks show a discontinuity with displacement of the broken segment, whereas gaps show a discontinuity with no displacement and were scored in our studies only if the discontinuity was longer than the chromatid width. These are sometimes referred to as "non-displaced breaks". X-irradiation produces several lesions in DNA including double-strand breaks, single-strand breaks and base damage. Because each chromatid contains a single continuous molecule of double-stranded DNA, chromatid breaks and gaps represent unrepaired DNA strand breaks. These may arise directly from the x-irradiation or indirectly during repair processes.[6] They can be quantified in cells entering metaphase at intervals after x-ray exposure.

The frequencies of these aberrations in metaphase cells arrested by colcemid 0.5 to 1.5 h after x-ray exposure (100 R in air, 58-68 R in borosilicate culture vessels because of absorption by the glass) were at least 3- to 5-fold higher in all human tumor cells examined, regardless of histopathology or tissue of origin, than in normal skin fibroblasts.[7] These studies were then extended to skin fibroblasts, PHA-stimulated peripheral blood lymphocytes or lymphoblastoid cells from individuals with a genetic disorder or condition predisposing to a high risk of cancer. Each disorder has distinct clinical symptoms that , with the exception of Down syndrome, may be inherited as recessive or dominant traits, *e.g.*, ataxia telangiectasia (A-T), Bloom syndrome, dyskeratosis congenita, dysplastic nevus syndrome, Down syndrome, Fanconi anemia and xeroderma pigmentosum. In other genetic conditions, neoplasia is the sole feature, *e.g.*, hereditary cutaneous malignant melanoma, familial polyposis, Gardner's syndrome, Li-Fraumeni syndrome, retinoblastoma and Wilms' tumor. Cells from individuals with any one of these genetic disorders or conditions showed a significantly higher frequency of chromatid breaks and gaps when arrested at metaphase 0.5 to 1.5 h after x-irradiation than observed in normal controls tested in parallel.[8-12]

Kinetic analyses of the cellular responses to G_2 phase x-irradiation showed that during the first 0.5 h after irradiation, frequencies of chromatid breaks and gaps were similar in normal, cancer-prone and tumor cells. One exception was the response of cells from A-T patients which showed a higher level (Fig. 1). There was a precipitous decline in frequency in normal cells entering metaphase during the subsequent hour. This decline could be prevented by adding the DNA repair inhibitor 1-β-D-arabinofuranosylcytosine (ara-C) with colcemid after x-irradiation.[12-14] Ara-C blocks the repair synthesis. The decline thus appears to result from efficient repair of the radiation-induced DNA damage during G_2. Frequencies of aberrations persisted at the high level or increased in tumor and cancer-prone cells during the corresponding period. This increase presumably results from the accumulation of DNA strand breaks generated during incomplete or deficient repair processes.[6]

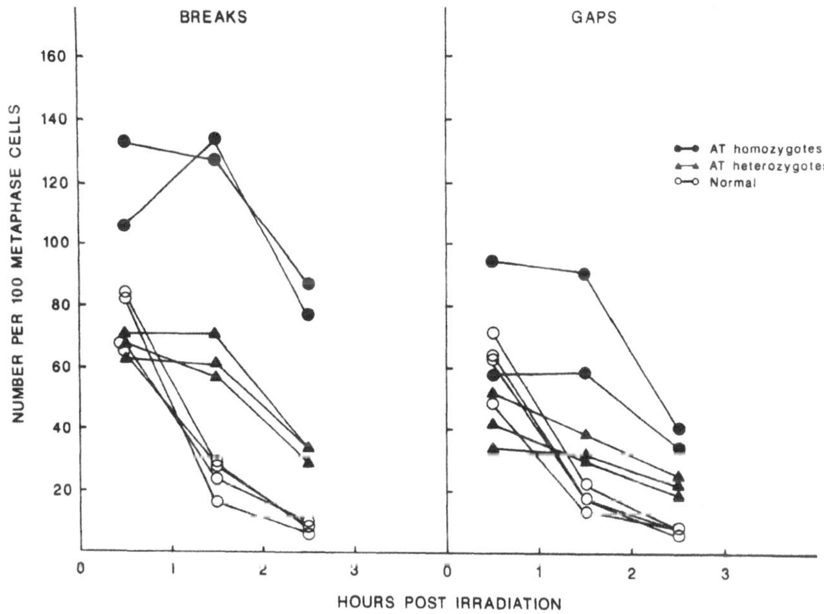

Figure 1. Chromatid breaks and gaps in PHA-stimulated peripheral blood lymphocytes from two A-T patients (homozygotes), three cancer-prone carriers of the A-T gene (heterozygotes) and four normal controls following x-irradiation with 58R during G_2 phase. Each determination is based on duplicate cultures processed for chromosome analysis at the times indicated.

Fibroblasts and tumor cells after low-level x-irradiation show a distinct G_2 mitotic block ~2 h before metaphase as determined by quantifying the numbers of metaphase cells in replicate cultures, irradiated and nonirradiated, at intervals after irradiation to establish their ratio.[7, 10, 15] The mean ratio approached zero at ~2 h after irradiation; such metaphase cells would have been in G_2 at the time of irradiation, ~2 h before metaphase. This perturbation of the cell cycle by x-irradiation could differentially affect the time available for repair during G_2 in normal, cancer-prone and tumor cells. However, the DNA repair deficiency occurs after the mitotic block, *i.e.*, from 0.5 to 1.5 h before metaphase. Furthermore, in six experiments comparing cancer-prone or tumor cells with normal cells, with respect to rate of progression after x-irradiation through G_2 to metaphase , no differences could be detected. Thus, there is no evidence of a difference in x-ray-induced delay between repair-efficient normal and repair-deficient tumor or cancer-prone cells, [7, 10 ,11, 15] (and unpublished).

In addition to cancer-prone and tumor cells cultured from *in vivo*, epithelial cells from normal human tissues were shown to acquire the G_2 repair deficiency after carcinogen-induced transformation to infinite lifespan in the case of human mammary epithelial cells,[16] or by transfection with pSV-3-*neo* or infection with adeno-12-SV40 hybrid virus in the case of human skin keratinocytes.[17] Furthermore, two lines of normal skin fibroblasts, transformed in culture by George Milo to anchorage independence but not infinite lifespan by chemical carcinogen treatment showed the enhanced susceptibility to light-induced chromatid breaks indicative of deficient repair.[18] In contrast to cells from primary or secondary cultures, the human epithelial cells after transformation to infinite lifespan acquired the repair deficiency and could then be transformed to malignant tumorigenic cells by infection with Kirsten murine sarcoma virus containing the *ras* oncogene,[16, 17] by ionizing radiation,[19] or by chemical carcinogens.[20] In a recent collaborative study with Boukamp and Fusenig (manuscript submitted for publication), we examined six clones of cells derived from a nontumorigenic line of human skin keratinocytes, HaCaT, spontaneously transformed to infinite lifespan. The clones were isolated after transfection of the parental line with the c-Ha-*ras* oncogene. All but one clone expressed the mutated p21 protein and grew as benign or malignant tumors in nude mice, thus representing two stages in carcinogenesis. The three benign hyperplastic clones showed efficient repair of the x-ray induced DNA damage, whereas the two clones that produced malignant tumors showed the G_2 repair deficiency. These results suggest that acquisition of deficient G_2 phase DNA repair in cultured human skin keratinocytes after *ras* oncogene transfection is required for progression from benign to malignant state. In all cases known to date, there appear to be two changes prerequisite for malignant transformation of human cells in culture, the acquisition of infinite lifespan and the G_2 phase DNA repair deficiency. Both of these have been shown to have a genetic basis.[21, 22]

Unrepaired strand breaks resulting from deficient repair during G_2 of damaged DNA lead to chromatid breaks with potential loss of acentric fragments during the subsequent mitosis. Strand breaks persisting from G_2 through mitosis into G_1, if not sealed, may rejoin at random to form interchanges, inversions, duplications, deletions and translocations with consequent gene rearrangements. The G_2 repair deficiency associated with human tumors, genetic predisposition to cancer and malignant transformation of human cells in culture thus provides a mechanism for the genetic and chromosomal alterations (such as gene mutations, chromosomal translocations and deletion of suppressor genes) known to be associated with the genesis of human cancer.

REFERENCES

1. M.J. Peak and J.G. Peak, Solar-ultraviolet-induced damage to DNA, *Photodermatology* 6:1 (1989).
2. B. Chance, H. Sies, and A. Boveris, Hydroperoxide metabolism in mammalian organs, *Physiol. Rev.* 59:527 (1979).
3. K.K. Sanford and V.J. Evans, A quest for the mechanism of "spontaneous" malignant transformation in culture with associated advances in culture technology, *J. Natl. Cancer Inst.* 68:895 (1982).
4. K.K. Sanford, R. Parshad, R. Gantt, and R.E. Tarone, A deficiency in chromatin repair, genetic instability and predisposition to cancer, *CRC Critical Reviews in Oncogenesis* 1:323 (1989).
5. R. Parshad, K.K. Sanford, and R. Gantt, G_2 chromatid radiosensitivity in relation to DNA repair and cancer susceptibility, *in*:: "The Eukaryotic Chromosome:Structural and Functional Aspects," R.C. Sobti and G. Obe, eds., Springer-Verlag, New York, pp. 175-183 (1991).
6. P.C. Hanawalt, P.K. Cooper, A.K. Ganesan, and C.A. Smith, DNA repair in bacteria and mammalian cells, *Annu. Rev. Biochem.* 48:783 (1979).
7. R. Parshad, R.R. Gantt, K.K. Sanford, and G.M. Jones, Chromosomal radiosensitivity of human tumor cells during the G_2 cell cycle period, *Cancer Res.* 44:5577 (1984).
8. K.K. Sanford, R. Parshad, R. Gantt, R.E. Tarone, G.M. Jones, and F.M. Price, Factors affecting and significance of G_2 chromatin radiosensitivity in predisposition to cancer, *Int. J. Radiat. Biol.* 55:963 (1989).
9. S. Takai, F.M. Price, K.K. Sanford, R.E. Tarone, and R. Parshad, Persistence of chromatid damage after G_2 phase x-irradiation in lymphoblastoid cells from Gardner's syndrome, *Carcinogenesis* 11:1425 (1990).
10. D.M. DeBauche, G.S. Pai, and W.S. Stanley, Enhanced G_2 chromatid radiosensitivity in dyskeratosis congenita fibroblasts, *Am. J. Human Genet.* 46:350 (1990).
11. R. Parshad, F.M. Price, K.F. Pirollo, E.H. Chang, and K.K. Sanford, Cytogenetic response to G_2 phase x-irradiation in relation to DNA repair and radiosensitivity in a cancer-prone family with Li-Fraumeni syndrome, *Radiation Res.*, in press.
12. K.K. Sanford, R. Parshad, F.M. Price, R.E. Tarone, and M.B. Schapiro, X-ray-induced chromatid damage in cells from Down syndrome and Alzheimer disease patients in relation to DNA repair and cancer proneness, *Cancer Genetics Cytogenet.*, in press.
13. H. Mozdarani and P.I. Bryant, Cytogenetic response of normal human and ataxia telangiectasia G_2 cells exposed to x-rays and ara-C, *Mutation Res.* 226:223 (1989).
14. R.D. Knight, R. Parshad, F.M. Price, R.E. Tarone, and K.K. Sanford, X-ray-induced chromatid damage in relation to DNA repair and cancer incidence in family members, *Intl. J. Cancer* 54:1 (1993).
15. R. Parshad, K.K. Sanford, and G.M. Jones, Chromatid damage after G_2 phase x-irradiation of cells from cancer-prone individuals implicates deficiency in DNA repair, *Proc. Natl. Acad. Sci. USA* 80:5612 (1983).
16. K.K. Sanford, F.M. Price, J.S. Rhim, M.R. Stampfer, and R. Parshad, Role of DNA repair in malignant neoplastic transformation of human mammary epithelial cells in culture, *Carcinogenesis* 13:1137 (1992).

17. R. Gantt, K.K. Sanford, R. Parshad, F.M. Price, W.D. Peterson, Jr., and J.S. Rhim, Enhanced G_2 chromatid radiosensitivity, an early stage in the neoplastic transformation of human epidermal keratinocytes in culture, *Cancer Res.* 47:1390 (1987).
18. R. Parshad, R. Gantt, K.K. Sanford, G.M. Jones, and R.F. Camalier, Light-induced chromatid damage in human skin fibroblasts in culture in relation to their neoplastic potential, *Int. J. Cancer* 28:335 (1981).
19. P. Thraves, Z. Salehi, and J.S. Rhim, Neoplastic transformation of immortalized human epidermal keratinocytes by ionizing radiation, *Proc. Natl. Acad. Sci. USA* 87:1174 (1990).
20. J.S. Rhim, J. Fujita, P. Arnstein, and S.A. Aaronson, Neoplastic conversion of human keratinocytes by adenovirus-12-SV40 and chemical carcinogens, *Science* 232:385 (1986).
21. O.M. Pereira-Smith and J.R. Smith, Evidence for the recessive nature of cellular immortality, *Science* 221:964 (1983).
22. K.K. Sanford, R. Parshad, E.J. Stanbridge, J.K. Frost, G.M. Jones, J.E. Wilkinson, and R.E. Tarone, Chromosomal radiosensitivity during the G_2 cell cycle period and cytopathology of human normal x tumor cell hybrids, *Cancer Res.* 46:2045 (1986).

41

CELL CYCLE REGULATION IN NORMAL VERSUS LEUKEMIC T CELLS

Toshio Nikaido[1], Koji Ono[1], Masuji Yamamoto[1], Toshiyuki Sakai[2], and Yasushi Magami [1]

[1] The Wistar Institute of Anatomy and Biology, 36th and Spruce Streets Philadelphia, PA 19104

[2] Department of Preventive Medicine, Kyoto Prefectural University of Medicine, Kyoto 602, Japan

INTRODUCTION

DNA replication is a central feature of the cell cycle, yet very little is known about the control of this process in eukaryotes. Genetic analysis of yeasts has defined proteins with potentially important roles in the control of DNA replication, e.g. CDC28 gene from *Saccharomyces cerevisiae* and its homolog cdc2 from *S. pombe*[1-4], for which human functional homologs exist[5-8]. A homolog of the cdc2 gene product, $p34^{cdc2}$, has been identified in *Xenopus* eggs as a component of the mitotic inducer maturation-promoting factor (MPF)[9, 10]. Active MPF is a complex of $p34^{cdc2}$ and cyclin B[11-14]. Cyclins are rapidly degraded as cells exit from metaphase[15] and therefore must be newly synthesized during interphase of each cell cycle so that cells can generate MPF and progress into mitosis[16, 17]. In addition to its role at metaphase, $p34^{cdc2}$ function is also required at the G1/S transition in yeast.

Recent evidence suggests a role for $p34^{cdc2}$ in the DNA replication of higher eukaryotes[18-20]. $p34^{cdc2}$ or a very closely related protein is involved in the initiation of DNA replication in *Xenopus* egg extracts[18]. In human cells, a factor termed RF-S has been purified from S phase cells, on the basis of its ability to activate DNA synthesis in G1 phase cell extracts, and shown to contain a human homolog of the $p34^{cdc2}$ kinase and cyclin A. Interestingly, addition of recombinant clam cyclin A to such G1 extracts was sufficient to activate DNA replication, suggesting that cyclin association with $p34^{cdc2}$ may be the rate-limiting step in the activation of the $p34^{cdc2}$ kinase at the G1/S transition[20].

Cyclin A is associated with $p34^{cdc2}$ and exhibits histone H1 kinase activity[21]; it arises earlier and is degraded before cyclin B in the cell cycle[22]. Cyclin A antisense RNA and antibody against cyclin A each inhibits DNA synthesis in mouse fibroblasts[23]. Cyclin A associates not only with cdc2, but also with the cdc2-related protein kinase, CDK2 (eg-1), and these complexes accumulate in the nucleus and are active during G1/S and G2 phase[24]. Two lines of evidence suggest a causal association between perturbations in cyclin A behavior and malignant transformation. First, the cyclin A gene is the site of integration of a fragment of the hepatitis B virus (HBV) genome in a human hepatocellular carcinoma[25]. Second, cyclin A is found associated with the adenovirus transforming protein E1A in adenovirus-transformed cells[24, 26]. Cyclin A was shown to interact with the E2F and the spacer region of p107, a protein that shares extended homology to pRB[27-30].

The Cell Cycle: Regulators, Targets, and Clinical Applications
Edited by V.W. Hu, Plenum Press, New York, 1994

Together, the evidence points to a critical role for p34^{cdc2} and cyclin A in tumorigenesis as well as in DNA synthesis. To determine whether the p34^{cdc2} and cyclin A genes are induced in normal T cells by IL-2, and how the genes might be activated in leukemic T cells, we analyzed transcriptional regulation of the p34^{cdc2} and cyclin A gene during T cell proliferation. We show that p34^{cdc2} and cyclin A gene expression is induced by IL-2 in normal human T cells. We also show that the RB suppresses the p34^{cdc2} expression, and that normal p53, but not mutant p53, suppresses cyclin A expression. We raise the possibility that the products of RB and p53 genes negatively regulate the expression of p34^{cdc2} and cyclin A genes.

MATERIALS AND METHODS

Reagents

The cell cycle inhibitor quercetin was purchased from Sigma Chemical Co. (St. Louis, MO) and prepared as 20 mM stock solution. Propidium iodide (Sigma) was dissolved in distilled water at 1 mg/ml. Plasmid DNAs were kind gifts from Drs. K. Itakura (ß2-microglobulin), J. Stein (histone H4), T. Hunter and J. Pines (cyclin A and B), P. Nurse (p34^{cdc2}), P. Hinds (p53), and P. Robbins (RB). Plasmid DNA of p53 was purchased from ATCC (Rockville, MD). Monoclonal antibodies for RB (Ab-1) and p53 (Ab-2) were purchased from Oncogene Science (Uniondale, NY).

Cell culture

Human peripheral blood lymphocytes were isolated from healthy donors by Ficoll-Hypaque density-gradient centrifugation as described[31]. Lymphocytes were further separated into sheep erythrocyte rosette-positive (T cells) and -negative populations. The T cells were cultured at a concentration of $1x10^6$/ml in RPMI 1640 supplemented with 10% heat-inactivated fetal bovine serum, 2 mM L-glutamine, and human purified recombinant IL-2 at 100 unit's /ml for 3 days in wells coated with anti-CD3 antibody. After anti-CD3 and IL-2 were removed, cell growth was arrested by cultivation for another 2 day. Arrested cells were restimulated with IL-2 for several days, harvested and characterized by flow cytometry. The human leukemic T cell line CEM, retinoblastoma cell line Y79 (purchased from ATCC) were grown in RPMI 1640 with 10% calf serum and 2 mM glutamine at 37°C in a humidified 5% CO2, 95% air atmosphere. For cell cycle blocking experiments, cells were seeded at 1×10^5 /ml. After 12 h, quercetin (70 μM) was added and the incubation continued for 14-16 h. Cells were collected by centrifugation and prepared for RNA or nuclear extraction, or resuspended in fresh medium with or without inhibitors.

RNA analysis

T cells were harvested and total RNA was prepared using the guanidine isothiocyanate (GIT)/ CsCl method[32]. Briefly, cells were lysed in GIT using a tissue homogenizer. The lysate was layered onto a CsCl gradient and spun in an ultracentrifuge. RNA in the pellets in the bottom of the tube was recovered by redissolving the pellet. Following quantitation of RNA by A^{260}, 10 μg per lane was size-fractionated by electrophoresis in 6% formaldehyde/ 1% agarose gels. Before RNA was transferred to Nytran nylon membranes (S and S; Keene, NH) in 10X SSC, gels were stained with ethidium bromide to verify that equal amounts of RNA were loaded per lane. Blots were baked at 80° C for 2 h before prehybridization and hybridization with random-primed ^{32}P-labeled DNA fragments (specific activity 2 to 10 x 10^8 cpm/μg). Hybridization signals were visualized by autoradiography using Kodak XAR-5 film with intensifying screens. The size of mRNA was estimated from the positions of 28S (5.1 kb) and 18S (2.0 kb) rRNA bands in ethidium-stained gels[33].

Cloning and DNA sequencing

A human T lymphocyte-derived genomic library using the lambda Dash phage vector was screened with 5' upstream DNA fragments of p34^{cdc2} or cyclin A cDNA. The lambda phage contained the genomic sequence corresponding to a 250-bp BamHI-KpnI fragment that is the most 5' upstream of p34^{cdc2} cDNA. This region of homology, contained in a 4.0-kb EcoRI fragment, was partially sequenced and was followed by an additional 11-kb sequence involved in the regulation of p34[cdc2] expression. The inserts from isolated phages were subcloned into the pBluescript plasmid vector, and the clones were used for partial DNA sequencing. A second phage contained the genomic sequence corresponding to a 200-bp AvaI-AvaI fragment that is the most 5' upstream of cyclin A cDNA. This region of homology, contained in a 3.8-kb XbaI-XhoI fragment, was partially sequenced and was followed by an additional 13-kb sequence involved in the regulation of cyclin A expression. Inserts from isolated phages were subcloned into the pBluescript plasmid vector and the clones were used for partial DNA sequencing[33].

Gel shift assay

Nuclear extracts were prepared from normal PHA- or IL-2-stimulated T cells and from quercetin-arrested and released leukemic cells as described[34]; the gel shift assay was performed according to a modification of the procedure described[35]. Radiolabeled double-stranded DNA was generated by annealing complementary oligomers with 5' overhangs and then radiolabeling with [^{32}P]ATP by T4 polynucleotide kinase. Radiolabeled oligonucleotides were incubated with cell nuclear extracts from normal and leukemic T cells.

Luciferase assay

A 2.0-kb EcoRI-StuI fragment containing the p34[cdc2] promoter region or a 1.0-kb HindIII-XhoI fragment containing the cyclin A gene promoter region was subcloned into the luciferase vector pXP1 plasmid DNA. In both deletion experiments, deleted fragments containing synthetic HindIII and BglII recognition sites at the 5' and 3' ends were amplified using the polymerase chain reaction method. Amplified fragments were digested with HindIII and BglII, and subcloned into the luciferase vector pXP1. Plasmid DNA (0.2-0.6 μg) was transfected into 5×10^6 of CEM or Y79 human retinoblastoma cells using the electroporator (Bio-Rad) at 250V, 960μF. Each transfection contained 2.5 μg of pcdc2-luc or pcyclin A-luc as a reporter, and 2.5 μg of one of the following plasmids; RB, p53 or their derivatives. pBluescript was added as a carrier to make total DNA to 30 μg. Cells were harvested after 24 hrs incubation and washed with phosphate-buffered saline (PBS) without calcium and magnesium twice. The cell pellets were lysed with 150 μl of cold lysis buffer (0.625% Triton X-100, 0.1 M K_2HPO_4, pH 7.8, and 1 mM dithiothreitol). Lysates were centrifuged at 13,000*g* for 5 min. Extract (50 μl) was subjected to measure luciferase activity with a Monolight 2001 luminometer (Analytical Luminescence, San Diego, CA) as described[36], and the activity was shown by raw light units (RLU) .

RESULTS

Induction of p34[cdc2] and cyclin A genes with IL-2 in normal human T cells

To establish an IL-2 inducible T cell system, we cultured T cells in wells coated with anti-CD3 antibody in the presence of human recombinant IL-2 for 3 days, removed the anti-CD3 and IL-2, and continued culture for another 2 day to arrest growth. Arrested cells were restimulated with IL-2 for several days, harvested and characterized by flow cytometry. Isolated cells represented a cell fraction enriched in T cells. Using this system, we examined the expression of several growth-related genes. Control cDNA probes for ß-2 microglobulin (ß2-m), histone H4 (H4), and c-*myb* detected only ß-2m, which is constitutively expressed in arrested cells, whereas H4 and c-*myb* were expressed in cells reactivated with IL-2. The mRNA of p34[cdc2] and cyclin A is also induced by IL-2 in normal human T cells (Fig. 1). Thus p34[cdc2] and cyclin A, like c-*myb*, are IL-2-inducible genes in T cells. Recently we demonstrated that quercetin reversibly blocks the cell cycle at a point 3-6 h before the start of DNA synthesis and that p34[cdc2] and cyclin A were constitutively expressed in human leukemic T cells CEM[37]. Expression of histone H4 and cyclin A was suppressed in CEM leukemic T cells blocked with quercetin (70 μM) for 14-16 h, while aphidicolin (5 μg/ml) and mimosine (200 μM) did not suppress expression of p34[cdc2] and cyclin A. These results suggest that cyclin A and p34[cdc2] are expressed at late G1 phase and deregulated in CEM cells[37] (Fig. 2).

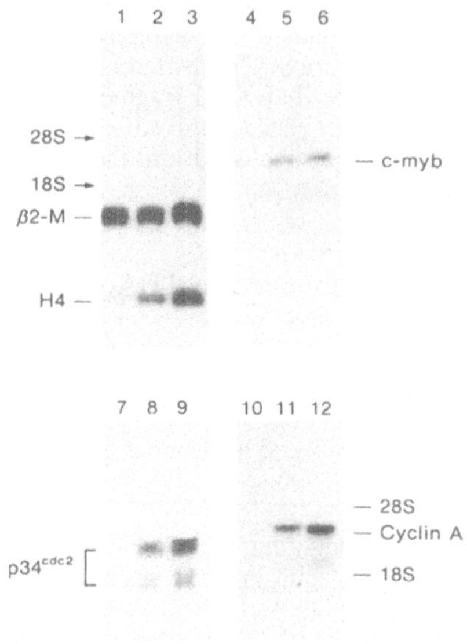

Figure 1. Induction of p34^{cdc2} and cyclin A mRNA in normal T cells. Human normal T cells were prepared and cultivated in wells coated with anti-CD3 antibody and in the presence of human purified recombinant IL-2 for 3 days. Anti-CD3 and IL-2 were removed (lanes 1, 4, 7, 10), cell growth was restimulated with IL-2 for 24 h (lanes 2, 5, 8, 11), and 36 h (lanes 3, 6, 9, 12), and cells were harvested and characterized. RNA (10 μg) was loaded in an agarose-formaldehyde electrophoresis gel and transferred to a nylon membrane filter as described in Materials and Methods. The filter was hybridized at various intervals after transfer with cloned DNA fragments containing histone 4 (H4), ß-2 microglobulin (ß2-m), *c-myb*, p34^{cdc2}, and cyclin A coding sequences.

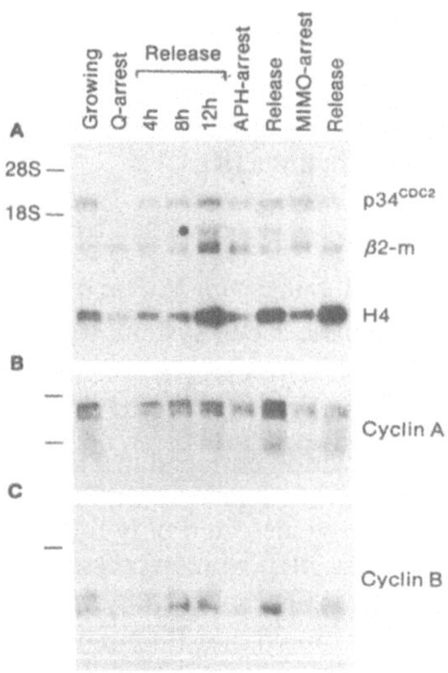

Figure 2. p34^{cdc2} and cyclin A are constitutively expressed in leukemic T cells and blocked with quercetin. Total RNA was extracted from leukemic T cells treated for 15 h with quercetin (**Q**), mimosine (**MIMO**), or aphidicolin (**APH**) then cells are released for 4, 8, 12h. Growing cells were left untreated in the left lane (**Growing**). The filter was hybridized at various intervals after transfer with cloned DNA fragments containing the histone 4 (H4), p34^{cdc2}, ß-2 microglobulin (**A**), cyclin A (**B**) and cyclin B (**C**) coding sequences. The cells were harvested for arrested cells. The remaining cells were released from them by centrifuging the cells, and resuspending in fresh medium. Cells released in the fraction were collected several hours after the change of medium.

Isolation and characterization of the p34^{cdc2} and cyclin A promoters

To determine whether differential regulation of the p34^{cdc2} and cyclin A gene promoter regions underlies the deregulated expression of these genes in leukemic as compared with normal T cells, we isolated and characterized the promoters of these genes to assess regulation at the transcriptional level. A human T lymphocyte-derived genomic library using the lambda Dash phage as a vector was screened with a 5' upstream DNA fragment of p34^{cdc2} cDNA. Sequence comparison between this cDNA and its respective genomic clone showed that the 5' untranslated region is encoded by two exons. DNA sequencing of the 5' upstream p34^{cdc2} fragment showed that the promoter lacks a typical TATA sequence. Primer extension experiments served to localize the site of transcription initiation. This putative promoter region contains DNA sequences that are potential recognition sites for E2F-like transcription factor, four inverted Sp1, inverted c-*myb*, and ATF/CREB-like. Because the E2F binding site is recognized by RB through E2F, we refer to this site as E2F/RB. Partial DNA sequencing of clones derived from a second phage containing the genomic sequence corresponding to 5' upstream region of cyclin A cDNA also indicated the absence of a typical TATA sequence in the promoter region and the presence of potential recognition sites for p53, Sp1, and ATF/CREB[38].

DNA binding proteins of the p34cdc2 and cyclin A promoters

We analyzed the promoter binding protein during the cell cycle of normal and leukemic T cells using gel mobility shift assay. Nuclear extracts were prepared from freshly isolated normal T cells cultured in the absence or presence of PHA for 3 days, and from CEM leukemic T cells arrested with quercetin, and arrested and released. Nuclear extracts were then incubated with several oligonucleotides probes constructed to contain typical DNA binding sites. The oligonucleotide containing the E2F/RB site from the p34^{cdc2} promoter reacted strongly with nuclear extract from normal arrested T cells but showed only low level reactivity with the extract from PHA-stimulated cells. Reactivity was weak with nuclear extract from arrested leukemic T cells and completely absent with nuclear extract from growing leukemic T cells (Fig. 3A). Further analysis for the possible involvement of RB in the observed reactivities using an anti-RB monoclonal antibody indicated the complete elimination of any binding product upon addition of the antibody, whereas addition of control CD11a monoclonal antibody did not alter the binding (Fig. 3B). These complexes were specifically inhibited by cold homologous oligonucleotides (Fig. 3C). Thus, the complex formed between E2F/RB binding site-containing oligonucleotides and the normal arrested T cell nuclear extracts contain the RB protein[38].

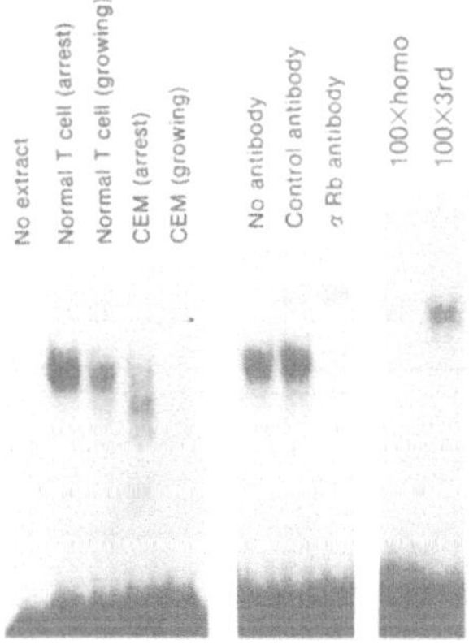

Figure 3: Gel shift assay of a sequence including the E2F/RB sites from the p34^{cdc2} promoter. (**A**): The 24bp oligonucleotides including the E2F/RB binding site from the p34cdc2 promoter were reacted with nuclear extracts prepared from normal and freshly isolated T cells (arrest) and PHA-stimulated T cells (growing), and from quercetin-arrested (arrest) and released (growing) CEM leukemic cells. (**B**): CD11a as a control or RB-specific monoclonal antibody was added to the reaction mixture. (**C**): Homologous oligonucleotides (100X homo) and third party oligonucleotides (100X 3rd), each in 100-fold excess, were added to the reaction mixture.

p34cdc2 and cyclin A promoter activities

To identify the essential promoter sequence of the 5' upstream DNA fragment of $p34^{cdc2}$, cells were transfected with $p34^{cdc2}$promoter-luciferase plasmid DNA constructed from the 2.0-kb EcoRI-StuI fragment containing the $p34^{cdc2}$ promoter region or with various deletion constructs, and assayed for luciferase activity 24 h later (Fig. 4A). Luciferase activity was sharply decreased when the deletion construct lacked the 102 bp containing the ATF binding site (131 and 15 RLU for constructs 1 and 2, respectively), whereas deletion of the 54 bp containing the *myb* binding site (construct 3) or of the region containing the Sp1 sites had only slight effect on luciferase activity. Deletion of all the known binding sites (construct 0) completely abrogated activity. These results suggest that these fragments have promoter activity ant that the 102 bp containing ATF contains potent positive regulatory elements. Expression of the $p34^{cdc2}$ promoter in Y79 human retinoblastoma cells that lack the RB expression cotransfected with the $p34^{cdc2}$ promoter-luciferase construct and a plasmid expressing human RB expression vector (phu RB) was reduced to 25 % of the activity of the cotransfection with a vector lacking the RB insert, indicating that RB suppressed the $p34^{cdc2}$ gene in the cells (Table 1).

Similar experiments using cells transfected with a vector constructed from the 1.0-kb HindIII-XhoI fragment containing the cyclin A gene promoter region and pXP1 plasmid DNA containing the luciferase gene (Fig. 4B) revealed a little potent positive promoter activity from the 142-bp containing three Sp1 binding sites (compare constructs 1 and 6), whereas the 69 bp containing the ATF site had a very potent positive promoter activity (282 and 5.6 RLU for constructs 6 and 2, respectively). The 76 bp containing the fragment between the ATF and p53 binding sites had only weak activity (constructs 2 and 3). Deletion of all but the final Sp1 site (construct 0) resulted in abrogated activity. Cotransfection of CEM cells with the cyclin A promoter-luciferase construct and a human normal or mutant p53 expression vector showed that normal p53 had 13 or 20 % of the activity of the murine or human mutant p53, respectively, indicated that either human and murine normal p53 but not mutant p53 suppressed cyclin A gene expression (Table 1).

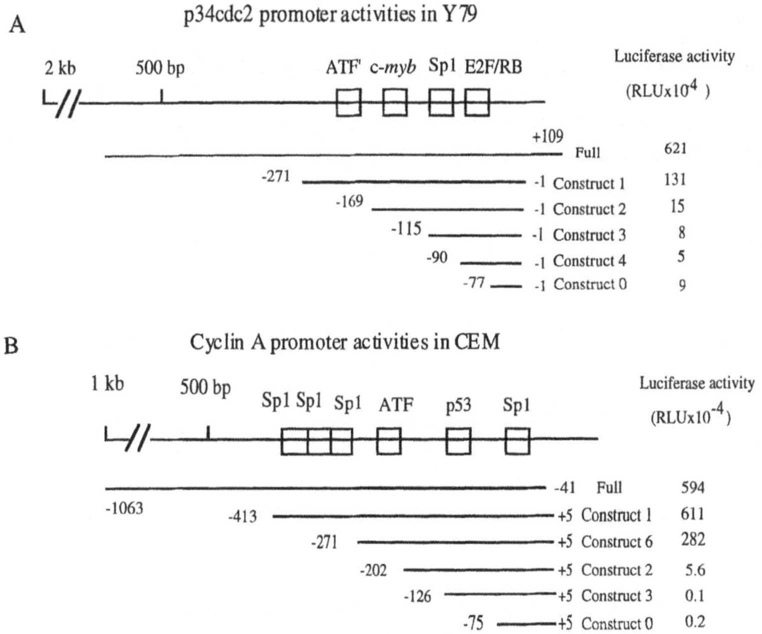

Figure 4: Assay of regulatory elements of the $p34^{cdc2}$ (A) and cyclin A (B) promoters.
A series of deletion constructs of $p34^{cdc2}$ and cyclin A promoters was transfected into human leukemic T cells CEM or retinoblastoma cells Y79, and luciferase activity, expressed as raw light units (RLU), was measured 1 day later. Two additional experimental runs gave similar results.

Table 1. Suppression of p34^{cdc2} and cyclin A gene expression by RB and p53

Promoter + Cotransfected plasmid	Luciferase activity (RLU x 10^{-4})	Suppression (%)
p34^{cdc2}-luciferase[a]		
pJ3W	137	100
phuRB[c]	34	25
pmRB[d]	141	103
cyclin A-luciferase[b]		
pCMV-neo	303	100
pC53SCX3.3 (mutant p53)[c]	309	102
pC53SN3.3 (normal p53)[c]	61	20
LTRp53CG val 135 (mutant p53)[d]	124	100
LTRp53CG ala 135 (normal p53)[d]	16	13

[a] Human retinoblastoma cells Y79 were plated ($5x10^6$) and transfected with 5μg of p34^{cdc2}-luciferase with RB expression vector or control vector lacking insert (7.5μg). Luciferase activity was assayed 24h later.

[b] Human leukemic cells CEM were plated ($5x10^6$) and transfected with 2.5μg of cyclin A-luciferase with 5.0 μg of p53 expression vector or control vector lacking insert. Luciferase activity was assayed 24h later.

[c] Human origin

[d] Mouse origin

DISCUSSION

Evidence has suggested a role for IL-2 in the cell cycle progression of T cells, but its precise pathway in this process has remained obscure. Most of the work to date has dealt with early cellular activation events associated with the G0/G1 transition. The present findings with freshly isolated T cells indicate that p34^{cdc2} and cyclin A messages are induced by IL-2. Previous studies have shown that antisense oligonucleotides or anti-cDNAs to p34^{cdc2} and cyclin A as well as to c-*myb* inhibit DNA synthesis[19, 23]. Although CEM leukemic T cells do not require IL-2 for proliferation, these cells express all three genes constitutively, perhaps as the cause of leukemogenic proliferation.

Our results suggest that the promoter region of the p34^{cdc2} gene contains binding sites for the E2F-like transcription factor, c-*myb*, Sp1, and ATF-like, and that the promoter region of the cyclin A gene contains binding sites for p53, Sp1, and ATF. Both promoters lack a typical TATA box, and contain potential binding sites (GC boxes) for transcription factor Sp1, all of which are characteristic features of housekeeping genes. However, the levels of both mRNAs are elevated during deregulated proliferation, and the DNA binding proteins that bind to the E2F/RB binding site fail to form complexes during the proliferation, suggesting that the DNA binding protein may influence the negative expression of the gene. In both cases, ATF or ATF-like binding sites have strong positive promoter activity. Thus, the interaction of specific DNA binding proteins with the sequences may represent a common mechanism by which IL-2 regulates gene expression.

Recently, it was shown that the RB protein products have a nuclear localization[7] and DNA binding activity *in vitro*, and form a complex with cyclin A and/or p34^{cdc2} [39]. It was also shown that RB associates with the transcription factor E2F *in vitro*[40-43] and that RB may suppress transcription at the E2F site[44, 45]. These findings point to E2F as a critical target for cellular regulatory interactions that control progression through the cell cycle. Our data indicated that the levels of the complex formed between the E2F/-like binding site of p34^{cdc2} and RB were decreased upon PHA stimulation at the G1/S boundary in normal T cells, and no complex formation was detected in leukemic T cells. In addition, the suppression of the p34^{cdc2} gene was detected by RB in Y79. Taken together, our results suggest that RB is involved in the transcription regulation of the p34cdc2 promoter through the direct interaction with a putative E2F site. Consistent with our *in vitro* data of RB binding to the E2F site, Dalton[45] has recently reported that the RB gene product downregulates p34cdc2 promoter activity in ICRF fibroblasts, although RB protein binding to the E2F/RB site was not analyzed[45]. We suggest that the RB gene product downregulates the p34cdc2 promoter by binding to the E2F/RB site. The fact that leukemic T cells arrested with quercetin, a kinase inhibitor, accumulates the RB complex raises the possibility that quercetin directly or indirectly inhibits the phosphorylation of RB.

Like RB, p53 gene products have a nuclear localization, DNA binding activity *in vitro* [46-48], and are restricted to a putative active form in G1. Thus, the p53 factor might also be a critical target for cellular regulatory interactions that control cell cycle progression. The experiments in which CEM were cotransfected with the cyclin A promoter-luciferase constructs together with a plasmid expressing p53 indicated that normal p53, but not mutant p53, suppresses cyclin A expression.

Our data suggest that RB and p53 gene products regulate p34cdc2 and cyclin A gene expression, respectively. Further information will come through identification of the proteins that regulate RB and p53 during cell cycle progression. Obvious candidates include G1 cyclins and serine-/threonine-specific kinases that are known to be phosphorylated during G1 and at the G1/S-phase boundary. The direct connection between p34cdc2 or cyclin A and a tumor suppressor gene product raise new questions about the role of cell cycle genes in tumorigenesis, in particular, the regulation of cell cycle genes in tumor cell differentiation and the mechanisms of tumor suppression.

SUMMARY

Analysis of transcriptional regulation of the p34cdc2 and cyclin A genes during T cell proliferation indicated that expression of both genes is induced by interleukin-2 (IL-2) in normal human T cells and deregulated in leukemic T cells. DNA sequencing data suggest that the promoter region of the p34cdc2 gene contains binding sites for the E2F-like transcription factor, c-*myb*, Sp1, and ATF, and that the promoter region of the cyclin A gene contains binding sites for p53, Sp1, and ATF. Cotransfection of Y79 human retinoblastoma cells with a p34cdc2 promoter-luciferase expression vector and a plasmid expressing the retinoblastoma gene (RB) indicated that the RB suppresses p34cdc2 expression. Cotransfection of CEM human leukemic T cells with a cyclin A promoter-luciferase expression vector and a plasmid expressing the normal or mutant p53 indicated that only the normal p53 suppresses cyclin A expression. Formation of complexes between the E2F-like binding site of the p34cdc2 promoter and the RB gene product by PHA stimulation in normal T cells and was completely absent in leukemic T cells. These data raise the possibility that RB regulates p34cdc2 expression and that p53 regulates cyclin A expression.

ACKNOWLEDGMENTS

We are grateful to Dr. Tom Sorger for critical reading of the manuscript, and to Drs. P. Nurse for p34cdc2 cDNA, T. Hunter and J. Pines for cyclin A/B cDNA, K. Itakura for b2-m cDNA, J. Stein for histone H4 cDNA, G. Trinchieri for anti-CD3, and Takeda Co. Ltd. for recombinant IL-2.

LITERATURE CITED

1. L. H. Hartwell, R. K. Mortimer, J. Culotti, and M. Culotti, Genetic control of the cell division cycle in yeast: V. Genetic analysis of *cdc* mutants. Genetics, *74:* 267-286 (1973).
2. M. Lorincz, and S. I. Reed, Primary structure homology between the product of yeast cell division control gene CDC28 and vertebrate oncogenes. Nature, *307:* 183-185 (1984).

3. P. Nurse, P. Thuriaux, and K. Nasmyth, Genetic control of the cell division cycle in the fission yeast *Schizosaccharomyces pombe*. Gen. Genet., *146:* 167-178 (1976).
4. J. Hindley, and G. A. Phear, Expression of the cloned genes encoding the putrescine biosynthetic enzymes and methionine adenosyltransferase of *Escherichia coli (speA, speB, speC and metK)* . Gene, *31:* 129-134 (1984).
5. D. Beach, B. Durkacz, and P. Nurse, Functionally homologous cell cycle control genes in budding and fission yeast. Nature, *300:* 706-709 (1982).
6. R. Booher, and D. Beach, Site-specific mutagenesis of cdc2+, a cell cycle control gene of the fission yeast *Schizosaccharomyces pombe*. Mol. Cell. Biol., *6:* 3523-3530 (1986).
7. M. G. Lee, and P. Nurse, Complementation used to clone a human homologue of the fission yeast cell cycle control gene cdc2. Nature, *327:* 31-35 (1987).
8. C. Wittenberg, , and S. I. Reed, Conservation of function and regulation within the cdc28/*cdc2* protein kinase family: characterization of the human cdc2Hs protein kinase in *Saccharomyces cerevisiae*. Mol. Cell. Biol., *9:* 4064-4068 (1989).
9. J. Gautier, C. Norbury, M. Lohka, P. Nurse, and J. Maller, Purified mutation-promoting factor contains the product of a *Xenopus* homolog of the fission yeast cell cycle control gene cdc2+. Cell, *54:* 433-439 (1988).
10. W. G. Dunphy , L. Brizuela, D. Beach , and J. Newport, The *Xenopus* cdc2 protein is a component of MPF, a cytoplasmic regulator of mitosis. Cell, *54:* 423-431 (1988).
11. M. Lohka , M. K. Hayes , and J. L. Maller, Purification of maturation-promoting factor, an intracellular regulator of early mitotic events. Proc. Natl. Acad. Sci. USA, *85:* 3009-3013 (1988).
12. G. Draetta , F. Luca , J. Westendorf, L. Brizuela , J. Rudeman , and D. Beach, cdc2 protein kinase is complexed with both cyclin A and B: evidence for proteolytic inactivation of MPF. Cell, *56:* 829-838 (1989).
13. J. Labbe, J. Capony, D. Cavadore, J. Derancourt, M. Kaghad, J. Lelias, A. Picard, and M. Doree, MPF from starfish oocytes at first meiotic metaphase is a heterodimer containing one molecule of cdc2 and one molecule of cyclin B. EMBO J., *8:* 3053-3058 (1989).
14. J. Gautier, J. Minshull, M. Lohka, M. Glotzer, T. Hunt, and J. L. Maller, Cyclin is a component of MPF from *Xenopus*. Cell, *60:* 487-494 (1990).
15. T. Evans, E. T. Rosenthal, J. Youngblom, D. Distel, and T. Hunt, Cyclin: a protein specified by maternal mRNA in sea urchin eggs that is destroyed at each cleavage division. Cell, *33:* 389-396 (1983).
16. J. Minshull, J. Blow, and T. Hunt, Translation of cyclin mRNA is necessary for extracts of activated *Xenopus* eggs to enter mitosis. Cell, *56:* 947-956 (1989).
17. A. W. Murray, and M. W. Kirschner, Cyclin synthesis drives the early embryonic cell cycle. Nature, *339:* 275-280 (1989).
18. J. J. Blow, and P. Nurse, A cdc2-like protein is involved in the initiation of DNA replication in *Xenopus* egg extracts. Cell, *62:* 855-862 (1990).
19. Y. Furukawa, H. Piwnica-Worms, T. J. Ernst, Y. Kanakura, and J. D. Griffin, *cdc2* gene expression at the G1 to S transition in human T lymphocytes. Science, *250:* 805-808 (1990).
20. G. D'Urso, R. L. Marraccino, D. R. Marshak, and J. M. Roberts, Cell cycle control of DNA replication by a homologue from human cells of the $p34^{cdc2}$ protein kinase. Science *250:* 786-791 (1990).
21. J. Minshull, R. Golsteyn, C. S. Hill, and T. Hunt, The A- and B-type cyclin associated cdc2 kinase in *Xenopus* turn on and off at different time in the cell cycle. EMBO J., *9:* 2865-2875 (1990).
22. C. F. Lehner, and P. H. O'Farrell, The roles of *Drosophila* cyclins A and B in mitotic control. Cell, *61:* 535-547 (1990).
23. F. Girard, U. Strausfeld, A. Fernandez, and N. J. C. Lamb, Cyclin A is required for the onset of DNA replication in mammalian fibroblasts. Cell, *67:* 1169-1179 (1991).
24. J. Pines, and T. Hunter, Human cyclin A is adenovirus E1A-associated protein p60, and behaves differently from cyclin B. Nature, *346:* 760-763 (1990).
25. J. Wang, X. Chenivesse, B. Henglein, and C. Brechot, Hepatitis B virus integration

in a cyclin A gene in a hepatocellular carcinoma. Nature, *343:* 555-557 (1990).
26. A. Giordano, P. Whyte, E. Harlow, B. R. Jr. Franza, D. Beach, and G. A Draetta, 60 kd cdc2-associated polypeptide complexes with the E1A proteins in adenovirus-infected cells. Cell, *58:* 981-990 (1989).
27. M. Mudryj, S. H. Devoto, S. W. Hiebert, T. Hunter, J. Pines, and J. R. Nevins, Cell cycle regulation of the E2F transcription factor involves an interaction with cyclin A. Cell, *65:* 1243-1253 (1991).
28. M. E. Ewen, Y. Xin, J. B. Lawrence, and D. M. Livingston, Molecular cloning, chromosomal mapping, and expression of the cDNA for p107, a retinoblastoma gene-related protein. Cell, *66:* 1155-1164 (1991).
29. M. E. Ewen, B. Faha, E. Harlow, and D. M. Livingston, Interaction of p107 with cyclin A independent of complex formation with viral oncoproteins. Science, *255:* 85-87 (1992).
30. B. Faha, M. E., Ewen, L-. H. Tsai, D. M. Livingston, and E. Harlow, Interaction between human cyclin A and adenovirus E1A-associated p107 protein. Science, *255:* 87-90 (1992).
31. T. Matsuyama, P. Anderson, J . F. Daley, S. F. Schlossman, and C. Morimoto, $CD4^+CD45R^+$ cells are preferentially activated through the CD2 pathway. Eur. J. Immunol., *18:* 1473-1476 (1988).
32. W. Gubler, and B. J. Hoffman, A simple and very efficient method for generating cDNA libraries. Gene, *25:* 263-269 (1983).
33. T. Nikaido, D. Bradley, and A. B. Pardee, Molecular cloning of transcripts that accumulate during late G1 phase in cultured mouse cells. Exp. Cell Res., *192:* 102-109 (1991).
34. J. D. Dignam, R. M. Lebovitz, and R. G. Roeder, Accurate transcription initiation by RNA polymerase II in a soluble extract from isolated mammalian nuclei. Nucl. Acids Res.,*11:* 1475-1489 (1983).
35. R. W. Carthew, L. A. Chodosh, and P. A. Sharp, An RNA polymerase II transcription factor binds to an upstream element in the adenovirus major late promoter. Cell, *43:* 439-448 (1985).
36. T. Sakai, N. Ohtani, T. L. McGee, P. D. Robbins, and T. P. Dryja, Oncogenetic germ-line mutations in Spl and ATF sites in the human retinoblastoma gene. Nature, *353:* 83-86 (1991).
37. M. Yoshida, M. Yamamoto, and T. Nikaido, Quercetin arrests human leukemic T cells in late G1 phase of the cell cycle. Cancer Res. *52:* 6676-6681 (1992).
38. M. Yamamoto, M. Yoshida, K. Ono, T. Fujita, N. Fujita-Ohtani, T. Sakai, and T. Nikaido, p34cdc2 and cyclin A promoters contain RB or p53 binding sites. (submitted, 1993).
39. Q. Hu, C. Bautista, G. M. Edwards, D. Defeo-Jones, R. E. Jones, and E. Harlow, Antibodies specific for human retinoblastoma protein identify a family of related polypeptides. Mol. Cell. Biol., *11:* 5792-5799 (1991).
40. S. P. Chellappan, S. Hiebert, M. Mudryj, J. M. Horowitz and J. R. Nevins, The E2F transcription factor is a cellular target for the RB protein. Cell, *65:* 1053-1061 (1991).
41. P. Raychaudhuri, S. Bagchi, S. H. Devoto, V. B. Kraus, E. Moran, and J. R. Nevins, Domains of the adenovirus E1A protein required for oncogenic activity are also required for dissociation of E2F transcription factor complex. Genes Dev., *5:* 1200-1211 (1991).
42. T. Chittenden, D. M. Livingston, and J. W. G. Kaelin, The T/E1A-binding domain of the retinoblastoma product can interact selectively with a sequence-specific DNA-binding protein. Cell, *65:* 1073-1082 (1991).
43. L. R. Bandara, and N. B. La Thangue, Adenovirus E1A prevents the retinoblastoma gene product from complexing with a cellular transcription factor. Nature, *351:* 494-497 (1991).
44. S. W. Hiebert, M. Blake, J. Azizkhan, and J. R. Nevins, Role of E2F transcription factor in E1A-mediated trans activation of cellular genes. J. Virol., *65:* 3547-3552 (1991).

45. S. Dalton, Cell cycle regulation of the human cdc2 gene. EMBO J., *11:* 1797-1804 (1992).
46. S. E. Kern, K. W. Kinzler, A. Bruskin, P. N. Friedman, C. Prives, and B. Vogelatein, Sequence-specific binding of p53 to DNA. Science, *252:* 1708-1711 (1991).
47. J. Bargonetti, P. N. Friedman, S. E. Kern, B. Vogelstein, and C. Prives, Wild-type but not mutant p53 immunopurified protein bind to sequence adjacent to the SV40 origin of replication. Cell, *65:* 1083-1091 (1991).
48. W. S. El-Deiry, S. E., Kern, J. A. Pietenpol, K. W. Kinzler, and B. Vogelstein, Definition of a consensus binding site for p53. Nature Genetics, *1:* 45-49 (1992).

42

CELLS UNDERGOING HIV ENVELOPE-MEDIATED PROGRAMMED DEGENERATION ACCUMULATE IN G2/M PHASE

Huan Tian[1], Dan Hartmann[2], Larry Wahl[3], Eileen Donoghue[1], Clare McGowan[4], Jeffrey Cossman[2], Paul Russell[4], Lawrence Samelson[5], and David I. Cohen[1]

[1]Laboratory of Immunoregulation
NIAID, NIH, Bethesda, MD 20892
[2]Department of Pathology, Georgetown University
School of Medicine
Washington, DC 20007
[3]Laboratory of Immunology
NIDH, NIH, Bethesda, MD 20892
[4]Department of Molecular Biology
Scripps Research Institute, La Jolla,CA 92037
[5]CBMB, NICHD, NIH
Bethesda, MD 20892

INTRODUCTION

Acquired immunodeficiency syndrome (AIDS) is a complex disease process induced by human immunodeficiency virus (HIV-1) infection.[1] Although the linkage between HIV-1 infection and the development of AIDS has been established for a decade,[2] the molecular and biochemical basis for the profound and irreversible depletion of helper $CD4^+$ T cells that follows HIV infection and paralyzes the immune system is not understood. A number of mechanisms have been proposed to account for $CD4^+$ T killing by HIV, including the direct lysis of virally-infected cells, and the functional disruption of uninfected cells through an interaction with viral proteins.[1,3,4] A recent hypothesis has proposed that, in HIV-infected individuals, there reemerges a cell death program normally utilized by immature T cells during development in response to specific stimuli accounting for both the early qualitative and late quantitative $CD4^+$ T cell defects associated with AIDS.[5]

The process of programmed cell death is an active suicide mechanism involved in a spectrum of biological processes, reflecting normal physiological homeostasis and development.[6-10] During thymocyte maturation in the immune system, programmed cell death plays a major role in negative selection of the T cell repertoire resulting in the clonal deletion of autoreactive T cells and to the establishment of self-tolerance.[11,13-17] This cell suicide program requires cellular activation[11,13-17] and the initiation of new RNA and protein synthesis[12,20,22], but not tyrosine phosphorylation[32]. An endogenous endonuclease is activated in dying cells resulting in a characteristic fragmentation of cellular DNA into nucleosomes[11]. In several systems, although not universally, it has been established that activation-induced death of thymocytes occurs at an early G1/S phase cell cycle block[18,19,20].

In this report, we summarize data establishing that HIV can induce a form of programmed cell death in $CD4^+$ T cells which is distinguished by cell cycle arrest at G2/M phase. Cells undergoing this HIV-directed suicide program have extensive hyperphosphorylation of human p34cdc2, the key mitotic regulator of G2/M phase cell cycle progression.[21] Cells degenerating of the cytopathic effect which characterizes HIV-1 accumulate cyclin B protein which serves as an additional indicator of G2/M cell cycle arrest. The HIV-mediated form of programmed cell death differs from types of programmed cell death associated with normal B cell and T cell selection. The inhibitor of protein synthesis, cycloheximide, blocks T cell programmed cell death during negative selection,[20,22] but cycloheximide increases HIV-induced T cell killing. Similarly, cyclosporin A[18,19,23] and actinomycin D[12] inhibit developmental programmed cell death of thymocytes, while neither affect HIV-induced cell death. Taken together, these data establish that HIV-induced cell killing represents a novel form of programmed cell death involving cellular activation and cell cycle arrest at G2/M phase.

RESULTS

Compared to the Programmed Cell Death Associated With B Cells and Thymocytes, HIV-Mediated Cell Killing Involves a Novel Suicide Program

HIV-induced cell killing was studied by isolating DNA from the cytoplasm of HIV-1 (LAV) infected peripheral blood T cells following lysis in dilute Triton (0.2%). DNA fragmentation was determined by gel electrophoresis (Figure 1) in a standard assay for programmed cell death. While uninfected PBLs have mainly high molecular weight DNA by this analysis (Figure 1, lane 1), DNA from HIV-infected cells was fragmented into low molecular weight material (Figure 1, lane 2). A $CD4^+$ T cell tumor line, Jurkat, transfected to express HIV envelope glycoproteins gp120/gp41 (Jenv) showed a similar laddered fragmentation of cellular DNA upon initiation of HIV-directed cell killing (data not shown). These studies established that HIV-directed cell killing was a form of programmed cell death.

To further investigate the mechanism of HIV-mediated programmed cell death (PCD), we evaluated several agents for their ability to inhibit HIV-envelope (HIVenv)-mediated cell killing in a model system where killing is initiated by combining equal numbers of a T cell line transfected to stably express HIVenv glycoproteins (Jenv) with a strongly $CD4^+$ T cell line (Jurkat) in the same culture. In the absence of any inhibitors, extensive cytopathicity and cell death occurs in this system between 8 and 24 hours. Herbimycin A, a protein tyrosine kinase inhibitor,[25] attenuated the HIV-induced cytopathicity, while it has been reported not to effect T cell selection[32]. Inhibitors of protein or mRNA synthesis block PCD in thymocytes,[12,20,22] but when HIVenv cytopathicity was studied in the Jenv/Jurkat co-culture system, the protein synthesis inhibitor, cycloheximide (50ng/ml), was found to increase cytopathicity (80 syncytia treated, 53 syncytia untreated, 51% increase). The inhibitor of

Table 1. Comparison of the effects of several inhibitors on HIVenv-mediated cytopathicity and on T cell selection.

Inhibitor	Function	Outcome HIV PCD	Outcome T Cell Selection
Herbimycin	pTK	↓	0
Cycloheximide	Protein Synthesis	↑	↓
Actinomycin D	RNA Synthesis	0	↓
Okadaic Acid	pp2A	↓	?

↓ = Decreased ↑ = Increased 0 = No Effect

mRNA synthesis, actinomycin D (10mcg./ml) had no effect (Table 1). The immunosuppressant cyclosporin A can largely block activation-induced T cell death,[18,19,23] but it had no effect on the HIV-mediated cell death program in our model system. Interestingly, okadaic acid, an inhibitor for protein phosphatase 2A, which has also been reported to promote cell cycle progression at least in part through activation of cdc25,[33] could dramatically inhibit the cytopathicity induced by HIVenv (Table 1).

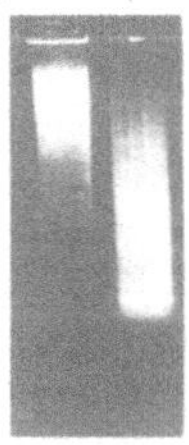

Figure 1. DNA fragmentation from normal PBLs infected with HIV-1 LAV. HIV-1 LAV-infected normal PBLs (lane 2) were harvested at day 7 and were lysed in hypotonic lysis buffer (10 mM Tris, 1 mM EDTA, 0.2% Triton X-100). Lysates were extracted twice with phenol before precipitating with ethanol. After digestion with 20 ug/ml RNase A, the resulting DNA was fractionated on a 1.5% agarose gel. Lane 1 shows DNA extracted from uninfected normal PBLs.

HIV-Induced Cell Killing Results in Tyrosine Hyperphosphorylation of cdc2, but not cdk2, with Accumulation of Arrested Cells in G2/M Phase

During HIVenv-mediated cytopathicity induced either by exposure of $CD4^+$ cells to Jenv or by HIV infection, a 34 KD protein became tyrosine hyperphosphorylated which was identified as the cyclin dependent kinase, p34cdc2, a universal regulator of mitotic cell cycle (Figure 2, lanes 4 - 6). The identity of this protein was established by its high affinity to the yeast suc1 product, p13, through its immunoprecipitation by carboxy terminal antibodies specific for human cdc2, and by electrophoretic mobility assay (Figure 2). Western blots of whole cell lysates from the co-culture of Jenv and Jurkat cells triggered to undergo HIVenv cytopathicity were analyzed with antibodies specific to human p34cdc2 and p33cdk2 (Figure 2). A constant level of cdk2 protein was observed during the killing process, with no evidence for changes in cdk2 phosphorylated forms which migrate more quickly than unphosphorylated cdk2 (Figure 2, lanes 1-3). Immunoblot analysis of cdc2, however, revealed increased amounts of a slightly slower migrating form after 24 hours of HIVenv cytopathicity,

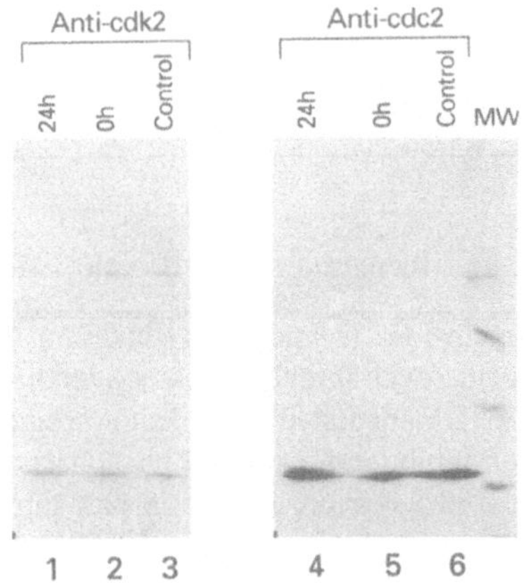

Figure 2. Co-cultivation of HIVenv-expressing line (Jenv) and $CD4^+$ Jurkat cells induces the hyperphosphorylation of cdc2, but not cdk2. Jurkat cells and Jenv were co-cultured for the indicated length of time. Samples were harvested by lysing with protein lysis buffer (50 mM Tris pH 7.5, 1% NP-40, 150 mM NaCl, 5 mM EDTA, 1 mM Na Ortho Vanadate, protease inhibitors). In the control samples (lane 3 and 6), Jurkat cells and HIVenv cells were lysed first with the protein lysis buffer, then equal amount of lysates were mixed. Protein levels were normalized in all samples before separating by 5-15% polyacrylamide gel electrophoresis. The levels of cdk2 and cdc2 were determined on immunoblots incubated with specific polyclonal antibodies to the carboxy terminals of cdk2 (lanes 1-3, UBI, Lake Placid, NY) or cdc2 (lanes 4-6, GIBCO BRL, Gaithersburg, MD).

representing the hyperphosphorylated form of p34cdc2 (Figure 2, lane 4). This slower mobility species of p34cdc2 was not observed at the beginning of the co-culture (Figure 2, lane 5). Phosphoamino acid analysis in our laboratory has confirmed the identity of this slower migrating band as hyperphosphorylated p34cdc2. These data indicated that p34cdc2, but not cdk2, was phosphorylated after co-culture of Jenv and Jurkat, consistent with the cells being arrested at the G2/M cell cycle checkpoint.

Since accumulation of intracellular cyclin B protein occurs only from late S through M phase,[34] its detection has been employed as a sensitive and specific indicator of G2/M phase. To verify that cells undergoing HIVenv-directed cytopathicity were arrested at the G2/M checkpoint, dying cells were analyzed for total cyclin B by protein immunoblot and for cellular cyclin B protein accumulation by staining with anti-cyclin B antibodies. These experiments showed that cyclin B levels increased in the model system of HIV cytopathicity, and that cells displaying the morphologic balloon degeneration characteristic of HIV cytopathicity had dramatically high levels of

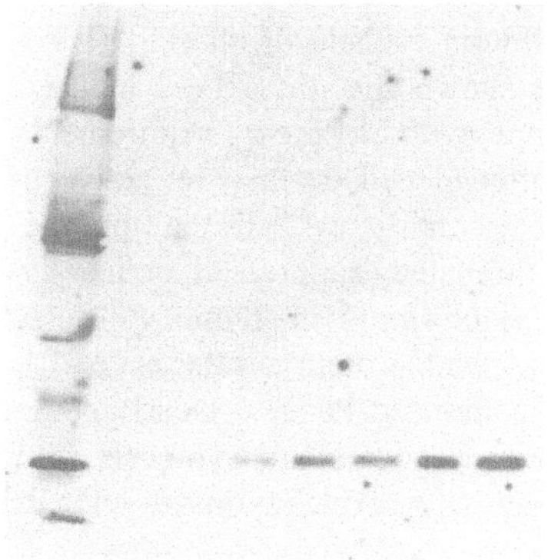

Figure 3. Association of hyperphosphorylation of p34cdc2 with HIV-mediated cytopathicity. Co-cultures of Jenv and Jurkat cells were harvested at the indicated time points by lysing with protein lysis buffer. p13 affinity precipitates were separated on a 5-15% polyacrylamide gel. Tyrosine phosphate content of p13-associated proteins was determined by immunoblot using a polyclonal rabbit anti-phosphotyrosine antibody.

intracytoplasmic cyclin B protein. Importantly, cyclin B antibody staining proved sensitive enough to detect accumulated cyclin B protein in the large, balloon degenerating cells occurring during primary infection of normal human peripheral blood T cells with clinical specimens of pathogenic HIV-1 virus. These studies verified that cells dying of HIV-induced killing had the characteristics of cells arrested at the G2/M checkpoint, including hyperphosphorylated p34cdc2 and accumulated cyclin B protein. Moreover, it was shown that this mechanism is operative in primary infections with clinical HIV-1.

The Kinetics of cdc2 Tyrosine Phosphorylation is Closely Associated with the Appearance of HIVenv-Induced Cytopathicity

Since p34cdc2 was tyrosine hyperphosphorylated during HIVenv-mediated cell killing, an obvious question concerned whether the kinetics of this tyrosine phosphorylation tracked the cytopathicity and cell death which occurred in the Jenv exposed co-cultures. At different time points following initiation of cytopathicity, endogenous cdks were specifically precipitated with p13-bound beads prior to immunoblot analysis with polyclonal rabbit anti-phosphotyrosine antibody (Figure 3). Only a single migrating species of cdk, corresponding to p34cdc2, became tyrosine phosphorylated. The tyrosine phosphate content of cdc2 (Figure 3) closely paralleled the appearance of HIVenv-induced cytopathicity, which dramatically appeared and progressed between 8 and 24 hours of culture (not shown).

DISCUSSION

In this report, we have demonstrated that the HIV-mediated cytopathic effect involves a novel cell death program. This program is characterized by cell cycle arrest at the G2/M checkpoint, fragmentation of cellular DNA, and cell death. HIV infection or exposure to HIV envelope proteins gp120/gp41 delivers suicide signals to T cells, resulting in the tyrosine hyperphosphorylation of the cyclin dependent kinase p34cdc2.

The hyperphosphorylation of p34cdc2 is closely associated with HIV-1-induced T cell killing. As a consequence of HIV exposure, normal cell cycle progression is blocked at the G2/M cell cycle checkpoint, leading to accumulation of intracytoplasmic cyclin B protein and to the bizarre, balloon cellular degeneration characteristic of cells dying during infections with pathogenic HIV-1.

Utilizing several inhibitors capable of altering HIVenv-mediated T cell killing in our model system, we have shown that the HIV-1-induced cytopathic mechanism is a novel T cell killing program with different requirements from developmental PCD (negative selection). Programmed cell death is an active process naturally occurring in a wide variety of physiological conditions.[6-10] In the immune system, thymocytes carrying self-reactive receptors are eliminated, suppressed or inactivated.[11-17] This classical type of programmed cell death, as observed in thymocyte selection, is characterized by its dependence on de novo RNA and protein synthesis, and by morphological features of apoptosis.[12] In contrast, HIV-1-induced PCD, as clearly demonstrated by experiments with actinomycin D, did not involve new RNA synthesis and was actually enhanced by blockade of protein synthesis with cycloheximide. Furthermore, cyclosporin A, an immunosuppressant shown to inhibit thymocyte programmed cell death in intrathymic T cell development,[18,19,23] had no protective effect on HIVenv-induced cytopathicity. Suicide signal transduction via tyrosine phosphorylation and dephosphorylation appears to play a unique and key role in the HIVenv-induced cell death mechanism.[29] Inhibition of src-related and other tyrosine kinases by herbimycin A largely attenuates the HIV-mediated cytopathic effect.[29] Tyrosine hyperphosphorylation of p34cdc2 is a central biochemical characteristic of cells committed to HIV-initiated PCD, which has not been seen in other forms of T cell PCD. Treatment of cells with okadaic acid not only strongly inhibits HIV-induced cytopathicity, but also biochemically blocks the production of tyrosine hyperphosphorylated cdc2. Okadaic acid, a phosphatase 2A inhibitor, is known to promote the bypass of the G2/M cell cycle checkpoint and to downregulate phosphatases which inactivate cdc25. These observations suggest the possibility that okadaic acid functions to release HIV cytopathicity by inactivating phosphatase 2A leading to stable activation of tyrosine phosphatase cdc25. Active cdc25 dephosphorelates and activates p34cdc2.[33] Against this model, however, we have overexpressed cdc25 protein in T cells but found that its overexpression does not release HIV-generated cytopathicity. Thus the full biochemical consequences of the HIV-induced G2/M checkpoint arrest remain to be elucidated.

Unlike other systems of PCD where G1 phase checkpoint arrest is observed,[19-21,31] disruption of normal cell cycle progression by HIV results in cells arresting and dying at G2/M phase. These observations provide the first molecular understanding of events surrounding HIV-mediated PCD. Despite intensive investigation of the immunopathogenicity of HIV-1, further work is required for fully understanding the molecular mechanisms by which HIV-1 disrupts immune function. Our data strongly support the contention that CD4 T lymphocyte depletion and dysfunction involves a novel cell death program initiated following exposure of cells to HIV envelope and most likely involving the triggering of specific cell surface receptors including, but not limited to, CD4. The finding that HIV utilizes an unusual form of PCD not employed by T cells during normal development could have important implications on the design of therapeutic agents capable of specifically interfering with the HIV-initiated cell death process.

REFERENCES

1. A.S. Fauci. The immunodeficiency virus: infectivity and mechanism of pathogenicity. Science 239:617 (1988).

2. F. Barre-Sinoussi, J.C. Chermann, F. Rey, M.T. Nugere, S. Chamaret, J. Gruest, C. Dauguet and C. Axler-Blin. Isolation of a T lymphotropic retrovirus from a patient at risk for acquired immunodeficiency syndrome (AIDS). Science 220:868 (1983).

3. J. Habeshaw, E. Hounsell and A. Dalgeish. Dose the HIV induce a chronic graft-versus-host-like disease? Immunol. Today 13:207 (1992).

4. M.L. Gougeon, V. Colizzi and L. Montagnier. AIDS Res. Hum. Retroviruses 9:287 (1993).

5. H. Groux, G. Torpier, D. Monte, Y. Mouton, A. Capron and J.C. Ameisen. Activation-induced death by apoptosis in $CD4^+$ T cells from human immunodeficiency virus-infected asymptomatic individuals. J. Exp. Med. 175:331 (1992).

6. R.E. Ellis, J. Yuan and H.R. Horvitz. Mechanisms and functions of cell death. Annu. Rev. Cell Biol. 7:663 (1991).

7. J.W. Sauders. Death in embryonic systems. Science 154:604 (1966).

8. J.R. Hinchliffe. Cell death in embryogenesis, in "Cell Death in Biology and Pathology," I.D. Bowen and R.A. Lockshin, ed., Chapman and Hill, New York (1981)

9. W.M. Cowan, J.W. Fawcett, D.D.M. O'leary and B.B. Stanfield. Repressive events in neurogenesis. Science 225:1258 (1984)

10. R.A. Lockshin. Cell death in metamorphosis. In "Cell Death in Biology and Pathology," I.D. Bowen and R.A. Lockshin, ed., Chapman and Hill, New York.

11. C.A. Smith, G.T. Williams, R. Kingston, E. J. Jenkinson and J.J.T. Owen. Antibodies to CD3/T-cell receptor complex induce death by apoptosis in immature T cells in thymic cultures. Nature 337:181 (1989)

12. H.R. MacDonald and R.K. Lees. Programmed death of autoreactive thymocytes. Nature 343:642 (1990).

13. I. Nakashima, Y.H. Zhang, S.M.J. Rahman, T. Yosida, K.I. Isobe, L.N. Ding, T. Iwamoto, M. Hamaguchi, H. Ikezawa and R. Taguchi. Evidence of synergy between Thy-1 and CD3/TCR complex in signal delivery to murine thymocytes for cell death. J. Immunol. 147:1153 (1991)

14. A. Veillette, J.C. Zuniga-Pflucker, J.B. Bolen and A.M. Kruisbeek. Engagement of CD4 and CD8 expressed on immature thymocytes induces activation of intracellular tyrosine phosphorylation pathways. J. Exp. Med. 170:1671 (1989)

15. D.J. McConkey, P. Hartzell, J.F. Amador-Perez, S. Orrenius and M. Jondal. Calcium-dependent killing of immature thymocytes by stimulation via the CD3/T cell receptor complex. J. Immunol. 143:1801 (1989).

16. T.H. Finkel, M. McDuffie, J.W. Kappler, P. Marrack and J.C. Cambier. Both immature and mature T cells mobilize Ca2+ in response to antigen receptor cross-linking. Nature 330:179 (1987).

17. W.L. Havran, M. Poenie, J. Kimura, R. Tsien, A. Weiss and J.P. Allison. Expression and function of the CD3-antigen receptor on murine CD4+CD8+ thymocytes. Nature 300:170 (1987).

18. M. NMercep, P.D. Noguchi and J.D. Ashwell. The cell cycle block and lysis of an activated T cell hybridoma are distinct processes with different Ca2+ requirements and sensitivity to cyclosporine A. J. Immunol. 142:4085 (1989).

19. Y.F. Shi, B.M. Sahai and D.R. Green. Cyclosporine A inhibits activation-induced cell death in T-cell hybridomas and thymocytes. Nature 339:625 (1989).

20. D.S. Ucker, J.D. Ashwell and G. Nickas. Activation-driven T cell death. I. Requirements for de novo transcription and translation and association with genome fragmentation. J. Immunol. 143:3461 (1989)

21. G. Draetta and D. Beach. Activation of cdc2 protein kinase during mitosis in human cells: cell cycle-dependent phosphorylation and subunit rearrangement. Cell 54:17 (1988).

22. D.J. McConkey, P. Hartzell and S. Orrenius. Rapid turnover of endogenous endonuclease activity in thymocytes: effects of inhibitors of macromolecular synthesis. Arch. Biochem. Biophys. 278:284 (1990).

23. C.M. Zacharchuk, M. Mercep and J.D. Ashwell. Thymocyte activation and death: a mechanism for molding the T cell repertoire. Ann. New York Acad. Sci. 636:52 (1991).

24. Y. Tani, H. Tian, H. C. Lane and D.I. Cohen. Normal T cell receptor mediated signalling in T cell lines stably expressing HIV-1 envelope glycoproteins. In press.

25. M. Graber, C.H. June, L.E. Samelson and A. Weiss. The protein tyrosine kinase inhibitor herbimycin A, but not genistein, specifically inhibits signal transduction by the T cell antigen receptor. Intl. Immunol. 4:1201 (1992).

26. D.J. McConkey, P. Hartzell, M. Jondal and S. Orrenius. Inhibition of DNA fragmentation in thymocytes and isolated thymocyte nuclei by agents that stimulate protein kinase C. J. Biol. Chem. 264: 13399 (1989).
27. C.M. Zacharchuk, M. Mercep, P.K. Chakraborti, S.S. Simons Jr. and J.D. Ashwell. Programmed T lymphocyte death: cell activation and steroid-induced pathways are mutually antagonistic. J. Immunol. 145:4037 (1990).
28. A.M. Zubiaga, E. Munoz and B.T. Huber. IL-4 and IL-2 selectively reuse Th cell subsets from glucocorticoid-induced apoptosis. J. Immunol. 149:107 (1992).
29. D.I. Cohen, Y. Tani, H. Tian, E. Boone, L.E. Samelson and H.C. Lane. Participation of tyrosine phosphorylation in the cytopathic effect of human immunodeficiency virus-1. Science 256:542 (1992).
30. C. Jessus and D. Beach. Oscillation of MPF is accompanied by periodic association between cdc25 and cdc2-cyclin B. Cell 68:323 (1992).
31. L.A. Sabourin and R.G. Hawley. Suppression of programmed death and G1 arrest in B-cell hybridomas by interleukin-6 is not accompanied by altered expression of immediate early response genes. J. Cell. Physiol. 145:564 (1990).
32. K. Nakayama and D.Y. Loh. No requirement for p56lck in the antigen-stimulated clonal deletion of thymocytes. Science 257:94 (1992).
33. L. Gautier, M.J. Solomon, R.N. Booher, J. Fernando Bazan and M.W. Kirschner. cdc25 is a specific tyrosine phosphatase that directly activates p34cdc2. Cell 67:197 (1991).
34. J. Pines and T. Hunter. Isolation of a human cyclin cDNA: Evidence for cyclin mRNA and protein regulation in the cell cycle and for interaction with p34cdc2. Cell 58:833 (1989).

PART VII

CLINICAL APPLICATIONS

43

THE INVOLVEMENT OF THE CELL CYCLE IN APOPTOSIS

Alan Eastman, Michael A. Barry, and Catherine Demarcq

Department of Pharmacology and Toxicology
Dartmouth Medical School
Hanover, NH 03755

APOPTOSIS IS A PRODUCT OF MULTIPLE PATHWAYS

The term apoptosis was proposed in 1972 to describe dying cells that undergo characteristic morphological changes that were distinctly different from necrosis (Kerr et al., 1972; reviewed in Wyllie et al., 1980). In necrosis, cells swell leading to rupture of plasma and organelle membranes, release of hydrolytic enzymes, and loss of organized structure. In contrast, apoptosis was recognized as reduction in cell size and therefore it was originally termed "shrinkage necrosis." It was believed that apoptosis was involved in tissue homeostasis so the new name was intended to emphasize the normal balance between cell replication (mitosis) and cell death (apoptosis). The earliest morphological change observed in a cell dying by apoptosis is chromatin condensation to the periphery of the nucleus, followed by nuclear and cytoplasmic blebbing, cell shrinkage, and eventual loss of membrane integrity. Biochemically, the earliest event observed is DNA digestion in the internucleosome spacer region. These events occur late in the pathway of cell death, but understanding what regulates them will lead to an understanding of the upstream events that directly cause cell death (Figure 1). One frequently cited requirement for apoptosis is new protein synthesis, although the identity of the critical protein(s) is unknown. These proteins are frequently thought of as lethal, but results discussed below will present an argument that they may be normal cell cycle regulatory proteins, and that their expression leads to passage of a cell to a phase of the cell cycle at which they can undergo apoptosis.

Apoptosis is observed during normal developmental and regulatory processes, and is frequently thought of as a programmed event, hence the use of the phrase "programmed cell death." Perhaps the most dramatic example of this programming occurs in the nematode *C. elegans* in which 131 cells die at precise times during development of the mature organism. Another example occurs in the developing human embryo where many more neurons are made than are eventually required; the survivors are those cells that receive neurotrophic growth factors from the target cells they innervate. There are two general initiators of apoptosis in these physiological conditions. Activation-induced apoptosis results from stimulation of specific receptors such as T cell receptor mediated deletion of self recognizing

The Cell Cycle: Regulators, Targets, and Clinical Applications
Edited by V.W. Hu, Plenum Press, New York, 1994

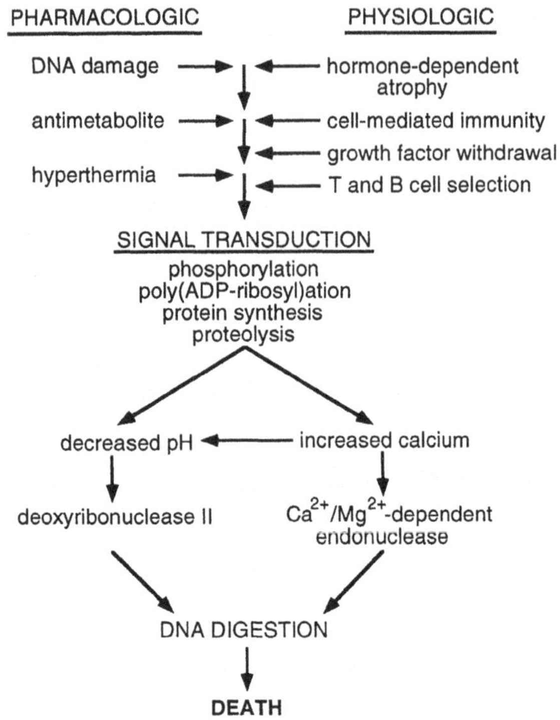

Figure 1. A summary of cellular events potentially associated with apoptosis. From multiple initiators of apoptosis, a variety of signalling events have been suggested to occur. These signals merge in a common pathway in which an endonuclease is activated and the characteristic DNA digestion and morphology of apoptosis is observed. Two candidate endonucleases are shown, each activated by different signals. The significance of these endonucleases is discussed in the text.

immune cells, thyroxine-mediated regression of the tadpole tail, or glucocorticoid-mediated death of immature thymocytes. Inactivation-induced apoptosis is probably the most common form of cell death which results from the removal of growth factors and the turning off of associated signal transduction pathways. An obvious example of inactivation as an initiating event is the requirement for serum in most cell culture. An important exception is in cancer where oncogenic changes frequently lead to reduced growth factor requirement.

The above examples of programmed cell death have often been considered cell suicide. However, there are many other agents that can kill cells, and the characteristic morphology of apoptosis is also observed. In these cases, cell death can be considered an unprogrammed event and the cells could more appropriately be said to be murdered. Cytotoxic T cells certainly murder their target cells, although they presumably do it by turning on a late stage in the target cell's endogenous apoptotic pathway. Similarly, an ever growing list of cytotoxic agents are reported to induce apoptosis as a consequence of interaction at many primary targets.

One important consideration that distinguishes programmed from unprogrammed apoptosis is the phase of the cell cycle involved. Hence, activation of a glucocorticoid receptor or removal of a growth factor lead to apoptosis at the G_0/G_1 phase. In contrast, unprogrammed cell death can result from damage elicited at any phase of the cell cycle. As an example, we have shown that the topoisomerase inhibitor etoposide induces apoptosis at two different stages of the cell cycle; an initial rapid phase of apoptosis is observed of cells in S phase at the time of insult, and a much slower phase in which cells progress to G_2 where they die over the following 48 h (Barry et al., 1993). The presence of multiple ways to kill a cell, both programmed and unprogrammed, suggest the existence of a variety of different signal transduction pathways, all of which eventually impact upon some common steps that result in the appearance of apoptotic cell death. This discussion of terminology in apoptosis is the subject of a recent review (Eastman, 1993).

Studies on cancer have provided apoptosis with perhaps its major impetus. The oncogene *bcl-2*, originally isolated from a chromosome break point in follicular B-cell lymphoma, was known for several years to protect cells from a variety of stresses (Vaux et al., 1988), but it was not until it was shown to prevent apoptosis that interest grew (Hockenbery et al., 1990). The same is also true for the oncogene c-*myc* which was known to kill cells when over-expressed (Wurm et al., 1986), but interest was stimulated by the observation that over-expression of c-*myc* killed cells by apoptosis (Evan et al., 1992). The tumor suppressor gene p53 has also been shown to induce apoptosis (Yonish-Rouach et al., 1991). It becomes important to determine whether the expression of these genes is related to unique or multiple pathways to apoptosis. The *bcl-2* gene, whose function remains elusive, can protect cells from activation-induced and inactivation-induced apoptosis as well as from most anticancer agents (Siegel et al., 1992; Miyashita and Reed, 1992). An antisense oligonucleotide to c-*myc* has been shown to suppress cell death in an activation-induced pathway but its contribution to other pathways has not yet been established (Shi et al., 1992). Over-expression of c-*myc* appears to lead to apoptosis independent of the cell cycle phase suggesting that cells can only tolerate expression of c-*myc* in the G_1 phase (Askew et al., 1991). To understand how these genes are involved in apoptosis requires a knowledge of the signal transduction pathways essential for apoptosis. To gain insight into this area, we began by analyzing apoptotic cells, and then proceeded back toward the initiating stimuli. In so doing we have established a model that may explain the interaction between oncogenes and apoptosis.

ENDONUCLEASES INVOLVED IN APOPTOSIS

The degradation of chromatin DNA into nucleosome fragments has become synonymous with apoptosis. There are several reports of apoptosis detected morphologically without DNA digestion, but this might be explained by limited DNA digestion that did not resolve into detectable DNA fragments (Ucker et al., 1992; Oberhammer et al., 1991; Zakeri et al., 1993). Certainly, digestion as far as nucleosome-length fragments can be considered "over-kill," but determining the signals for DNA digestion arguably provide the best insight into the onset of apoptosis.

Mammalian cells contain a variety of endonucleases, any of which could be involved in apoptosis (reviewed in Eastman and Barry, 1992). It has become highly cited that the endonuclease involved in apoptosis is Ca^{2+}/Mg^{2+}-dependent. However, early articles only hypothesized this because the Ca^{2+}/Mg^{2+}-endonuclease was observed in thymocytes undergoing apoptosis (Cohen and Duke, 1984). Thymocytes are known to contain other endonucleases, any of which could theoretically be involved in apoptosis (Nikonova et al., 1982). There are also other reasons to doubt at least the universal involvement of a Ca^{2+}/Mg^{2+}-endonuclease: many cells do not appear to contain it, and there are many reports of undetectable increases in Ca^{2+} which is thought to be the trigger for its activation. Even in those systems in which the endonuclease is observed, the increases in Ca^{2+} remain far below those required for in vitro activity.

While attempting to purify an endonuclease from Chinese hamster ovary cells that was potentially responsible for apoptosis, we found an endonuclease active only at acidic pH. Upon characterization, it was identified as deoxyribonuclease II (Barry and Eastman,

1993). Its maximal activity was observed at pH 5, but some activity was obtained up to almost pH 7. Apoptosis could be induced in the CHO cells by directly reducing the intracellular pH with a proton ionophore. This led to experiments to determine whether changes in Ca^{2+} or pH occurred in cells as they underwent cytotoxin-induced apoptosis.

Apoptosis was induced in human promyelocytic leukemia HL60 cells by the topoisomerase inhibitor etoposide (Barry et al., 1993). These cells began to digest their DNA after about 4 h. Cells were loaded with either a Ca^{2+}- or pH-sensitive fluorescent dye, and then analyzed by flow cytometry for changes in these two parameters. No change in calcium was observed. However, concurrent with the onset of DNA digestion, the cells underwent acidification to about pH 6.4. The flow cytometer was used to sort the cells on the basis of intracellular pH, and only the acidic cells showed the morphology and DNA digestion characteristic of apoptosis. These results were consistent with deoxyribonuclease II as the endonuclease responsible for the apoptotic DNA digestion.

Ongoing studies are addressing whether acidification occurs during other incidences of apoptosis. We have been working with an interleukin-2-dependent cytotoxic T lymphocyte cell line CTLL-2. Upon withdrawal of IL-2, cells accumulate in the G_1 phase where they begin to digest there DNA after about 16 h. Concurrent with this DNA digestion, we have observed both increases in Ca^{2+} and decreases in pH. Cells were incubated with EGTA to remove extracellular Ca^{2+}, and this was shown to inhibit the increase in intracellular Ca^{2+} following removal of IL-2. However, chelation of Ca^{2+} did not protect the cells, rather acidification was still observed and the cells still digested their DNA. These results are again consistent with deoxyribonuclease II being responsible for the DNA digestion associated with apoptosis.

REGULATION OF INTRACELLULAR pH

The acidic shift observed during apoptosis in the experiments described above was not a continuum of cells progressively more acidic, rather there were two clear populations, one at normal pH, the other at about pH 6.4 (Barry et al., 1993). To understand the origin of this acidification it is important to emphasize that the cells are still metabolically viable. First, the acidic cells are able to take up fluorescent dye, deesterify it and then retain it, thereby demonstrating membrane integrity. The cells also excluded trypan blue. Second, these cells showed no increase in intracellular Ca. To maintain low Ca against a Ca gradient requires metabolic activity. A metabolically inactive cell or necrotic cell would have intracellular pH and Ca equal to the extracellular medium. These cells therefore maintain an electrochemical gradient across the cytoplasmic membrane. Under conditions of a normal electrochemical gradient, selective inhibition of pH regulation would cause pH to drop 1 pH unit (Madshus, 1988). Therefore, the large acidification in the apoptotic cells could be due to selective inhibition of pH regulation without interference with the homeostasis of other ions.

Cells regulate their pH through a variety of ion transport mechanisms. These regulators include Na^+/H^+ antiporters, ATP-driven H^+-pumps, and several bicarbonate exchangers; the latter influence intracellular H^+ concentration by altering the intracellular concentration of HCO_3^-. Certain specialized cells also regulate pH through other exchangers such as a H^+/K^+ exchanger or a lactate-H^+ symporter.

The Na^+/H^+ antiporter is a major H^+-extruding mechanism which is driven by the inward-directed Na^+ gradient. This antiporter has a high affinity for H^+ at pH 6, but is inoperative at neutral pH. However, a wide variety of external signals including growth factors enhance the affinity of the antiporter for H^+ at neutral pH leading to alkalinization of the cells. This modulation of intracellular pH is brought about by phosphorylation of the antiporter (Sardet et al., 1989; Sardet et al., 1990). In the presence of bicarbonate, cells can use alternate mechanisms to regulate their intracellular pH. However, the Na^+/H^+ antiporter has been shown to be important for tumor growth (Rotin et al., 1989) perhaps because bicarbonate transporters alone are unable to regulate the degree of acidification that may occur during growth of a tumor in vivo, a much more stressful environment than in cell culture. Further evidence for the importance of pH regulation for tumor growth comes from experiments in which a yeast H^+/ATPase was expressed in NIH/3T3 cells. This led to increased intracellular pH and transformation of the cells (Gillies et al., 1990). Another oncogene Ha-*ras*, has also been shown to activate the Na^+/H^+ antiporter (Maly et al., 1989), which would suggest that it may also be able to suppress cell death.

The interaction of growth factors with their receptors leads to activation of a number of different signal transduction pathways (Figure 2). For example, phosphatidylinositol-3-kinase (PI3K) and ras-GTPase activating protein (GAP) activate pathways that lead to changes in nuclear transcription events such as up-regulation of c-*fos*, c-*jun* and c-*myc*, as well as phosphorylation of the tumor suppressor proteins Rb and p53 (Aaronson, 1991). A third signal transduction pathway involves the activation of phospholipases such as phospholipase C (PLC) which produces diacylglycerol, an activator of protein kinase C (PKC). Activation of PKC can also cause induction of nuclear events, but more importantly for the current discussion is its mediation of the phosphorylation of several ion transport proteins including the Na^+/H^+ antiporter (Sardet et al., 1990). Direct activation of PKC with phorbol esters, which also causes phosphorylation of the Na^+/H^+ antiporter, can also overcome apoptosis induced by removal of growth factors (Swann and Whitaker, 1985). Accordingly, the PKC pathway, and its mediation of phosphorylation of the Na^+/H^+ antiporter appears to be critical for cell survival. Hence the available information is consistent with apoptosis being induced by intracellular acidification that can be prevented by alteration in the capacity of cells to regulate pH changes. One caveat is the possibility that the Na^+/H^+ antiporter may be phosphorylated by PKC-independent pathways (Maly et al., 1989) suggesting multiple survival strategies for a cell.

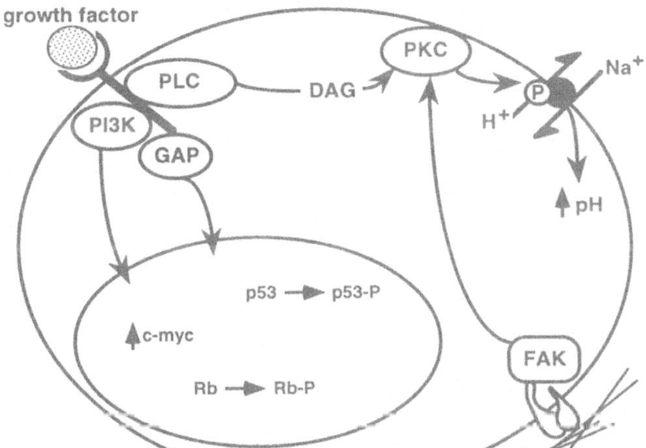

Figure 2. Schematic representation of growth factor-mediated signal transduction pathways in a cell, emphasizing those pathways that stimulate nuclear events, and those that activate survival pathways such as through PKC-mediated phosphorylation of membrane ion transporters. The abbreviations for the components shown are described in the text.

This model also addresses the susceptibility of certain cells to undergo apoptosis upon over-expression of c-*myc* or wild-type p53. These are nuclear oncogenic events which lead to apoptosis in the absence of activation of a survival pathway such as mediated by PKC. For example, the induction of apoptosis by both c-*myc* and p53 can be suppressed by growth factors (Evan et al., 1992; Yonish-Rouach et al., 1991). Furthermore, tumor cells over-expressing c-*myc* frequently appear to have mutated or deleted p53, suggesting that p53 may be a mediator of the c-*myc*-induced apoptosis. Therefore, deletion of p53 may be one way to prevent apoptosis induced by over-expression of c-*myc*.

APOPTOSIS INDUCED BY CISPLATIN

Our investigations into apoptosis originated from studies aimed at explaining how a drug such as cisplatin kills cells. Cisplatin is an effective drug for a variety of human tumors. It kills cells as a consequence of its interaction with DNA where it produces both DNA-intrastrand and DNA-interstrand cross-links (Eastman, 1987). This binding to DNA is not itself sufficient to cause cell death, rather cells are much more sensitive if cycling, and passage through S phase appears critical, at least at pharmacologically relevant drug concentrations. Damaged cells progress through the cell cycle, arrest at G_2, and subsequently die (Sorenson and Eastman, 1988a,b). As the cells die, they exhibit characteristics of apoptosis (Sorenson et al., 1990; Barry et al., 1990). However, our previous experiments did not adequately resolve the order of events that occur as a cell dies. We therefore investigated these phenomena in a well synchronized population of cells as described below (Eastman et al., 1993; Demarcq et al., 1993).

These experiments have used a DNA repair-deficient Chinese hamster ovary cell line, UV41. This cell line was specifically chosen so that the level of DNA damage would remain constant throughout the time course of the experiments. Furthermore, this facilitated the use of a very low concentration of cisplatin, which although killing all the cells, was inadequate to cause inhibition of DNA synthesis. Accordingly, the damaged cells arrested and died synchronously.

Cells were synchronized at the G_1/S border by mitotic shake followed by aphidicolin. Upon release from aphidicolin, undamaged cells progressed immediately through S phase, and through mitosis by 12 h (Figure 3). Some synchrony was retained through a second cycle. These cells were analyzed for their $p34^{cdc2}$ synthesis and phosphorylation and were shown to hyperphosphorylate $p34^{cdc2}$ and then rapidly dephosphorylate it just prior to entry into mitosis.

Synchronized cells were incubated with cisplatin for 1 h in early S phase. These cells progressed to the G_2 phase at the same rate as undamaged cells, however, they remained in G_2 for up to 16 h. These cells also contained the hyperphosphorylated form of $p34^{cdc2}$ thereby demonstrating no inhibition in synthesis or phosphorylation of the protein. The arrest was caused by an inability to dephosphorylate $p34^{cdc2}$, a function assigned to the cdc25C phosphatase (Hoffmann et al., 1993). At later time periods (24 h) cells eventually dephosphorylated $p34^{cdc2}$ and entered mitosis. However, during mitosis, the chromosomes segregated abnormally leading to G_1 progeny with unequal numbers of chromosomes. Apoptotic events occurred at even later times and apparently as a consequence of this lethal mitosis.

Cells arrested in G_2 following cisplatin could be forced to undergo mitosis by the addition of 5 mM caffeine; within 2 h $p34^{cdc2}$ was dephosphorylated and mitosis was completed by 4 h. Under these conditions, the chromosomes also segregated unequally and apoptosis occurred much sooner. In contrast, the onset of apoptosis can be delayed by inhibiting protein synthesis with cycloheximide. This occurs because protein synthesis is

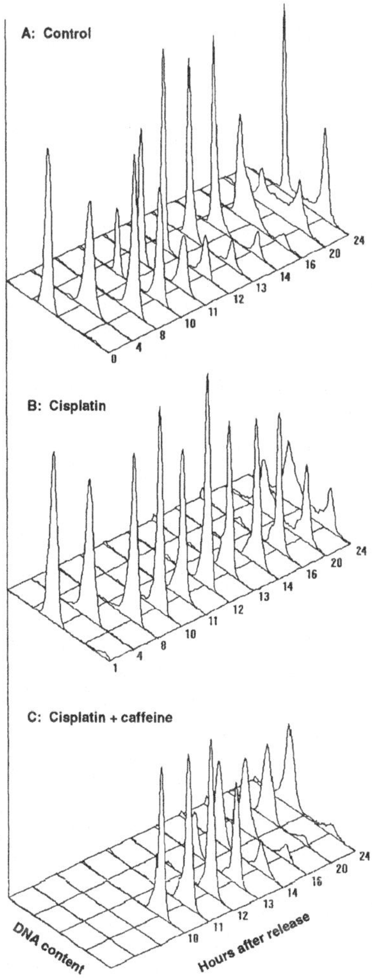

Figure 3. Effect of cisplatin on cell cycle progression in synchronized CHO/UV41 cells. (A) Following synchronization, cells were allowed to progress through the cycle and were harvested at the indicated times and analyzed for DNA content by flow cytometry. (B) Cells were damaged with cisplatin for 1 h in early S phase immediately upon release from synchronization. Progression through the cell cycle was arrested prior to mitosis. (C) Cells damaged with cisplatin as above were also incubated with 5 mM caffeine at 10 h which forced them to immediately enter mitosis.

required for normal passage through mitosis and without the appropriate proteins, cells can not undergo a lethal mitosis. These results are consistent with the premise that mitosis is an essential step in the lethality and that a G_2 arrest can prevent cell death, or in this model in which there is no DNA repair, G_2 arrest can delay cell death.

Following a lethal mitosis, cells began to float off the culture dish. This is a common observation when cells die but its relevance to apoptosis has been generally over-looked, in part because the majority of investigations on apoptosis have been performed in suspension cultures. Only the detached cells contained the DNA digestion characteristic of apoptosis. We surmise that the extracellular matrix, in this case the culture dish, acts like a growth factor (Figure 2). Extracellular matrix proteins such as fibronectin, which communicate through integrin-type receptors have been shown to activate a signal transduction pathway through a focal adhesion kinase (FAK; Zachary and Roxengurt, 1992). The extracellular matrix can also activate the Na^+/H^+ antiporter leading to intracellular alkalinization (Schwartz et al., 1991). Therefore, loss of attachment is likely to inactivate the Na^+/H^+ antiporter leading to intracellular acidification and activation of DNase II. This is also what is observed in vivo where a cell is seen to shrink away from its neighbors, and this loss of contact could lead in a similar manner to apoptotic DNA digestion.

In summary, these experiments have established a sequence of events that occur during the induction of apoptosis by cisplatin. Damaged cells progress to and arrest in G_2 due to suppression of $p34^{cdc2}$ kinase activity. Upon subsequent dephosphorylation of $p34^{cdc2}$, cells undergo an aberrant mitosis with unequal chromosome segregation. In the subsequent G_1 phase, the cells detach from the extracellular matrix and undergo apoptotic DNA digestion. It is hypothesized that DNA digestion results from loss of a signal transduction pathway originating from the extracellular matrix which leads to intracellular acidification and subsequent activation of DNase II.

SUMMARY

Apoptosis can occur at various phases of the cell cycle. In the case of growth factor withdrawal, cells die at the G_1 phase and the pathway of apoptosis may utilize various oncogenes or tumor suppressor genes. Cancer cells modify these pathways to facilitate cell survival. Cytotoxins can induce apoptosis at many phases of the cell cycle. Examples described here are of etoposide-induced apoptosis occurring in S phase cells, and cisplatin-induced apoptosis occurring in G_1 phase after cells have progressed through a lethal mitosis. In each case, we have observed intracellular acidification consistent with activation of DNase II. These results suggest that altered pH regulation is an important component of apoptosis.

Cells have developed survival pathways: activation of PKC is probably just one such pathway. We suggest that critical molecules in these survival pathways are ion regulatory proteins, presumably at the cell surface, but also potentially at the nuclear membrane. As an example the Na^+/H^+ antiporter can be activated by both PKC-dependent and independent pathways. Perhaps genes such as bcl-2 that protect cells are involved in regulation of these ion transporters. One example to support this hypothesis is the E5 oncoprotein of bovine papilloma virus that binds to the H^+/ATPase (Goldstein et al., 1991).

There are multiple initiators of apoptosis, and multiple pathways culminating in some common final steps. Previous hypotheses for induction of apoptosis have focussed predominantly on Ca^{2+} as a critical ion in these final steps. However, the involvement of Ca^{2+} is certainly not ubiquitous. The results presented here suggest that intracellular pH regulation may be much more important. This hypothesis fits with our current knowledge of intracellular signalling mechanisms, as well as with the current understanding of the functions of oncogenes and tumor suppressor genes.

ACKNOWLEDGMENTS

This study was supported by NIH Grant CA50224 and Cancer Center Support Grant CA23108. Michael Barry was supported by a National Cancer Institute Training Grant CA09658.

REFERENCES

Aaronson, S.A., 1991, Growth factors and cancer, *Science* 254:1146.

Askew, D.S., Ashmun, R.A., Simmons, B.C., and Cleveland, J.L., 1991, Constitutive c-*myc* expression in an IL-3-dependent myeloid cell line suppresses cell cycle arrest and accelerates apoptosis, *Oncogene* 6:1915.

Barry, M.A., Behnke, C.A., and Eastman, A., 1990, Activation of programmed cell death (apoptosis) by cisplatin, other anticancer drugs, toxins and hyperthermia, *Biochem. Pharmacol.* 40:2353.

Barry, M.A. and Eastman, A., 1993, Identification of deoxyribonuclease II as an endonuclease involved in apoptosis, *Arch. Biochem. Biophys.* 300:440.

Barry, M.A., Reynolds, J.E., and Eastman, A., 1993, Etoposide-induced apoptosis in human HL-60 cells is associated with intracellular acidification, *Cancer Res.* (in press).

Cohen, J.J. and Duke, R.C., 1984, Glucorticoid activation of a calcium-dependent endonuclease in thymocyte nuclei leads to cell death, *J. Immunol.* 132:38.

Demarcq, C., Creswell, D., and Eastman, A. (1993). Involvement of $p34^{cdc2}$ kinase in cisplatin-induced G_2 arrest and apoptosis in Chinese hamster ovary cells, submitted.

Eastman, A., 1987, The formation, isolation and characterization of DNA adducts produced by anticancer platinum complexes, *Pharmac. Ther.* 34:155.

Eastman, A. and Barry, M.A., 1992, The origins of DNA breaks: a consequence of DNA damage, DNA repair or apoptosis, *Cancer Invest.* 10:229.

Eastman, A., 1993, Apoptosis: a product of programmed and unprogrammed cell death, *Toxicol. Applied Pharmacol.* in press.

Eastman, A., Barry, M.A., Creswell, D., and Demarcq, C., 1992, Cytotoxicity as a consequence of DNA damage, *in*: "DNA Repair Mechanisms," V.A. Bohr, K. Wassermann, K.H. Kraemer, and J.H. Theysen, eds., Munksgaard, Copenhagen.

Evan, G.I., Wyllie, A.H., Gilbert, C.S., Littlewood, T.D., Land, H., Brooks, M., Waters, C.M., Penn, L.Z., and Hancock, D.C., 1992, Induction of apoptosis in fibroblasts by c-myc protein, *Cell* 69:119.

Gillies, R.J., Martinez-Zaguilan, R., Martinez, G.M., Serrano, R., and Perona, R., 1990, Tumorigenic 3T3 cells maintain an alkaline intracellular pH under physiologic conditions, *Proc. Natl. Acad. Sci. USA.* 87:7414.

Goldstein, D.J., Finbow, M.E., Andresson, T., McLean, P., Smith, K., Bubb, V., and Schlegel, R., 1991, Bovine papillomavirus E5 oncoprotein binds to the 16K component of vacuolar H^+-ATPases, *Nature* 352:347.

Hockenbery, D., Nunez, G., Milliman, C., Schreiber, R.D., and Korsmeyer, S.J., 1990, Bcl-2 is an inner mitochondrial membrane protein that blocks programmed cell death, *Nature* 348:334.

Hoffmann, I., Clarke, P.R., Marcote, M.J., Karenti, E., and Draetta, G., 1993, Phosphorylation and activation of human cdc25-C by cdc2/cyclin B and its involvement in the self-amplification of MPF at mitosis, *EMBO J.* 12:53.

Kerr, J.F.R., Wyllie, A.H., and Currie, A.R., 1972, Apoptosis: a basic biological phenomenon with wide-ranging implications in tissue kinetics, *Br. J. Cancer* 26:239.

Madshus, I.H., 1988, Regulation of intracellular pH in eukaryotic cells, *Biochem. J.* 250:1.

Maly, K., Uberall, F., Loferer, H., Doppler, W., Oberhuber, H., Groner, B., and Grunicke, H.H., 1989, Ha-*ras* activates the Na^+/H^+ antiporter by a protein kinase C-independent mechanism, *J. Biol. Chem.* 264:11839.

Miyashita, T. and Reed, J.C., 1992, Bcl-2 gene transfer increases relative resistance of S49.1 and WEHI7.2 lymphoid cells to cell death and DNA fragmentation induced by glucocorticoids and multiple chemotherapeutic drugs, *Cancer Res.* 52:5407.

Nikonova, L.V., Nelipovich, P.A., and Umansky, S.R., 1982, The involvement of nuclear nucleases in rat thymocyte DNA degradation after γ-irradiation, *Biochim. Biophys. Acta* 699:281.

Oberhammer, F., Bursch, W., Parzefall, W., Breit, P., Stadler, M., and Schulte-Hermann, R., 1991, Effects of transforming growth factor β on cell death of cultured rat hepatocytes, *Cancer Res.* 51:2478.

Rotin, D., Steele-Norwood, D., Grinstein, S., and Tannock, I., 1989, Requirement of the Na^+/H^+ exchanger for tumor growth. , *Cancer Res.* 49:205.

Sardet, C., Franchi, A., and Pouyssegur, J., 1989, Molecular cloning, primary structure, and expression of the human growth factor-activatable Na^+/H^+ antiporter, *Cell* 56:271.

Sardet, C., Counillon, L., Franchi, A., and Pouyssegur, J., 1990, Growth factors induce phosphorylation of the Na^+/H^+ antiporter, a glycoprotein of 110 kD, *Science* 247:723.

Schwartz, M.A., Lechene, C., and Ingber, D.E., 1991, Insoluble fibronectin activates the Na/H antiporter by clustering and immobilizing integrin $\alpha_5\beta_1$, independent of cell shape, *Proc. Natl. Acad. Sci. USA.* 88:7849.

Shi, Y., Glynn, J.M., Guilbert, L.J., Cotter, T.G., Bissonette, R.P., and Green, D.R., 1992, Role for c-myc in activation-induced apoptotic cell death in T cell hybridomas, *Science* 257:212.

Siegel, R.M., Katsumata, M., Miyashita, T., Louie, D., Greene, M.I., and Reed, J.C., 1992, Inhibition of thymocyte apoptosis and negative selection in *bcl-2* transgenic mice, *Proc. Natl. Acad. Sci. USA.* 89:7003.

Sorenson, C.M. and Eastman, A., 1988a, Influence of *cis*-diamminedichloroplatinum(II) on DNA synthesis and cell cycle progression in excision repair proficient and deficient Chinese hamster ovary cells, *Cancer Res.* 48:6703.

Sorenson, C.M. and Eastman, A., 1988b, Mechanism of *cis*-diamminedichloroplatinum(II)-induced cytotoxicity: role of G_2 arrest and DNA double-strand breaks, *Cancer Res.* 48:4484.

Sorenson, C.M., Barry, M.A., and Eastman, A., 1990, Analysis of events associated with cell cycle arrest at G_2 and cell death induced by cisplatin, *J. Nat. Cancer Inst.* 82:749.

Swann, K. and Whitaker, M., 1985, Stimulation of the Na/H exchanger of sea urchin eggs by phorbol ester, *Nature* 314:274.

Ucker, D.S., Obermiller, P.S., Eckhart, W., Apgar, J.R., Berger, N.A., and Meyers, J., 1992, Genome digestion is a dispensable consequence of physiological cell death mediated by cytotoxic T lymphocytes, *Mol. Cell. Biol.* 12:3060.

Vaux, D.L., Cory, S., and Adams, J.M., 1988, *Bcl*-2 gene promotes haemopoietic cell survival and cooperates with c-*myc* to immortalize pre-B cells, *Nature* 335:440.

Wurm, F.M., Gwinn, K.A., and Kingston, R.E., 1986, Inducible overproduction of the mouse c-myc protein in mammalian cells, *Proc. Natl. Acad. Sci. USA.* 83:5414.

Wyllie, A.H., Kerr, J.F.R., and Currie, A.R., 1980, Cell death: the significance of apoptosis, *Int. Rev. Cytol.* 68:251.

Yonish-Rouach, E., Resnitzky, D., Lotem, J., Sachs, L., Kimchi, A., and Oren, M., 1991, Wild-type p53 induces apoptosis of myeloid leukaemic cells that is inhibited by interleuken-6, *Nature* 352:345.

Zachary, I. and Roxengurt, E., 1992, Focal adhesion kinase ($p125^{FAK}$): a point of convergence in the action of neuropeptides, integrins and oncogenes, *Cell* 71:891.

Zakeri, Z.F., Quaglino, D., Latham, T., and Lockshin, R.A., 1993, Delayed internucleosomal DNA fragmentation in programmed cell death, *FASEB J.* 7:470.

44

CELL CYCLE REGULATION AND THE CHEMOSENSITIVITY OF CANCER CELLS

Kurt W. Kohn, Patrick M. O'Connor and Joany Jackman

Laboratory of Molecular Pharmacology
Developmental Therapeutics Program
Division of Cancer Treatment
National Cancer Institute
Bethesda, Maryland 20892

INTRODUCTION

Current strategies of cancer chemotherapy and drug development focus on the concept of targets of drug attack leading to cytotoxicity that, for reasons unknown, can be selective against malignant cells. Some classical targets of anti-cancer drug action include DNA, enzymes of nucleic acid metabolism, topoisomerases, and microtubules (table 1). Although most of the classical anti-cancer drugs interfere in some way with DNA replication or chromosome segregation, it is not known how these actions lead to antitumor activity. Combinations of drugs are often employed in therapy, but the regimens are usually designed empirically without clear knowledge of mechanisms.

Recently elucidated details of cell cycle regulation, as well as of the functions of oncogenes and tumor suppressor genes, have given new impetus to the earlier suggestion that vulnerability of tumors to drugs may arise from regulatory defects existing in certain tumor cell types (Pardee 1987). The outlines of new therapeutic possibilities can now be discerned that would take advantage of specific cell cycle checkpoint defects. The general form of such therapy could be a sequence of two drug regimens: (1) challenge of the defective checkpoint by means of a drug that would activate a step logically upstream from the defect, followed by (2) a cytotoxic drug to kill cells that are not arrested at the checkpoint. In effect, such strategies would selectively protect normal cells against toxicity while regulation-defective tumor cells would remain vulnerable.

This type of stratagem is attractive because it aims at the core of what makes cells malignant: specific defects in growth control. The logic would be that the regulatory defects that are necessary for malignancy would confer drug sensitivity. By linking drug sensitivity with malignancy, two notorious conundrums of cancer chemotherapy -- cytogenetic heterogeneity and drug-resistance -- might be evaded. Figure 1 outlines the concept. The road to malignancy begins with defective function of a regulatory gene, such as an oncogene or tumor suppressor gene. The consequent defect in cell cycle control leads to chromosome instability which increases the probability of acquisition of other genetic defects in the progression to malignancy. Defective controls in sensing DNA damage

The Cell Cycle: Regulators, Targets, and Clinical Applications
Edited by V.W. Hu, Plenum Press, New York, 1994

Table 1. Classical anti-cancer drug targets.

Target	Mode of action	Drugs
DNA	intercalation	doxorubicin amsacrine ellipticine
DNA	crosslinking	cisplatin nitrogen mustards mitomycin
DNA	incorporation	thioguanine arabinosylcytosine
DNA	free radical attack	ionizing radiation bleomycin
enzymes	inhibition of nucleic acid metabolism	methotrexate 5-fluorouracil 6-mercaptopurine
topoisomerases	stabilization of DNA cleavable complexes	etoposide (VP-16) doxorubicin amsacrine ellipticine camptothecin
tubulin	disruption of microtubules	vinblastine taxol

would enhance both chromosome instability and drug sensitivity, the ratio of these effects depending on whether the DNA damage is mainly single base damage leading to mutation or more serious lesions, such as crosslinks, which are more often lethal. Drugs acting by other mechanisms, such as mitotic inhibition or interference with nucleic acid syntheses, might also act in generally this way.

Some of the known or suspected cell cycle checkpoints responding to DNA damage are diagrammed in figure 2. The best understood, to which we will pay most attention here, are the G1 commitment to S which corresponds to Pardee's restriction point R, and the G2/M checkpoint where cultured cells treated with DNA damaging drugs commonly are observed to become arrested. We have undertaken two types of investigation. The first aims to determine how the G2/M checkpoint detects DNA damage, and the second investigates the behavior of different human tumor cell types with respect to both G1/S and G2/M checkpoint controls. Before summarizing our recent findings, we will comment on some aspects of the G2/M checkpoint mechanism as it is currently understood.

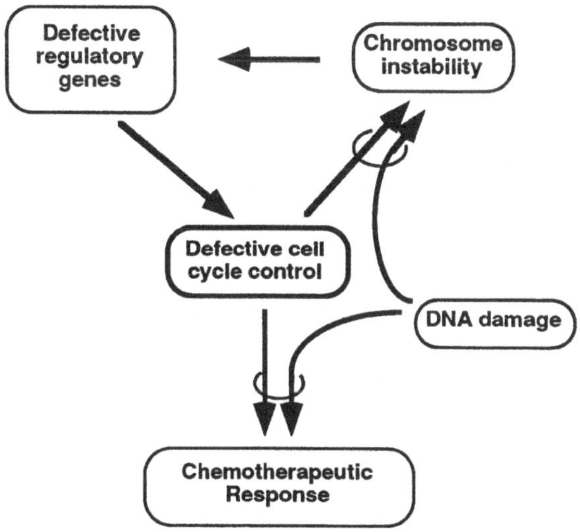

Figure 1. Role of cell cycle control in carcinogenesis and chemotherapy. Repairable DNA damage, such as single base damage, is more mutagenic than lethal and would favor the accumulation of genetic defects leading to malignancy. Other types of DNA damage, such as crosslinks or double-strand breaks, are more difficult to repair and favor cell killing that, if selective, could produce a therapeutic response.

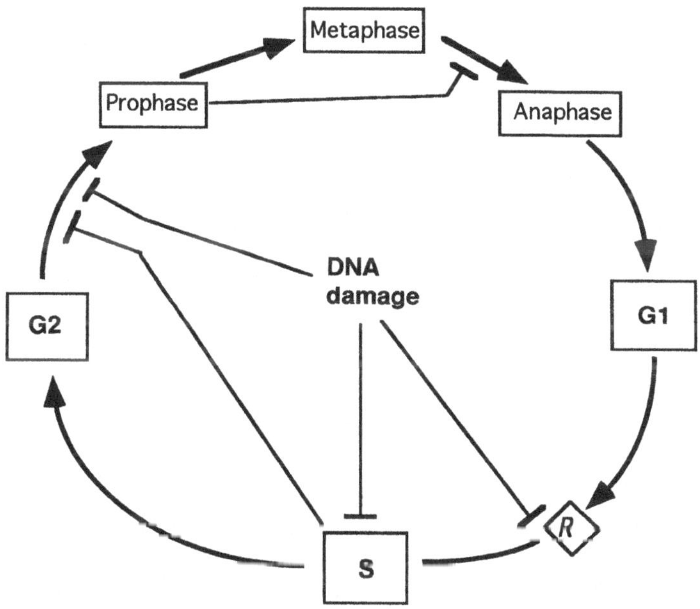

Figure 2. Map of cell cycle checkpoints. Detection of DNA damage causes delay in (1) commitment to S at restriction point R, (2) onset of prophase events at the G2/M boundary, and (3) progress through S phase due, perhaps, to control of replicon initiation. Also indicated are (4) the physiologic control by incomplete replication upon prophase initiation and (5) the requirement of a properly formed metaphase spindle for the onset of anaphase.

THE G2/M CHECKPOINT MECHANISM

Cells that have been treated with any of a variety of DNA-damaging drugs often continue slowly through S phase and then arrest in G2. Figure 3 outlines how this may occur. The presence of DNA damage is in some way signalled to the protein kinase/phosphatase system (consisting of cyclin B, cdc2, cdc25 and wee1) which may directly initiate the prophase events of chromosome condensation and nuclear membrane breakdown. The signal is probably conveyed via the RCC1/ran system (Dasso 1993) which also monitors DNA replication. While DNA damage is present or DNA replication is incomplete, RCC1/ran somehow blocks the cyclin B-cdc2 switch that initiates prophase.

The functional integrity of the cyclin B-cdc2 switch is becoming evident (figure 4). The switch consists, at least, of cyclin B-cdc2 and cdc25C which constitute a positive feedback system to bring about the rapid activation of cdc2. The switch is OFF when cdc2 is phosphorylated at Tyr15 and Thr14 and cdc25C is under-phosphorylated. The system switches to ON when cdc2 becomes dephosphorylated due to the action of phosphorylated cdc25C and cdc25C becomes phosphorylated due to the action of dephosphorylated cdc2. In order to allow the switching reaction to start, a small background process is necessary, and this could be provided by the weak phosphatase activity of under-phosphorylated cdc25C, or possibly by cdc25A or B. The wee1 side of the system (figure 4) may also be part of the switch, but the details of its role, relative to that of cdc25C, remain to be explored. The switch might be turned towards ON by nim1 which inactivates wee1 by phosphorylation and turned towards OFF by phosphatases that dephosphorylate wee1-P and/or cdc25C-P. The state of the switch would depend also on the concentrations of cyclin B, cdc2 and cdc25C.

The integrity of a switch involving cdc2, cdc25C and wee1 would imply that the phosphorylation states of these components should always change in synchrony under a variety of conditions, including drug responses.

The experiments described in the next section define some aspects of the G2 arrest state induced by DNA damage. Our eventual goal is to trace back the causal chain of regulatory steps to the point where the reactions that control the cell cycle responses to DNA damage can be identified.

CHARACTERIZATION OF THE G2/M ARREST POINT IN BURKITT'S LYMPHOMA CELLS TREATED WITH NITROGEN MUSTARD (HN2)

The most commonly reported cell cycle effect of DNA damage on cultured tumor cells is arrest in G2 phase. We have previously explored the association of cdc2 with G2 arrest (O'Connor et al., 1992). More recently, we have extended these studies in Burkitt's lymphoma CA46 cells synchronized near the G1/S border by double aphidicolin block and then treated with HN2 for 30 minutes (O'Connor et al., 1993). Constituents of the cyclin B and cyclin A systems were assayed as a function of time following treatment. Nocodozole was added after the HN2 treatment in order to prevent cells that enter mitosis from moving on into the next cell cycle. This was useful mainly for the control cell comparison, since most of the HN2-treated cells arrested in G2. The results show that HN2-induced G2 arrest occurred prior to the activation of the cyclin B-cdc2 switch: dephosphorylation of Tyr15 and Thr14 of cdc2 had not yet occurred, and cyclin B and cdc2-associated kinase activities were still low (figure 5A and B). Consistent with the hypothesis of an integrated switch, cdc25C in the G2-arrested cells had not become hyperphosphorylated, and its associated phosphatase activity remained low (figure 6). We are now testing the state of wee1.

In contrast to the cyclin B-cdc2 switch, the cyclin A-cdk2 system was still active in the G2 arrested cells: cyclin A protein levels, as well the kinase activities associated with cyclin A and cdk2, remained elevated. In control cells, cyclin A levels and the associated kinase activities declined at about the time that the cdc2 and cyclin B-associated kinase activities were rising. The HN2-induced

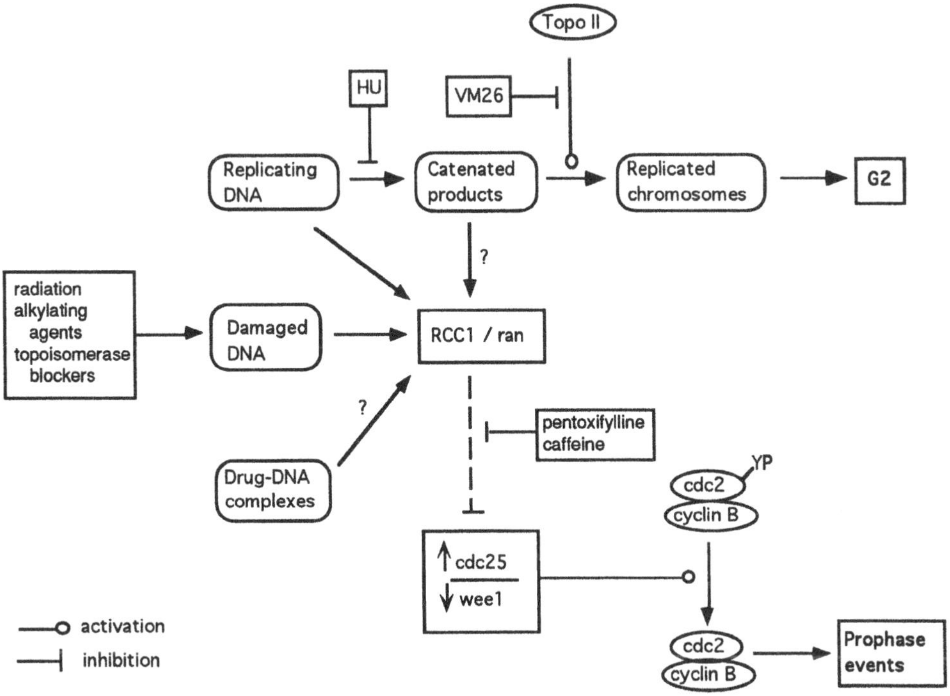

Figure 3. Outline of how the initiation of prophase events is inhibited by DNA damage or by incompletely replicated chromosomes. The controls appear to converge on the RCC1/ran system towards a final common pathway leading to the cyclin B-cdc2 switch. It is not known exactly what signals this system detects; it may by the presence of replication forks or catenated chromosomes. The points of action of several classes of drugs are suggested. Caffeine and pentoxifylline block these controls, suggesting that they may act along the final common pathway. Stimulatory effects on a reaction are represented by circles, inhibitory effects by bars. (HU, hydroxyurea; VM26, etoposide.)

G2 arrest therefore precedes the switching off of the cyclin A/cdk2 system and the turning on of the cyclin B-cdc2 system.

During S phase, cyclin A kinase activity is associated solely with cdk2, but during mitosis, it is found associated with both cdk2 and cdc2. In the HN2-treated G2 arrested cells cyclin A kinase activity was associated almost solely with cdk2, showing that the arrest point precedes the activation of cyclin A-cdc2 complexes (O'Connor et al 1993).

RELATIONSHIP BETWEEN p53 FUNCTION AND THE ABILITY OF BURKITT'S LYMPHOMA CELLS TO ARREST IN G1

The role of p53 in DNA-damage-induced G1 arrest was examined in 17 Burkitt's lymphoma cell lines provided by Drs. Ian Magrath and Kishore Bhatia of the Pediatric Oncology Branch of the National Cancer Institute. The status of the p53 gene of these lines had been determined, but the lines were sent to us coded, so that we were blinded of this information until we had made our inferences regarding G1 arrest. Cell doubling times ranged from 19 to 24 hours. Ability to arrest in G1

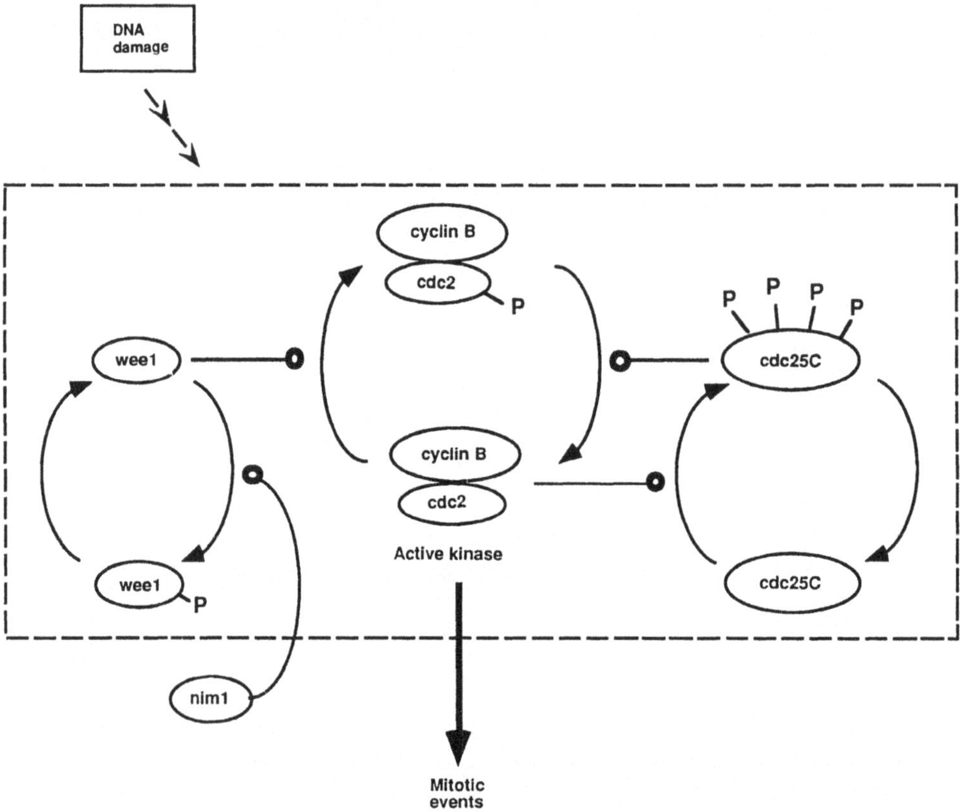

Figure 4. Cyclin B-cdc2 switch control of the G2/M transition. The cdc2 and cdc25C species may form a positive feedback couple that could rapidly turn the switch ON. The indicated phosphorylations inhibit the cdc2 kinase and enhance the activity of the cdc25C phosphatase. The excitation level of the switch may be controlled by nim1 which inactivates wee1 by phosphorylation. A phosphorylation cycle involving position Thr161 of cdc2 may also be important, but is only now becoming clarified and is not included in this scheme.

was determined by subjecting the cells of 6.3 Gy of γ-rays and then letting the cells grow for 16 hours, at which time cell cycle distribution was determined by flow cytometry. The cells lines were judged to show 3 types of responses: (1) strong G1 arrest (at least 60% of the original G1 population), (2) little or no G1 arrest (less than 25% of the original G1 population, and (3) intermediate response. Cell responses were unaffected when 0.4 μg/ml nocodozole was added after irradiation in order to block progression beyond metaphase. Consistent results were also obtained when the radiation dose was doubled.

Five of the cell lines showed strong G1 arrest, and contained only normal p53 genes. Eight of the lines showed little or no G1 arrest, and these had only mutant p53 genes. The remaining 4 lines showed intermediate responses; two of these were heterozygous, containing both mutant and normal p53 genes.

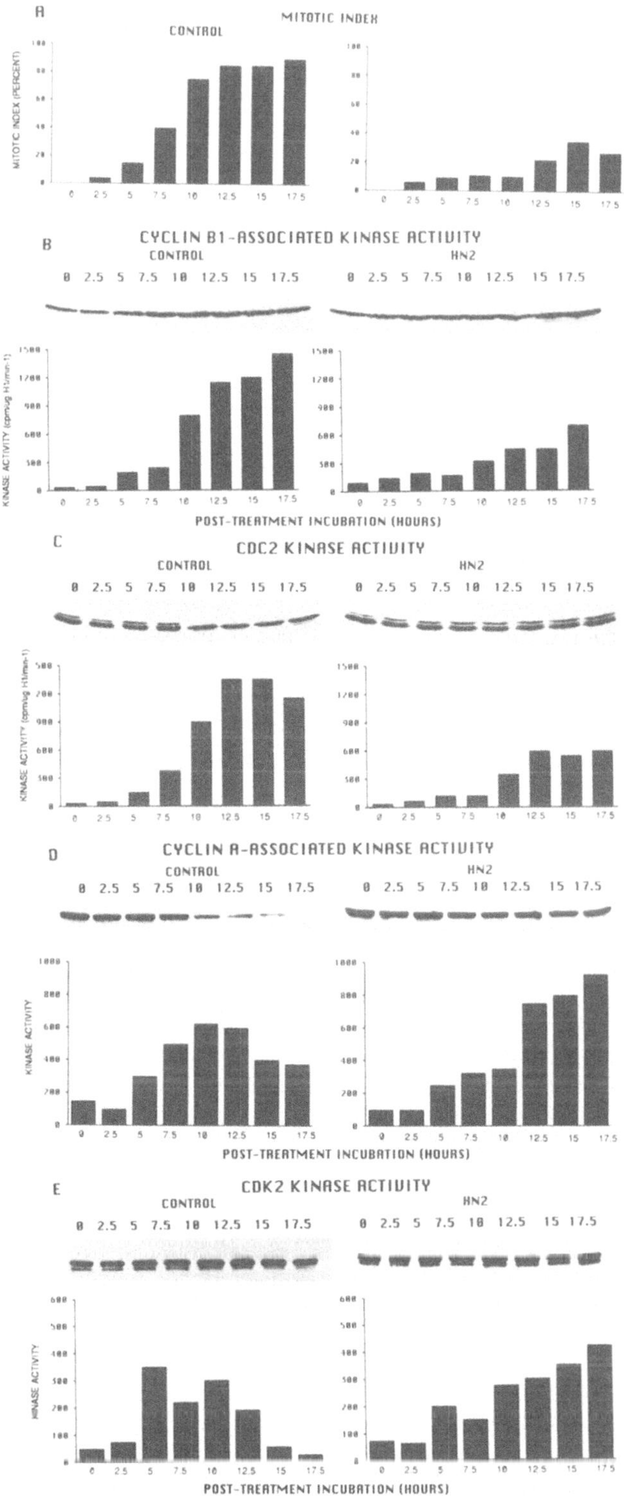

Figure 5.
Behavior of cyclins A and B and their associated kinaseses HN2 treated CA46 cells arrest synchronized near the G1/S boundary by a double aphidicolin and then treated with 0.5µM HN2 for 30min. The cells were then incubated in the presence of 0.5µg/ml nocodozole for the indicated number of hours. (A)mitotic index. (B)*Above:* cyclin B1 immunoblots. *Below:* H1 kinase activity in cyclin B1 immunoprecipitates. (C) *Above:* cdc2 immunoblots. *Below:* H1 kinase activity in the cdc2 immunoprecipitates. (D) *Above:* cyclin A immunoblots. *Below:* H1 kinase activity in cyclin A immunoprecipitates. (E) *Above:* cdk2 immunoblots. *Below:* H1 kinase activity in the cdk2 immunoprecipitates.

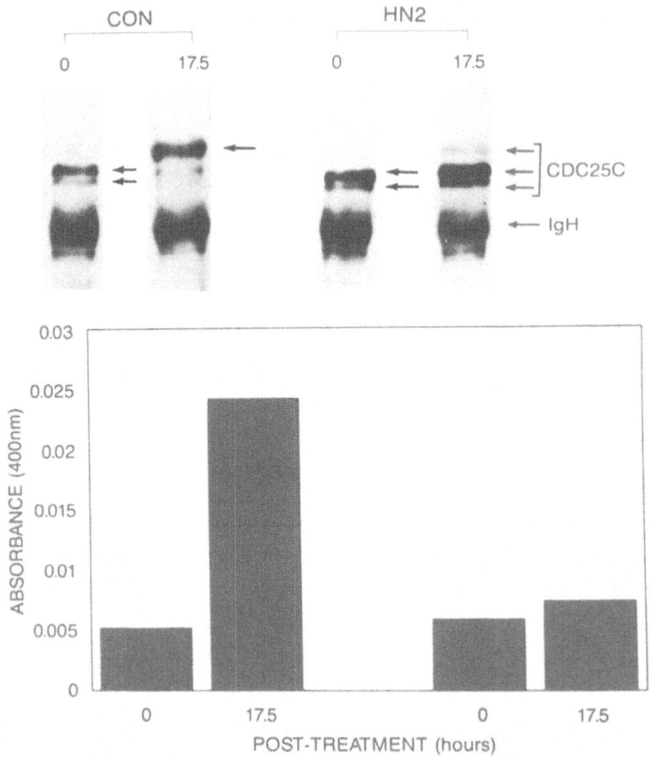

Figure 6. Behavior of cdc25C phosphatase in cells treated as in figure 5 (but using 0.7 μM HN2). *Above:* Immunoelectrophoretic blots. *Below:* phosphatase activity in cdc25C immunoprecipitates (p–nitrophenylphosphate substrate).

The other 2 intermediate lines had only normal p53 genes; preliminary results suggest that these may differ from the other normal-p53 lines in showing less p53 protein accumulation in response to radiation. The other normal-p53 lines all showed strong p53 protein accumulation responses.

Although most of the mutant p53 lines had constitutively high p53 protein content, two mutant lines showed low p53 protein content that became enhanced following radiation. It seems therefore that the lifetime of most of the mutant proteins was enhanced, as expected, but that in some cases a lack-of-function mutation did not prolong the lifetime of the protein.

This assay may be useful as a test of p53 function with respect to its ability to affect cell cycle progression in DNA-damaged cells.

G2/M CHECKPOINT CONTROL DEFECTS IN COLON CARCINOMA CELL LINES

We are currently testing the fidelity of cell cycle checkpoint controls in human tumor cell lines of the NCI's cell screen panel (provided by Drs. Anne Monks and Nicholas Scudiero). So far, 6 colon carcinoma cell lines were tested using the same protocols as described above for the Burkitt's lymphoma cells. Doubling times ranged from 18 to 34 hours. Experiments in the presence of nocodozole showed that none of the colon lines arrested significantly in G1 following radiation. We do not as yet know the state of the p53 genes in these lines. Interestingly, 4 of the lines showed marked differences between the presence or absence of nocodozole, in that there was a large G1 component in the absence but not in the presence of nocodozole. No such differences occurred in the case of the lymphoma lines. Since nocodozole prevented the appearance of the G1 population, we conclude that the G1 population derived from progress of cells through mitosis, suggesting defective G2/M checkpoint control in these lines.

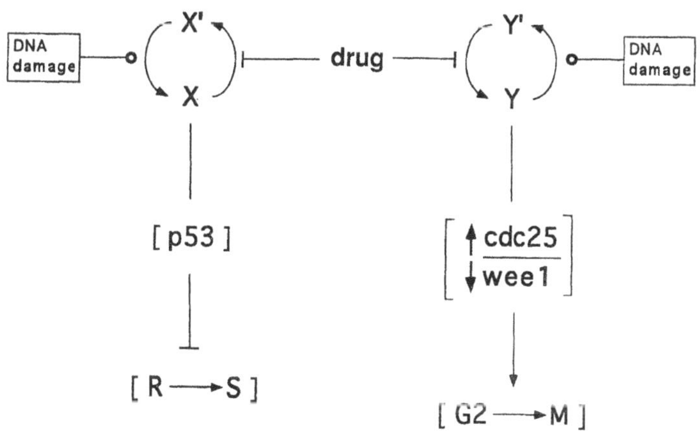

Figure 7. Outline of therapeutic strategies to take advantage of defective control of G1/S (left) or G2/M (right). The unknown control network upstream of the p53-dependent events is represented hypothetically by X, X', and the network upstream of cdc25/wee1 is represented by Y, Y'. DNA damage is represented as stimulating a reaction step. A drug that remains to be identified or developed is represented as inhibiting a reaction step. When the upstream reactions become known, specific enzymes could be used to test for candidate drugs.

PROSPECT FOR THERAPY BASED ON DEFECTIVE CHECKPOINT CONTROL

A form of therapy based on checkpoint control could be comprised of two steps: (1) challenge of the integrity of a checkpoint, followed by (2) selective killing of cells that pass through to a subsequent cell cycle phase. It may soon be possible to formulate this idea in terms of detailed molecular events. Schemes for therapy based on defective control at G1/S or G2/M are outlined in figure 7. For G1/S control, a pathway of response to DNA damage is known to depend on the function of p53. Since a large fraction of human malignant tumors are defective in p53 function, a therapy that takes advantage of this defect could have considerable practical impact. The control of G2/M passes via cdc25 or wee1 to the cdc2 switch and often arrests cells containing DNA damage in G2. Our findings with colon cancer cells suggest that some malignant cells may be defective in this checkpoint.

For both schemes, it is obviously important to understand the controlling molecular network that operates logically upstream of p53 or cdc25/wee1. In figure 7, these networks, hopefully soon to be elucidated, are represented for illustrative purpose in the simplest possible hypothetical form. In present therapies, DNA damage seems to serve the dual functions of challenging the checkpoint and also killing the cells that pass through. We would predict that these ends could be achieved with non-DNA damaging therapy. When the controlling reactions are known, it should become possible to identify a specific step – undoubtedly involving a protein kinase or phosphatase – such that a specific inhibitor of that step would cause normal cells to arrest in G1 or G2. Neoplastic cells that do not arrest could then be killed by available poisons specific for S phase or mitosis. Normal cells would be protected while cycling tumor cells are killed by a drug regimen that need not be very toxic and that could be repeated many times for effective therapy.

This approach would have to be individualized. It requires that tests become available to identify the pertinent molecular defects in clinical tumors, so that patients could be selected for therapy appropriate to the defect.

Although variant tumor cells could still become resistant by actively exporting or metabolizing the drugs, they probably could not easily overcome the checkpoint defect without ceasing to be malignant. Therefore the tumor would remain fundamentally vulnerable and sensitive to modified drugs that are not exported or metabolized.

Certainly there will be additional problems in developing this kind of therapy. For example, p53-defective tumor cells could still escape by arresting in G2. This might be prevented by means of a pentoxifylline-like drug that would specifically prevent G2 arrest. The detailed form of the therapy must await a fuller characterization of the role of p53 in the control of cell cycle events -- and apoptosis (Lane, 1992).

REFERENCES

Dasso, M., 1993, RCC1 in the cell cycle: the regulator of chromosome condensation takes on new roles, *TIBS* 18:96

Lane, D.P., 1992, p53, guardian of the genome, *Nature* 358: 15.

O'Connor, P.M., Ferris, D.K., Pagano, M., Draetta, G., Pines, J., Hunter, T., Longo, D.L., and Kohn, K.W., 1992, G2 delay induced by nitrogen mustard in human cells affects cyclin A/cdk2 and cyclin B1/cdc2-kinase complexes differently, *J. Biol. Chem.* 268: 8298.

O'Connor, P.M., Ferris, D.K., White, G.A., Pines, J., Hunter, T., Longo, D.L., and Kohn, K.W., 1992, Relationships between cdc2 kinase, DNA crosslinking and cell cycle perturbations induced by nitrogen mustard, *Cell Growth & Diff.* 3: 43.

Pardee, A.B., 1987, Molecules involved in proliferation of normal and cancer cells. *Cancer Res.* 47: 1488.

45

MULTIPLE CELL CYCLE CHECKPOINT OVERRIDE AND ITS POTENTIAL FOR BINARY TUMOR THERAPY

Robert L. Margolis and Paul R. Andreassen

Institut de Biologie Structurale
41 avenue des Martyrs
38027 Grenoble cedex 1, France

ABSTRACT

The progression of the cell to the next stage of its cycle is regulated by checkpoints, which assure that prerequisite events have been completed before the next cell cycle stage ensues. The overall purpose of checkpoints is to ensure fidelity in DNA replication and in cell division. We have found that checkpoint override is lethal to mammalian cells. We propose, therefore, that it may be possible to destroy tumor cells by induction of checkpoint override. Previously, the purine analogue 2-aminopurine (2-AP), a specific protein kinase inhibitor, had been shown to cause S-phase arrested cells to inappropriately enter mitosis (Schlegel et al., 1990). We then found that 2-AP causes BHK cells in drug induced mitotic arrest to inappropriately exit mitosis (Andreassen and Margolis, 1991). As the drug showed capacity to advance cells past checkpoints at two distinct parts of the cell cycle, this result suggested that there is an underlying common factor responsible for the various inhibitory controls of the cell cycle. We therefore have tested the ability of 2-AP to inappropriately advance the cell cycle following blockage with a variety of stage specific inhibitors. We have obtained the striking result (Andreassen and Margolis, 1992) that 2-AP causes BHK cells to override every cell cycle checkpoint examined, regardless of whether the arrest point is in G_1, S phase, G_2 or mitosis. Several of the inhibitors that we have used to induce cell cycle arrest are used therapeutically for cancer treatment. We have now extended the results of multiple checkpoint override to CHO cells, and have examined the effects of checkpoint override on cell viability. Neither the cell cycle arrest drugs, nor 2-AP, is lethal to CHO cells during short exposure. However, induction by 2-AP of inappropriate exit from arrest in interphase is quickly lethal to the cell. Our results therefore suggest that a binary therapy approach using a cell cycle arrest drug in combination with a drug such as 2-AP will have a synergistic destructive effect on tumors. Death following override from interphase stages ensues rapidly by induction of DNA cleavage, which in BHK cells gives the internucleosomal cleavage pattern characteristic of apoptosis.

RESULTS

The eukaryotic cell cycle is composed of discrete stages. The major events of the cycle are the replication of the chromatin and the subsequent cell division. Replication occurs in S-phase, and cell division occurs after DNA replication is completed, in mitosis (or M-phase). There are additionally two gaps in the cycle during which events that are preparative for S-phase or mitosis occur. These are designated G_1 (which occurs between mitosis and the subsequent S-phase), and G_2 (lying between S-phase and the next mitosis). When a cell has withdrawn from the cell cycle into a quiescent state, it is said to be in G_0.

The progression of the cell to the next stage of its cycle is under the control of factors that act as checkpoints to assure that the previous stage has been completed before the subsequent stage ensues (Murray, 1992; Hartwell and Weinert, 1989). Such checkpoints occur to ensure the fidelity of chromosomal segregation to daughter cells (Hartwell and Weinert, 1989). The cell contains exquisitely sensitive feedback control circuits that can, for example, prevent exit from S-phase when a fraction of a percent of DNA remains unreplicated (Dasso and Newport, 1990) and can block advance into anaphase in mitosis until all chromosomes are aligned on the metaphase plate (Rieder and Alexander, 1990). The nature of these checkpoints and how they act to block cell cycle progression is unknown.

Various mutants have been isolated which escape specific cell cycle control circuits and progress inappropriately to the subsequent cell cycle stages. They include cdc2 wee1 double mutants (Russell and Nurse, 1987) wee1 mik1 double mutants (Lundgren et al., 1991) and rad9 (Weinert and Hartwell, 1988) in yeast, bimE7 in *Aspergillis* (Osmani et al., 1988), and the RCC1 mutant tsBN2 in mammalian BHK cells (Nishitani et al., 1991), all of which exhibit an uncoupling of entry into mitosis from the completion of DNA replication. Also, mutations in *S. cerevisiae*, BUB (Hoyt et al., 1991) and MAD (Li and Murray, 1991), have been isolated that fail to arrest in mitosis with microtubule destabilizing drugs. It is significant in terms of the rationale to be developed here that most, if not all, checkpoint override events that have been induced in lower eukaryotic mutants are lethal (for example Russell and Nurse, 1987; Lundgren et al., 1991; Li and Murray, 1991).

In addition, drug treatments such as the combined exposure to the DNA replication inhibitor hydroxyurea and caffeine or to hydroxyurea and the phosphatase inhibitor okadaic acid can cause normal mammalian cells to enter mitosis without completing S-phase (Schlegel and Pardee, 1986; Yamashita et al., 1990).

The purine analogue 2-aminopurine (2-AP), a specific protein kinase inhibitor (Farrell et al., 1977; Mahadevan et al., 1990), has been shown to cause S-phase arrested cells to inappropriately enter mitosis (Schlegel et al., 1990). Recently, we have found that it also causes BHK cells in mitotic arrest to rapidly exit mitosis (Fig. 1; Andreassen and Margolis, 1991). These findings showed that 2-AP has the capacity to advance cells inappropriately past checkpoints at two distinct parts of the cell cycle. Therefore, the evidence suggested that there might be an underlying common factor responsible for the various inhibitory controls of the cell cycle. We therefore tested the capacity of 2-AP to inappropriately advance the cell cycle following cell blockage with a variety of stage specific inhibitors.

These experiments yielded a striking result. 2-AP causes BHK cells to override every drug induced cell cycle block point examined, regardless of whether the arrest point is in G_1, S-phase, G_2, or mitosis (Andreassen and Margolis, 1992).

It also appears evident that BHK cells exposed continuously to 2-AP alone may exit S-phase without completion of replication, and can exit mitosis without metaphase, anaphase or telophase events (Andreassen and Margolis, 1992). We conclude that, in addition to overriding checkpoints in the cell cycle created by drug blockage, 2-AP has the capacity in some cells to cause a failure of normal checkpoint controls that control passage through stages of the cell cycle. Thus far only BHK cells have been found susceptible to spontaneous override of both S-phase and mitosis in the presence of purine analogues.

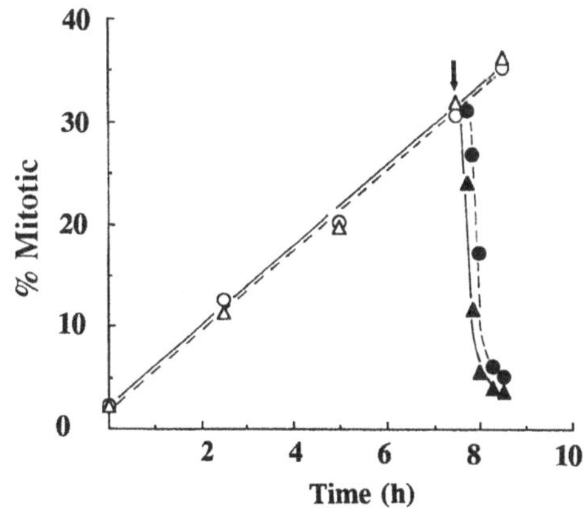

Figure 1. Override of a mitotic block is induced by 2-AP. A randomly cycling BHK cell population was treated with mitotic inhibitors. The mitotic index, as determined by immunofluorescence microscopy of dispersion of lamin B and as confirmed by propidium iodide assay of chromatin condensation, increases in the presence of 0.06 μg/ml nocodazole (open circles) or 5 μM taxol (open triangles) applied at time Ø. Addition of 10 mM 2-AP (arrow) induces rapid exit of blocked cells from mitosis [nocodazole + 2-AP, closed circles; taxol + 2-AP, closed triangles].

Other cells, such as NIH 3T3, HeLa and CHO, show no spontaneous premature exit from S-phase, although CHO and HeLa have thus far shown the ability to exit mitosis without completing metaphase, anaphase or telophase events during exposure to 2-AP.

As a result of exposure to the drug 2-AP, the cell does not escape a metabolic block and revert to a normal cell cycle. The metabolic block remains in effect, but the cell progresses inappropriately to subsequent stages of its cycle. For example, when the cell is blocked in G_1 by mimosine, 2-AP causes the cell to arrive in mitosis at the appropriate time despite the fact the drug blockage has prevented DNA replication from occurring (Fig. 2; Andreassen and Margolis, 1992).

The molecular control of cell cycle checkpoints is not well understood at present. The most complete current information relates to the transition from G_2 into mitosis. Progression from G_2 into mitosis is controlled by the state of phosphorylation of $p34^{cdc2}$ on a tyrosine residue (tyr-15) (Gould and Nurse, 1989; Krek and Nigg, 1991). Mutation at the tyrosine phosphorylation site on $p34^{cdc2}$ causes premature entry into mitosis, without prior completion of DNA synthesis (Krek and Nigg, 1991). Furthermore, the function of the checkpoint that yields G_2 blockage in response to γ-irradiation requires the function of wee1 (Rowley et al., 1992), a protein kinase that phosphorylates $p34^{cdc2}$ at tyr-15 (Parker et al., 1992). Override of hydroxyurea block by mutation of RCC1 requires active cdc25, the tyrosine phosphatase that dephosphorylates $p34^{cdc2}$ at tyr-15 (Seki et al., 1992). Given our results, we now have cause to believe an underlying commonality of checkpoint regulation exists throughout the cell cycle, perhaps at the level of a specific 2-AP sensitive protein kinase (Andreassen and Margolis, 1992).

Our observations on the effect of purine analogues on checkpoint override may have consequences for developing a new approach to tumor therapy. The basic thesis is that checkpoint override induced by the combination of two drugs is a lethal event. The two drugs individually may be relatively innocuous, one being an agent that would temporarily arrest the cell in its cycle, the other being an agent that would override the arrest but would be without effect on unarrested cells.

Two of the inhibitors that we have used to induce cell cycle arrest in the published study (VM-26 and taxol) are used therapeutically for cancer treatment (Tummarello et al., 1990; Rowinsky et al., 1990). In CHO cells, VM-26 is not of itself lethal during short exposure. However, inappropriate exit from an arrested state, induced by 2-AP, is ultimately lethal for the cell. Therefore, our results suggest that "binary" therapy, using a drug such as VM-26 in combination with 2-AP or another purine analogue will cause inappropriate escape from cell cycle blockage, and may have a synergistic destructive effect on tumors. In contrast, taxol alone has proven to be highly toxic to the cells studied, even in brief exposure. Our recent

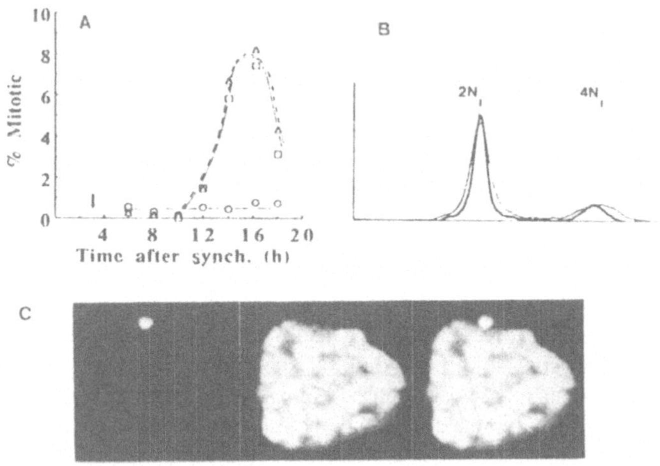

Figure 2. 2-AP overrides mimosine dependent late-G_1 blockage. (A) BHK cells were synchronized by shake-off detachment of nocodazole-arrested mitotic cells and replated in the absence of nocodazole. (A) The mitotic index is shown for cells exposed to 200 μM mimosine alone (circles) or to 200 μM mimosine plus 10 mM 2-AP (triangles), or left untreated (squares). The point of drug addition is indicated by an arrow. Mitotic index was determined as in Figure 1. (B) Flow cytometric analysis of cells treated with mimosine alone (light line) or mimosine plus 2-AP (heavy line) at 18h after shakeoff demonstrates the blockage of DNA replication in both populations. (C) Immunofluorescence image of typical mitotic cell in mimosine plus 2-AP. Images of a small intensely staining mitotic aster detected with anti-tubulin antibody (left), condensed mitotic chromatin stained with propidium iodide (middle), and the merged image (right) are shown.

results indicate that checkpoint override occurs when cells blocked in mitosis by chemotherapeutic agents of the Vinca alkaloid family are exposed to purine analogues.

It is important to point out that the cause of death appears to be through mechanisms related to cell cycle checkpoint override. We have determined that cells blocked in G_0 by serum starvation, as well as cells terminally exited from the cell cycle (senescent cells, in this instance) are not adversely influenced by binary drug treatment. This aspect is important, as it suggests that only cells that are actively traversing a cell cycle or otherwise vulnerable to checkpoint override are susceptible to lethal action.

A limitation of using cell cycle arrest drugs for chemotherapy is that the effectiveness of specific drugs varies with the particular tumor type. Because 2-AP triggers override of checkpoints imposed by administration of a variety of cell cycle arrest drugs, the binary treatment approach could have unprecedented versatility in augmenting treatment for a wide range of tumor types. We would not argue that 2-AP would be the drug of choice for tumor treatment, but that 2-AP serves as a model for the development of binary tumor therapies based upon compounds with the capacity to override cell cycle checkpoints. Cell cycle arrest agents which have been used for tumor chemotherapy are listed by their cell cycle point of arrest in Table 1.

Table 1. Cell Cycle Arrest Points of Selected Chemotherapeutic Agents*

S-phase	G2	Mitosis
Hydroxyurea Methotrexate	Alkylating agents: Cyclophosphamide Carmustine Epipodophyllotoxins: Teniposide (VM-26) Etoposide (VP-16) Cisplatin	Taxol Vinca Alkaloids: Vinblastine Vincristine

*from Goodman and Gilman's *The Pharmacological Basis of Therapeutics*, 1990.

In the ideal situation, chemotherapeutic agents that act to arrest the cell cycle would be used in a dosage range that is specific for tumor cells. In chemotherapy, side effects partly result from the arrest and death of non-target cells, and are exacerbated by the length of the course of drug administration required for chemotherapy.

An advantage of the binary treatment that we propose would be to allow a much shorter treatment regimen, and thereby lessen side effects.

Additionally, we have recently found that different mammalian cells vary greatly with respect to capacity for override of specific cell cycle arrest points. A pattern emerges in which all cells thus far examined override late cycle arrest (arrest induced in G_2 or mitosis), but a smaller subset of cells is capable of overriding early cell cycle arrest (G_1 or S phase). In the limited set of cells thus far observed, there appears to be a correlation of the ability to override cycle arrest with the early expression of cyclin B. These observations open the possibility that the induction of override of early cell cycle arrest may be an event specific for tumor cells, and that a binary drug therapeutic regimen may be highly selective for tumor cells.

This possibility is augmented by the very interesting observations recently made with respect to cyclin B expression in human tumor cells. In a recent study that may be of importance with respect to the therapy regimen proposed here, it has been found that human breast tumor cells universally exhibit expression of cyclin B early in the cell cycle when compared with normal tissue controls (Keyomarsi and Pardee, 1993).

We have conducted an analysis of the effect of 2-AP on override of cell cycle arrest points and induction of death in CHO cells. We chose CHO cells because they exhibit override from arrest points early in the cell cycle, but do not spontaneously exit S-phase with incomplete DNA replication. The CHO cells thus remain euploid through prolonged exposure to 2-AP. After prolonged exposure to 2-AP, CHO cells reinitiate normal proliferation upon release from the drug.

CHO cells arrested for 12 hours with mimosine (G_1 arrest), or with hydroxyurea or aphidicolin (S phase arrest), will recover and proliferate normally when released from the drug inhibition. As stated above, these cells tolerate, without apparent harm, exposure to 2-AP alone over the same time course. If, however, the cells are preblocked with one of these early cycle inhibitors and then exposed to 2-AP, the effect of double drug exposure is the rapid and nearly quantitative death of the treated cells.

Double drug treatment by the combination of VM-26 (teniposide), which induces G_2 arrest, with 2-AP also appears effective at inducing cell death. When CHO cells are induced to override mitotic arrest, the cells have thus far not exhibited greater sensitivity to double drug treatment than to exposure to the mitotic arrest agent alone. Short term exposure to mitotic arrest drugs such as taxol appears to ultimately induce cell death. There is no strong augmentation of effect with the addition of 2-AP.

An additional interesting aspect of the binary drug treatment effect is that CHO cells are as liable to die from tandem exposure to the two drug agents as from exposure to the two drugs together. Thus, for example, cells arrested in S-phase with aphidicolin may be released from aphidicolin into 2-AP. The two drugs are never present together at the same time. Nonetheless, a synergistic action occurs and the cells proceed to their death. This effect may be due to an advancement past a critical checkpoint during recovery from the first drug, with a lethal result. From the point of view of potential clinical applications, this effect is important, as it establishes that time of exposure to either drug may be kept minimal, and that it is not necessarily critical that there be a high titer of both drugs in the system simultaneously.

The mechanism of cell death appears to be by apoptosis. When cell death is evident as determined by microscopy, DNA samples extracted from these cells give "ladders" of DNA typical of apoptotic mechanisms (Zheng et al., 1991; Peitsch et al., 1993). The mechanism by which binary drug treatment can induce apoptotic cell death following checkpoint override is currently being studied. We propose that override from cell cycle arrest severely compromises the cell, since different elements of its cycle are out of synchrony. For example, DNA may be partially replicated (aphidicolin block) but the cell is in mitosis, or topoisomerase II has not resolved DNA winding after the completion of replication (teniposide block) but the cell enters mitosis. There would appear to be a feedback control system that reads these extreme flaws in the cell cycle and consequently induces cell death.

We have tested a variety of purine analogues to determine which of these induces checkpoint override. Thus far, the effective inducers of override fall into a discrete class. They include 2-aminopurine, 6-dimethylaminopurine and 2,6-diaminopurine. These congeners all have basic substitutions in either the 2 or 6 position. No purine analogue has been found effective if it did not contain a modification of this sort. No compound with acidic modifications of the 2 or 6 positions induces override (compounds tested include 2-chloroadenosine, 6-thioguanine and mercaptopurine).

The purine analogue 2,6-diaminopurine was used in clinical trials for cancer chemotherapy in the 1950s (Burchenal et al., 1951). Since patients were able to tolerate high doses of the drug during repeated short exposures, it may prove to be effective in a binary therapy approach. The drug doses used in these clinical trials were quite high, approaching concentrations we have found effective in vitro.

FUTURE DIRECTIONS

We have established that a binary drug approach to tumor therapy might ultimately be a viable possibility. We believe that this is the first demonstration that synergistic drug treatment may be applied at a series of cell cycle checkpoints to achieve override and cell death. The approach has a rational basis in that neither drug alone is capable of effecting substantial damage, but the two together in the proper combination with respect to the time of application will be highly effective at causing the death of vulnerable cells. The concept works with cells in culture, but there is much that must be done to demonstrate that it is feasible and practicable in real life situations.

One of the directions to be taken is to screen for other agents that are effective inducers of checkpoint override and that stand a reasonable chance of being tolerated by human subjects at effective doses. These agents need not be limited to the purine analogue class of compounds. We have established a drug screen procedure that can be used to detect other agents effective at inducing cell cycle checkpoint override. Importantly, we have established that the principle of this approach is sound, and we have shown that screening procedures exist that can be used to search for other effective compounds. We are currently assaying other compounds that might reasonably be expected to affect checkpoint override, based on their reported properties.

Another important aspect to be explored is whether tumor cells are especially vulnerable to override of checkpoints early in the cell cycle. If it is true that early expression of cyclin B is common among and restricted to tumor cells, and that its presence is indicative of permissiveness for checkpoint override at the point of arrest, then tumor cell death by checkpoint override would be, in principle, highly selective.

There are many questions of a basic biological nature that arise from the observations we have made. Is cyclin B acting as a permissive agent to allow the cell to proceed inappropriately to mitosis, or is it merely indicative of the overall advanced state of the cell with respect to the full complement of elements required for progression in the cycle? Is the induction of apoptosis universal for override from all points in the cell cycle, and does the resulting apoptosis occur only once a particular control point has been passed in the next cell cycle? If so, what is the molecular nature of this control?

Finally, of course, we intend to apply our observations as quickly as possible to animal model systems to determine if and under what circumstances a binary drug treatment regimen might be effective in eliminating tumors.

REFERENCES

Andreassen, P.R., and Margolis, R.L., 1991, Induction of partial mitosis in BHK cells by 2-aminopurine, *J. Cell Sci.* 100:299.

Andreassen, P.R., and Margolis, R.L., 1992, 2-Aminopurine overrides multiple cell cycle checkpoints in BHK cells, *Proc. Natl. Acad. Sci. USA* 89:2272.

Burchenal, J.H., Karnofsky, D.A., Kingsley-Pillers, E.M., Southam, C.M., Laird-Myers, W.P., Escher, G.C., Craver, L.F., Dargeon, H.W., and Rhoads, C.P., 1951, The effects of the folic acid antagonists and 2,6-diaminopurine on neoplastic disease, *Cancer* 4:549.

Dasso, M., and Newport, J.W., 1990, Completion of DNA replication is monitored by a feedback system that controls the initiation of mitosis in vitro: studies in Xenopus, *Cell* 61:811.

Farrell, P.J., Balkow, K., Hunt, T., and Jackson, R.J., 1977, Phosphorylation of initiation factor eIF-2 and the control of reticulocyte protein synthesis, *Cell* 11:187.

Gilman, A.G., Rall, T.W., Nies, A.S., and Taylor, P., "Goodman and Gilman's The Pharmacological Basis of Therapeutics," eighth edition, Pergamon Press, New York (1990).

Gould, K.L., and Nurse P., 1989, Tyrosine phosphorylation of the fission yeast cdc2 protein kinase regulates entry into mitosis, *Nature* 342:39.

Hartwell, L., and Weinert, T.A., 1989, Checkpoints: controls that ensure the order of cell cycle events, *Science* 246:629.

Hoyt, M.A., Totis, L., and Roberts, B.T., 1991, *S. cerevisiae* genes required for cell cycle arrest in response to loss of microtubule function, *Cell* 66:507.

Keyomarsi, K., and Pardee, A.B., 1993, Redundant cyclin overexpression and gene amplification in breast cancer cells, *Proc. Natl. Acad. Sci. USA* 90:1112.

Krek, W., and Nigg, E.A., 1991, Mutations of $p34^{cdc2}$ phosphorylation sites induce premature mitotic events in HeLa cells: evidence for a double block to $p34^{cdc2}$ kinase activation in vertebrates, *EMBO J.* 10:3331.

Li, R., and Murray, A.W., 1991, Feedback control of mitosis in budding yeast, *Cell* 66:519.

Lundgren, K., Walworth, N., Booher, R., Dembski, M., Kirschner, M., and Beach, D., 1991, Mik1 and wee1 cooperate in the inhibitory tyrosine phosphorylation of cdc2, *Cell* 64:1111.

Mahadevan, L.C., Wills, A.J., Hirst, E.A., Rathjen, P.D., and Heath J.K., 1990, 2-Aminopurine abolishes EGF- and TPA-stimulated pp33 phosphorylation and c-fos induction without affecting the activation of protein kinase C, *Oncogene* 5:327.

Murray, A.W., 1992, Creative blocks: cell-cycle checkpoints and feedback controls, *Nature* 359:599.

Nishitani, H., Ohtsubo, M., Yamashita, K., Iida, H., Pines, J., Yasudo, H., Shibata, Y., Hunter, T., and Nishimoto, T., 1991, Loss of RCC1, a nuclear DNA-binding protein, uncouples the completion of DNA replication from the activation of cdc2 protein kinase and mitosis, *EMBO J.* 10:1555.

Osmani, S.A., Engle, D.B., Doonan, J.H., and Morris, N.R., 1988, Spindle formation and chromosome condensation in cells blocked at interphase by mutation of a negative cell cycle control gene, *Cell* 52:241.

Parker, L.L., Atherton-Fessler, S., and Piwnica-Worms, H., 1992, p107wee1 is a dual-specificity kinase that phosphorylates $p34^{cdc2}$ on tyrosine 15, *Proc. Natl. Acad. Sci. USA* 89:2917.

Peitsch, M.C., Polzar, B., Stephan, H., Crompton, T., Robson-MacDonald, H., Mannherz, H.G., and Tschopp, H., 1993, Characterization of the endogenous deoxyribonuclease involved in nuclear DNA degradation during apoptosis (programmed cell death), *EMBO J.* 12:371.

Rieder, C.L., and Alexander, S.P., 1990, Kinetochores are transported poleward along a single astral microtubule during chromosome attachment to the spindle in newt lung cells, *J. Cell Biol.* 110:81.

Rowinsky, E.K., Cazenave, L.A., and Donehower, R.C., 1990, Taxol: a novel investigational antimicrotubule agent, *J. Natl. Cancer Inst.* 82:1247.

Rowley, R., Hudson, J., and Young, P.G., 1992, The wee1 protein kinase is required for radiation-induced mitotic delay, *Nature* 356:353.

Russell, P., and Nurse, P., 1987, Negative regulation of mitosis by $wee1^{+}$, a gene encoding a protein kinase homolog, *Cell* 49:559.

Schlegel, R., Belinsky, G.S., and Harris, M.O., 1990, Premature mitosis induced in mammalian cells by the protein kinase inhibitors 2-aminopurine and 6-dimethylaminopurine, *Cell Growth Diff.* 1:171.

Schlegel, R., and Pardee, A.B., 1986, Caffeine-induced uncoupling of mitosis from the completion of DNA replication in mammalian cells, *Science* 232:1264.

Seki, T., Yamashita, K., Nishitani, H., Takagi, T., Russell, P., and Nishimoto, T., 1992, Chromosome condensation caused by loss of RCC1 function requires the cdc25 protein that is located in the cytoplasm, *Mol. Biol. Cell* 3:1373.

Tummarello, D., Guidi, F., Torresi, U., Dazzi, C., and Cellerino, R., 1990, Teniposide (VM-26) as second-line treatment for small cell lung cancer *Anticancer Res.* 10:397.

Weinert, T.A., and Hartwell, L.H., 1988, The RAD9 gene controls the cell cycle response to DNA damage in Saccharomyces cerevisiae, *Science* 241:317.

Yamashita, K., Yasuda, H., Pines, J., Yasumoto, K., Nishitani,H., Ohtsubo, M., Hunter, T., Sugimura, T., and Nishimoto, T., 1990, Okadaic acid, a potent inhibitor of type 1 and type 2A protein phosphatases, activates cdc2/H1 kinase and transiently induces a premature mitosis-like state in BHK21 cells, *EMBO J.* 9:4331.

Zheng, L.M., Zychlinsky, A., Ojcius, D.M., and Young, J.D.-E., 1991, Extracellular ATP as a trigger for apoptosis or programmed cell death, *J. Cell Biol.* 112:279.

46

RADIATION INDUCED G2 DELAY AND MITOTIC CYCLIN EXPRESSION

W. Gillies McKenna[1] and Ruth J. Muschel[2]

[1]Department of Radiation Oncology
[2]Department of Pathology and Laboratory Medicine
University of Pennsylvania
Philadelphia, PA. 19104

INTRODUCTION

Many studies have demonstrated that oncogene transfection affects the radiation sensitivity of primary rat embryo fibroblasts. Our results indicate that the expression of H-ras plus v-myc oncogenes confers radiation resistance to a much greater extent than the expression of either gene alone. We have further shown that the radioresistant phenotype is accompanied by a prolonged G2 delay. This is consistent with the hypothesis that the extent of this delay is an important determinant of radiation sensitivity. The study of mitotic cyclin expression during the progression of cells through G2 and M after irradiation has also revealed several possible mechanisms for induction of a G2 arrest by radiation. We found that radiation affected the control of cyclin B levels both at the mRNA level as evidenced by decreased cyclin B mRNA expression after irradiation in S, and at the protein level as demonstrated after irradiation in G2. We also found that cyclin A expression was affected quite differently. Cyclin A expression was unaffected in its timing by irradiation and in fact levels of cyclin rose to supranormal levels during the period of time that cyclin B was depressed. The relationship of cyclin B expression to cyclin A levels is consistent with a link in the regulation of their degradation after M. It remains to be determined how these mechanisms might be differentially regulated in radioresistant and sensitive cells.

THE RAT EMBRYO CELL MODEL

Primary rat embryo cells (REF) have been used extensively as a model for cellular transformation with oncogenes (1-4). The use of these cells provides a system in which the parental cells are uniform in genetic composition, in which the rate of spontaneous transformation is low. The use of these cells also reduces the possibility that the effects seen after transfection with oncogenes are influenced by changes resulting from prolonged *in vitro* passage of the parental cell line. Primary REF cells are transformed at a low frequency by activated ras oncogenes (1). However, when the ras oncogene is introduced simultaneously with a cooperating oncogene such as v-myc or the adenovirus E1A gene, transformation occurs at a much higher frequency (3,5). Transfection with cooperating oncogenes by themselves does not transform rat embryo cells, but can result in immortalization of these usually senescent cells (6). The cell lines we obtained from transfection with H-ras or H-ras + v-myc are for the most part diploid with little karyotypic heterogeneity (2,4). These cells have been used in the studies described below to determine the effects of altered oncogene expression on radiation resistance.

The radiosensitivity of cells transfected with oncogenes

FitzGerald and co-workers (7) first suggested that the ras oncogene may induce radioresistance when they showed that NIH 3T3 cells transformed with the N-ras oncogene were more radioresistant than their untransformed parent cells when tested at a dose rate of 200 cGy/min. Sklar (8) correlated increased radioresistance in tissue culture with activation of other forms of the ras oncogene in a number of NIH 3T3 cell transfectants. Other transforming oncogenes have also been shown to confer radiation resistance. Kasid et al. (9) showed that transfection of an activated raf-1 gene into NIH3T3 cells induced radioresistance, a result confirmed by Chang et al. (10).

We examined REF which had been transformed by either H-ras alone or H-ras plus v-myc for survival by clonogenic assay after varying doses of X-rays. The results obtained with these cells were compared to results obtained with primary REF cells and with two cell lines immortalized by transfection with myc genes (11). Primary and immortalized rat embryo cells were sensitive to ionizing radiation with a D_0 of approximately 1.00 Gy. The H-ras plus v-myc transformants were markedly more resistant to radiation than the parent primary cells or than the immortalized rat embryo cells. The mean D_0 of this group of cells and their subclones was approximately 1.9 Gy. The difference in radiosensitivity between the H-ras plus v-myc transformants and the control cells was significant at the $p < 0.01$ level when the two groups were compared using Student's T test. Visual inspection of the curves as well as fitting of the data to the linear quadratic equation indicated that there was enhanced survival at both the low doses in the shoulder region, as well as at higher doses in the H-ras plus v-myc transformants. The rare transformants obtained by transfection with H-ras alone without any introduced cooperating oncogene also had enhanced radioresistance as compared to the controls, but to a lesser degree and this increase did not greatly affect the shoulder region. From these data we concluded that introduction of the H-ras plus v-myc oncogenes into rat embryo cells could result in transformants with significantly greater resistance to X-rays than either the parental cells or myc immortalized cells. These results for rat embryo cells have been confirmed by Ling (12), and more recently by Pardo and coworkers (Pers. comm.), but they are not generalizable to all other cell lines.

Damage induction and repair in radiosensitive and resistant lines

In order to elucidate the mechanism(s) responsible for the increased radiation resistance of H-ras + myc transformed REF, we compared DNA double strand break (dsb) induction, and DNA dsb rejoining in a radiation sensitive c-myc immortalized cell line, mycREC, (D_0 = 1.1 Gy); and in a radiation resistant, H-ras plus v-myc transformed cell, 3.7, (D_0 = 1.9 Gy). This work was carried out in collaboration with Dr. George Iliakis at Thomas Jefferson University (13). Breakage and rejoining were measured using pulse field gel electrophoresis. Cells were irradiated in the exponential phase of growth and the amount of DNA dsb present was quantified by measuring the fraction of the total DNA released into the gel (FAR) from agarose plugs in which the cells were embedded. Similar values of FAR were measured for both cell lines at equal X-ray doses. The results we obtained suggested a similar induction of DNA dsb per gray per dalton in both the sensitive and resistant cell line. Repair of DNA dsb was measured after exposure to 40 gray of X-rays. Both cells displayed a fast and a slow component of rejoining. The fast component had a t_{50} of approximately 12 minutes and the slow component a t_{50} of about 3 hours. These values were the same for both cell lines. The results indicate that the increased radioresistance of 3.7 cells, as compared to that of mycREC cells (MR), cannot be attributed to a decrease in the induction of DNA dsb per gray per dalton. Although small differences were observed in the dose response curves for DNA dsb induction in the two cell lines, these could be explained by small differences in the distribution of cells in the cell cycle in each population. Since the kinetics of DNA dsb repair were also similar in mycREC and 3.7 cells, it can also be hypothesized that the increased radioresistance of the latter cells does not derive from changes that modify repair capacity as measured by this technique.

The results indicate that the increased resistance to ionizing radiation of 3.7 cells are probably not due to modulations in the induction and repair of DNA dsb. This suggested that alterations of radiation induced delays in the progression through the cycle might underlie the observed differences in radiosensitivity between mycREC and 3.7 cells since these alterations could lead to differences in the fixation of radiation induced damage. These results do not suggest that alterations in DNA dsb induction or repair are unimportant determinants of radiosensitivity. These are clearly important in other systems and probably in at least some human tumors, but our results imply that because these factors are identical for both of the cell lines we have studied that other mechanisms must predominate in our system.

Radiation induced G_2 delay and DNA repair

Ionizing radiation has been shown to induce perturbations in the cell cycle of most eukaryotic cells. After irradiation, cells undergo a division delay which is reflected by increased time spent in the G2 portion of the cell cycle (14,15). It is thought that this G2 delay contributes to the ability of the cells to survive irradiation. In yeast the checkpoint for G2 arrest was first defined by the rad 9 mutant. The rad 9 mutation is characterized by increased sensitivity to ionizing radiation due to the inability to arrest in G2 in response to DNA damage (16). Induction of a G2 delay in these mutants through the use of a microtubule inhibitor results in a partial recovery of radiation resistance (16,17). The mammalian counterpart to this gene has not yet been identified. Mammalian cells deficient in G2 arrest also show an increased sensitivity to ionizing radiation. Some cell lines derived from patients with ataxia telangectasia are highly sensitive to ionizing radiation and do not undergo a G2 delay after radiation exposure (18,19). Baby hamster Kidney derived tsBN2 cells which Nishimoto and his group isolated as temperature sensitive for DNA replication, undergo mitotic events even without completion of DNA replication. These cells contain a point mutation in the RCC1 gene (20,21). The RCC1 gene is thought to be part of a complex that regulates cdc2 histone H1 kinase activity through a GTPase cycle (22). Incubation of tsBN2 cells at the non-permissive temperature after irradiation results in decreased survival relative to cells maintained at the permissive temperature (23). Additional evidence for the role of the G2 delay in DNA damage repair comes from studies using caffeine. This drug, which renders cells more sensitive to irradiation, also reduces or abolishes the radiation-induced G2 delay (24,25). Thus the G2 delay in response to DNA damage appears to be important in enhancing survival.

Cell cycle control proteins involved in the G2/M transition

To attempt to understand the mechanisms underlying the G2 delay, we can make use of recent discoveries elucidating the biochemical mechanisms controlling the cell cycle. The progression through G2 and M in the cell cycle has been shown to be regulated by p34-cdc2 kinase activity (26). The p34-cdc2 kinase is activated by two obligatory and independent pathways, one is through binding to cyclin B and the other is through the phosphorylation state of p34-cdc2. Although p34-cdc2 protein is present at essentially equal levels throughout the cell cycle in growing cells, its kinase activity only appears in G2/M. The p34-cdc2 is active as a kinase only when bound to a cyclin (27). Mitotic cyclin levels rise at the S/G2 transition. As cyclin B levels rise, it is found in a complex with p34-cdc2. Thus one level of regulation of activation of the kinase activity of p34-cdc2 is controlled by the increased levels of the cyclin B. While the binding of cyclin B to p34-cdc2 is necessary for its activation, it is not sufficient. The subsequent dephosphorylations of p34-cdc2 are also required(28,29).

In human cells two mitotic cyclins called cyclin A and B have been identified. The mRNA levels of both cyclin A and B rise in Hela cells at late S phase and peak in G2 (30,31). The rise in mRNA levels appears to be controlled at least in part at the transcriptional level, but may also occur posttranscriptionally. In Hela cells cyclin B complexes with p34-cdc2 while cyclin A is also found in a complex with both p34-cdc2 and a different molecule, cdk2, which shares some antigenic determinants with p34-cdc2 (31,32,33). Cyclin B is known to bind to p34-cdc2 but not to any other protein. The actual role of both of the cyclins in human cells is not yet defined, but in Drosophila it has been demonstrated that cyclin A and B are both required for completion of the cell cycle (34). The role of cyclin A in cell cycle progression is more complex than cyclin B as it contributes to the regulation of both S and G2/M. Cyclin A levels rise throughout S while cyclin B levels remain low. Cyclin A is required, however, both for DNA replication and for the G2/M transition (34,35). Because the cyclins play such a pivotal role in cell cycle regulation, the cyclin cdc2 system was a prime candidate for factors regulating the radiation induced G2 delay.

The activity of p34-cdc2 is also regulated by its phosphorylation state. In G1 p34-cdc2 is not phosphorylated, but early in G2 p34-cdc2 is phosphorylated at threonine 161, a phosphorylation which may increase its affinity for cyclin B (36). Upon binding of cyclin B to p34-cdc2, the threonine 14 and tyrosine 15 residues are phosphorylated. Both of these phosphorylations inhibit the kinase activity of p34-cdc2 (37,38). In yeast the tyrosine phosphorylation is mediated by kinases, wee1 or mik1 (39). Final activation of the cyclin B-p34-cdc2 complex is dependent upon dephosphorylation of threonine 14 and tyrosine 15 which appears to be mediated by the cdc25 phosphatase(40).

At the completion of mitosis,both cyclin A and B are destroyed. If mutations which prevent that destruction are introduced, progression through mitosis into G1 is not accomplished (41). The destruction of cyclin A precedes that of cyclin B and there is some evidence that the presence of cyclin B is required to initiate the destruction of cyclin A (42). Thus, both the increase and subsequent decrease in levels of the mitotic cyclins are required for the cells to transit through G2/M.

ALTERATIONS IN CYCLIN EXPRESSION IN HELA CELLS AFTER RADIATION

Because of the role cyclins play in cell cycle progression through G2/M, the effects of ionizing radiation on cyclin expression were examined (42, 43). Hela cells were synchronized using a double block, thymidine followed by ap' .dicolin, using minor modifications of the method of Heintz et al. (44). Cells were then released from the block and irradiated in S phase. Survival curves for Hela cells at this time of the cell cycle have been extensively reported by Tolmach's laboratory and clonogenic survival fractions are about 0.5% at a dose of 10 Gy (45). However, the irradiated cells will continue to divide for several rounds before loss of viability is noted and indeed cell numbers rose as expected during the experiment. Thus, cells are viable throughout the experiment. In the unirradiated controls, the population of cells began to enter G2 phase at 9 to 12 hours and had exited from M phase at 15 hours. The levels of cyclin B mRNA, as evaluated by RNA blotting, rose at 9 hours, peaked at 12 hours coincident with the greatest percentage of cells in G2/M phases and fell as the cells re-entered G1 phase. In contrast, cells irradiated at the start of S phase did not exit from M phase until after 24 hr. Thus, the irradiated cells remained in G2 for at least 9 h longer than the controls at this dose. These cells also remained in S several hours longer, an expected result since all phases of the cell cycle are somewhat delayed by radiation (except perhaps M) but the lag in G2 was by far the longest. Nonetheless, in spite of more than 80% of the cells in the population accumulating in G2, cyclin B mRNA remained at low levels. Thus, the delay in G2 is associated with an absence of accumulation of cyclin B mRNA when Hela cells are irradiated in early S phase. Cyclin B protein as detected in western blot analysis was also delayed after irradiation in S. In later experiments, we showed that the decrease in cyclin B mRNA and protein was dose dependent and could be detected at from 2 to 4 Gy (43).

Effects of radiation on cyclin A expression

Cyclin A expression after irradiation in S was also examined (43). Cyclin A mRNA and protein levels rise at the expected time and continue to rise to levels greater than those found in the control cells. This contrasts with cyclin B mRNA and protein levels which remain depressed after irradiation in S. After the cells begin to exit from the block and enter G1, the cyclin A mRNA and protein levels fall. These findings are consistent with the hypothesis formulated by Ruderman (42), that cyclin B is responsible for triggering the destruction of cyclin A. Kufe et al. has found similar results in other human cell lines but have also found that at higher doses, 20 Gy, cyclin A mRNA levels may also be depressed (47).

These results further indicate that irradiation has some specificity in its effects on both cyclin B and mRNA and protein levels since cyclin A expression was not decreased. Although some genes have been found to be induced in cells after irradiation, including members of the heat shock gene family, metallothioneins, c-fos, c-jun, the transcription factor kB, type I collagenase, b-polymerase and a series of genes called DDI(DNA damage inducible) which were identified because of their increase in level after DNA damage, others have been shown to be unaffected (48,49,50,51,52). Only cyclin B has been specifically reported to be depressed at the doses considered here. Holland et al. (53) examined the patterns of protein synthesis after irradiation in G2 using one dimensional gel electrophoresis and found only one band at Mr 170,000 to be reduced, the others were not changed. Boothman et al. (54) examined protein synthesis using two dimensional gel electrophoresis after irradiation in plateau phase. While the great majority of proteins appeared unaltered, eight spots increased in intensity and two decreased. Thus, the effects on cyclin B mRNA and protein are not a general consequence of irradiation on protein or RNA synthesis and the pattern of their altered expression, while not unique is unusual.

The effect of irradiation in G2 was also examined (43). Hela cells were synchronized and irradiated in G2 phase with 6 Gy of X-rays. In the irradiated cells cyclin B mRNA rose at the same time as in the control cells, but the cyclin B protein dropped to very low levels even though cyclin B mRNA levels were rising at this point. The levels of cyclin B protein in the irradiated cells did not begin to rise rapidly until after 6 hours, corresponding to the exit of the cells from the radiation-induced G2 block. Thus, cyclin B levels were also depressed after irradiation in G2, but apparently through a different mechanism than after irradiation in S when both mRNA and protein levels were depressed. This indicates that a posttranscriptional mechanism must operate after radiation in G2.

These data suggest that there is a cellular mechanism which recognizes DNA damage and transduces this signal to eventually result in altered cyclin expression. This mechanism cannot yet be identified because the controls which allow cells to recognize DNA damage and the signals which transduce this recognition are not well understood. The data presented here suggest that one target of the mechanisms regulating G2 transition after DNA damage might involve redundant controls on the levels of cyclin B mRNA.

REFERENCES

1. D.A. Spandidos and N.M. Wilkie, Malignant transformation of early passage rodent cells by a single mutated human oncogene. Nature 310, 469-475 (1984).
2. R.J. Muschel, K. Nakahara, E. Chu, R. Pozzatti and L.A. Liotta, Karyotypic analysis of diploid or near diploid metastatic Harvey ras transformed rat embryo fibroblasts. Cancer Res. 46, 4104-4108 (1986).
3. R. Pozzatti, R. Muschel, J. Williams, R. Padmanabhan, B. Howard, L. Liotta and G. Khoury, Primary rat embryo cells transformed by one or two oncogenes show different metastatic potentials. Science 232, 223-227 (1986).
4. W.G. McKenna, K. Nakahara and R.J. Muschel, Site-specific integration of H-ras in transformed rat embryo cells. Science 241, 1325-1328 (1988).
5. H.E. Ruley, Adenovirus early region 1A enables viral and cellular transforming genes to transform primary cells in culture. Nature 304, 602-606 (1983).
6. H.L. Land, L.F. Parada and R.A. Weinberg, Cellular oncogenes and multistep carcinogenesis. Science 222, 771-778 (1983).
7. T.J. FitzGerald, C. Daugherty, K. Kase, L.A. Rothstein, M. McKenna and J.S. Greenberger, Activated human N-ras oncogene enhances X-irradiation repair of mammalian cells in vitro less effectively at low dose rate. Am. J. Clin. Onc. 8, 517-522 (1985).
8. M.D. Sklar, The ras oncogenes increase the intrinsic resistance of NIH 3T3 cells to ionizing radiation. Science 239, 645-647 (1988).
9. U.N. Kasid, A. Pfeifer, R.R. Weichselbaum, A. Dritschilo and G. Mark, The raf oncogene is associated with a radiation resistant human laryngeal cancer. Science 237, 1039-1041 (1987).
10. E.H. Chang, K.F. Pirollo, Z.Q. Zou, Cheung, H.Y., E.L. Lawler, R. Garner, E. White, W.B. Bernstein, J.W.J. Faumeni and W.A. Blattner, Oncogenes in radioresistant, noncancerous skin fibroblasts from a cancer-prone family. Science 237, 1036-1039 (1987).
11. W.G. McKenna, M.C. Weiss, B. Endlich, C.C. Ling, V.I. Bakanauskas, M.L. Kelsten and R.J. Muschel, Synergistic effect of the v-myc oncogene with Hras on radioresistance. Cancer Research 50, 97-102 (1990).
12. C.C. Ling and B. Endlich, Radioresistance induced by oncogenic transformation. Radiat. Res. 120, 267-279 (1989).

13. G. Iliakis, L. Metzger, R.J. Muschel and W.G. McKenna, Induction and repair of DNA double strand breaks in radiation-resistant cells obtained by transformation of primary rat embryo cells with the oncogenes H-ras and v-myc. Cancer Res. 50, 6575-6579 (1990).
14. M.M. Elkind, A. Han and K. Volz, Radiation response of mammalian cells grown in culture. IV. Dose dependence of division delay and post-irradiation growth of surviving and non-surviving Chinese hamster cells.. J. Natl. Cancer Inst. 30, 705-711 (1963).
15. J.S. Bedford and J.B. Mitchell, Mitotic accumulation of HeLa cells during continuous irradiation. Radiat. Res. 70, 173-180 (1977).
16. Weinert T. A. and Hartwell L. H., The RAD9 gene controls the cell cycle reponse to DNA damage in Saccharomycer cerevisiae. Science 241, 317-322 (1988).
17. T.A. Weinert and L.H. Hartwell, Characterization of RAD9 of saccharomyces cerevisiae and evidence that its function acts posttranslationally in cell cycle arrest after DNA damage.. Mol. Cell Biol. 10, 6554-6564 (1990).
18. H. Nagasawa and J.B. Little, Comparison of kinetics of X-ray induced cell killing in normal, ataxia telangectasia and hereditary retinoblastoma fibroblasts. Mutat. Res. 109, 297-308 (1983).
19. F. Zampetti-Bosseler and D. Scott, Cell death, chromosome damage and mitotic delay in normal human, ataxia telangectasia and retinoblastoma fibroblasts after X-irradiation. Int. J. Radiat. Biol. 39, 547-558 (1981).
20. M. Ohtsubo, R. Kai, N. Furuno, T. Sekiguchi, M. Sekiguchi, H. Hayashi, K. Kuma, T. Miyata, S. Fukushige, T. Murotsu, K. Matsubara and T. Nishimoto, Isolation and characterization of the active cDNA of the human cell cycle gene (RCC1) involved in the regulation of onset of chromosome condensation. Genes Develop. 1, 585-593 (1987).
21. S. Uchida, T. Sekiguchi, H. Nishitani, K. Miyauchi, M. Ohtsubo and T. Nishimoto, Premature chromosome consdensation is induced by a point mutation in the hamster RCC1 gene. Mol. Cell. Biol. 10, 577-584 (1990).
22. T. Nishimoto, S. Uzawa and R. Schlegel, Mitotic checkpoints. Curr. Opinions Cell. Biol. 4, 74-79 (1992).
23. H. Sasaki and T. Nishimoto, Chromosome condensation may enhance X-ray-related cell lethality in a temperature-sensitive mutant (tsBN2) of baby hamster kidney cells (BHK21). Radiat. Res. 109, 407-418 (1987).
24. P.M. Busse, S.K. Bose, R.W. Jones and L.J. Tolmach, The action of caffeine on X-irradiated HeLa cells. II. Synergistic lethality.. Rad. Res. 71, 666-677 (1977).
25. H. Sasaki and T. Nishimoto, X-ray related potentially lethal damage expressed by chromosome condensation and the influence of caffeine. Rad. Res. 120, 72-82 (1989).
26. P. Nurse, Universal control mechanism regulating onset of M-phase. Nature 344, 503-508 (1990).
27. E.A. Nigg, P. Gallant and W. Krek, Regulation of p34cdc2 protein kinase activity by phosphorylation and cyclin binding. CIBA Foundation Symp. 170, 72-84 (1992).
28. J. Gautier, T. Matsukawa, P. Nurse and J. Maller, Dephosphorylation and activation of Xenopus p34cdc2 protein kinase during the cell cycle. Nature 339, 626-9 (1989).
29. A.O. Morla, G. Draetta, D. Beach and J.Y.J. Wang, Reversible tyrosine phosphorylation of cdc2: dephosphorylation accompanies activation during entry into mitosis.. Cell 58, 193-203 (1989).
30. J. Pines and T. Hunter, Isolation of a human cyclin cDNA: evidence for cyclin mRNA and protein regulation in the cell cycle and for interaction with p34cdc2. Cell 58, 833-846 (1989).
31. J. Pines and T. Hunter, Human cyclin A is adenovirus E1A-associated protein p60 and behaves differently from cyclin B. Nature 346, 760-763 (1990).
32. B. Faha, M.E. Ewen, L.H. Tsai, D.M. Livingston and E. Harlow, Interaction between human cyclin A and adenovirus E1A-associated p107 protein. Science 255, 87-90 (1992).
33. L.-H. Tsai, E. Harlow and M. Meyerson, Isolation of the human cdk2 gene that encodes the cyclin A- and adenovirus E1A-associated p33 kinase. Nature 353, 174-177 (1991).

34. C.F. Lehner and P.H. O'Farrell, The roles of Drosophila cyclins A and B in mitotic control. Cell 61, 535-547 (1990).
35. F. Girard, U. Strausfeld, A. Fernandez and N.J.C. Lamb, Cyclin A is required for the onset of DNA replication in mammalian fibroblasts. Cell 67, 1169-1179 (1991).
36. K.L. Gould, S. Moreno, D.J. Owen, S. Sazer and P. Nurse, Phosphorylation at Thr 167 is required for Schizosaccharomyces pombe p34cdc2 function. EMBO J. 10, 3297-3309 (1991).
37. G. Draetta and D. Beach, Activation of cdc2 protein kinase during mitosis in human cells: cell cycle-dependent phosphorylation and subunit rearrangement. Cell 57, 393-401 (1989).
38. K.L. Gould and P. Nurse, Tyrosine phosphoryation of the fission yeast cdc2+ protein kinase regulates entry into mitosis. Nature 342, 39-45 (1989).
39. K. Lundgren, N. Walworth, R. Booher, M. Dembski, M. Kirschner and D. Beach, mik1 and wee1 cooperate in the inhibitory tyrosine phosphorylation of cdc2. Cell 64, 1111-1122 (1991).
40. A.W. Murray, M.J. Solomon and M.W. Kirschner, The role of cyclin synthesis and degradation in the control of maturation promoting factor activity. Nature 339, 280-286 (1989).
41. F.A. Luca, E.K. Shibuya, C.D. Dohrmann and J.V. Ruderman, Both cyclin A delta 60 and B delta 97 are stable and arrest cells in M phase, but only cyclin B delta 97 turns on cyclin destruction. EMBO J. 10, 4311-4320 (1991).
42. R.J. Muschel, H.B. Zhang, G. Iliakis and W.G. McKenna, Cyclin B expression in HeLa cells during the G2 block induced by ionizing radiation. Cancer Res. 51, 5113-5117 (1991).
43. R.J. Muschel, H.B. Zhang and W.G. McKenna, Differential effect of ionizing radiation on the expression of cyclin A and cyclin B in Hela cells. Cancer Res. 53, 1128-1135 (1993).
44. N. Heintz, H.L. Sive and R.G. Roeder, Regulation of human histone gene expresssion: kinetics of accumulation and changes in the rate of synthesis and in the half-lives of individual histone mRNAs during the HeLa cell cycle. Mol. Cell. Biol. 3, 539-550 (1983).
45. L.J. Tolmach and P.M. Busse, The action of caffeine on X-irradiated HeLa cells. IV. Progression delays and enhanced cell killing at high caffeine concentrations. Rad. Res. 82, 374-392 (1980).
46. R. Datta, R. Hass, H. Gunji, R. Weichselbaum and D. Kufe, Down-Regulation of cell cycle control genes by ionizing radiation. Cell Growth and Diff. 3, 637-644 (1992).
47. A.J.J. Fornace, B. Zmudzka, M.C. Hollander and S.H. Wilson, Induction of beta-polymerase mRNA by DNA-damaging agents in Chinese hamster ovary cells. Mol. Cell. Biol. 9, 851-853 (1989).
48. A.J.J. Fornace, D.W. Nebert, M.C. Hollander, J.D. Leuthy, M. Papathanasiou, J. Fargnoli and N.J. Holbrook, Mammalian genes coordinately regulated by growth arrest signals and DNA-damaging agents. Mol. Cell. Biol. 9, 4196-4203 (1989).
49. M.C. Hollander and A.J.J. Fornace, Induction of fos RNA by DNA-damaging agents. Cancer Res. 49, 168-92 (1989).
50. M. Sherman, R. Datta, D. Hallahan, R. Weichselbaum and D. Kufe, Ionizing radiation regulates expression of the c-jun protooncogene. Proc. Natl. Acad. Sci. USA 87, 5663-5666 (1990).
51. M.A. Brach, R. Hass, M. Sherman, H. Gunji, R. Weichselbaum and D. Kufe, Ionizing radiation induces expression and binding expression and binding activity of the nuclear factor kB. J. Clin. Invest. 88, 691-695 (1991).
52. J.M. Holland, W.D. Wright, R. Higashikubo and J.L. Roti Roti, Effects of irradiation on nuclear protein synthesis in G2 phase of the cell cycle. Radiat. Res. 122, 197-208 (1990).
53. D.A. Boothman, I. Bouvard and E.N. Hughes, Identification and characterization of X-ray-induced proteins in human cells. Cancer Res. 49, 2871-2878 (1989).

47

CYCLIN B DEGRADATION AS A TARGET OF ANTIPROLIFERATIVE DRUG ACTION

Steven W. Sherwood, Robert D. Simoni and Robert T. Schimke

Department of Biological Sciences
Stanford University, Stanford CA, 94305

INTRODUCTION

Mitotic Control and Drug Sensitivity in Mammalian Cells

For a number of years we have been examining the genotoxic and cytotoxic effects of antiproliferative drugs as well as cytotoxic treatments such as u.v. irradiation and hypoxia in cultured mammalian cells. The approach we have taken is to examine cellular response to treatment rather than drug-target interaction, and in particular, we have been interested in determining the relationship of treatment-induced cell cycle perturbation to cytotoxicity and genotoxicity (1-5).

Studies by Kung, et al, demonstrated the existence of important differences between "standard" mammalian cell lines in a mitotic checkpoint regulating transit through mitosis (6). The relationship between disturbances in cell cycle control and the genotoxicity and cytotoxicity of inhibitors of DNA synthesis had been examined in earlier studies but not directly linked to mitosis (1-5). The experiments of Kung, et al examined the propensity of different cell types to continue cell cycle progression in the absence of a mitotic spindle (induced by colcemid, a model anti-tubulin drug). It was found that Chinese hamster ovary cells (CHO) readily enter a tetraploid cell cycle when spindle formation is blocked, while HeLa S3 cells are permanently arrested in a diploid pseudometaphase state. HeLa cells will remain arrested in this state in the presence of drug until they undergo apoptosis. The mitotic behavior of these two cell lines correlates with biochemical differences in the way that MPF activity (measured as histone H1-kinase activity) and cyclin B degradation are regulated in the absence of a spindle; CHO cells undergo cyclin B degradation and inactivation of MPF activity when maintained in colcemid while in HeLa cells, cyclin B is not degraded and H1 kinase activity remains elevated for the duration of colcemid treatment. The cell cycle checkpoint identified in these experiments appears to be analogous, if not homologous, to the mitotic checkpoint in yeast identified as "mitotic-arrest deficient" (mad) mutants (7-9). In mammalian cells these differences in the regulation of mitosis were described as reflecting varying degrees of "stringency"; HeLa S3 (and the majority of human cell lines examined) display "stringent", and CHO (and many rodent cell lines) a relatively "relaxed" mitotic control (6). Relative to “stringent” HeLa S3 cells, CHO cells are less sensitive to the cytotoxic effects of colcemid and other anti-spindle drugs (6,7).

The “relaxed” mitotic control observed in CHO cells is defined by continued cell cycle progression in the presence of colcemid at a concentration sufficient to completely inhibit spindle formation. CHO AA8 cells "blocked" in mitosis with high concentrations of colcemid are actually only delayed for 2-4 hrs before exiting mitosis (measured by decline in cyclin B1 content and histone H1 kinase activity, as well as morphologically using chromatin decondensation and nuclear membrane reformation as criteria and entering a tetraploid G1 (6). This "relaxed" mitotic control is interpreted to indicate that interfering with microtubule-driven mitotic processes imposes little feedback (or feed-forward) control on cell cycle progression in such cells (6, 10). Inhibition of cytokinesis by cytochalasin also induces a short transient delay in cell cycle progression, but not a cell cycle “block” (10).

Anti-Mitotic Effects of ALLN

To better understand the regulation of mitosis in CHO cells, we sought other means of inhibiting mitotic progression in these cells. The report that the cell permeant neutral cysteine protease inhibitor acetyl-leucyl-leucyl-norleucinal (ALLN, also called calpain I inhibitor) inhibited 3-hydroxy-3-methyl-glutaryl-coenzyme A-reductase (HMG CoA-reductase) degradation in-vivo (11) prompted us to examine the possibility that ALLN might reduce the rate of cyclin B degradation. . A mitotic arrest phenotype of undefined mechanism has been reported for the epoxide-based cysteine protease inhibitor E64d (12), and ALLN inhibits cyclin B degradation in a cell-free clam oocyte system (13). ALLN also was reported to be cytotoxic to CHO cells (LD_{50}=2 uM, ref 14).

Addition of ALLN to asynchronously growing CHO cells results in a complex cell cycle arrest phenotype in which cells arrest at two distinct points, the G1/S boundary and the metaphase-anaphase transition. Cells in S-phase at the time of drug exposure are not blocked but are slowed in in S-phase progression. Where a cell becomes blocked is determined in part by the concentration of inhibitor and, importantly, by the cell’s cell cycle position at the time it "sees" the inhibitor (15).

Examination of the mitotic arrest induced by ALLN shows that ALLN directly or indirectly inhibits cyclin B proteolysis. CHO cells exposed to ALLN after they have passed G1/S but before attaining mitosis proceed to metaphase where they undergo formation of a normal-appearing bipolar spindle, nuclear envelope breakdown, chromosome condensation and congression to the metaphase plate (15). Specifically at the point that cells would proceed into anaphase they become arrested. Histone H1 kinase activity remains elevated during the arrest and cyclin B levels actually continue to increase beyond levels normally occurring during mitosis (Fig. 1,). Chromosome condensation continues during the block and chromosomes become "hypercondensed" while still aligned on the metaphase plate. After a delay of approximately 12-14 hrs, cells undergo a rapid and chaotic "mitosis".

The aberrant mitosis occurring after prolonged ALLN arrest results from the spindles becoming multipolar. While these spindles retain a polarized character, their multipolar organization results in cell divisions which generate multiple genetically unbalanced daughter cells of different size. This “mitosis” is accompanied by rapid disappearance of cyclin B and the reformation of the nuclear envelope as well as chromatin descondensation in the daughter cells, and thus we conclude that the cells exit from mitosis and do not simply undergo a terminal cytolytic process (Fig 2, ref 15). However, while the daughter cells are "normal" G1 cells with respect to gross morphology, they are extremely unbalanced genetically and subsequently die in what we refer to as "pseudo-G1" (16).

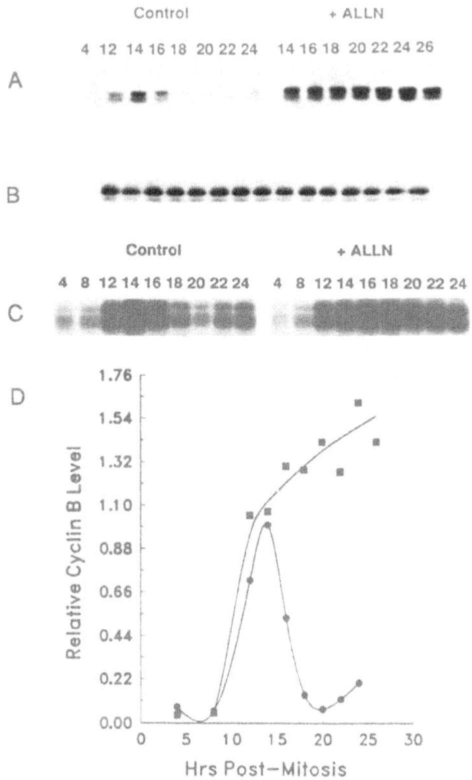

Figure 1. Level of cyclin B1 protein and histone H1 kinase activity in CHO cells arrested in mitosis with ALLN. Cells synchronized by mitotic shakeoff were exposed continuously to 40 ug/ml ALLN beginning 12 hrs after mitosis (cells in late S/G2). Cell extracts were made at the hours indicated and total protein (100 ug/lane) analysed by SDS-PAGE followed by immunoblotting with anti-cyclin B1 antibody (panels A). Densitometric analysis of cyclin B content for two experiments is summarized in panel D. Cell extracts were also analysed for histone H1 kinase activity (panels C). Panel B is the filter shown in A reprobed for the 30 Kd subunit of anti-calpain II as a loading control.

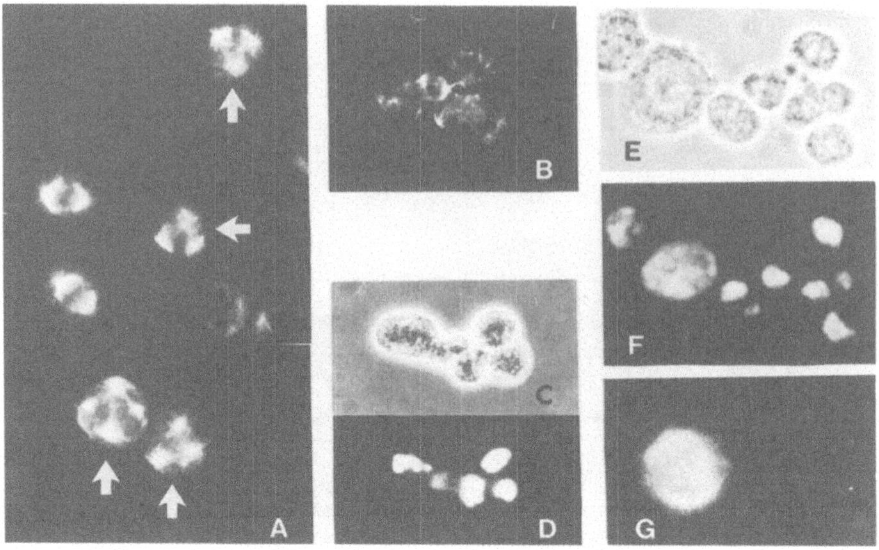

Figure 2. Mitotic phenotype of ALLN-treated CHO cells. Panel A, Anti-tubulin immunofluorescence showing initially normal bipolar spindles which become multipolar during ALLN-blocked mitosis. Panels B-G show daughter cells of aberrant "mitosis" produced after ALLN-induced mitotic delay. Panel B, anti-tubulin immunofluorescence. Panel C,D, phase image and Hoechst 33242 stained mitotic cell. Panels E-F show phase, Hoechst and anti-cyclin B1 immunofluorescence image of a prophase cells and telophase daughter cells of aberrant mitosis.

The reason CHO cells finally "escape" the ALLN-induced arrest is unclear. It is known that ALLN is a substrate for the MDR1 efflux pump (11, 14) and also that it is detoxified by an aldo-keto reductase (14). Hence, it may be that the intracellular concentration of active peptide declines over time. We have not examined this possibility closely.

The spindle abnormalities which develop as CHO cells are blocked at metaphase with ALLN closely resemble those occurring when a non-degradable cyclin B gene is expressed in HeLa cells (17), as well as when cells are pharmacologically slowed through mitosis (18-20). Aberrant mitosis of this nature is the cytotoxic endpoint of a variety of agents we have examined, and we believe it may represent a general aspect of antiproliferative drug toxicity, at least under certain conditions of drug exposure (SWS, RTS, in preparation). Because we have observed this phenomenon associated with the toxicity of anti-tubulin drugs as well as inhibitors of DNA synthesis and cis-platinum, we are examining this process in more detail to determine whether the supernumerary spindle poles result from centrosomal replication (a G1 event) occurring during the arrest, or whether this type of aberration reflects "mechanical" disruption of already formed spindle poles.

The mitotic block induced in CHO cells by ALLN is clearly correlated with delayed degradation of cyclin B and resembles the phenotype observed in cells expressing a transfected non-degradable cyclin B. Data showing that in-vivo proteolysis of HMG-CoA reductase is affected by ALLN suggests that ALLN is not a only an inhibitor of a cyclin B protease (11,14). However, some degree of substrate specificity is indicated by the studies of Luca and Ruderman (13) which showed that in clam oocyte extracts, ALLN inhibits cyclin B degradation but not the degradation of other proteins detectable in this system. Evidence that calpastatin is not involved with the degradation of cyclin B in an in-vitro embryo system suggests that ALLN's effect on cyclin B degradation may be independent of calpain (21), although calpain has been implicated in normal mitotic function (22). Thus, the proteolytic process which is involved in cyclin B degradation and inhibited by ALLN remains unidentified.

The effect of ALLN at the G1/S transition suggests that this agent might also inhibit cyclin proteolysis at other cell cycle transition points. While the events of cyclin turnover at this transition are much less well understood in mammalian cells than those occurring at mitosis, the G1/S effects of ALLN are quite pronounced and in fact this transition point is more sensitive to ALLN (i.e. requires a lower ALLN concentration to be observed) than is the metaphse-anaphase transition (15). We are examining the possibility that the rate of degradation of other cyclins might be affected by ALLN in a manner which leads to the observed cell cycle blocks.

The effects of ALLN on mitosis are clearly correlated with a strong inhibition of cyclin B degradation. Whether this reflects direct or indirect effects of ALLN on cyclin B proteolysis remains to be determined. It seems clear, however, that ALLN interferes with mitosis in CHO cells in a way which affects mitotic feedback controls in a manner distinct from those revealed by cellular response to anti-tubulin drugs. It is important to note that whereas the drug-sensitivity (measured as clonogenic survival) of anti-tubulin anti-mitotic agents is lower in "relaxed" mitotic control cells such as CHO than in more "stringent" cells (eg. HeLaS3), the cytotoxicity of ALLN, which exerts an anti-mitotic effect indirectly of spindle formation, is similar in both cell lines (15). Because the studies of clonogenic survival which we have conducted utilize asynchronous cultures, it is possible that the double cell cycle effect of ALLN may contribute to it's overall toxicity.

Our results with ALLN suggest that targeting regulatory processes associated with cell cycle progression may present new opportunities for antiproliferative drug development and use. Because phase-specific proteolysis of specific cyclin proteins is critical in mitotic progression and may also be important in the transitions between other cell cycle phases (particularly G1-S), protease inhibitors may represent an important class of agent from the standpoint of their utility as probes of cell cycle proteolytic processes as well as potentially useful therapeutic agents. It is very interesting in this regard that there are reports showing that protease inhibitors inhibit oncogene transformation in cultured cells (22-24), alter the bleomycin sensitivity of carcinoma cells independent of bleomycin metabolism (25) and display some efficacy as anti-carcinogens (26,27). Whether these effects of protease inhibitors are related to the cell cycle effects of ALLN we have described remains to be determined.

ACKNOWLEDGEMENTS

We thank the members of our laboratories for helpful discussions. This work was supported by grants to RTS (NIH CA 16318) and RDS (HL 26502).

REFERENCES

1. Hill, A.B. and Schimke, R.T. 1985. Increased gene amplification in L5178Y mouse lymphoma cells with hydroxyurea-induced chromosomal aberrations. Cancer Res. 45:5050-5057.
2. Hoy, C.A., Rice, G.C., Kovacs, M., and Schimke, R.T. 1987. Overreplication of DNA in S-phase Chinese hamster ovary cells after DNA synthesis inhibition. J. Biol. Chem. 262:11927-11934.

3. Sherwood, S.W., Daggett, A.S., and Schimke, R.T. 1987. Interaction of hyperthermia and metabolic inhibitors in the induction of chromosome damage in Chinese hamster ovary cells. Cancer Res. 47:3584-3588.
4. Sherwood, S.W., Schumacher, R.I. and Schimke, R.T. 1988. Effect of cycloheximide on development of methotrexate resistance in Chinese hamster ovary cells treated with inhibitors of DNA synthesis. Mol. Cell. Biol. 8:2822-2827.
5. Kung, A.L., Zetterberg, Z., Sherwood, S.W. and Schimke, R.T. Cytotoxic effects of cell cycle phase specific agents: Result of cell cycle perturbation. Cancer Res. 50:7307-7317.
6. Kung, A.L., Sherwood, S.W. and Schimke, R.T. 1990 Cell line specific differences in the control of cell cycle progression in the absence of mitosis. Proc. Nat. Acad. Sci. USA 87:9553-9557.
7. Rieder, C.L. and Palazzo, R.E. 1992. Colcemid and the mitotic cycle. J. Cell Sci. 102:387-392.
8. Hoyt, M.A. , Totis, L. and Roberts, B.T. 1991. S. cerevisiae genes required for cell cycle arrest in response to loss of microtubule function. Cell 66:507-518.
9. Li, R. and Murray, A.W. 1991. Feedback control of mitosis in budding yeast. Cell 66:519-531.
10. Schimke, R.T., Kung, A.L., Rush, D.F., and Sherwood, S.W. 1'991. Differences in mitotic control among mammalian cell lines. In, **The Cell Cycle.** Cold Spring Harbor Symp. Quant, Biol. LVI:417-425.
11. Inoue, S., Bar-Nun, S., Roitleman, J., and Simoni, R.D. 1991. Inhibition of degradation of 3-hydroxy-3-methylglutaryl-coenzyme A reductase in vivo by cysteine protease inhibitors. J. Biol. Chem. 266:13311-13317.
12. Shoji-Kasi, Y., Senshu, M., Iwashita, S., and Imahori, K. 1988. Thiol specific protease inhibitor E64d arrests human epidermoid carcinoma A431 cells at mitotic metaphase. Proc. Nat. Acad. Sci. USA 85:146-150
13. Luca, F.A., and Ruderman, J.V. 1989. Control of programmed cyclin degradation in a cell-free system. J. Cell Biol. 109:1895-1909.
14. Inoue, S., Sharma, R.C., Schmke, R.T. and Simoni, R.D. 1993. Cellular detoxification of tripeptidyl aldehydes by an aldo-keto reductase. J. Biol. Chem. 268:5894-5898.
15. Sherwood, S.W. Kung, A.L., Roitelman, J., Simoni, R.D. and Schimke, R.T. 1993. In-vivo inhibition of cyclin B proteolysis and induction of cell cycle arrest in mammalian cells by the neutral cysteine protease inhibitor N-acetyl-leucy-leucyl-norleucinal. Proc. Nat. Acad. Sci. USA 90:3353-3357.
16. Sherwood, S.W., and Schimke, R.T. 1993 Induction of apoptosis by cell cycle phase specific drugs. In, **Apoptosis**. 6th Pezcoller Foundation Symposium.
17. Gallant, P. and Nigg, E.A. 1992. Cyclin B2 undergoes cell cycle dependent nuclear translocation and when expressed as a non-destructible mutant causes mitotic arrest in HeLa cells. J. Cell Biol. 1217:213-224.
18. Sluder, G., and Begg, D.A. 1985. Control mechanisms of the cell cycle: Role of spatial arranmgement of spindle components in the timing of mitotic events. J. Cell Biol. 97:877-886.
19. Mazia, D. 1987. The chromosome cycle and the centrosome cycle in the mitotic cycle. Int. Rev. Cytol. 100: 49-92.
20. Kreyer, G., Ris, H., and Borisy, G.G. 1984. Centriole distribution during tripolar mitois in Chinese hamster ovary cells. J. Cell biol. 98:2222-2229.
21. Watanabe, N., Vande Woude, G.F., Ikawa, Y., and Sagata, N. 1989. Nature (London) 342:505-511.
22. Schollmeyer , J.E. 1988. Calpain II involvement in mitosis. Science 240:911-913.
23. Garte, S.J., Currie, D.D., and Troll, W. 1987. Inhibition of H-ras oncogene transformation of NIH 3T3 cells by protease inhibitors. Cancer Res. 47:3159-3162
24. Cox, L.R., Jotz, W., Troll, W., aaand Garte, S.J. 1991. Antipain-induced suppression of oncogene expression in H-ras transformed NIH 3T3 cells. Cancer Res. 51:4810-4814
25. Billings, P.C., and Habres, J.M. 1992. A growth-related protease activity that is inhibited by the anticarcinogenic Bowman-Birk protease inhibitor. Proc. Nat. Acad. Sci. USA 89:3120-3124
26. Jitesh, P., Jani, J.S., Mistry, G.M., Lazlo, J.S., G., and Sebti, S.M.. 1992. In-vivo sensitization of human lung carcinoma to bleomycin by the cysteine protease inhibitor E-64d. Oncology Res. 4(2):59-64.
27. Hozumi, M., Ogawa, M., Sugimura, T., Takeguchi, T., and Umezawa, H. 1972. Inhibition of tumorigenesis in mouse skin by leupeptin, a protease inhibitor from Actinomyces. Cancer Res. 32:1725-1728
28. Troll, W., Fenkel, K., Weisner, R. 1984. Protease inhibitors as anticarcinogens. J. Nat. Cancer Inst. 73:1245-1250

48

CLINICAL RELEVANCE OF DNA PLOIDY AND CELL CYCLE PHASES IN TRANSITIONAL CELL CARCINOMA OF THE RENAL PELVIS AND URETER: A STUDY BY MEANS OF STATIC DNA-CYTOPHOTOMETRY

Hussain-Al-Abadi, Reinhard Nagel

Department of Urology
Clinical Cytology and DNA Cytophotometry
Rudolf Virchow Medical Center
Free University of Berlin
Berlin, Germany

ABSTRACT

In 72 patients with urothelial carcinoma of the renal pelvis or ureter the ploidy and counts of cell cycle phases in the tumor were analyzed by means of single cell DNA cytophotometry with the intention of finding new prognostic factors in addition to those already known (stage and grade). Follow-up ranged from 1 to 10 years. The results of the DNA analyses were related to the tumor categories, histopathological grading and clinical course. Malignancy grade 1 tumors showed DNA frequency peaks in the diploid range, while tumors assessed as malignancy grade 2 showed heterogeneous DNA distribution patterns. Malignancy grade 3 tumors exhibited 71% aneuploid and 29% tetraploid DNA values.

DNA histograms also show the distribution of the cell cycle phases in a cell population measured. The individual tumor cell nuclei are assigned to the different phases of the cell cycle (G0/G1, S phase and G2/M phases) according to the DNA content. Our results show that the prognosis for patients with more than 50% diploid cells in G0/G1 phase is better than for those with a lower percentage of cells. As a rule, a high percentage of G0/G1 cells is a sign of a slow growth rate of the tumor. The differences in survival times associated with the rates of G0/G1 cells and S and G2/M cells are not only evident in the patients with malignancy grade 1 but also in those with malignancy grades 2 and 3.

There was also a positive correlation between pT category and DNA ploidy. The cell lines were aneuploid in 38% of the patients with stage pT1 tumors, 56% with stage pT2 tumors and almost 85% with stage pT3, N+ tumors. A significant correlation was found between the results of DNA cytophotometry and the clinical course of the disease. Patients with diploid tumor cell nuclei had no metastases and no local tumor progression for up to 10 years, whereas patients with aneuploid tumor cell nuclei suffered metastasis and no local tumor progression within 24 to 36 months. The patients died of the tumor 36 months after primary diagnosis on the average.

The determination of tumor ploidy and tumor cell cycle phases in urothelial carcinoma of the renal pelvis and ureter by means of DNA cytophotometry yields valuable prognostic information.

INTRODUCTION

The histomorphological classification of tumors of the urothelium is a difficult task even today, mainly because there is no grading system which affords reliably reproducible results (1 - 3). Urothelial tumors that are not highly differentiated constitute a heterogeneous group with overlapping histological characteristics, clinical courses and biological behaviors (1, 2). Far more accurately than is possible with visual morphology, the different malignancy potentials of the tumor cell nuclei can be evaluated with the aid of DNA cytophotometry of the actual tumor cell (4 - 8), and the progression of these tumors can thus be predicted more exactly. The combined information obtained by the two methods - morphology and DNA cytophotometry - provides valuable prognostic indicators which have an influence on treatment and follow-up (4 - 9).

The heterogeneity of tumor cells has recently received great interest by clinicians because of its significant interrelation with the course of the tumor before and during treatment and because of its value as a prognostic indicator. This fact has also been pointed out by several authors of studies on malignant tumors in a wide variety of organs (4 - 8, 10 - 13).

In the present prospective study, the ploidy, DNA heterogeneity and the counts of cell-cycle phases in the individual malignancy grades were investigated by single-cell DNA absorption cytophotometry with the aim of finding new cell-related prognostic factors to complement those conventionally employed (stage and grade).

PATIENTS AND METHODS

Between February 1982 and December 1991, 72 patients underwent radical nephroureterectomy with a bladder cuff and lymphadenectomy for carcinoma of the renal pelvis and ureter. The average age of these patients (38 women and 34 men) was 67 years, with a range of 43 to 85 years. Patients with some other malignant tumor were excluded from the study. Of the patients 52 (72%) were diagnosed as having a tumor of the renal pelvis and 20 (28%) as having a tumor of the ureter (see table 1). Abuse of analgesics was reported by 15% of the patients during a 10 to 25-year period and 97% reported recurrent macroscopic hematuria (for an average of 5 to 34 months) before hospitalization. Follow-up examinations included exfoliative urine cytology every 6 weeks; sonography, cystoscopy. and lavage cytology and, if necessary, also DNA cytophotometry every three months, chest radiography, voiding urogramm and, if necessary, also computer tomography every six months.

Immediately after removing the tumor-bearing kidney, the renal pelvis was dissected and one or several wedge-shaped segments were excised from the tumor, the number depending on the size of the tumor. Several cytological preparations were made of the sagittal cut surface of these sections and of several macroscopically normal areas of the renal pelvis or ureter and fixed with Merckofix Spray (Merck). Some of the preparations were stained with Papanicolaou's stain for cytomorphological evaluation. These specimens were later compared with the final histological results. For the determination of DNA the preparations were stained by Feulgen's reaction (30 minutes' hydrolysis in 5 N HCl at room temperature). Leukocytes from peripheral blood from healthy donors treated in the same manner as the test material were used as reference preparations for the determination of DNA diploidy.

As a rule, approximately 100 tumor cell nuclei were measured, occasionally more. The DNA content of these nuclei was determined by means of single-cell absorption photometry.

The fine-scanning procedure with a Leitz MPV-2 cytophotometer was used to measure the cell nuclei. The scanning stage, measuring procedure and recording of the absorption were controlled by a digital pdp 8a process computer. The respective total extinction values measured were printed together with the mean value, standard deviation, variance and coefficient of variation.

Table 1. Clinical, pathological and DNA cytophotometric parameters in 72 patients with transitional cell carcinoma of the renal pelvis and ureter.

	No. Pts. (%)
Renal pelvis	52 (72)
Ureter	20 (28)
Multiplicity:	
Singel	54 (75)
Multiple	18 (25)
Grade:	
1	17 (24)
2	36 (50)
3	19 (26)
Stage:	
1	32 (45)
2	19 (26)
3	21 (29)
DNA ploidy:	
Diploid	23 (32)
Tetraploid	23 (32)
Aneuploid	26 (36)
Analgesics abuse (phenacetin)	10 (15)
Status:	
Alive	41 (57)
Dead	31 (43)
Cause of death:	
Transitional cell Ca	26 (36)
Other	5 (7)

The results are printed out in the form of a histogram in which the number of tumor cell nuclei measured is marked as 'n' on the ordinate, and the relative DNA content of the individual cell populations, given in relative c and expressed in arbitrary units, is plotted against the abscissa. Figures 1a-d show typical DNA histograms. The tumors exhibit either a diploid or several aneuploid or tetraploid DNA cell lines.

RESULTS

Tumor stage - histological and cytomorphological classification

The dissemination of the tumor was determined according to the tumor, nodes and metastasis classification recommended by the International Union Against Cancer (14). Of the patients 32 (45%) had stage pT1 pNO, 19 (26% stage pT2 pNO, 19% stage pT3 pNO and 7 (10% stage pT3 pN+ disease (table 1). The tumors were classified histologically according to the guidelines issued by the World Health Organization (15), whereas the evaluation of cellular anaplasia in the cytological preparations was assessed according to the classification suggested by Mostofi et al and Bennington and Beckwith (16). The histological malignancy grade 2 was found to prevail at 50%, in the entire case material, followed by malignancy grades 3 (26%) and 2 (24%, table 1).

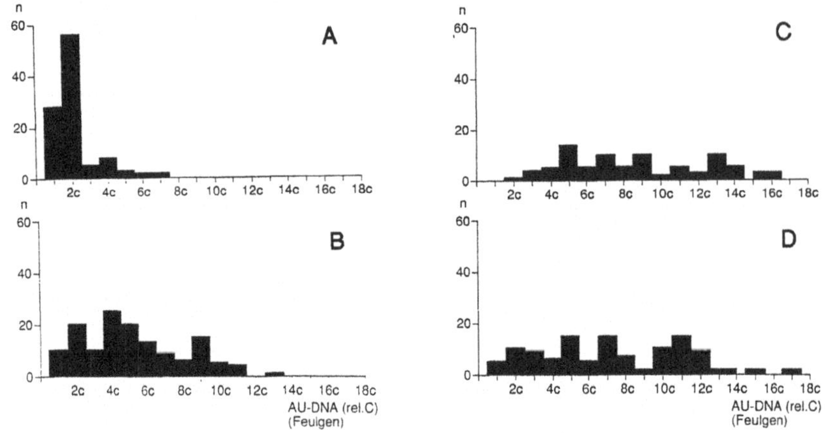

Fig. 1. ***A***, malignancy grade 1. DNA histogram with frequency peak in diploid range and scattered values up to 7c. ***B***, malignancy grade 2. DNA histogram with broad heterogeneous distribution ranging from 2c to 13c. ***C***, malignancy grade 3. DNA histogram with heterogeneous distribution shows frequency peaks in 5c, 7c, 9c, 13c, and widely scattered values up to 16c. ***D***, DNA histogram of lymph node metastasis shows aneuploid DNA distribution. *AU-DNA (rel. C)*, arbitrary units.

DNA Ploidy and Grade of Malignancy

A positive correlation was found between the cellular anaplasia of the tumor cells and their DNA content. A total of 117 DNA histograms, 11 of which were histograms of lymph node metastases, was analyzed. In the 17 patients (24%) with grade 1 tumors, 88% of the tumor cell nuclei measured exhibited diploid (2c), 5% tetraploid (4c) and 7% aneuploid DNA distribution (Fig. 1a and 2).

The 36 patients with grade 2 tumors showed heterogeneous DNA distribution patterns, with DNA frequency peaks in the diploid, tetraploid and aneuploid ranges, although the carcinomas exhibited the same morphological degree of differentiation. The DNA content measured in the tumor cell nuclei was in the diploid range in 23%, in the tetraploid range in 35%, in the aneuploid range in 42%. (Figs 1b and 2)

The DNA content measured in the tumor cell nuclei of the 19 patients with grade 3 tumors was tetraploid in 29% and aneuploid in 71% as demonstrated by the ploidy distribution (Figs. 1 and 2).

DNA Ploidy and Pathological Tumor Stage

There was also a positive correlation between DNA ploidy and tumor stage. The DNA cell lines were aneuploid in 38% of the patients with stage T1 tumors, 56% of the patients with stage T2 tumors and almost 85% of the patients with stage T3 tumors, with or without lymph node infiltration (Fig. 3).

DNA Ploidy and Abuse of Analgesics

It is noteworthy that the nuclear DNA distribution measured in the 11 patients with carcinoma of the renal pelvis or ureter who had reported phenacetin abuse was aneuploid. In

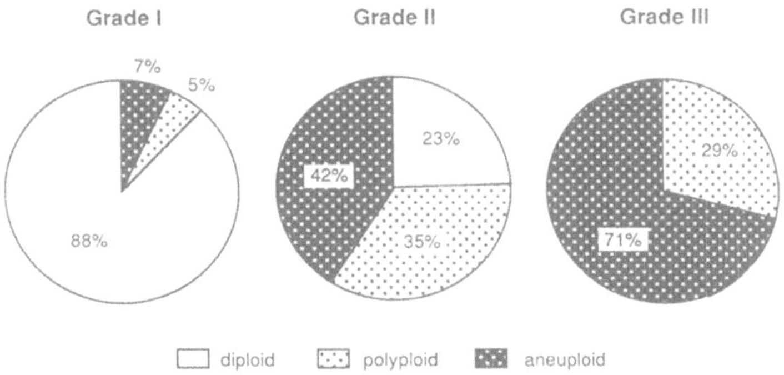

Fig. 2. Correlation of DNA ploidy and grades of malignancy in 72 patients with cacinoma of renal pelvis and ureter.

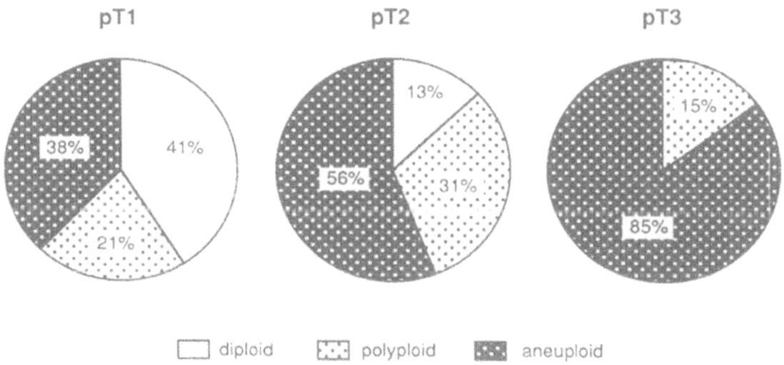

Fig. 3. Correlation of DNA ploidy and histological staging of urothelialcarcinoma of renal pelvis and ureter in 72 patients.

addition to multifocal tumor dissemination, these patients exhibited severe cellular anaplasie even at 6 cm from the tumor. These findings were established by scrape cytology and confirmed by the broad DNA distributions demonstrated by DNA cytophotometry (Fig. 1c and 2). 8 of the 11 patients had DNA distribution patterns with several cell lines.

DNA Ploidy and Multiplicity of the Tumor

Of the 26 patients whose tumors showed DNA aneuploidy 18 (69%) exhibited multifocal spreading (table 2). In these patients, cytology even showed microscopic changes indicative of carcinoma in situ in the urothelium in the immediate vicinity of the tumor.

Table 2. Multiplicity distribution of transitional cell carcinoma of the renal pelvis and ureter in 18 patients

Multiplicity	No. Patients	(%)
Renal pelvis	5	(28)
Ureter	3	(17)
Renal pelvis + bladder	4	(22)
Bilateral renal pelvis	4	(22)
Bilateral renal pelvis + ureter + bladder	2	(11)

DNA Ploidy and Metastatic Dissemination

A total of 23 patients with diploid DNA distribution patterns (DNA frequency peak in the 2c range) had no metastases during the 10-year observation period and none died of carcinoma. Of the patients with aneuploid tumors, however, and especially patients with reported phenacetin abuse, 84% suffered tumor progression with metastases to other organs (that is retroperitoneal lymph nodes in 43%, lungs in 32% and bones in 26%) within 28 to 36 months and they died of the tumors.

Table 3. Correlation of nuclear DNA patterns clinical course in 72 patients with transitional cell carcinoma of the renal pelvis and ureter.

DNA Histogram Pattern	No. Pts.	Died of Metastatic Transitional Cell Ca No. (%)
Diploid	23	–
Tetraploid	23	4 (17)
Aneuploid	26	22 (84)

Prognostic Significance of DNA Ploidy

The DNA ploidy and cell line heterogeneity were significant for the prognosis. Patients with aneuploid DNA distribution patterns died sooner than patients with diploid cell lines. Of the patients with diploid or tetraploid DNA distribution patterns 5 (7%) died of cardioplumonary disease: 3 had diploid and 2 had tetraploid DNA distribution patterns. Of the 26 patients with aneuploid DNA distribution patterns 22 (84%) died of metastasis of the primary disease. Two patients had metastases in the lymph nodes, bones and lungs within 16 and 21 months after tumor nephrectomy but they have been clinically stable with methotrexate, vinblastine, doxorubicin and cisplatin chemotherapy (17). Two patients were clinically stable while exhibiting neither metastases nor tumor progression at 11 and 17 months postoperatively. The remaining 41 patients (57%) have shown no signs of progression of the primary disease (table 3).

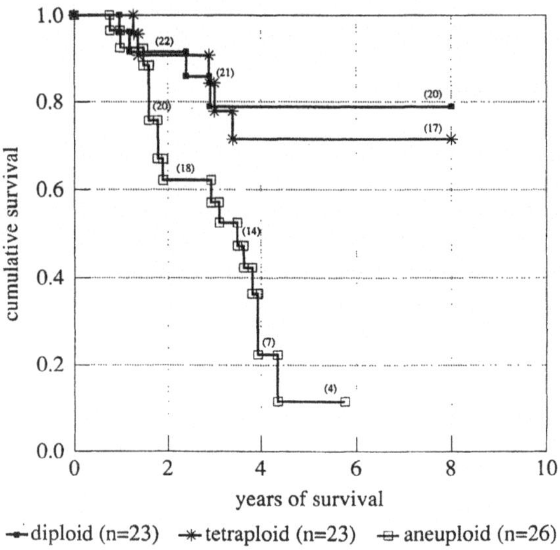

Fig 4. Survival times of 72 patients with transitional cell carcinoma grouped according to DNA ploidy: diploid versus tetraploid ($p>0.5$) versus aneuploid ($p<0.001$) .

DISCUSSION

According to Bergkvist et al, (1) the stage and pathohistological grading of a tumor do not suffice to judge its biological aggressiveness in respect of recidivation, invasion and metastatic growth. Since histological grading is subjective and not uniform (2, 3), it is necessary to assess the grade of malignancy and tumor stage objectively by means of prognostic methods of investigation (4, 5, 6, 10, 12).

Presently there are 2 different methods of measuring the nuclear DNA content of malignant and benign tumors: flow-through cytophotometry, which is the most common method and single cell cytometry.

Researchers often employ auto-radiographic examinations to assess the proliferate activity of a tumor. This method is not suitable for clinical application, though, on account of the procedure it involves.

This is the first prospective study on urothelial carcinomas of the renal pelvis and ureter in which DNA analyses have been conducted by single-cell cytophotometry over a follow-up period of 10 years.

Our results show that the DNA ploidy, tumor heterogeneity and the distribution of the individual cell cycle phases of the tumors are independent factors which are of considerable value for predicting patient's survival times.

Detailed investigations have shown that nuclear DNA analysis can contribute to the existing clinical and morphological parameters by the prognostically supplementary information it affords. In this context we would like to mention our previous studies on 329 patients with prostatic carcinoma, 112 patients with renal cell carcinoma and on 127 patients with urothelial carcinoma of the urinary bladder conducted over a follow-up period of 9 years (6, 18, 19).

The results of the study on 409 patients with mammary carcinoma conducted by "Fallenius (1987)" (20) underline the prognostic significance of DNA ploidy.

Other investigations based on flow-through cytophotometry point to a correlation between the histological grade of malignancy of a tumor and the DNA content of the tumor cell nuclei (4, 7, 8).

In the past few years various working groups have reported that it is possible to measure DNA by flow-through cytophotometry in cell material embedded in paraffin. In 1988, a research team at the Mayo Clinic employed this method in 119 cases of urothelial carcinoma of the renal pelvis and confirmed the prognostic relevance of DNA ploidy that we had shown using single-cell cytophotometry (13).

In our study we found an aneuploidy rate of 7% in highly differentiated urothelial tumors and one of 71% in poorly differentiated tumors. Other authors who employed flow-through cytophotometry to investigate tumors of the lungs found aneuploidy rates of up to 85% (21).

A negative turn of prognosis in the individual case can be deduced from a broad scattering of DNA values as manifest in the histogram. A similar conclusion can be drawn from the ratio of diploid cells to aneuploid cells. This ratio is particularly important for carcinomas with grade 2 malignancy which are uniform with regard to morphology but heterogeneous with respect to malignancy potentials.

DNA histograms show also the distribution of the cell cycle phases in a cell population measured. The individual tumor cell nuclei are assigned to the different phases of the cell cycle (G0/G1 phase, S phase and G2/M phase) according to their DNA content. Changes in cellular genetics result in corresponding changes in the DNA histograms (12, 13). The prognosis for patients with more than 88% diploid cells (G0/G1 phase) is better than for those with a lower percentage of diploid cells. As a rule, a high percentage of G0/G1 cells is a sign of a slow growth rate of the tumor. The differences in survival times associated with the rates of G0/G1 and proliferating cells (S and G2/M) are not only evident in the group of patients with malignancy grade 1 but also in the ones with malignancy grades 2 and 3.

The patients with nephropathy caused by analgesics abuse who also showed multifocal tumor dissemination exhibited high DNA values with predominantly aneuploid DNA distribution and several DNA cell lines in DNA cytophotometry. Prolonged phenacetin abuse presumably provokes aggressive cytochemical and morphological changes in the epithelial cells which lead to the development of a biologically highly aggressive tumor.

CONCLUSIONS

DNA histograms define the biological characteristics of a tumor more precisely than the criteria by which the clinical status of the disease is judged. Only in the case of grade 1 and grade 3 tumors, the histological grading correlates with the nuclear DNA content. The prognosis for grade 1 tumors is good whereas it is very unfavorable for grade 3 tumors. For both groups (patients with grade 1 and grade 3 tumors) DNA ploidy affords no additional prognostic information. Grade II tumors, on the other hand, are very heterogeneous in respect of DNA ploidy although they exhibit the same histomorphological degree of differentiation. These tumors can be sub classified in aneuploid (biologically aggressive) and diploid or tetraploid (biologically less aggressive) tumors with the aid of DNA cytometry.

Analyses of DNA ploidy, tumor cell heterogeneity and cell cycle phases by means of DNA cytophotometry afford additional valuable information as to prognosis and treatment.

Acknowledgements

We thank Mrs. Ikonija Rotter for her excellent technical assistance and Prof. Küster for helping with the statistics.

LITERATURE

1. A. Bergkvist, A. Ljungkvist, and G. Moberger: Classification of bladder tumours based on the cellular pattern. Preliminary report of a clinical-pathological study of 300 cases with a minimum followup of eight years. *Acta Chir. Scand.* 130: 371 (1985)
2. C. Busch, A. Engberg, B.J. Norlen, and B- Stenkvist: Malignancy grading of epithelial bladder tumours. Reproducibility of grading and comparison between forceps biopsy, aspiration biopsy and exfoliative cytology. *Scand. J. Urol. Nephrol.* 11: 143 (1977)
3. H. Zincke: Treatment modalities in superficial bladder cancer. *Seminar in Urology,* 1: (Suppl. 1) 3 (1989)
4. F.A. Klein, H.W. Herr, P.C. Sogani et al.: Detection and follow-up of carcinoma of the urinary bladder by flow cytometry. *Cancer* 50: 389-395 (1982)
5. H. Al-Abadi, R. Nagel: Prognostische Bedeutung von Ploidie und proliferativer Aktivität beim lokal fortgeschrittenen Prostatakarzinom. *Akt. Urol.* 4: 182 (1988)
6. H. Al-Abadi, R. Nagel: Prognostic Relevance of ploidy and proliferative activity of ploidy and proliferative activity of Renal Cell Carcinoma. *Eur. Urol.* 15: 271 (1988)
7. P.J. Oljans, and H.J. Tanke: Flow cytometric analysis of DNA content in bladder cancer: Prognostic value of the DNA index with respect to early tumour recurrence in 62 tumours. *World J. Urol.* 4: 205 (1986)
8. B. Tribukait, H. Gustafson, and P.L. Esposti: The significance of ploidy and proliferation in the clinical and biological evaluation of bladder tumours: A study of 100 untreated cases. *Brit. J. Urol.* 54: 130 (1982)
9. R. Nagel, H. Al-Abadi, M. Petkovic, and M. Uhlig: Therapie der Urotheltumoren, *Z. Urol. Nephrol.* 82: 91 (1989)
10. G. Auer, E. Eriksson, E. Azevedo, T. Caspersson, and A. Wallgren: Prognostic significance of nuclear DNA content in mammary adenocarcinomas in humans. *Cancer Res.* 44: 394 (1984)
11. L.L. Vindelov, H.H. Hansen, I.B.J. Christensen, M. Spang-Thomsen, F.R. Hirsch, M. Hansen, and N.L. Nissen: Clonal heterogeneity of small cell anaplastic carcinoma of the lung demonstrated by flow cytometric DNA analysis. *Cancer Res.* 40: 4295 (1980)
12. W.M. Murphy, R.W. Chandler, and R.M. Trafford: Flow Cytometry of deparaffinized nuclei compared to histological grading for the pathological evaluation of transitional cell carcinoma. *J. Urol.* 135: 694 (1986)

13. M.L. Blute, K. Tsushima, G.M. Farrow, T.M. Therneau, and M.M. Lieber: Transitional cell carcinoma of the renal pelvis: Nuclear desoxyribonucleic acid ploidy studied by flow cytometry. *J. Urol.* 140: 944 (1988)
14. UICC Union International Contre le Cancer: TNM classification of malignant tumours, 3rd ed. (International Union Against Cancer), Geneva (1978)
15. F.K. Mostofi, L.H. Sobin, and H. Torloni: Histological Typing of Urinary Bladder Tumors.International histological classification of tumours 10. WHO, Geneva (1973)
16. J.L. Bennington, and J.B. Beckwith: Tumours of the kidney, renal pelvis and ureter, *in:* Atlas of Tumour Pathology, Washington D.C.: Armed Forces Institute of Pathology, 2nd series, fasc. 12, 308 (1975)
17. C.N. Sternberg, A. Yagoda, H.I. Scher, R.C. Watson, H.W.Herr, M.J. Morse, P.C. Sogani, E.D. Vaughan, N. Blander, L.R. Weiselberg, N. Geller, P.S. Hollander, R. Cipperman, W.R. Fair, and F. Whitmore: M-VAC (Methotrexate, Vinblastine, Doxorubicin and Cisplatin) for advanced transitional cell carcinoma of the urothelium. *J. Urol.* 139: 461-469 (1988)
18. H. Al-Abadi, V. Borgmann, and R. Nagel: Urothelkarzinom der Harnblase: Einzelzell-DNS-Zytometrie und ihre klinische Bedeutung. *Verh. Dtsch. Ges. Urol.* 42 (1990)
19. H. Al-Abadi, V. Borgmann, R. Friedrichs, and R. Nagel: Nuclear deoxyribonucleic acid analysis: Relevance of ploidy, DNA heterogeneity and phases of the cell cycle: An 8-year study of 329 patients with prostatic carcinoma. *J. Urol.* 143: Suppl. 326 A, 522 (1990)
20. A. Fallenius: DNA Content and Prognosis in Breast Cancer. Thesis. Karolinska Institute, Stockholm (1986)
21. M. Volm, J. Mattern, J. Sonka, K. Wayss, P. Drings, and L. Vogt-Moykopf: Prognostic value of ploidy and proliferative activity in non-small cell lung carcinoma. *Tumor-Diagnostik Ther. 6:* 8 - 13 (1985)

INDEX

The manufacturer's authorised representative in the EU is Springer Nature Customer Service Centre GmbH, Europaplatz 3, 69115 Heidelberg, Germany. If you have any concerns regarding our products, please contact ProductSafety@springernature.com

Printed and bound by CPI Group (UK) Ltd, Croydon, CR0 4YY

28/07/2026

02179215-0009